NATURAL MINERAL NANOTUBES

Properties and Applications

NATURAL MINERAL NANOTUBES

Properties and Applications

Edited by

**Pooria Pasbakhsh, PhD, and
G. Jock Churchman, PhD**

Apple Academic Press Inc. | Apple Academic Press Inc.
3333 Mistwell Crescent | 9 Spinnaker Way
Oakville, ON L6L 0A2 | Waretown, NJ 08758
Canada | USA

©2015 by Apple Academic Press, Inc.

First issued in paperback 2021

Exclusive worldwide distribution by CRC Press, a member of Taylor & Francis Group

No claim to original U.S. Government works

ISBN 13: 978-1-77463-367-0 (pbk)
ISBN 13: 978-1-77188-056-5 (hbk)

Library and Archives Canada Cataloguing in Publication

Natural mineral nanotubes: properties and applications / edited by Pooria Pasbakhsh, PhD, and G. Jock Churchman, PhD.

Includes bibliographical references and index.
ISBN 978-1-77188-056-5 (bound)
1. Nanotubes. 2. Nanocomposites (Materials). I. Churchman, G. Jock, author, editor II. Pasbakhsh, Pooria, author, editor

TA418.9.N35N38 2015   620.1'15   C2014-907857-9

Library of Congress Cataloging-in-Publication Data

Natural mineral nanotubes : properties and applications / Pooria Pasbakhsh, PhD, and G. Jock Churchman, PhD, editors.

pages cm
Includes bibliographical references and index.
ISBN 978-1-77188-056-5 (alk. paper)
1. Mineralogy. 2. Nanotubes. 3. Nanostructures. 4. Halloysite. I. Pasbakhsh, Pooria, editor. II. Churchman, G. J., editor.

QE369.P49N38 2015   553.6--dc23   2014045760

Apple Academic Press also publishes its books in a variety of electronic formats. Some content that appears in print may not be available in electronic format. For information about Apple Academic Press products, visit our website at **www.appleacademicpress.com** and the CRC Press website at **www.crcpress.com**

ABOUT THE EDITORS

Pooria Pasbakhsh

Pooria Pasbakhsh, PhD, has been a senior lecturer at Monash University, Selangor, Malaysia, since 2010, with a specialty in polymer nanocompocites and nanomaterials. He obtained his Bachelor and Master degrees from the Iran University of Science and Technology and Khaje Nasireddin Toosi University of Technology, Tehran, Iran, in Ceramic Engineering and Material Science and Engineering, respectively and he recieved his PhD from University of Science Malaysia in 2010. He published several journal and conference papers on characterization and applications of halloysite nanotubes. Since 2007 his research has been supported by the Ministry of Higher Education of Malaysia; Monash University; University of Science Malaysia; and Geological Survey of South Australia Resources and Energy Group, where he did a short postdoctoral program on characterization of various halloysite nanotubes from different origins at the University of Adelaide, South Australia. His current research interests concern the preparation, characterization and modeling of bionanocomposites for packaging, medical and self-healing applications. In 2013 he completed a graduate certificate at Monash University on Higher Education where he focused on the effect of different teaching techniques on the learning of undergraduate students in engineering material.

G. Jock Churchman, PhD

G. Jock Churchman is Adjunct Senior Lecturer in Soils at the University of Adelaide and part-time Associate Professor in the Centre for Environment Risk Assessment and Remediation at the University of South Australia. He obtained degrees in Chemistry from Otago University in his native New Zealand. He studied the physical chemistry of halloysite for his PhD under a fellowship from the New Zealand pottery and ceramics industry and carried out research for this industry for a short time before beginning a 2-year postdoctoral fellowship in soil science at the University of Wisconsin, USA. He has continued trying to understand why halloysite contains interlayer water (his specific PhD thesis topic) for all of his career, while also pursuing many other research topics on clays, especially in soils. He was employed in the New Zealand Soil Bureau, Department of Scientific and Industrial Research, for 16 years and in CSIRO (Commonwealth Scientific and Industrial Research Organisation) Division of Soils (later Land and Water) for 14 years, and has held visiting fellowships in soil science at Reading University, England and the University of Western Australia. He is a former Editor (now Emeritus) of the journal of *Applied Clay Science.* In 2005, he completed a BA (Hons) in philosophy from Flinders University of South Australia with a thesis on the philosophical status of Soil Science.

DEDICATION

From Pooria: To my family, for their unfailing love, loyalty, support, and motivation; without their sentiment chemistry the life would be too hollow for me.

From Jock: To Jan—who has lived with halloysite all her married life—for her support and encouragement.

CONTENTS

Contents

LIST OF CONTRIBUTORS

Elshad Abdullayev
Applied Minerals Inc., Eureka, UT 84628, Email: hovsan@gmail.com

Carola Aguzzi
Department of Pharmacy and Pharmaceutical Technology, School of Pharmacy, University of Granada, Campus of Cartuja, Granada 18071 s/n, Spain

Ana C. S. Alcântara
Instituto de Ciencia de Materiales de Madrid, ICMM-CSIC, Cantoblanco, c/Sor Juana Inés de la Cruz 3, Madrid 28049, Spain, Email: anaclecia@gmail.com

Peng Ao
Department of Materials Science and Engineering, Jinan University, Guangzhou 510632, PR of China

Patricia Aparicio
Departmento de Cristalografía Mineralogía y Química Agrícola, Facultad de Química, Universidad de Sevilla, Professor García González 1, Seville 41012, Spain, Email: paparicio@us.es

Pilar Aranda
Instituto de Ciencia de Materiales de Madrid, ICMM-CSIC, Cantoblanco, c/Sor Juana Inés de la Cruz 3, Madrid 28049, Spain, Email: aranda@icmm.csic.es

C. Bailly
Bio- and Soft Matter (BSMA), Institute of Condensed Matter and Nanosciences (IMCN), Université catholique de Louvain (UCL), Croix du Sud 1, Louvain-la-Neuve B-1348, Belgium

Pilar Cerezo
Department of Pharmacy and Pharmaceutical Technology, School of Pharmacy, University of Granada, Campus of Cartuja, Granada 18071 s/n, Spain

G. Jock Churchman
School of Agriculture, Food and Wine, The University of Adelaide, Australia, Email: jock.churchman@adelaide.edu.au; and Centre for Environmental Risk Assessment and Remediation (CERAR) CRC for Contamination Assessment and Remediation of the Environment (CRC CARE), University of South Australia, University Boulevard, Mawson Lakes SA 5095, Email: jock.churchman@unisa.edu.au

Margarita Darder
Instituto de Ciencia de Materiales de Madrid, ICMM-CSIC, Cantoblanco, c/Sor Juana Inés de la Cruz 3, Madrid 28049, Spain, Email: darder@icmm.csic.es

Gustave Kenne Dedzo
Department of Chemistry and Center for Catalysis Research and Innovation, University of Ottawa, 10, Marie-Curie, Ottawa (Ont), Canada K1N6N5, Email: kennegusto@yahoo.fr

Christian Detellier
Department of Chemistry and Center for Catalysis Research and Innovation, University of Ottawa, 10, Marie-Curie, Ottawa (Ont), Canada K1N6N5, Email: dete@uottawa.ca

Francisco M. Fernandes

Sorbonne Universités, UPMC Univ Paris 06, UMR 7574, Chimie de la Matière Condensée de Paris, F-75005, Paris, France; CNRS, UMR 7574, Chimie de la Matière Condensée de Paris, Paris F-75005, France, Email: francisco.fernandes@upmc.fr

Saverio Fiore

Institute of Methodologies for Environmental Analysis, CNR, Tito Scalo, Potenza, Italy, Email: saverio.fiore@cnr.it

Emilio Galán

Departmento de Cristalografía Mineralogía y Química Agrícola, Facultad de Química, Universidad de Sevilla, Professor García González 1, Seville 41012, Spain, Email: egalan@us.es

Kheng Lim Goh

School of Mechanical and Systems Engineering, Newcastle University, NE1 7RU, United Kingdom, NU International Singapore Pte Ltd, Singapore 569830, Email: kheng-lim.goh@newcastle.ac.uk

Kavitha Govindasamy

School of Engineering, Monash University Malaysia, Jalan Lagoon Selatan, Bandar Sunway, Selangor 47500, Malaysia, Email: Kavitha@monash.edu

Stephen Guggenheim

Department of Earth and Environmental Sciences, University of Illinois at Chicago, Chicago, Illinois, 60607 USA, Email: xtal@uic.edu

F. Javier Huertas

Instituto Andaluz de Ciencias de la Tierra, CSIC-University of Granada, Armilla, Granada, Spain

Marie-Claude Jaurand

Director of Research at the Institut National de la Santé et de la Recherche Médicale (INSERM), Email: marie-claude.jaurand@inserm.fr

John L. Keeling

Geological Survey of South Australia; Department for Manufacturing, Innovation, Trade, Resources and Energy, South Australia, Adelaide 5000, South Australia, Australia, Email: John.Keeling@sa.gov.au

Benoit Lecouvet

Bio- and Soft Matter (BSMA), Institute of Condensed Matter and Nanosciences (IMCN), Université Catholique de Louvain (UCL), Croix du Sud 1, box L7.04.02, B-1348 Louvain-la-Neuve, Belgium, Email: benoit.lecouvet@uclouvain.be

Mingxian Liu

Department of Materials Science and Engineering, Jinan University, Guangzhou 510632, PR of China Email: liumx@jnu.edu.cn

Binghong Luo

Department of Materials Science and Engineering, Jinan University, Guangzhou 510632, PR of China

Yuri Lvov

Louisiana Tech University, Ruston, LA 71272, Email: ylvov@coes.latech.edu

Ravi Naidu

Centre for Environmental Risk Assessment and Remediation (CERAR), University of South Australia, Building X, Mawson Lakes, SA 5095, Australia, and Cooperative Research Centre for Contamination Assessment and Remediation of the Environment (CRC CARE), P.O. Box 486, Salisbury, SA 5106, Australia, Email: ravi.naidu@crccare.com

Antonio Nieto-Camacho

Instituto de Química, Universidad Nacional Autónoma de México, México

B. Nysten

Bio- and Soft Matter (BSMA), Institute of Condensed Matter and Nanosciences (IMCN), Université catholique de Louvain (UCL), Croix du Sud 1, box L7.04.02, B-1348 Louvain-la-Neuve, Belgium, Email bernard.nysten@uclouvain.be

Pooria Pasbakhsh

School of Engineering, Monash University Malaysia, Jalan Lagoon Selatan, Bandar Sunway, Selangor 47500, Malaysia, Email: pooria.pasbakhsh@monash.edu, ppooria@gmail.com

Chai Siao Peng

School of Engineering, Monash University Malaysia, Jalan Lagoon Selatan, Bandar Sunway, Selangor 47500, Malaysia, Email: chai.siang.piao@monash.edu

Qi Peng

Department of Materials Science and Engineering, Jinan University, Guangzhou 510632, PR of China

Manuel Pozo

Ph.D, Department of Geology and Geochemistry. Faculty of Science. Autonomous University of Madrid. Cantoblanco, Madrid 28049, Spain, Email: manuel.pozo@uam.es

María Teresa Ramírez-Apán

Instituto de Química, Universidad Nacional Autónoma de México, México

Eduardo Ruiz-Hitzky

Instituto de Ciencia de Materiales de Madrid, ICMM-CSIC, Cantoblanco, c/ Sor Juana Inés de la Cruz 3, Madrid 28049, Spain, Email: eduardo@icmm.csic.es

Binoy Sarkar

Centre for Environmental Risk Assessment and Remediation (CERAR), University of South Australia, Building X, Mawson Lakes, SA 5095, Australia, and Cooperative Research Centre for Contamination Assessment and Remediation of the Environment (CRC CARE), P.O. Box 486, Salisbury, SA 5106, Australia, Email: Binoy.Sarkar@unisa.edu.au

Javiera Cervini Silva

Departamento de Procesos y Tecnología, Universidad Autónoma Metropolitana, Artificios No. 40, 6 Piso, Col. Miguel Hidalgo, Delegación Álvaro Obregón, C.P. 01120 México, D. F., Email: jcervini@correo.cua.uam.mx

Rangika Thilan De Silva

School of Engineering, Monash University Malaysia, Jalan Lagoon Selatan, Bandar Sunway, Selangor 47500, Malaysia, Email: rangika.desilva@monash.edu

Vahdat Vahedi

School of Engineering, Monash University Malaysia, Jalan Lagoon Selatan, Bandar Sunway, Selangor 47500, Malaysia, Email: vahdat.vahedi@monash.edu

César Viseras

Department of Pharmacy and Pharmaceutical Technology, School of Pharmacy, University of Granada, Campus of Cartuja, Granada 18071 s/n, Spain; Andalusian Institute of Earth Sciences, CSIC-University of Granada, Avda. de Las Palmeras 4, Armilla (Granada) 18100, Spain, Email: cviseras@ugr.es

Bernd Wicklein
Department of Materials and Environmental Chemistry, Stockholm University, Stockholm 106 91, Sweden, Email: bernd.wicklein@mmk.su.se
Changren Zhou
Department of Materials Science and Engineering, Jinan University, Guangzhou 510632, PR of China

LIST OF ABBREVIATIONS

AA	allyl alcohol
AFM	atomic force microscopy
AGE	allylglycidylether
AIBN	azobisisobutyronitrile
BBT	(2-benzoxazolyl) thiophene
BPO	benzoyl peroxide
BSP	biphasic calcium phosphate
CEC	cation exchange capacity
CMC	carboxymethylcellulose
CNTs	carbon nanotubes
CRTA	controlled rate thermal analysis
CT	compact tension
CTE	coefficient of thermal expansion
CTMAB	cetyl trimethylammonium bromide
DCM	dichloromethane
DGEBA	bisphenol a diglycidyl ether
DMA	dynamic mechanical analysis
DMF	dimethyl formamide
DMF	dimethyl formamide
DS	diclofenac sodium
DSC	differential scanning calorimetric
EDS	energy dispersive analysis system
ENR	epoxidized natural rubber
EPD	electrophoretic deposition
EPDM	ethylene propylene diene monomer
FDA	food and drug administration
FTIR	fourier transform infrared
GMA	glycidylmethacrylate
GMS	γ-glycidoxypropyltrimethoxy silane
GTA	glutaraldehyde
HA	haemagglutinin
HNTs	halloysite nanotubes
HPMC	hydroxypropylmethylcellulose
HRP	horseradish peroxidase
IBU	ibuprofen
ILSS	interlaminar shear strength

KPS	potassium persulfate
LbL	layer-by layer
MCC	microcrystalline cellulose
MICINN	ministerio de ciencia e innovación
MMT	montmorillonite
MPO	mieloperoxidase
MPS	methacryloxypropyl trimethoxysilane
MTT	tetrazolium bromide
NAD	nicotinamide adenine dinucleotide
NR	natural rubber
OCD	open channel defect
ODP	octadeylphosphonic acid
O-MMT	organo-modified montmorillonite
PAA	polyacrylic acid
PAAm	polyacrylamide
Paly	palygorskite
PC	polycarbonate
PCB	printed circuit boards
PCL	polyprolactone
PCOM	phase-contrast optical microscopy
PGA	polyglycolide
PGS	poly (glycerol sebacate)
PLA	poly (lactic acid)
PLGA	poly lactic-co-glycolic acid
PMMA	poly methyl methacrylate
PP	polypropylene
PPA	phenylphosphonic acid
PS	polystyrene
PVA	polymer polyvinyl alcohol
PVDF	polyvinylidene fluoride
QACs	quaternary ammonium compounds
RH	resorcinol and hexamethylenetetramine
SA	sodium alginate
SAED	small angle electron diffraction
SBR	styrene-butadiene rubber
SDS	sodium dodecyl sulfate
SEM	scanning electron microscopy
SEP	sepiolite
TD	thermal degradation
TEM	transmission electron microscopy
TGA	thermogravimetric analysis
Tpa	tonnes per annum
WG	wheat gluten

XRF   X-ray fluorescence
xSBR   carboxylated styrene butadiene rubber

LIST OF SYMBOLS

F	applied force
I	cross-sectional moment of inertia
E	elastic modulus
G	shear modulus
A	cross-sectional area
f_s	shape factor
X_{min}	a constant value
X_{min} and X_{max}	minimum and maximum values of the geometrical parameter
E and V	modulus and volume fraction
H and C	halloysite and chitosan
τ_y	yield stress in shear
L"	embedded length of the upper half of the HNT

LIST OF REVIEWERS

Elshad Abdullayev, USA
Volker Altstädt, Germany
Maria Franca Brigatti, Italy
G. Jock Churchman, Australia
Fernanda Cravero, Argentina
Rangika Thilan De Silva, Malaysia
Hélio Anderson Duarte, Brazil
Kheng Lim Goh, Singapore
Saied Hojati, Iran
Abul Huq, USA
Bill Jaynes, USA
Selahattin Kadir, Turkey
John L. Keeling, Australia
Stefano Leporatti, Italy
Benoit Lecouvet, Belgium
Mingxian Liu, PR China
David Lowe, New Zealand
Pooria Pasbakhsh, Malaysia
Binoy Sarkar, Australia
Javiera Cervini-Silva, Mexico
Mercedes Suárez, Spain
Ian Wilson, UK
Hüseyin Yalçin, Turkey

INTRODUCTION

Pooria Pasbakhsh, G. Jock Churchman

"If you don't plow the soil, it's going to get so hard nothing grows in it. You just plow the soil of yourself to find the quest for truth within yourself. You just get moving and you allow yourself to move around, and then you will see the benefit"… Rumi (Jalāl ad Dīn Muhammad Balkhī, 1207–1273)[1]

Since 29 December,1959, when Richard Feynman delivered an after-dinner lecture at the annual meeting of the American Physical Society and expressed that "there's plenty of room at the bottom," nanotechnology, the science and technology of working with the smallest possible materials, has raised hopes for the future to create new opportunities such as in telecommunication and to overcome challenges encountered in various areas such as medicine, engineering, agriculture, physics, chemistry, etc.

Recently there has been increasing interest in new research and industrial applications for the natural mineral nanotubes such as palygorskite, sepiolite, and halloysite due to their tubular morphology, nanoscale diameters, and their specific chemistry. These natural nanotubes are considered as clay minerals and are found in different geographical and geological environments in the USA, China, Spain, New Zealand, Australia, Turkey, Iran, Algeria, among other countries.

khâk, literally meaning "soil"; Khaki (UK/'kɑːkiː/, US/'kækiː/, in Canada/'kɑrkiː/[[1]]) is a color, a light shade of yellow-brown similar to tan or beige. *Khaki* is a loan word incorporated from Hindustani ख़ाकी and Urdu یکاخ (both meaning "soil-colored") and is originally derived from the Persian: یکاخ [xɒː'kiː], which came to English from British India via the British Indian Army since 1848 (Wikipedia.com; www.merriam-webster.com, Accessed Jan 2014).

This book tries to introduce its readers to the major types of natural mineral nanotubes which are providing increasing insights and benefits to nanotechnology-related subjects for application in various areas such as engineered plastics, medicine, the environment, agriculture, and chemistry.

---

[1]The translation of this quote by Rumi was selected from three different resources (Discourses of Rumi (or Fihi mafih) by, A. J. Arberry, Omphaloskepsis, Ames, Iowa; Dr Fatemeh Keshavarz's talk on http://www.onbeing.org/program/ecstatic-faith-rumi/189 and http://www.chooseyourmetaphor.com/soil-as-metaphor/by Daneil Goldsmith).

The book has been divided into 10 parts containing 26 chapters overall. In the first chapter, Steve Guggenheim introduces the structures, chemistries, and textures of chrysotile, halloysite, imogolite, and allophane as well as palygorskite-sepiolite phyllosilicates. In the first part of his chapter he describes a general classification scheme commonly used to understand the connections between the structural components. In Part II of this book there are three chapters which focus on the identification and nomenclature of halloysite, palygorskite-sepiolite and chrysotile, respectively, in a historical perspective. In Chapter 2, Jock Churchman explains the chemistry, morphology and definition of halloysite nanotubes (HNTs) and he discusses the relationship between halloysite and kaolinite. In Chapter 3, by Emilio Galán and co-authors, and Chapter 4, by Saverio Fiore and F. Javier Huertas, these aspects are discussed for the palygorskite-sepiolite group and for chrysotile, respectively. Part III of this book is devoted to the geology, mineralogy, and occurrences of these natural mineral nanotubes. John Keeling, in Chapter 5, describes the major halloysite deposits in the world including those of Northland New Zealand, the Dragon mine in Utah and the Biga Peninsula in Turkey, other deposits in China, Thailand, Algeria, and Japan, as well as a unique type of halloysite with very uniform tubular shapes and high surface area (~76 m^2/g) from Camel Lake in South Australia. Further, in Chapter 6 in the same section, Emilio Galán and Manuel Pozo report that economical deposits of sepiolite and palygorskite have complex geological origins and although their mineralogy and geological settings are different, genetic integrative models can nonetheless establish useful patterns for exploration purposes. They report that the annual tonnages of mined palygorskite and sepiolite are estimated to be 1,300,000 and 850,000 tons, respectively and they identify the USA and Spain as the largest producers of palygorskite and sepiolite, respectively, accounting in turn for about 75 and 95 percent of the world's annual production. In Chapter 7, Saverio Fiore explains that the recent discoveries of seafloor serpentinites and the unique mechanical and physical properties of chrysotile have given rise to a renewed scientific interest in serpentine minerals, and more specifically chrysotile, including for the earth and environmental sciences, geomicrobiology, and materials science.

In Parts IV to IX, the book presents some recent and interesting applications of these natural mineral nanotubes. In Chapter 8 polymer nanocomposites reinforced by halloysite nanotubes and their different applications are reviewed by Vahdat Vahedi and Pooria Pasbakhsh. They show that the lower cost, unique chemical structure and nanotubular shapes of halloysite, alongside their easier processability and the generally superior mechanical properties of their composite products especially at higher loadings compared to carbon nanotubes and montmorillonite, make halloysite a material of special interest to the polymer composite industry. Part V brings four chapters related to the usage of natural mineral nanotubes in biocomposite materials. In the first of these chapters, Eduardo Ruiz-Hitzky and co-authors describe how bionanocomposites based on sepiolite and palygorskite- "exotic clay minerals"- are combined with polysaccharides, proteins and other biomolecules (e.g., phospholipids). These materials give rise to biodegradable, biocompatible, and bio-mimetic platforms which can introduce specific functionalities enabling their uses in various applications. In Chapter 10, Rangika De Silva and coauthors review the possible applications of halloysite nanotubes in reinforcing Polylactic Acid (PLA) bionanocomposites for packaging applications, while biomedical applications of Poly (lactic-co-glycolic acid) biocomposites with halloysite are discussed in Chapter 11 by Mingxian Liu and colleagues in the context of a research study. They show that while these nanocomposites can have good biocompatibility and sustained drug release abilities in addition to their promising applications in tissue engineering and as absorbable antiadhesive membranes, one of the major obstacles to their use is the possible non-biodegradation of HNTs *in vivo*; this shows that there is a long way to go before their practical application in clinical treatments. In the last chapter of this section Kavitha Govindasamy and coauthors present and review chitosan-halloysite nanoubes bionanocomposites as new types of materials. Both chitosan and halloysite are abundantly available and obtainable at low cost and both are biocompatible so this makes their composites attractive for industries like packaging and for medicine.

At the current stage of progress in and usage of polymer-natural mineral nanotube composites, a key question which remains unanswered is that of being able to determine the separate mechanical properties of the individual nanotubes in order to assess their contribution to the final products as polymer composites. In Part VI there are two chapters which show how the elastic modulus of halloysite nanotubes can be calculated and modeled. Benoit Lecouvet (in Chapter 13) and Kheng Lim Goh (in Chapter 14), together with their coauthors, respectively approach this challenge experimentally (by using AFM) and mathematically.

Four chapters in Part VII focus on the procedures for modification of different mineral nanotubes, as well as the effect of modified mineral nanotubes on their performance in their applications and on their final products. In Chapter 15, Vahdat Vahedi and Pooria Pasbakhsh review the research progress in the functionalization of halloysite nanotubes. Gustave Kenne Dedzo and Christian Detellier, in coauthoring Chapter 16, review and discuss the reactive sites, mechanisms involved and strategies for the modification of sepiolite and palygorskite as well as some examples of applications of these nanomaterials. Binoy Sarkar and Ravi Naidu show, in Chapter 17 show that organopalygorskites can find potential applications to remediate contaminated waters and soils due to the easier and relatively inexpensive preparation methods required compared to those for other commercially available materials. In research work described in Chapter 18, Vahdat Vahedi and coauthors discuss how and why the quantity of hydroxyl groups on the surface of halloysite nanotubes can affect their efficiency for modification as well as the final performance of polymer composites.

Returning to the applications of mineral nanotubes again in Part VIII, three chapters therein describe the uses of natural mineral nanotubes in industrial and agricultural applications. In Chapter 19 Abdullayev and Lvov review some very promising prospects for the use of HNTs in commercial applications, including as nanoscale additives in polymers, as nanocontainers and for controlled release such as in anticorrosion and antimolding protection, flame-retardancy, dentistry, cosmetics, catalysts for hydrocarbon cracking, and many others. Benoit Lecouvet reviews some very novel and interesting industrial, environmental and biomedical applications of halloysite nanotubes in Chapter 20. In Chapter 21, Gustave Kenne Dedzo and Christian Detellier report that the external surfaces of sepiolite and palygorskite can undergo chemical modifications or they can be modified by metals or enzymes. They further report that, although it is difficult to encapsulate the reagents and control the reaction conditions in their inner tunnels, this approach cannot be dismissed, with future work being aimed at controlling the environment in the tunnels of these nanotubes.

Other interesting applications of natural mineral nanotubes that are being developed currently are found in medical topics and some of these are covered in Part IX, comprising three chapters. Elashad Abdullayev from an industrial perspective reports in Chapter 22 on some novel and fascinating uses of HNTs such as for the encapsulation of biologically active agents into the HNTs' lumen structure, as halloysite-polymer scaffolds in tissue engineering and as substrates for the immobilization of enzymes and these same aspects for sepiolite, palygorskite, imogoilite, and again for halloysite are reviewed and discussed in Chapter 23 by César Viseras and coauthors. Chapter 24 sees Javiera Cervini-Silva and coauthors presenting a research study on anti-inflammatory properties of different halloysites which shows that there is a direct relationship between the surface area of the halloysites and their anti-inflammatory activity and cytotoxicity. In the final chapter of Part IX, Marie-Claude Jaurand interestingly compares some possible health effects of natural mineral nanotubes with those of carbon nanotubes, which is described extensively for its health effects. She shows that both CNTs and asbestos produce oxidative stress, inflammation, activation of signalling pathways, and genotoxicity with the longest particles being the most dangerous to health.

The final chapter of this book (26) in Part X consists of a broad editorial review and summary of the field covered in the book, including data on the current trends in research and applications of the natural mineral nanotubes. Just as Feynman in 1959 pointed to "the problem of manipulating and controlling things on a small scale," we also believe that manipulating and controlling the structure (including their lumen space) and chemistry (including the nature of surfaces and of lumen functional groups) of natural mineral nanotubes can enable these unique nanomaterials to provide advanced applications in medicine, agriculture, the environment, and engineered plastics.

PART I
THE MAJOR NANOTUBULAR MINERALS
AND THEIR STRUCTURES

CHAPTER 1

PHYLLOSILICATES USED AS NANOTUBE SUBSTRATES IN ENGINEERED MATERIALS: STRUCTURES, CHEMISTRIES, AND TEXTURES

STEPHEN GUGGENHEIM

CONTENTS

1.1 INTRODUCTION

Naturally occurring phyllosilicates (silicate-based layer structures) are structurally diverse and some of these phases have the ability to form nanotube and tunnel forms. These tubes and tunnels are mesoporous because the tube center and the tunnels can potentially encapsulate ions and molecules. Chrysotile, ideally $Mg_3Si_2O_5(OH)_4$, is a common asbestos mineral that is nanotubular in form with potentially deleterious biological effects when inhaled, but the mineral has been considered as a nanowire substrate and synthetic analogs may avoid the human health issues (e.g., Roveri et al., 2006). In contrast, the nanotube mineral halloysite, ideally $Al_2Si_2O_5(OH)_4$ when H_2O is not part of the structure, is of low toxicity and is potentially useful in many biologically relevant applications, such as in drug delivery systems (e.g., Veerabadran et al., 2007) and delivery systems of insecticides, herbicides, etc. (Lvov et al., 2008). The tunnel-like structures of palygorskite-sepiolite minerals are not believed to have health hazards, have excellent absorptive, rheological and catalytic properties, and are used in many industrial applications including nanomaterials (e.g., Galán, 1996, Álvarez et al., 2011). Imogolite and allophane (both Al-rich silicates) and hisingerite (an Fe-rich silicate) are nanotubular minerals that are poorly crystalline, and the structures are therefore poorly known. Duarte et al. (2012) have reviewed computer-modeling studies of imogolite, halloysite, and chrysotile.

This chapter, which emphasizes (experimental) structural studies, is an effort to provide an introduction for the researcher with a chemical or materials science background, but with limited mineralogical background; it is not meant to be an all-inclusive review. Recent reviews are referred in the text at appropriate locations, and they cover other aspects of the literature in great detail. The reader is encouraged to consult with these papers or original sources if more background is desired. This chapter is organized in the following way: this section provides a general background, a description of structural components, a general classification scheme, and a description of structural parameters commonly used to understand connectivity between the structural components. The remaining sections cover four nanomaterial substrates: chrysotile, halloysite, imogolite and allophane, and palygorskite-sepiolite. Parts of this chapter have been presented previously in Guggenheim (2011) and Guggenheim and Krekeler (2011).

In this chapter, wherever tube fibers are described relative to unit cell dimensions of the layers, the layer-stacking direction is the c axis repeat distance. The fiber axis is generally not defined as the c axis because for the layer minerals described here, the fiber axis is usually along X or Y within the layer. A useful construct is to consider the structure much like a carpet, with the layers being analogous to the carpet when used as a floor covering (the c axis, more precisely c^*, is perpendicular to the floor and the a and b axes are within the plane of the carpet), but when the carpet is rolled for storage the axis of the scroll is either X or Y. The symbols X and Y are equivalent to the [100] and [010] directions, respectively.

1.1.1   THE IDEALIZED PHYLLOSILICATE MODEL AND BASIC NOMENCLATURE

The phyllosilicate structure involves a series of *planes* of atoms, each plane defined as a single set of atoms stretching in two dimensions. Planes may be a set of atoms (e.g., oxygen anions) or dissimilar atoms (e.g., Si and Al). Three planes, an anion O plane, a plane with tetrahedrally coordinated cations, and an O + OH plane, form a tetrahedral sheet. The oxygen anion plane, which forms the base of the model, is often referred to as the basal plane. The tetrahedral (*T*) cations are commonly Si and Al, which are considerably smaller in size than oxygen anions, and they sit nestled in the "dimple" formed from three adjacent basal oxygen atoms (where atoms, to a first approximation, are considered as hard spheres). To complete a fourfold coordination about the Si, a fourth oxygen anion, the apical oxygen anion, sits adjacent to the Si and off the basal plane. Thus, the tetrahedral sheet is ideally comprised of corner-sharing tetrahedra which form hexagonal-like rings extending in two dimensions as shown in Fig. 1a. The OH is located in the same plane as the apical oxygen anions and is considered a part of the tetrahedral sheet, but the OH is not a part of any tetrahedron. Instead, the OH group is in the center of the hexagonal ring.

The top part of the tetrahedral sheet, the O + OH plane, is also the bottom part of the octahedral sheet, and thus this plane is a common junction between the two sheets. The ideal octahedral sheet (Figure 1.1b) differs for different types of phyllosilicates. For example, for an ideal model of the serpentine and kaolin mineral groups, the octahedral sheet consists of the O + OH plane, a cation plane with sixfold coordinated cations (*M*), and a terminating plane of OH groups. The *M* cations (e.g., Mg, Fe, Al) sit in the "dimple" formed by two apical oxygen atoms and the adjacent OH group on the top part of the tetrahedral sheet, and this arrangement forms half of the sixfold coordination of the *M* cations. The remaining portion of the coordination unit involves the terminating plane of OH groups, with the *M* cation closest packed with OH groups by residing in the "dimples" formed from three adjacent OH groups. Each *M* cation is coordinated to two O atoms and four OH groups. This layer, the *1:1 layer*, has a 1:1 ratio of tetrahedral sheets to octahedral sheets (Figure 1.1c). For the remaining types of phyllosilicate minerals, a *2:1 layer* is the basis of the structure, which is formed where a second tetrahedral sheet is inverted relative to the first, such that O + OH planes occur on both sides of the *M* atom plane (Figure 1.1d). In this case, each *M* cation is coordinated to four O atoms and two OH groups, and the OH groups are well within the layer (instead of on the top surface as in 1:1 layers). The construction of the model defines the different portions of the phyllosilicate structure precisely with increasing thicknesses: planes, sheets, and layers, and these terms should not be used interchangeably.

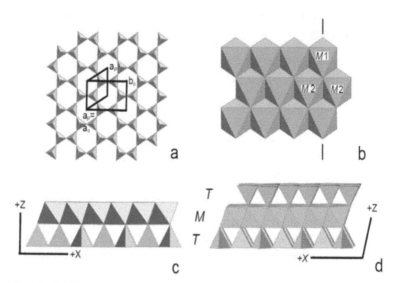

FIGURE 1.1 Different parts of the phyllosilicate structure are illustrated. A hexagonal network of silicate or aluminosilicate tetrahedra (e.g., Si cations coordinated by four oxygen anions) are shown in (a) and this is commonly referred to as the "tetrahedral sheet." An "octahedral sheet" (six coordinated M cations with anions forming octahedral polyhedra) is shown in (b); the dark line represents a mirror plane separating the two M2 sites, with the M1 site on the mirror plane. In (a), two unit cells are shown, one is an orthohexagonal cell (a_o, b_o) and the other is a primitive hexagonal cell. If distortions occur, the hexagonal cell may be referred to as being "pseudohexagonal." The orthohexagonal cell is C centered. Most phyllosilicate structure descriptions use the orthohexagonal cell for convenience of comparison. Two types of layers are illustrated, in (c) the 1:1 layer is shown and in (d) the 2:1 layer is illustrated, with labels of T = tetrahedral sheet and M = octahedral sheet. Note that the common plane of junction between the tetrahedral and octahedral sheets involves the apical oxygen anions of the tetrahedra, and it is this common junction that is involved in minimizing misfit between the two component sheets (After Bailey, 1980, from Guggenheim, 2011, reproduced with the kind permission of the European Mineralogical Union).

A completed model (Figure 1.2) involves the placement of material between the layers (in the interlayer) for many 2:1 layer structures. For both 1:1 and 2:1 structures, the placement of another layer above the first establishes a layer sequence, and this is referred to as stacking. Layer structures, like the phyllosilicates, have a variety of structures that result from different layer-stacking sequences. The term *polytypism* refers to different layer-stacking sequences within a mineral species, and the layers are compositionally similar. With regard to interlayer materials, cations, cation-hydrate complexes, hydroxide sheets, polar molecules, or no interlayer material can occur. Where interlayer material is present, this material has a net positive charge or is polar. Control of what enters the interlayer is dependent on bulk chemistry and the charge on the layer (layer charge). The net charge on the layer is either zero or negative. For 2:1 phyllosilicates, the talc-pyrophyllite group has no net layer charge and no interlayer material, and the layers are bonded by van der Waals interactions. For micas, an interlayer cation resides in the silicate ring, with a univalent cation in the true micas (and the layer charge of −1) and a divalent cation in the brittle micas (and the layer charge of −2). The vermiculites and smectites, sometimes considered as separate groups and sometimes not, have a layer charge of approximately −0.2 to −0.6 (smectite) and −0.6 to −0.85 (vermiculite). Both have cations + H_2O molecules in the interlayer, with the cations offsetting the net negative charge on the layer and the H_2O variable in content, depending on the cation and the activity of H_2O in the environment around the particles. Because of the ability of vermiculite and smectite to expand owing to the variable interlayer H_2O, these clay minerals are referred to as "swelling clays." Finally, the chlorites consist of the 2:1 layer and a positively charged interlayer of ideal composition of $(R^{2+}, R^{3+})_3(OH)_6$, where R is a cation. A combination of hydrogen bonds and electrostatic interactions link the interlayer to the 2:1 layer in chlorites. In general, 1:1 phyllosilicates also have both hydrogen bonding and electrostatic interactions linking the layers, but unlike the 2:1 layer structures, the layers do not have a significant net layer charge.

In the ideal case, a trioctahedral sheet forms where only divalent cations reside in all possible M site positions. The smallest structural unit contains three octahedral cations and a trioctahedral arrangement indicates a charge of $3 \times 2^+ = 6^+$ electrostatic valence charges. A dioctahedral sheet forms where one of the three sites is vacant, and to maintain charge balance, the remaining M sites of the smallest structural unit must be filled with trivalent cations, hence $2 \times 3^+ = 6^+$ charges. In general, a group (Figure 1.2) is defined on the basis of both the type of layer and the layer charge, subgroups are based on the presence of trioctahedral or dioctahedral varieties, and chemical composition determines differences among species. The octahedral sites are usually designated as $M1$, which resides on the mirror plane in ideal symmetry (in micas, $C2/m$: $a \approx 5.4$ Å, $b \approx 9.5$ Å, $c \approx 10.0$ Å, $\beta \approx 100°$) for a one-layer structure, and two $M2$ sites, which are related by reflection across the mirror plane. The $M1$ is the *trans* site because OH groups are located in opposing corners of the octahedron, whereas the $M2$ sites are *cis* sites because these groups are at adjacent corners.

The 1:1 layer structure is noncentric because the layer is composed of a single tetrahedral sheet residing on one side of the octahedral sheet. The serpentine minerals are trioctahedral and, for the ideal one-layer platy structure (e.g., lizardite), there is only one symmetry-unique octahedral site and one tetrahedral site. In contrast, the platy dioctahedral kaolin minerals [kaolinite, nacrite, dickite, and halloysite (7 Å)] are polymorphs, and they are more complex primarily because of the arrangement of the octahedral vacant site. Important substrates for nanocomposite materials include a fibrous and nonplaty serpentine mineral, chrysotile, and the dioctahedral kaolin mineral, halloysite, also a fibrous and nonplaty form. Both chrysotile and halloysite are discussed in more detail below.

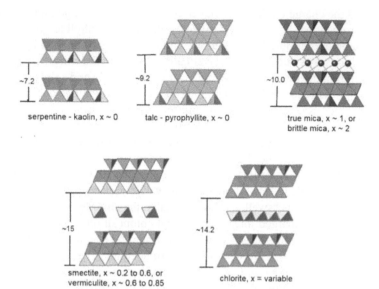

FIGURE 1.2 Five basic phyllosilicate groups are shown. The classification is based on the type of layer (1:1 or 2:1) and the charge per formula unit, x. The value x is the layer charge given as a positive number (= interlayer charge). For serpentine-kaolin and talc-pyrophyllite, the assigned names are given as trioctahedral members, followed by the dioctahedral members. The other groups also have trioctahedral and dioctahedral members, but they are not included as part of the group name. Two mica groups (true micas and brittle mica), the smectite group, and the vermiculite group are separated based on layer charge also. The interlayer material may be cations as in the micas, hydrate complexes as found in smectite and vermiculite, and hydroxide sheets, which are found in the chlorites. The $d(001)$ values (given in Å) found to the left of each illustration are the typical layer to layer spacing. For the cases of smectite and vermiculite, each of which may swell upon exposure to humidity or other changes in environmental conditions, these values are very rough approximations depending on conditions. Many phyllosilicate structures used as nanotube substrates do not fall directly into these groups, but are related. For example, halloysite is classified as belonging to the 1:1 phyllosilicates (serpentine-kaolin group), but halloysite has variable numbers of H_2O molecules in the interlayer (After Bailey, 1980, from Guggenheim, 2011, reproduced with kind permission of the European Mineralogical Union.)

1.1.2   TETRAHEDRAL/OCTAHEDRAL SHEET MISFIT

Because there is a common junction of O + OH between the tetrahedral and octa-hedral sheets, these two sheets must compensate for a lateral misfit between the tetrahedral sheet and the octahedral sheet (Figure 1.3). In ideal planar structures, a relatively larger tetrahedral sheet can reduce its lateral dimensions via tetrahedral rotation, the in-plane rotation of adjacent tetrahedra in opposite directions around the silicate ring, which is measured by the tetrahedral rotation angle, α. This adjustment is common in most phyllosilicates, and this is one reason why so many different compositions of phyllosilicates can exist. Although tetrahedral rotation occurs in kaolinite, it does not appear to occur in halloysite, and the reason for this is discussed in more detail below. For cases where the lateral dimensions of the tetrahedral sheet are smaller than the lateral dimensions of the octahedral sheet, the mechanisms to compensate for misfit are considerably more complex and may involve perturbations that include tetrahedral inversions, incomplete tetrahedral sheets, etc. (see Guggenheim and Eggleton, 1988 for a review). In chrysotile, the mechanism to relieve misfit is an out-of-plane tilting of tetrahedra such that the apical oxygen anion distance of the tetrahedral sheet can increase to form a common junction with the octahedral sheet. The result is the scroll or cylinder form.

There are other ways to reduce misfit, but the structural results are minimal and it is not always clear if they result only from misfit issues or other factors. The tetrahedral sheet may decrease or increase its lateral dimensions by adjusting the three apical oxygen-T-basal oxygen angles (one of the angles, designated as τ, is shown in Figure 1.3). An increase in τ produces a decrease in the lateral dimensions of the tetrahedral sheet. The angle ψ (Figure 1.3) is defined as the body diagonal of an octahedron and the vertical (relative to the basal plane). Thus, an increase in ψ results in an elongation of the octahedron and an increase in the lateral dimension of the octahedral sheet. (Portions of this section were updated from Guggenheim, 2011, and reprinted with the kind permission of the European Mineralogical Union.)

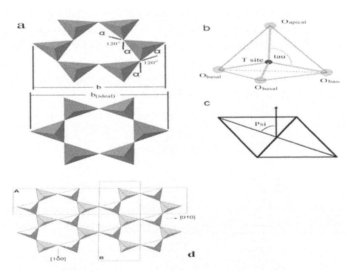

FIGURE 1.3 Illustration of three parameters that describe structural distortions. (a) Description of how the tetrahedral sheet may reduce its lateral dimensions by tetrahedral rotation; the tetrahedral rotation angle, α, is given as the angular deviation from 120° of adjacent tetrahedra divided by 2. The resultant rotation produces a ditrigonal silicate ring. In Part (b), the tau (τ) value is the O_{apical}-T-O_{basal} angle and it may be either an individual angle

FIGURE 1.3 *(Caption continued)*

or the average of the three O_{apical}-T-O_{basal} angles in a tetrahedron. The ideal value is 109.47°. The value increases with an increase in tetrahedral height and a reduction in width. In Part (c), the psi (ψ) value is described as the angle between the vertical and the body diagonal. Part (d) shows how tetrahedral rotation affects the linkage of chains within the tetrahedral sheet. For example, one chain is segmented (but continuous) and occurs along the [010] direction (labeled A), whereas the other chain is only slightly kinked and continuous along the [100] direction (labeled B, and often referred to as a "pyroxene" chain). The amount of kinking is related to α; at α 0, the pyroxene chain is fully extended (*cf.* Part (a)). (Modified from Guggenheim and Krekeler, 2011, and reproduced with the kind permission of Elsevier Press.)

1.2   CHRYSOTILE

The magnesium-rich serpentines occur as three general structural forms: planar structures with platy morphology (lizardite, Figure 1.4); alternating wave structures (antigorite), which are platy also; and roll structures (chrysotile) forming fibrous morphology. Magnesium-rich lizardite has a near-end-member composition of $Mg_3Si_2O_5(OH)_4$, whereas the antigorite structure, which commonly has structural adjustments at wave and half-wave boundaries (see below), is near $Mg_{28}Si_2O_5(OH)_{36}$ for a common form. Half waves involve 1:1 layers with tetrahedra tilting to form a curvature of the sheet with a radius of near 70 Å, which is similar to the roll radius in chrysotile. Linking half waves in antigorite produces a reversal in direction of the tetrahedral apices so that the tetrahedral sheet coordinates to an adjacent octahedral sheet. Chrysotile is not spatially homogeneous in its properties (nor is halloysite) and does not satisfy the conditions as required for the Phase Rule, and thus chrysotile is metastable relative to antigorite and lizardite (see Evans, 2004 for an in-depth discussion of stability relationships, possible parameters to control the radial dimensions, and effects of stress, pores, and modes of occurrence).

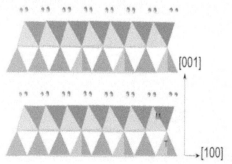

FIGURE 1.4 The projection of the platy serpentine mineral, lizardite (Data from Mellini et al., 2010), down the [010] direction using the CrystalMaker (v. 2.7) visualization program. M = octahedral site, T = tetrahedral site. The spheres in the interlayer are H+ ions that serve to link the 1:1 layers via O-H–O bonds across the interlayer.

1.2.1 UNIT CELL

Chrysotile $Mg_3Si_2O_5(OH)_4$ occurs as a monoclinic cell (clinochrysotile with the fiber axis as X) and as orthorhombic unit cells (orthochrysotile, fiber axis along X, and parachrysotile with the fiber axis along Y) (Whittaker, 1953, 1956a, b, c). Clinochrysotile is usually the dominant form of chrysotile, followed by or-thochrysotile, although the latter may occasionally occur as the dominant form, and parachrysotile is rare. Two dominant types of layer-stacking sequences occur within the fibers of clinochrysotile: chrysotile-$2M_{cl}$ ("2" indicates two-layer repeat, M = monoclinic, the subscript "c" indicates cylindrical, and the subscript "1" distinguishes the structure from other two-layer monoclinic cylindrical-type polytypes; older literature uses "Or" instead of "O" to designate orthorhombic) with cell parameters of $a \sim 5.34$, $b \sim 9.25$, $c \sim 14.65$ Å, $\beta \sim 93.3°$, and chrysotile-$1M_{cl}$ (Zvyagin, 1967) with a cell of $a \sim 5.35$, $b \sim 9.25$, $c \sim 7.33$ Å, $\beta \sim 94.2°$. The orthochrysotile polytype, chrysotile-$2O_{cl}$, has cell parameters of $a \sim 5.34$, $b \sim 9.25$, $c \sim 14.63$ Å, $\beta \sim 90°$, and parachrysotile is a two-layer orthorhombic structure also, with $a \sim 5.3$, $b \sim 9.24$, $c \sim 14.7$ Å, $\beta \sim 90°$. The orthorhombic forms are comprised of two layers oriented such that the upper layer is rotated by 180° relative to the lower layer. Bailey (1969) has shown how to derive these orthorhombic polytypes by simple alternations of oc-tahedral set (I, II) positions from layer to layer, which closely simulates how atoms would attach themselves to the surfaces of a crystal during crystallization.

1.2.2 TEXTURE

In general, roll structures of chrysotile have been shown by electron microscopy (Figure 1.5) to have curved layer walls that form cylinders, spirals, and cone-in-cone morphologies (Yada, 1967, 1971, 1979; see also Veblen and Buseck, 1979). Although many tubes are hollow, ultrasonic treatment is capable of extracting tube fillings, which are apparently amorphous and of low density (Martinez and Comer, 1964). Fiber diameters are variable, but common values occur from an inner diam-eter of 70-80 Å to an outer diameter of 220–270 Å (Yada, 1971). Very thick fibers (>350 Å) appear to be a result of multiple-step growth mechanisms which involve dislocations oriented in two directions, radial and axial to the fiber (Jagodzinski and Kunze, 1954; Yada, 1971). Whittaker (1956a) described the diffraction effects of a clinochrysotile fiber that showed an offset of the 601 reflection. On the basis of this overlap, the fiber could be described as helical with a one unit-cell offset along the fiber axis implying an average radius of 84 Å. In selected-area electron-diffraction patterns (Yada, 1971), cylinders show 4.5-Å fringes (from imaging the 020 reflec-tion) parallel to the fiber axis, but these fringes are inclined commonly from 2 to 3° and, less commonly up to 10°, in helical forms, along with splitting of $0k0$ and $hk0$ reflections from diffraction originating from both sides of the helix. Toman and Frueh (1968) found that clinochrysotile samples from 15 sites have nearly identical average lengths, diameters of 234 ± 1 Å, and an average wall thickness of 155 ±

1 Å by using the positions and widths of X-ray reflections. Falini et al. (2004) reported synthetic stoichiometric samples with a core diameter of 70 ± 10 Å, an outer maximum diameter of 490 ± 10 Å, and a wall thickness of 70 Å. Parachrysotile forms as a helix with a mean radius of ~97 Å (Whittaker, 1956c). Whittaker (1956b) found that the orthochrysotile he studied was cylindrical with a nonhelical, circular cross-section. Calculated values of fiber sizes from X-ray diffraction techniques are consistent with direct observation by electron microscopy.

Some lizardite platelets have been observed to form nanotube-like structures at their edges (Wicks and Chatfield, 2005; Wicks, 1986), with the fiber axis along X (or pseudo-X) of the parent lizardite (Wicks and Chatfield, 2005). Unlike chrysotile, the scrolls do not appear to have a central core, but instead have "no visible core or a tapered central core, commonly with bubble-like structures" often ending in a secondary scroll at 60° to the fiber axis and with fragments of the parent platelets. The scrolls are mechanically weak, are often torn, are more stable in the electron beam of the TEM than chrysotile, and produce selected area electron-diffraction patterns sometimes similar to lizardite, but commonly more complex and quite different than chrysotile. Insufficient detail of the internal stresses in lizardite is available to determine the cause of these scrolls (Wicks and Chatfield, 2005), and they may be related to sample preparation.

A nonfibrous, splintery serpentine better described as lath-like and sometimes referred to as Povlen chrysotile, or more commonly as polygonal serpentine (Zussman et al., 1957; Krstanovic and Pavlovic, 1967; Cressey and Zussman, 1976; Cressey, 1979), occurs naturally and is often comprised of a cylindrical core and polyhedral overgrowths of platy lizardite radiating around the core. The core may also be composed of planar layers, antigorite and partially curved layers, and groups of fibers. Recently, this composite form has been studied intensely (Cressey et al., 2008, 2010; Logar and Mellini, 2009; Grauby and Baronnet, 2009; Andréani et al., 2008; Baronnet et al., 2007), but it is not commonly used as nanocomposite substrates and is not discussed further.

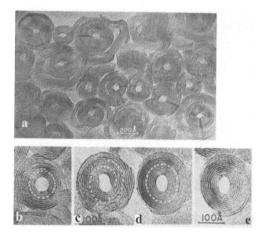

FIGURE 1.5   *(Continued)*

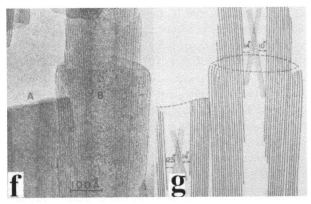

FIGURE 1.5   Transmission electron microscope image (in Part (a)) showing lattice fringes of approximately closest-packed chrysotile scrolls. The texture here suggests a stepwise growth mechanism. Parts (b) and (c) show sample coils of chrysotile with one-unit and multiunit wall thicknesses, respectively. Parts (d) and (e) show parts of a through-focus series of sample cylinders of chrysotile. Parts (f) and (g) illustrate a cone-in-cone structure for a chrysotile particle. All parts of the figure are looking down the fiber direction, except for Parts (f) and (g), which are along the fiber (From Yada, 1971, reproduced with the kind permission of the International Union of Crystallography, IUCr).

1.2.3   STRUCTURE

The present day understanding of the chrysotile structure relies on the early work of Whittaker (1956a, b, c) using fiber bundles and single-crystal, two-dimensional refinement procedures. The chrysotile-$2M_{cl}$ structure was refined using (010) projections in plane group $p111$ to $R = 0.14$ and the chrysotile-$2O_{cl}$ refinement was achieved in $p11g$ at $R = 0.21$, both using film data (mostly integrated by microdensitometer). The significance of this work was that the microtexture of the fibers could be incorporated in the interpretation of the structure as derived from the X-ray data. However, the relatively high R factors suggest that the details of the structure require additional confirmation. This texture information is lost when X-ray powder data from samples are processed, and thus more modern powder analyses (e.g., Falini et al., 2004; Leoni et al., 2004 as described below) are describing the structure of stacking associated with the fiber walls, rather than the microtexture information relating to the details of curvature. Use of the transmission electron microscope (TEM) provides information about the microtexture (e.g., Yada, 1971) and selected area diffraction can provide stacking information (e.g., Dódony and Buseck, 2004), but electron-diffraction data have not been used to provide an atomic-coordinate refinement.

The structure of clinochrysotile is not based on the layer-to-layer hydrogen bonding network that was used to theoretically derive the planar serpentine polytypes. The curving of the layers in chrysotile prevents the normal formation of hydrogen bonding (Wicks and Whittaker, 1975). The linkage of adjacent layers as described by Whittaker (1956a) appears related to steric packing effects, where a corrugated or buckled basal oxygen anion surface fixes the layers along the [100] direction parallel to the fibers. In projection, basal oxygen anions, O(1), of an upper layer is between parallel rows of OH groups on the top surface of the adjacent layer below. Two basal oxygen anions, O(1) and O(2), are not coplanar; the O(2) anion, which is withdrawn into the layer, is separated from O(1) by 0.2 Å. The O(1) anion is offset away from the layer. In projection, O(2) nearly falls directly over the OH groups, possibly indicating that hydrogen bonding may occur here (Wicks and O'Hanley, 1988). The rows of OH groups extend parallel to the [010] direction and follow the curvature of the layers at $a/2$ intervals along the fiber axis, with $\pm a/2$ disorder common. The disorder in position of $a/2$ and the lack of a hydrogen bonding network owing to layer curvature produces disorder along the [010] direction between layers. The formation of the β angle of ~93.3° also indicates how the basal oxygen anion surface of one layer is keyed to the OH-group surface of the adjacent layer. The upper layer is shifted by 0.417 Å parallel to the fiber axis to maintain the steric stacking and to obtain the 93.3° β angle value. Wicks and O'Hanley (1988) suggested that this shift may be related to hydrogen bonding, although the tendency for hydrogen bonding is very limited. There is also a shift of $\delta = 0.101$ Å caused by distortions that occur in every layer, possibly owing to tetrahedral rotation (Wicks and O'Hanley, 1988), but in opposing directions so that it does not affect the β angle, although a two-layer structure results (polytype symbol $2M_{c1}$ describes this structure). A second two-layer monoclinic structure is theoretically possible, but has not been observed. Because the two-layer aspect is based on relatively minor distortions, the X-ray reflections that define the structure as two layers are very weak. The chrysotile-$1M_{c1}$ structure has the layer shifts of 0.417 Å as the two-layer structure, but the shift of $\delta = 0.101$ Å occurs in the same + direction between layers to produce a β angle of near 94.1°. Bailey (1980) noted the similarity of this angle to the β angle of a one-layer chrysotile described by Shitov and Zvyagin (1966) involving the supplement β angle near 106.5°, thereby suggesting their equivalence. If the shift of $\delta = 0.101$ Å occurs in the direction between layers throughout, a β angle of near 92.5° is produced and the polytype is assigned the $1M_{c2}$ symbol. The chrysotile-$1M_{c1}$ structure is very rare.

In addition to the buckled basal oxygen anion surface of the layers, Wicks and Whittaker (1975) found that the Mg cation planes are slightly buckled, although Wicks and O'Hanley (1988) noted that this buckling is not related to the O(1) and O(2) displacements of 0.2 Å along the stacking direction around the fiber. In chrysotile, the plane of the Mg cations is shifted from the center of the octahedral sheet and away from the tetrahedral sheet by 0.07 Å, and a similar shift occurs in lizardite (0.08, 0.09 Å; based on refinements by Mellini 1982; Mellini and Zanazzi, 1987).

Also, the thicknesses of the tetrahedral sheet (2.1–2.4 Å in $2M_{c1}$ and 2.13 Å in $2O_{c1}$) and the octahedral sheet (2.08 Å in both polytypes) can be compared to powder data results (Table 1.1). Table 1.1 shows octahedral sheet thicknesses in chryso-tile of 1.81–2.00 Å, which are comparable to lizardite (tetrahedral sheet thickness, 2.21–2.19 Å; octahedral sheet thickness, 2.12–2.16 Å).

TABLE 1.1 Structural parameters for platy serpentine (lizardite) and chrysotile. The lizardite structure of Mellini et al. (2010) is near-end-member composition and thus with tetrahedral/octahedral sheet misfit approaching that of chrysotile

Parameter	Lizardite (Data from Mellini et al., 2010)		Chrysotile (Data from Leoni et al., 2004; Falini et al., 2004	
	KG-2	KG-3	Falini et al.	Leoni et al.
α (°)[a]	0.84	0.72	-10.75	-1.21
Sheet thickness				
octahedral (Å)	2.115	2.155	1.813	1.999
tetrahedral (Å)[b]	2.208	2.194	2.434	2.051
Interlayer separation (Å)	2.967	2.952	3.155	3.124
τ (°)[c]	111.30	110.90	T1: 109.6; T2: 109.2	T1: 105.2; T2=T3=T4: 105.3
ψ (°)[d]	59.09	58.65	M1: 63.12; M2: 63.28;	M1=M2=M3=M4: 61.36;
			M3: 63.03	M5=M6: 61.39

[a] α 0.5 [120° mean $O_b - O_b - O_b$ angle], minus indicates basal oxygen atoms of upper layer move toward O (of hydroxyl group) of adjacent layer to achieve more favorable H bonding
[b] Tetrahedral thickness includes "internal" OH anion within silicate ring
[c] Defined in Figure 1.3, average of three values
[d] Defined in Figure 1.3, cos ψ (octahedral sheet thickness)/[2(size of M site)]

Falini et al. (2004) used synthetic chrysotile tubes to partially refine the structure by Rietveld techniques in Cc symmetry without using parameters that relate to the curved cylindrical effect (texture). Thus, the agreement factors were only fair at R_{wp} = 0.260 and R_p = 0.234. They refined the cell parameters to a = 5.340(10), b = 9.241(10), c = 14.689(20) Å, and β = 93.66(3)°, which are close to the $2M_{c1}$ values of Whittaker (1956a). Using a modified Rietveld refinement program capable of including stacking disorder information and limited curvature-related information, Leoni et al. (2004) partially refined models using similar materials described by

Falini et al. (2004). They found a more reasonable fit by approximating random stacking disorder along Y for a six-layer model, and by developing an "idealized" single layer. The reported atomic coordinates are of an ideal orthogonal layer that, if randomized appropriately, yielded $R_{wp} = 0.139$ and $R_{exp} = 0.023$ and orthogonal cell parameters, which are consistent with the Rietveld results. Because all refinement programs involve minimizing the fit between calculated and observed data, the resulting model in a Rietveld refinement may have systematic errors from the adjustment of the model to any texture information contained in the powder data. Dódony and Buseck (2004) simulated chrysotile fibers with a 323-Å diameter down the X fiber axis using a lizardite starting model to obtain simulated X-ray and electron-diffraction patterns for all the cylindrical chrysotile polytypes for comparison to observed data. A TEM image of a single fiber, Fourier transformed to obtain a diffraction pattern, showed that the pattern consisted of streaking indicating random stacking. Other fibers showed similar disorder. Phyllosilicate samples often occur having disorder stacking whereas other samples may show stacking order, and thus a general conclusion that all fibers have random stacking would not be appropriate.

In contrast to the two-layer clinochrysotile structure that involves a distortion that is relatively small, the two-layer structure of orthochrysotile is produced by the alternation of the occupancy between two sets of octahedral sites. One of these two octahedral sets of sites is found in one layer and the other set of sites is found in an adjacent layer, to complete a two-layer structure. Both sets of sites can potentially exist within a layer to maintain closest packing but only one set can be occupied at a time (mixing of sets within a layer is prohibited owing to cation-to-cation repulsions). The resulting diffraction pattern shows strong reflections indicative of the strong two-layer character. In this structure, $2O_{c1}$, successive layers are shifted by 0.417 Å parallel to the fiber axis, but the distortional shifts of magnitude 0.101 Å are negative. A model, not yet found, is designated $2O_{c2}$, which is similar to $2O_{c1}$ but with distortional shifts of $+\delta$ between every layer. Yada (1979) resolved the Mg and Si positions in chrysotile-$2M_{c1}$ and chrysotile-$2O_{c1}$ by TEM and confirmed the generalized aspects of the two stacking sequences. Dódony and Buseck (2004), using a TEM image taken perpendicular to the fiber axis, showed orthorhombic symmetry with spacings that differ from $2O_{c1}$, indicating a different two-layer cylindrical polytype.

Parachrysotile is very rare and has not been well studied. Like in clinochrysotile, there are basal oxygen anion corrugations but at intervals of $nb/6$ along the fiber axis (Y), probably with considerable disorder. Alternate layers may be displaced by $b/12$ shifts (Whittaker, 1956c). Yada and Tanji (1980) used high-resolution TEM observations to confirm the stacking as a two-layer structure.

1.2.4   CHEMISTRY

Chernosky (1975) has shown in synthesis studies that lizardite (platy) serpentine coexists with chrysotile in the range of $Mg_3Si_2O_5(OH)_4$ to $Mg_{2.875}Al_{0.125}(Si_{1.875}Al_{0.125})$

$O_5(OH)_4$, although the individual product particles were not analyzed. The chrysotile fibers are shorter, but thicker and less abundant as Al content increases. Bailey (1980) concluded that chrysotile can occur with small amounts of Al and platy serpentines can exist without Al, and Wicks and Plant (1979) determined that most chrysotile samples contain less than 0.9 wt % Al_2O_3. By changing the temperature of crystallization, Chernosky (1975) was able to take chrysotile tubes and lizardite plates at a composition of $Mg_3Si_2O_5(OH)_4$ synthesized at 431°C (P = 2 kbar) and recrystallize the assemblage to only lizardite plates at 413°C (P = 2 kbar). Using SiO_2 gels and cryptocrystalline $Mg(OH)_2$ reactants in a highly basic environment, Jancar and Suvorov (2006) examined the synthesis parameters and physical properties of the resultant tubes, with the lengths of tubes affected by the temperature of crystallization and duration of the experiment, but the average diameter of the tubes is unaffected. Helical structures appear to form from plates. In a kinetic study at 300°C and Si/Mg molar ratio of 0.67, Lafay et al. (2013) found that reaction steps could be described with a serpentine precursor of a flake-like morphology forming initially within two hours, nucleation and growth of chrysotile by dissolution of the precursor from 3 to 8 h, and Ostwald ripening (small chrysotile crystals dissolved and redeposited on larger crystals) from 8 to 30 h.

Wicks and Plant (1979) found that iron (calculated as FeO from electron microprobe analyses) may be present in natural chrysotile at <6 wt %. Rozenson et al. (1979) and Blaauw et al. (1979) concluded from Mössbauer analyses that Fe^{2+} only occupies the octahedral sites and that Fe^{3+} may also occur in tetrahedral coordination, but when it does, the substitution is very limited. O'Hanley and Dyar (1998), also using Mössbauer techniques, confirmed the prior work and further showed that the proportion of tetrahedral Fe^{3+} to octahedral Fe^{3+} was inversely related so that a nearly constant total Fe^{3+} is maintained. They suggested that lizardite and chrysotile are not polymorphic in natural systems because chrysotile has greater Fe^{2+} and tetrahedral Al. In addition, chrysotile has fewer tetrahedral Fe^{3+} cations and H^+ vacancies than lizardite. Fe-doped (0.29 to 1.37 wt %) chrysotile confirmed that iron can enter both the octahedral and tetrahedral sites, and there is an apparent flattening of the octahedral sheet (Foresti et al., 2005a). Viti and Mellini (1997) identified a compositional gap in a serpentinite at Elba, Italy, between lizardite and chrysotile, with lizardite having greater Al and Fe than chrysotile, which also suggests that the two are not true polymorphs in natural systems. A natural Ni-rich clinochrysotile, pecoraite has been described by Faust et al. (1969), and synthesized by Perbost et al. (2003) and Korytkova et al. (2005). Ni substitution reduces the length and diameter of the nanotubes. Noll et al. (1960) and Jasmund and Sylla (1970) have synthesized Co^{2+} and Fe^{2+} forms of clinochrysotile. Ti-substituted (0.25–11.50 wt % Ti) chrysotile increases the outer diameter of the nanotubes and reduces the length of the fiber (Foresti et al., 2005b). Trace amounts (<200 μg/g) of boron have been shown to occur in chrysotile, probably in the tetrahedral site (Pabst et al., 2011).

1.2.5 WHY ROLLING OF THE LAYERS OCCURS

Chernosky (1975) calculated that the ideal composition of zero misfit for Al-bearing, Mg-rich serpentine based on $b_{tet} = b_{oct}$ is: $Mg_{2.7}Al_{0.3}(Si_{1.7}Al_{0.3})O_5(OH)_4$, (at 11 wt % Al_2O_3) but it is clear (see above) that both platy and rolled forms can exist over a significant range in composition . Wicks and Plant (1979) found that most chrysotile contained less than 0.9 wt % Al_2O_3, and Wicks and O'Hanley (1988) estimated an upper limit of 3.7 wt % of Al_2O_3 $[Mg_{2.9}Al_{0.1}(Si_{1.9}Al_{0.1})O_5(OH)_4]$ from both natural and synthesis studies.

Wicks and O'Hanley (1988) noted that a cylindrical-like structure as found in chrysotile does not fully compensate for misfit between the tetrahedral and octahedral sheets. For a helical structure, each particle will have only one layer position at the ideal radius of curvature to compensate for the structural mismatch in lateral dimensions. Yada (1971) found outer diameters commonly at 220–270 Å, whereas Whittaker (1957) calculated the ideal radius at 88 Å (diameter = 176 Å) for rolled structures with a X fiber axis. This suggests that large volumes of the fiber are either overcompensating or undercompensating for misfit. As the fiber radius increases past the ideal value and the curvature becomes less efficient at minimizing misfit, greater compression occurs in the octahedral sheet along Y. Even at the ideal radius, though the layer curvature will reduce the effect of misfit in the roll direction (Y), there is no minimizing of misfit along the direction perpendicular to the roll curvature and parallel to the fiber axis (X). Wicks and O'Hanley (1988) noted that, owing to layer curvature in one primary direction, fibers have variable misfit relief along Y but not along X and perhaps this misfit produces a tetrahedral sheet deformed to compensate for the unequal stresses. Therefore, to compensate, they speculated that the silicate rings become somewhat elongated and do not remain ditrigonal.

Perbost et al. (2003) systematically examined cation-size differences between the tetrahedral and octahedral sheets by synthesizing serpentine phases of composition $Ni_3(Si,Ge)_2O_5(OH)_4$ with morphologies of plates and tubes (curved features). Temperature (200°C), period of crystallization (9 days), and $P(H_2O)$ = 16 bar were kept constant, and the experiments generally produced a uniform product. At tetrahedral compositions of $(Si_{1.50}Ge_{0.50})$ to $(Si_{2.0}Ge_{0.0})$, curved structures formed. Products at tetrahedral compositions of $(Si_{1.28}Ge_{0.72})$ consisted of both plates and tubes. The radius of curvature at 0 percent Ge was 10 nm, which increased (to about 40 nm) with increasing Ge. The curvature is independent of crystallite thickness. At Ge compositions greater than $(Si_{1.50}Ge_{0.50})$, planar structures formed with plates of hexagonal or triangular (001) faces. Calculations using an elastic "thin-plate" analysis to relate the amount of curvature were similar to the experimental data. The analysis used a surface stress on one side of the thin plate to represent the tetrahedral sheet and another surface stress on the opposing side for the octahedral sheet. The difference in the two stresses can be used to calculate the surface radius for a single serpentine layer. The analysis assumed no layer-to-layer interactions, although such interactions would be expected from hydrogen bonding, even if hydrogen bonding is limited, and used the available elastic constants of a 2:1 layer (which would differ considerably from a 1:1 layer). In contrast, platy (lizardite) structures have been structurally refined with X-ray data at end-member compositions (Mellini et al., 2010), although the sample occurs as polyhedral spheres comprised of radial arrays of plates.

1.3 HALLOYSITE

Halloysite occurs with many morphologies. In general terms, the particles appear as plates, or curved 1:1 layers along one axis (e.g., tubes, cylinders, spirals), or partially curved as 1:1 layers along multiple axes (e.g., spheres); see Figure 1.6. Joussein et al. (2005), in a review of halloysite, documented ten different morphologies from the literature (e.g., fiber, disk, crumpled lamellar, lath, etc.) where these additional morphologies are subsets of the basic three. Bailey (1990) suggested that a fourth morphology should be considered as a basic form, a prismatic form where flattened faces occur on curved regions (often on spherical forms and tubular forms). Tubes are not always hollow, but chemical treatments, such as hot 2 percent Na_2CO_3 soaking (Kirkman, 1977), can apparently attack the inside solid portions before tube walls.

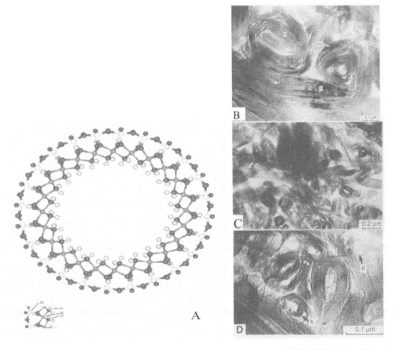

FIGURE 1.6 *(Continued)*

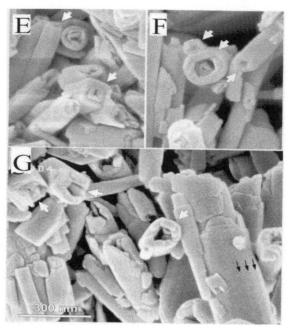

FIGURE 1.6 Computer simulation (Part (A)) of the halloysite structure showing an idealized cylindrical form in cross section. (B) shows TEM image of kaolinite (k) with halloysite (h) and halloysite spirals with some straight sections of kaolinite-like material with a mottled diffraction contrast (mc). (C) and (D) show a variety of halloysite tubes, polygonal shapes, and longitudinal sections; labels in (D) are s = spiral, and d = electron-beam damage. Parts (E), (F), and (G) are field-emission scanning electron microscope (FE-SEM) images of prismatic halloysite. White arrows indicate cross sections and *black arrows* in lower right of (G) indicate facets on the cylinder. (Part (A) reprinted from Guimarães et al., 2010, copyright 2010, with permission of the American Chemical Society; Parts (B), (C), and (D) from Robertson and Eggleton, 1991, reproduced with kind permission of The Clay Minerals Society, publisher of *Clays and Clay Minerals*; and Parts Є, (F), and (G) from Kogure et al., 2013, reproduced with kind permission of the Mineralogical Society of America).

1.3.1   CELL PARAMETERS AND HYDRATION STATE

The 1:1 layers, regardless of morphology present, may be described based on hydration state of the interlayer and resulting c-cell axial length, and this produces two common structural forms of halloysite. The more hydrous form, referred to in the early literature as "halloysite" and "endellite," and later as "halloysite (10 Å)" or variations thereof (e.g., halloysite-10 Å, 10Å-halloysite, etc.), has unit-cell parameters of a ~5.1, b ~8.9, c ~10.25 Å (basal spacing, $c\sin\beta$ ~ 10.1 Å), and β ~100°, whereas the less hydrated form, known earlier as "metahalloysite" and more recently as "halloysite (7 Å)" or variations thereof, has unit-cell parameters of a ~5.1, b ~8.9, c ~7.3 Å ($c\sin\beta$ ~ 7.2 Å), and β ~ 96.5°. Only the terms halloysite

(7 Å) and halloysite (10 Å) should be used. The chemical composition of halloysite (7 Å) is near that of kaolinite, $Al_2Si_2O_5(OH)_4$, and that for halloysite (10 Å) is $\sim Al_2Si_2O_5(OH)_4$ $2H_2O$. Dehydration of the 10-Å form by gentle heating is readily achieved and produces a 7.2-Å spacing, and dehydration is irreversible. Halloysite (10 Å) must be stored in water to prevent (partial) dehydration. For a given ambient relative humidity, a steady state may be approached with spacings between the two end members (Churchman and Carr, 1972; Churchman et al., 1972), indicating a more general and ideal chemical composition of $\sim Al_2Si_2O_5(OH)_4$ nH_2O, where n may vary from 0 to 2.

1.3.2 STRUCTURE

X-ray powder data (Brindley and Robinson, 1948) usually shows diffraction hk bands suggesting completely random stacking in halloysite (7 Å), although superposed sharper reflections on these bands may be observed in "well crystallized" samples. Investigation of individual particles by electron diffraction indicates that at least some stacking order is present (Honjo and Mihama, 1954; Honjo et al., 1954), and thus a highly disordered two-layer structure can be described by considering the superposed peaks over the diffraction bands. Kohyama et al. (1978) found similar electron-diffraction peaks for halloysite (10 Å), also indicating sequences of two-layer stacking in this halloysite. Using prismatic material, Chukhrov and Zvyagin (1966) proposed a two-layer monoclinic cell (β_{ideal} = 97°) with a space group of Cc for halloysite (7 Å) based on calculated intensity comparisons to observed data over a previously described (Honjo et al., 1954) triclinic cell. The former corresponds to the $2M_1$ polytype of Bailey (1969) in his description of trioctahedral polytypes. Bailey indicated that in the resultant monoclinic cell, the vacant octahedral site is located on the pseudomirror plane of each layer (the derivation of this two-layer structure involves alternating the vacant site across the pseudomirror plane in adjacent layers followed by choosing the monoclinic unit cell so that the vacant site is on the pseudomirror plane) (1990). Bailey (1990) noted that the small increase in β angle from the 7-Å form ($\beta \sim 96.5°$) to the 10-Å form ($\beta \sim 100°$) indicates a similarity in the "partly ordered" layer-stacking sequence. Little is known about the location of the interlayer H_2O in the 10-Å form. Using a low-electron dose, high resolution TEM, and tubular halloysite (7 Å), Kogure et al. (2011) found one-layer packets, and they suggested that the orientation of layers may be indicative of partial hydrogen bonding between the layers (e.g., as found in kaolinite). The stacking sequence appears to be 1:1 layers with layer displacements of -a/3-b/3 and parallel to [110], which differs from that of kaolinite. Using prismatic material of halloysite (7 Å), Kogure et al. (2013) found powder X-ray diffraction data consistent with the two-layer model of Chukhrov and Zvyagin (1966). Corresponding high resolution TEM data of Kogure et al. (2013) showed that the two-layer stacking model is only a local structure observed for six or seven layers in a multisequence set of disordered layers, and these results are consistent for samples showing small X-ray maxima superposed over diffraction bands. An extended abstract (Zhang et al., 2012) reported on the Rietveld analysis of halloysite (7 Å). As noted by the authors, the sample was chosen because the volume of material was believed sufficient, but the halloysite was poorly crystalline showing a limited number of broad maxima, and the resulting model has unreasonable interatomic distances. Walker and Bish (1992) warned users of the Rietveld method of the dangers of using low-quality data containing limited structural information; convergence may appear reasonable although the results may not be correct owing to the limited data.

The atomic positions for interlayer H_2O are unknown, although Hendricks and Jefferson (1938) suggested a model where the H_2O is ice-like. Using kaolinite starting material and intercalating H_2O between the 1:1 layers by using DMSO (dimethylsulfoxide) as an initial expanding agent with ammonium fluoride (which replaces at least some OH with F), Costanzo et al. (1982, 1984) proposed a model and Lipsicas et al. (1985) confirmed the model using nuclear magnetic resonance (NMR) where there are two components of H_2O: one where the molecule resides within the ditrigonal silicate ring ("hole water") and one referred to as "associated water" where the H_2O is more mobile. The use of the term "water" instead of H_2O is unfortunate because water is a phase with specific bulk properties that would not apply to interlayer H_2O molecules. The proposed model indicates that the $H_2O_{(hole)}$ is hydrogen bonded to the 1:1 layer, whereas the $H_2O_{(assoc)}$ is strongly hydrogen bonded to other $H_2O_{(assoc)}$ within a discontinuous plane in the interlayer, but only weakly bonded to the 1:1 layer. If only $H_2O_{(hole)}$ is present, the $c\sin\beta$ spacing is 8.4–8.6 Å, and a spacing of 10 Å occurs if both types are present, half as $H_2O_{(hole)}$ and half as $H_2O_{(assoc)}$. Costanzo and Giese (1985) related this model to the dehydration of halloysite. They suggested that the bonding strengths of $H_2O_{(hole)}$ and $H_2O_{(assoc)}$ to the 1:1 layer overlap in energy because of stacking defects and defects from the curvature of the layers. Thus, the basal spacing continuously changes from 10 Å to 7.2–7.9 Å upon dehydration. This interpretation does not preclude interstratifications of the 10 and 7-Å forms upon partial dehydration (e.g., Churchman et al., 1972), but it does suggest that a (metastable) 8.4–8.6 Å spacing may have to be considered as well when modeling diffraction data. It is also unclear if the arrangement of H_2O in a partially fluorine-exchanged phase corresponds directly to halloysite. Using a self-consistent, charge density-functional tight-binding method analysis and a starting model of a two-dimensional, single "halloysite" layer without H_2O, ideal-nanotube models were constructed by Guimarães-Enyashin et al. (2010) to obtain electronic and mechanical properties of the different nanotube models. The Young's modulus was smaller for the single-walled nanotubes than for carbon nanotubes, and all tubes are insulators. Simulated diffraction patterns for the nanotubes were generally good in comparison to halloysite.

The ability to convert kaolinite to an expanding phyllosilicate with a basal spacing similar to halloysite (7 Å) suggests that there is no way to differentiate between the two phases, unless the history of the sample is known. One common working hypothesis is that if the phase can be hydrated with an increase in the basal spacing, then it is halloysite, but dehydrated halloysite shows nonreversible swelling and thus, the issue remains. Others have suggested that X-ray or electron optical data showing a two-layer structure are evidence for a halloysite over kaolinite (e.g., Giese, 1988, p. 54), although polytype differences are not, in themselves, a method for identification of mineral species. Intercalation tests have been used also because polar compounds intercalate into halloysite more readily than into kaolinite (Churchman and Carr, 1975). Halloysite expands to 11 Å by intercalation treatment of hydrazine, followed by water, and then glycerol treatment (Range et al., 1969), and Churchman et al. (1984) used a formamide treatment to expand halloysite (see a comparison of techniques in Churchman, 1990). Churchman and Gilkes (1989) found that dehydrated halloysite tubes do not always intercalate formamide, and Janik and Keeling (1993) suggested a Fourier transform-infrared, partial least-squares test. Morphology of the particles and details of chemical composition may be useful in identifying halloysite; these interrelated properties are also important in understanding why different halloysite morphologies exist and how they compare to chrysotile (see below). It is cautioned, however, that bulk analyses may not reflect the actual chemical composition of the halloysite grains because these natural materials often contain other fine-grained phases (e.g., iron oxides, oxyhydroxides, poorly crystalline material such as allophane, etc.), either as separate grains or within a halloysite particle.

1.3.3   CHEMISTRY AND TEXTURE

Morphology of halloysite appears to be related to both crystallization conditions, which can often be surmised from geological occurrences (e.g., Joussein et al., 2005), and chemical composition (e.g., as compiled by Bailey, 1990). Halloysite has been reported to have increasing (tetrahedral site) Al substitution for Si (e.g., Tazaki, 1981) from spheres (apparent limited curvature of the layers), to short tubes, to long tubes (with apparent greatest curvature). In addition, Fe^{3+} is commonly observed in halloysite in (octahedral) substitutions for Al. Bailey (1990) summarized the reports (e.g., Noro, 1986) of iron and morphological consequences: maximum Fe_2O_3 amounts (>3.1 wt. %) produce platy morphologies, spherical and short-tube halloysites occur where iron content is between about 1– 3.1 wt. %, and long-tube halloysites exist where Fe_2O_3 is <1 wt. %. Thus, the least curvature of the layers occurs with maximum Fe^{3+}, where the iron is in octahedral sites as determined from ESR (e.g., Nagasawa and Noro, 1987a, b) and Mössbauer studies (Quantin et al., 1984), and greatest curvature occurs where Fe substitutions are very limited. However, Singh and Gilkes (1992) examined (Al + Fe^{2+}) content and found only very small, if any, differences between tubes, lath-like, and platy crystals. With regard to crystallization conditions, Tomura et al. (1985), in synthesis studies involving the direct dissolution/precipitation mechanism of spherical and platy kaolinite, found that spherical "kaolinite" is favored at high degrees of supersaturation (of calcined silica and alumina, where Al/Si = 1), whereas platy kaolinite is favored at low degrees of saturation. The study did not adequately establish if the spherical forms were kaolinite vs. halloysite. However, they did suggest that the results are analogous to the occurrence of spherical halloysite forming from volcanic glass.

White et al. (2012) examined the effect of Ge doping in the synthesis of a platy morphology ($\approx$ kaolinite) and a tube form ($\approx$ halloysite) by hydrothermal techniques at pH = 2 and a temperature of 220°C. Heterogeneity of the products, presumably because some Ge remained dissolved in solution, was observed and thus the actual composition of the products is only approximately known. Multilayered (25–40 layers) nanotubes [of $Al_2(Si_{2-x}Ge_x)O_5(OH)_4$, $x \leq 0.004$] with an inner diameter of 50–150 Å, an outer diameter of 400–600 Å, and some lengths of >5,000 Å, were obtained at 0.2 Ge/(Si + Ge) molar fraction starting material after 7 days. The fiber axis of the tubes is parallel to Y of the unit cell. A molar ratio of 0.5 produces single-wall nanotubes believed to be near $Al_2Si_xGe_{2-x}O_5(OH)_4$ ($x \sim 0.65$) in composition, with an inner diameter of 9–12 Å, an average outer diameter of 30 Å, and lengths of 200–300 Å. A low molar ratio (<0.1) forms partially curved plates of kaolinite. Spheroidal amorphous nanoparticles of an average diameter of 50–100 Å were present in all the experiments.

The substitution of Al for Si requires charge balance elsewhere in the structure, and cation exchange values from bulk material suggest that exchangeable cations are present, primarily Na, K, and Ca in very small numbers, <0.04 atoms per $O_5(OH)_4$ (Bailey, 1990; Weiss and Russow, 1963). Cation exchange implies a cause for hydration behavior. However, measured cation exchange capacities (7-Å form: 5–10 meq/100g; 10-Å form: 40–50 meq/100g, Grim, 1953; unspecified: 6–50; 33–48; 60; 58; 2–22 meq/100g, Kerr, 1951; Olivieri, 1961; Kunze and Bradley, 1964; Wada and Mizota, 1982; Noro, 1986, respectively) have values that vary greatly and all bulk-property measurements are suspect because of possible impurities (see, e.g., Eggleton et al. 1991, where smectite layers were found to terminate kaolinite (001) surfaces). Newman et al. (1994), using ^{27}Al NMR techniques, showed that tetrahedral Al cannot be used to distinguish between kaolinite and halloysite, with tetrahedral Al content of <1 percent. Natural ferric kaolin analogous to halloysite, forming spherical or partly spherical shapes approximately 60–200 Å in diameter, is known as hisingerite (Eggleton and Tilley, 1998; Baker and Strawn, 2012).

1.3.4   KAOLINITE TO HALLOYSITE TRANSFORMATIONS

Robertson and Eggleton (1991), using TEM and SEM methods, showed that natural weathering of plates of kaolinite can produce attached (polygonal) spirals of halloysite. Mixed halloysite and kaolinite particles initially form parallel hollow tubes/spirals (no cones) with planar sides ("prismatic" forms) nearest the kink on the kaolinite basal plate surface where the coil is attached to the kaolinite. Presumably, hydration also begins, at least partially, at the kink. The planar regions represent "kaolinite relics" about 0.1–0.2 μm in length nearest the kink, whereas the interior of the spiral becomes more circular or oval owing to the loss of the rigidity of the kaolinite portions of the sections as they further hydrate. Cross sections of tubes with planar regions were considered by Dixon and McKee (1974) as evidence of tubes forming from thin plates of kaolinite that had rolled. The kaolinite relics of Robertson and Eggleton (1991) can probably be described alternatively as dehydrated forms. The spiral/prisms showed that initial cross sections were irregular three- to eight-sided shapes, with five- and six-sided shapes most common and with angles between two faces approaching 120° The long axis of the halloysite forms is parallel to the X axis of the kaolinite plate. Sheaves of halloysite form if pore spaces are not present and the chemical compositions of kaolinite and halloysite were indistinguishable but, if differences existed, they were at a scale finer than ~5,000 nm. These observations are consistent with the results of Newman et al. (1994) in that Al content cannot distinguish between kaolinite and halloysite.

Singh and Mackinnon (1996) transformed kaolinite to a hydrated form using repeated applications (~10 to ~35) of potassium acetate as the swelling agent. After 35 cycles, they found that cation exchange capacity increased to 32 meq/100 g (vs. 10 meq/100 g for the kaolinite), that the surface area was ~10 m^2/g (vs. 6 m^2/g for kaolinite), that suspensions were viscous and readily flocculated at pH 10, and the number of coarse particles decreased from a size of <15 μm in kaolinite to <1 μm after treatment. This treatment is not believed to affect the chemical composition of the products. Upon hydration, the products consisted of residual kaolinite plates and tubes, with the latter commonly elongated along Y and less commonly along X of the kaolinite substrate. The tube axis orientation is opposite from the natural samples examined by Robertson and Eggleton (1991). Tubes with stacking described as "well ordered" showed a two-layer sequence, whereas others with more considerable stacking disorder showed little or no periodicity. As with the natural material (Robertson and Eggleton, 1991), tubes were spirals with planar features (prismatic) on the outside and "smoothly curved" layers on the inside. Tube characteristics, such as length and stacking periodicity, are probably a function of the parent kaolinite. Tube diameters and wall thicknesses are variable and similar to natural tubes at diameters of 100–500 nm (Singh and Gilkes, 1992), with wall thicknesses of 10–50 nm (Singh and Mackinnon (1996). Inner diameters of natural halloysite (7 Å) tubes are 7–38 nm (Dixon and McKee, 1974) and 20–100 nm (Bates et al., 1950). Diameters of natural spherical forms are ~150 nm.

X-ray diffraction patterns of the kaolinite starting material as well as after 27 and 35 cycles of treatment are given by Singh and Mackinnon (1996). With 35 cycles of treatment, a 10-Å spacing had developed and a 7.9-Å peak had broadened and apparently had moved to 8.0 Å. Upon successive dehydration, the 8.0-Å peak eventually moved to 7.2 Å, which suggests that there are 7.2-Å and 10.0-Å interstratified spacings in the sample. Costanzo et al. (1984) and Singh and Mackinnon (1996) suggested that hydration and extreme stacking disorder occurs together because of the weakening of the interlayer hydrogen bonding.

Although a topotactic reaction is described in the transformation of kaolinite to halloysite, either experimentally or from studies involving natural textures, this process is not the only way to produce halloysite. Hydrothermal experiments by Tomura et al. (1985) for spherical kaolinite and observations from weathered rocks for spherical halloysite (summarized in Joussein et al., 2005) indicate that supersaturated solutions associated with volcanic ash and pumice, and volcanic glass in marine environments, commonly lead to the formation of spherical and spherical-like forms by direct precipitation. White et al. (2012) reported the synthesis of a variety of morphologies, including spheres, in hydrothermal reactions which presumably do not involve a topotactic reaction.

1.3.5 THE ORIGIN OF CURVATURE IN HALLOYSITE

1.3.5.1 WHY HYDRATION OCCURS?

Hydration in kaolinite will occur if the hydrogen bonding network (= layer-to-layer interactions) is sufficiently disrupted so that H_2O molecules may enter the interlayer to form halloysite. With the presence of H_2O in the interlayer to disrupt the hydrogen bonding network, the number of H_2O molecules (i.e., "hydration") is related to the degree of stacking disorder (Costanzo et al., 1984; Singh and Mackinnon, 1996). In controlled experiments, intercalation produces kaolinite hydrates either by DMSO/NH_4F applications to form a 11-Å hydrate (Costanzo et al., 1984), by KAc applications to form a 14-Å hydrate (Singh and Mackinnon, 1996), or by the intercalation of guest molecules (e.g., various alkylammonium cations) and simultaneous swelling by a solvent in a methoxy-modified kaolinite (Kuroda et al., 2011). Hydration increases with either greater expansion and/or by multiple cycles of intercalation, and thus the layer-to-layer interactions diminish and stacking disorder increases (Singh and Mackinnon, 1996).

All kaolin minerals (kaolinite, nacrite, dickite, halloysite) have layer stacking where Si ions of a 1:1 layer project directly onto two of the three octahedral sites, and this produces cation-cation repulsions if all sites are occupied as in trioctahedral 1:1 layers (Bailey, 1990). Because the kaolin minerals are dioctahedral and thus one octahedral site is vacant, the kaolin minerals minimize Coulombic repulsions as the vacant site removes the effect. Nonetheless, one Al-occupied site remains to interact with Si, and repulsive forces exist. In halloysite, these forces are further minimized by the shielding effect of H_2O molecules that enter into the interlayer and the increase in layer-to-layer distances. Thus, an increase in stability occurs for increasing hydration, and this may be a further initial driving force for H_2O to enter the interlayer as suggested by Bailey (1990). However, as layer stacking becomes more turbostratic and less regular, this driving force becomes less important, although water activity (e.g., relative humidity) becomes more important.

Natural samples have been clearly documented showing a similar transformation from platy kaolinite to halloysite tubes by a hydration process (Robertson and Eggleton, 1991; Singh and Gilkes, 1992). The process of hydration, however, must differ from synthesis experiments. Bobos et al. (2001) noted that changes in surface speciation may occur with changes in pH. For example, they suggested that surface OH groups can either ionize to OH_2 or O with acid/base reactions and with increasing pH, the Si tetrahedra can be affected. Interestingly, Ma and Eggleton (1999) noted that even apparently "pure" kaolinite particles may have either pyrophyllite or smectite terminating layers, indicating that different kaolinite-rich samples may behave quite differently under similar conditions.

1.3.5.2 STRUCTURAL REASONS FOR ROLLING?

Bailey (1990) and earlier workers invoked misfit between the tetrahedral and octahedral sheets and a common O,OH junction between the two semi-independent sheets to explain why rolling occurs in halloysite. For example, an isolated $Al(OH)_3$ octahedral sheet has a b-axis value of 8.64 Å in gibbsite and 8.67 Å in bayerite. Although a Si_4O_{10} tetrahedral sheet does not occur in isolation in nature, calculations using idealized bond lengths indicate that an isolated tetrahedral sheet would have a b-axis value of 9.15 Å. Bates et al. (1950) suggested that, in halloysite, the Si tetrahedra exist on the outside of the walls of the coil because the tetrahedral sheet is too large and the Al-rich octahedral sheet is too small to form a planar junction. Although there is evidence for control of morphology on the basis of misfit being relieved by either Al substitutions or Fe^{3+} substitutions tetrahedrally or by substitutions of larger Ge for Si, the data are not always consistent with the idea of minimizing the misfit between the component sheets for tubes of halloysite. In kaolinite, tetrahedral rotation occurs to minimize this misfit. Bailey (1990) hypothesized that tetrahedral rotation may be blocked in halloysite owing to $H_2O_{(hole)}$ and/or by exchangeable cations. The latter explanation requires charge balance to be offset by tetrahedral substitutions, which are not always present. The effects of minimizing misfit by tetrahedral substitutions may work to a limited degree in natural samples, but it is not a universal mechanism to explain the rolling of layers in halloysite.

Two important findings since 1990 have revised the explanation for why rolling occurs in halloysite: (1) kaolinite hydrates were prepared that induced layer rolling in kaolinite without effectively changing the composition of the layers (e.g., Singh and Mackinnon, 1996) and (2) the topotactic transformation of platy kaolin to halloysite tubes was described in detail from natural samples (e.g., Robertson and Eggleton, 1991). Singh (1996) pointed out that tetrahedral rotation reduces lateral dimensions of both the Si plane and basal-oxygen plane, whereas only the common junction (apical oxygen atoms and OH ions) needs to be affected. Repulsive forces will increase for those planes where Si–Si and O–O distances decrease. He calculated that repulsion effects are greater by 12 times for tetrahedral rotation when compared to repulsions generated by layer rolling. In contrast, because tetrahedral rotation does affect the basal oxygen plane, the oxygen atoms are better positioned to form hydrogen bonds in kaolinite, and this is the reason why the hydrogen-bond network forming layer-to-layer interactions requires disruption before layer rolling can proceed. This model also accounts for the irreversible nature of the dehydration of the 10-Å form by gentle heating to produce a 7.2-Å spacing. Because the tube radius changes as a function of the distances from the axis, Singh (1996) also noted that appropriate tetrahedral substitutions may compensate for misfit by limited tetrahedral rotation.

Although the natural samples appear to approximate a "solid state" reaction, the fact that studies involving natural samples describe rocks/soils undergoing weathering processes strongly indicates that surrounding fluids play an important, although perhaps limited, role in the reaction. Layer rolling occurs where hydration disrupts the layer-to-layer interactions from hydrogen bonds, and this is evident also by an increase in stacking disorder. The natural samples showed no evidence of a change in composition between kaolinite and halloysite and that the coils may incorporate platy kaolinite within the coil walls, thus explaining why some coils appear to have facets.

Kogure et al. (2013), in a study on prismatic halloysite (7 Å), suggested a somewhat different formation pathway for these halloysite prisms. As above, the process starts with a hydrated halloysite forming as a 10-Å form without hydrogen bonds across the interlayer, and rolling is initiated. The fiber axis length is along the Y direction, normal to the pseudomirror plane of the kaolinite 1:1 layer. The halloysite then dehydrates to develop a 7-Å form with limited hydrogen bonding between the layers. The hydrogen bonding is very limited because of the variable radius of the coil. As long-term crystal "ripening" proceeds, the coils become more prismatic with "sector domains" in which the 1:1 layers are flattened, and layers adjust themselves by $\pm b/3$ or $\pm b/6$ shifts similar to kaolinite to form hydrogen-bonding, layer-to-layer interactions. According to Kogure et al. (2013), the number of $+b/3$ shifts is approximately equal to the number of $-b/3$ shifts to maintain the external form of the prism, and the formation of regular interstratifications is generally promoted.

1.3.5.3   WHY THE PREFERENCE FOR SCROLLING IN ONE DIRECTION?

In general, the fiber axis for halloysite tubes is the Y axis (relative to kaolinite plates), although scrolling can occur less commonly along X. White et al. (2012) suggested a relationship between the rate of kaolinite crystallization because the platy particles are elongated along the [010] direction. For their compositions, White et al. argued that an "inclusion of a GeO_4 tetrahedron or Si(IV) vacancy and the distribution of stress in both directions would cause a stronger bending of the shorter side compared to the longer side." However, this suggestion is not universal for all halloysite and has limited applicability because of the synthetic Ge samples.

Perhaps a more fundamental and universal approach is to consider structural differences along the X and Y axes and the configuration of the chains that form the tetrahedral sheet. The tetrahedral sheet in kaolinite is ditrigonal rather than the ideal hexagonal array because the tetrahedral sheet is larger than the octahedral sheet and tetrahedral rotation occurs in kaolinite (see Figure 1.3d). Thus, the linkage of tetrahedra becomes different for the two directions, with a pyroxene-like chain extended along X and a more segmented (still continuous, however) chain along Y. The difference in chain configurations increases with increasing tetrahedral rotation. It is the difference in configuration of the two chains that is important because both chains can bend, although the bending along the Y axis is presumably favored.

1.4 IMOGOLITE AND ALLOPHANE

1.4.1 IMOGOLITE

Natural imogolite occurs in weathered volcanic ash, but may also be found in the clay-size fraction of podzolized soils and in pumice. Imogolite, which is very poorly crystalline, forms hollow tubes comprised of curved dioctahedral gibbsite-like $[Al_2(OH)_3]$ sheets linked at the inner part of the curved surface by O_3SiOH tetrahedra (Cradwick et al., 1972). Three oxygen ion apices of each tetrahedron replace three OH on one side of the gibbsite-like sheet, and the Si and the OH apex are located over the vacant site (Figure 1.7). Thus, the tetrahedra are not linked to other tetrahedra. The span of a tetrahedral edge of a Si-rich tetrahedron is about 2.7 Å, which is shorter than the O-O distance in gibbsite, and presumably this difference causes curling of the sheet (Figure 1.7). Cradwick et al. (1972) describe a unit cell of the natural nanotube having 10 Si atoms and 20 Al atoms. The OH group of the tetrahedral apex points into the tube interior. The nanotube walls are one layer thick. The chemical formula is $(OH)_3Al_2O_3SiOH$ with natural samples having the range of $Al_2(OH)_3(SiO_2)_{10-12}(H_2O)_{23-30}$ (Wada, 1977), although impurities may be at least partly responsible for the variability. Imogolite is characterized by the Si/Al ratio of near 0.5, a diffraction pattern of dried material (at 100°C) showing broad X-ray peaks at 23.0, 8.9, 6.6, and 5.2 Å, and the direct observation of nanotubes by transmission electron microscopy.

Farmer et al. (1983) described the dimensions of the natural nanotubes as having an inner diameter of ~10 Å and an outer diameter of 23–27 Å, and values are 10–15 percent greater for synthetic samples. Lengths can vary from nanometers to micrometers. Gustafsson (2001) showed that the surface charge is weakly positive on the outer walls whereas the inner surface is weakly negative, perhaps explaining why most imogolite nanotubes are found naturally arranged in closest packed aggregates. Yah et al. (2010), in a review paper, discussed several ways to surface modify imogolite to better develop integrated organic matrices with the material.

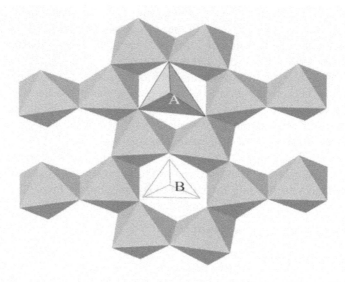

FIGURE 1.7 Illustration of the source of misfit in imogolite prior to rolling. The layer simulation incorporates the dioctahedral gibbsite structure (Saalfeld and Wedde, 1974) as the base structure, with Al present in the octahedra and octahedral corners of OH. One tetrahedron, labeled (A), is drawn larger than a Si-rich tetrahedron (and requires distortions to span a vacant octahedral site) to illustrate connectivity. Connectivity to the octahedral corners is achieved by replacing the OH ions on the basal plane of the tetrahedron with three O ions. The fourth corner of the tetrahedron, the apical corner, is an OH to achieve charge balance. A silicon tetrahedron (B) drawn to approximately the same scale as the octahedra, only partially fills the gap across the vacant octahedral site. Because Si tetrahedra do not span the octahedral vacancies completely, this misfit results in curling and scrolling. Note that the tetrahedra do not link to other tetrahedra and the sheet-like character results from the octahedral sheet. The apical OH of the tetrahedron points toward the interior of the resultant roll (see text for more details).

1.4.2 SYNTHESIS AND CHEMISTRY OF IMOGOLITE

Because of the poor crystalline nature of imogolite and difficulty in obtaining pure material, minor differences in chemical composition may not be significant. Most of what is known about the composition is based on synthesis work. Most synthesis studies involve the formation of Al-, Si-rich imogolite and ways to increase yield. These studies include that by Farmer and Fraser (1979) where they used $Al(ClO_4)_4$ and $Si(OH)_4$ in dilute solutions. They produced a proto-imogolite sol and then imogolite by heating to 100°C from dilute hydroxy aluminum and orthosilicic acid. Farmer et al. (1983) modified the procedure and carefully adjusted the pH to produce imogolite also, and they discussed ways to increase yield. This study was followed by Wada (1987) who used a similar procedure but at 25°C and exceptionally long aging times of 7 years. More recent studies include those by Ma et al. (2012), where they synthesized imogolite in polymer solutions, and Abidin et al. (2009), who used a silica gel source in an $AlCl_3$ solution for imogolite synthesis. Using natural and synthetic material and two relative humidity conditions, Bishop et al. (2013) determined infrared bands assignments for imogolite.

Using orthosilicic and orthogermanic acid solutions and $AlCl_3$, Wada and Wada (1982) synthesized imogolite with substitutions of Ge for Si to determine the effect of the substitution on the curvature of the nanotubes. The outside diameter of the nanotube increased to 33 Å with complete Ge substitution, in accord with the larger size of Ge over Si and the better fit of the O_3GeOH tetrahedra to the gibbsite-like sheet. The number of gibbsite-like unit cells around the nanotube circumference increases from 10 to 12 in the Si-rich form to 18 in the Ge-rich form. Levard et al. (2008) has increased the yield of Ge-rich imogolite products, and Maillet et al. (2010) has found evidence for double-walled Ge-rich imogolite nanotubes. Ookawa (2012) synthesized a Fe-rich imogolite at the atomic ratio of Fe/(Al + Fe) = 0.05 using $NaSiO_4$, $FeCl_3$, and $AlCl_3$, and found that the product had catalytic properties for liquid phase oxidation reactions for some hydrocarbons.

1.4.3   ALLOPHANE

Allophane is also poorly crystalline and shows curved walls but with a more spherical (diameter of 35– 50 Å, Kitagawa, 1971; Henmi and Wada, 1976) morphology than imogolite. Allophane has a chemical formula of $Al_2O_3(SiO_2)_{1.3-2.0} 2.5–3.0(H_2O)$. Several structural models have been proposed for allophane. Parfitt and Henmi (1980) suggested from infrared data that allophane is composed of imogolite-like units, and Parfitt et al. (1980) found that "proto-imogolite" allophane spheres form, with silica in the interior. They examined such a sphere with a 40-Å diameter and suggested that it was composed of 125 unit cells of imogolite. They suggested that the imogolite-type allophane would have an Al/Si ratio of near 0.5, but a second type of allophane has a ratio of near 1.0 and a halloysite-like structure. Bishop et al. (2013) provided infrared band assignments under two relative humidity conditions. Childs et al. (1990) described a silica-rich allophane with a partial octahedral sheet within a rolled tetrahedral "sheet." He et al. (1995), in an electron spectroscopy for chemical analysis (ESCA) and near NMR study, supported the idea that allophane is a phyllosilicate similar to a kaolin-type phyllosilicate. Small numbers of Fe ions have been found to substitute for Al in allophane (e.g., Kitagawa, 1973). Baker and Strawn (2012) determined from an Fe K-edge X-ray absorption fine structure (XAFS) study that the substitution occurs in octahedral sites only and that Fe clusters occur in small domains. In contrast to the model proposed for "proto-imogolite," they found that a montmorillonite-like (2:1 layer) structure model better fit their data for allophane. Because montmorillonite is a 2:1 layer and the octahedral sheet is held in tension between the opposing tetrahedral sheets, an unmodified layer such as that found in montmorillonite does not appear to be a crystal-chemical reasonable model for a curved structure. Baker and Strawn (2012) also suggested that the clustering of Fe may be the incipient formation of an iron-rich phase as the allophane ages.

1.5 PALYGORSKITE-SEPIOLITE

1.5.1 SPECIES AND NOMENCLATURE

The palygorskite-sepiolite group (Table 1.2) consists of palygorskite, sepiolite, falcondoite, kalifersite, loughlinite, raite, tuperssuatsiaite, yofortierite, windhoekite, and $\sim NaCa(Fe^{2+}, Al, Mn)_5[Si_8O_{19}(OH)](OH)_7\,5H_2O$. Except for palygorskite and its Mn analog yofortierite, which are dioctahedral, all the other species are trioctahedral. Although not a formal member of the group, intersilite has structural similarities to the group and is included in the table. The names, "attapulgite" (often used in industry), "hormite" (Robertson, 1962), and "palysepiole" (Ferraris et al., 1998), all of which have been used to describe either palygorskite or as an alternative to palygorskite-sepiolite, are not proper mineral names and should not be used in the scientific literature.

TABLE 1.2 Palygorskite-sepiolite group (Modified from Guggenheim and Krekeler, 2011).

Mineral	Formula	Reference
Falcondoite	$\sim(Ni_{8-y-z}R^{3+}{}_y\square_z)\,(Si_{12-x}R^{3+}{}_x)\,O_{30}\,(OH)_4$ $(OH_2)_4 \cdot R^{2+}{}_{(x-y+2z)2}\,(H_2O)_8$	modified from Springer (1976), see also Proenza et al. (2007)
Ferrisepiolite	$(Fe^{3+}Fe^{2+}Mg)_4(Si,Fe^{3+})_6O_{15}(O,OH)_2 \cdot$ $6H_2O$	Xiangping et al. (2013)
Intersilite	$(Na_{0.80}K_{0.45}\square_{0.75})Na_5Mn(Ti_{0.75}Nb_{0.25})$ $[Si_{10}O_{24}(OH)] \cdot (O,OH)(OH)_2 \cdot 4H_2O$	Yamnova et al. (1996)
Kalifersite	$(K,Na)_5Fe^{3+}{}_7\,(Si_{20}O_{50})\,(OH)_6 \cdot 12(H_2O)$	Ferraris et al. (1998)
Loughlinite	$\sim Na_4Mg_6\,(Si_{12}O_{30})(OH)_4\,(OH_2)_4$	Fahey et al. (1960), Kadir et al. (2002)
Palygorskite	$\sim(Mg_{5-y-z}R^{3+}{}_y\square_z)\,(Si_{8-x}R^{3+}{}_x)\,O_{20}\,(OH)_2$ $(OH_2)_4 \cdot R^{2+}{}_{(x-y+2x)/2}\,(H_2O)_4$	Bailey (1980)
Raite	$\sim Na_3Mn_3Ti_{0.25}\,(Si_8O_{20})\,(OH)_2 \cdot 10(H_2O)$	Pluth et al. (1997)
Sepiolite	$\sim(Mg_{8-y-z}R^{3+}{}_y\square_z)\,(Si_{12-x}R^{3+}{}_x)\,O_{30}\,(OH)_4$ $(OH_2)_4 \cdot R^{2+}{}_{(x-y+2z)2}\,(H_2O)_8$	Bailey (1980)

TABLE 1.2 *(Continued)*

Mineral	Formula	Reference
Tuperssuatsi-aite	$\sim Na_{1.87}Fe_{2.14}Mn_{0.48}Ti_{0.14}$ (Si_8O_{20}) $(OH)_2 \cdot n(H_2O)$	Cámara et al. (2002)
Un-named	$\sim NaCa(Fe^{2+},Al,Mn)_5$ $[Si_8O_{19}(OH)]$ $(OH)_7 \cdot 5H_2O$	Rastsvetaeva et al. (2012)
Windhoekite	$(Ca_{1.68}Mn_{0.32})Fe^{3+}_{2.96}(Si_{7.87}Al_{0.08})$ $O_{20}(OH)_4 \cdot 10H_{1.98}O$	Chukanov et al. (2012)
Yofortierite	$\sim (R^{2+},R^{3+},\square)_5 \, Si_8O_{20}(OH, H_2O)_2(H_2O)_7;$ where Mn^{2+} dominates	Hawthorne et al. (2013)

1.5.2 STRUCTURE OF THE PALYGORSKITE-SEPIOLITE MINERAL GROUP

Although single-crystal structure refinements of palygorskite and sepiolite are un-available owing to the difficulties in obtaining suitable crystals, powder data studies have provided important structural information. In addition, single-crystal X-ray re-finements of other members of the group are available and, along with comparative studies, an in-depth understanding of palygorskite and sepiolite has resulted. Early structural studies of palygorskite-sepiolite were done by Drits and Sokolova (1971), Christ et al. (1969), and Chrisholm (1992). In addition, Rietveld powder techniques were used by Artioli et al. (1994), Chiari et al. (2003), Giustetto and Chiari (2004), and Post and Heaney (2008). There are two known structural modifications, mono-clinic (*C2/m*) and orthorhombic (*Pbmn*), and both forms tend to be intergrown. See Bailey (1980) and Jones and Galán (1988) for summaries of earlier work and model evolution.

Each member of the palygorskite-sepiolite group has infinitely extending tetra-hedral sheets involving sixfold rings of tetrahedra, similar to the ideal phyllosilicates. Also similar to the ideal phyllosilicates, the tetrahedral sheets have a continuous basal oxygen-atom plane, but unlike the ideal phyllosilicates, the palygorskite-sepi-olite group has apical oxygen atoms pointing in opposing directions within a con-tinuous sheet (Figure 1.8). Each section of like-pointing tetrahedra forms a strip or ribbon pattern and each ribbon consists of a tetrahedral ring (or two pyroxene-like chains) in palygorskite and 1.5 rings (or three pyroxene-like chains) in sepiolite. The (octahedral) metal atoms, Mg or Al in palygorskite and sepiolite, are linked to six nearest-neighbor oxygen atoms, some of which are obtained from the apical oxygen atoms of tetrahedral ribbons pointing in one direction. Apical oxygen atoms of an opposing tetrahedral ribbon also form a portion of the octahedral coordina-tion, in addition to two OH groups (or by OH_2 groups in special cases, see below). Therefore, most octahedra are formed from two apical oxygen atoms obtained from one ribbon, two additional apical oxygen atoms from the opposing ribbon, and two

OH groups. In palygorskite and sepiolite, the octahedra, which are linked via edge sharing, form strips that are not continuous sheets. In sepiolite, the octahedral strips are eight octahedra wide, whereas strips that are five octahedra in width occur in palygorskite. The terminal anion at the edges of the octahedral strip involves four OH_2 groups per formula unit. Because these groups are well bonded to the octahedral metal cation and not isolated, they are not referred to as H_2O. The OH_2 groups are required for charge balance. Palygorskite has an octahedral sheet with one site vacant per five octahedra, and is dioctahedral (Martin-Vivaldi and Cano-Ruiz, 1955; Drits and Aleksandrova, 1966), based on chemical analyses and early structural analysis. Later, this was confirmed from Rietveld analysis (Post and Heaney, 2008; Giustetto and Chiari, 2004, Chiari et al., 2003; Artioli and Galli, 1994).

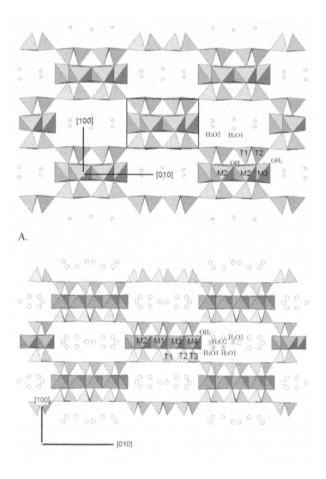

FIGURE 1.8   *(Continued)*

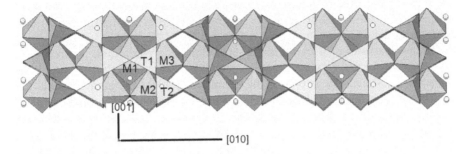

C.

FIGURE 1.8 Palygorskite (A) and sepiolite (B) structures projected along the [001] direction (along the plane of the sheets). A polysome is depicted as a box in the center of (A). The M octahedral cations, Mg and Al, are given as dark-fill octahedra and the T sites (Si) are shown as lighter-fill triangles. In projection (A), the M1 site (not shown) is in front, between the two labeled M2 sites; Figure C shows the site information in more detail. Zeolitic H_2O are given as open circles. In projection (B) for sepiolite, M2 resides behind M1, which is behind M3. Zeolitic H_2O are given as open circles and identified (labeled) with "x." Projection (C) is palygorskite projected down the [100] direction. Coordinates from Post and Heaney (2008) and Post et al. (2007) (From Guggenheim and Krekeler, 2011, reproduced with the kind permission of Elsevier Press).

A "polysome" is the section of the structure involving the combination of an octahedral strip and adjacent strips of tetrahedra, which resembles the 2:1 layer, although more limited in lateral extent. The spacing of the basal-oxygen-atom plane to basal-oxygen-atom plane is about 6.5 Å, which is similar to the analogous spacing of the 2:1 layer found in mica. Polysomes have been found in many mineral structures (e.g., Baronnet, 1997), where they may be modules that involve units of two or more building blocks within a structure. The compositions and structures of the different polysomes that form a modular structure may or may not differ from each other. For example, interstratifications of chlorite and lizardite of the same composition may be considered modular, as may interstratifications of muscovite and paragonite, which are both mica structures but with different compositions. Structures within the palygorskite-sepiolite mineral group may show variations in width of the polysomes, polysome omissions ("open channel" defects), and stacking errors, and these defects can be monitored by TEM.

Vacant regions, zeolitic H_2O, and exchangeable cations may reside in the channels formed at the edges of the octahedral strips in palygorskite and sepiolite. Exchange reactions with organic molecules are possible if the size of the organic cations is appropriate, as steric constraints control as to what can enter this channel. The indigo molecule (4.8×12.3 Å) in the channels of palygorskite or sepiolite form a synthetic pigment (Maya Blue), where the blue color remains highly stable for centuries, probably because of the protection of the structure around the channel.

The pigment was used by the Maya civilization because of its bright color. Larger molecules also may be adsorbed by the structure, but this is probably because of the existence of defects.

The two varieties of stacking, either monoclinic or orthorhombic, are related to the occupancy for the set of octahedral cations (Set I or Set II) in the 2:1 layer of the polysome; see Bailey (1980, p. 8) for the definition of these octahedral set positions. Set I or Set II occupancy is based on closest packing within the polysome, and this occupancy produces a shift or stagger, either $+c/3$ or $-c/3$, of the upper tetrahedral strip in the polysome relative to the lower tetrahedral strip. Where the direction of shift always remains the same, a monoclinic structure results (with an ideal ß = 105.2°), whereas if an alternation of shift directions occurs because Set I alternates with Set II in adjacent strips, an orthorhombic structure results (ß = 90°). Chrisholm (1992) proposed space groups and atomic coordinates for these derived models.

The most reasonable palygorskite Rietveld-based model, and one developed by using pure material, is obtained from Post and Heaney (2008). For monoclinic palygorskite, there are two sites for zeolitic H_2O, and these site locations are similar to those found by Giustetto and Chiari (2004). However, an additional H_2O site found by Giustetto and Chiari was not located by Post and Heaney (2008). Post and Heaney (2008) recognized that some H_2O-H_2O distances are too close (1.03 Å) for both sites to be occupied simultaneously, and some zeolitic H_2O is closely associated to the OH_2 anions of the octahedral strip. This suggests that the H_2O is weakly H bonded to the framework. The monoclinic polymorph differs from the orthorhombic palygorskite (Giustetto and Chiari, 2004) based on the zeolitic H_2O locations and a complex H-bonding scheme for each polytype was proposed. For sepiolite, Post et al. (2007) located four zeolitic H_2O sites; two sites are fully occupied, one is half occupied, and one is about a third occupied.

Results from Rietveld refinements are generally less precise than from single-crystal refinements (e.g., see Post and Bish, 1989). The lack of precision is observed in reported bond distances, with associated errors about a magnitude larger than those reported from a typical single-crystal refinement. However, accurate bond lengths and angles are an essential part of understanding a structure, including the determinations of site occupancy, distortions, and bonding character. Unlike most single crystal refinements, constraints are used in Rietveld refinements to obtain reasonable results, and these constraints often involve fixing (or limiting the variation of) atomic parameters. Constraining the atomic parameters to reasonable values, which was done to some extent for all sepiolite and palygorskite Rietveld refinements, may establish the size and shape of the coordination polyhedra for these complex materials, which is the fundamental reason for the refinement in the first place. Thus, care must be taken to avoid detailed interpretation of structural data derived from Rietveld X-ray analysis. However, unlike individual bond distances and angles, *average* polyhedral sizes derived from Rietveld refinements tend to be similar to those obtained from single-crystal data, even for polyhedra that are not constrained. Therefore, comparisons of generalized structural parameters, such as rotation angles, ψ values, and average τ values are probably more fruitful than direct comparisons of individual distances and angles. For example, the three O_{apical} - T - O_{basal} angles (the τ angles) describing the shape of a tetrahedron in the Giustetto and Chiari (2004) monoclinic palygorskite model varies between 83.1° and 119.8°. This is crystal chemically unreasonable because a large deviation from near 109.5° implies that the Si – O bond lacks significant covalent character (where overlapping electron orbitals in Si tetrahedra require near ideal angle values). In contrast, however, the *average* of the values, 106.9°, is reasonably close to the ideal value of 109.5°. Similar problems exist for the orthorhombic palygorskite model.

1.5.3 CHEMISTRY

Newman and Brown (1987) found that octahedral occupancy varied from $R^{3+}_{2.5}$ + $R^{2+}_{1.5}$ for Al-rich specimens to $R^{3+}_{0.5}$ + $R^{2+}_{4.25}$ for Mg-rich palygorskite, although impurities may be an issue. The occupancy of tetrahedral sites tends to be very Si rich, with Al substitutions limited from $Al_{0.12}$ to $Al_{0.66}$ per eight T sites. In a recent review of the chemical composition of palygorskite and sepiolite, Suárez and García-Romero (2011) found that Si ranges from 7.20 to 8.13 atoms per eight sites and octahedral cations range from 2.86 to 4.66 atoms per four sites, with most samples near four. Octahedral cations are Al (0.12–2.35 per four sites) and Mg (0.0–3.91 per four sites), with minor amounts of Fe^{3+} (<1.31 cations per four sites) and Ti (<0.17 cations per four sites). Exchangeable cations may occur (Ca, Na, K). Infrared spectroscopy (Serna et al., 1977) and infrared plus Mössbauer spectroscopy (Heller-Kalai and Rozenson, 1981) have shown that vacancies are ordered into M1 in palygorskite (where M1 is the central octahedral site on the special position), and that Mg (and Fe) preferentially orders into M3 (two sites away from the M1 site in the octahedral strip). Also using infrared analysis, Chryssikos et al. (2009) found that regions of the palygorskite structure were dioctahedral (with AlAlOH, $AlFe^{3+}OH$, $Fe^{3+}Fe^{3+}OH$ interactions) and trioctahedral (MgMgOH), with these interactions implying that Al and Fe^{3+} order into M2. However, Stathopoulou et al. (2011) found that these samples were intergrowths of sepiolite and palygorskite polysomes. Using Rietveld analysis and average bond lengths, Post and Heaney (2008) determined that vacancies ordered in M1, Al in M2, and Mg in M3. These results are consistent with other Rietveld refinement studies (Artioli and Galli, 1994; Chiari et al., 2003; Giustetto and Chiari, 2004) and with earlier spectroscopy studies. Suárez and García-Romero (2011) suggested that Fe^{3+} substitutes for Al in M2, with Fe_2O_3 content as high as 14.8 wt %. Krekeler et al. (2004) found small amounts of Ti (TiO_2 ≤3.5 wt %) and Mn.

Sepiolite shows limited, and randomly dispersed, octahedral vacancies for samples where R^{3+} content is high and with total octahedral occupancies from 7.0 to 8.0 (Newman and Brown, 1987), with cations of primarily Mg and minor Mn, Fe^{3+}, Fe^{2+}, and Al, although sepiolite is generally considered trioctahedral (e.g., Galán and Carretero, 1999). García-Romero and Suárez (2010) also concluded that substitution for Mg by Al (15–39 %) and Fe can occur along with vacancies. MgO content may vary between 18.58 to 30.57 wt. % and Al_2O_3 content may be to 8.35 wt. %. Titanium, potassium, or sodium is low with oxides of these elements at ≤1 wt. %. A Ni-rich sepiolite exists that appears to vary to falcondoite (Fal) based on a solid solution series from Fal_3 to Fal_{70} (Suárez and García-Romero, 2011). Suárez and García-Romero found total octahedral cations range from 6.7 to 8.03, which corresponds to 0.5 to 14 percent vacancies, with only minor amounts of Al (0.01 to 1.24 atoms per ~8 cations) and Fe^{3+} (0.01– 0.43 atoms per ~8 cations) and trace Ti (2011). Tetrahedral content has been summarized (Bailey, 1980) as from $(Si_{11.96}Al_{0.05})$ to $(Si_{11.23}Fe^{3+}_{0.53}Al_{0.24})$, although R^{3+} content may be as high as 1.3 atoms per 12 sites. Suárez and García-Romero (2011) updated the values to $Si_{11.16}$ to $Si_{12.05}$ with tetrahedral Fe^{3+} of <0.04, based on previously published data. Santaren et al. (1990) found a small substitution (1.3 wt.%) of F for OH (25% of the OH groups) in a sepiolite from Spain. Although early authors believed that a compositional gap existed between palygorskite and sepiolite, a continuous series actually exists if new data are included in the analysis (Suárez and García-Romero, 2011). Microtextural data (see below) suggest that polysome intergrowths are common, which may account for a compositional series.

1.5.4   WHY DO TETRAHEDRAL REVERSALS OCCUR IN THE PALYGORSKITE-SEPIOLITE GROUP MINERALS?

In an examination of structural parameters in the palygorskite-sepiolite mineral group, Guggenheim and Krekeler (2011) concluded that misfit is not a requirement to produce tetrahedral reversals in the mineral group, in general. For example, the average octahedral cation M–O bond distance vs. the average T–O bond distance in palygorskite (Post and Heaney, 2008) is similar to several (synthetic) mica structures, which do not have modulations. Misfit does occur in some species, for example, sepiolite shows slight misfit and raite (Table 1.2) has a more significant amount of misfit. However, another mechanism is required to explain why tetrahedral reversals occur for samples where misfit is minimal or nonexistent, and this mechanism is probably not a geometric one.

There are two important sets of crystallization parameters common to the members of the palygorskite-sepiolite group: the environment of formation and the activity of the components that affect the formation of these minerals. The environment of formation ranges from low-temperature aqueous solutions (e.g., salt lakes) to high-temperature hydrothermal (<350°C) conditions, and natural solutions tend to be alkali-rich with (Na + K)/Al > 1. Guggenheim and Krekeler (2011) described the environments of formation as a continuum where a common value of the activity of the components comprising the polysomes can occur. The octahedral strips in the mineral group are enriched by OH and OH_2, which is consistent with an aqueous environment with a high activity, a_{OH}. Other factors can affect a_{OH} values, such as

a_{alkali}, a_{SiO2}, and a_{Mg} and these parameters need to be explored to determine their effect on the formation of palygorskite and sepiolite.

1.5.5 MICROTEXTURE

Sorption behavior, solubility, density, and many other fundamental properties involved in environmental interactions and industrial applications of clay-sized materials are strongly related to structure, particle size, and microtexture. Microtextures involve the geometrical relationship between fine-grained components and include crystallinity (including defects, amorphous materials), grain size variations, and grain and crystal shapes within the material. In this section, palygorskite and sepiolite or related members of the group are considered, with an emphasis on defects that may affect chemical composition and sorption behavior. Defect types discussed below include stacking errors, variation in widths of polysomes, and omission of polysomes (Figures 1.9 and 1.10).

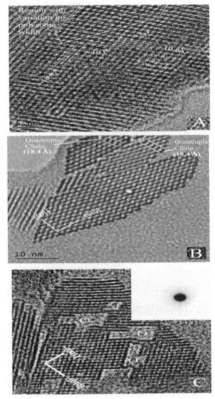

FIGURE 1.9 (A) TEM image along the [100] direction of yofortierite, the Mn-rich version of palygorskite. A grain boundary occurs in the lower right portion of the image. The extreme upper

FIGURE 1.9   *(Continued)*

left of the image is a beam-damaged particle in a different orientation from the central portion. Structural information is best illustrated in the upper central portion, with a parallelogram-like lattice and spacings of ~10.6 Å and angles of 73° and 103°. Widths of white regions are variable, which may represent widths of channels or polysomes. (B) TEM image along [100] of sepiolite (Helsinki) showing a strip of 18.4 Å consistent with a 4 chain width polysome. (C) Image near the [100] direction of yofortierite. This image has a rhombus-like lattice fringe contrast with a spacing of ~10.8 Å, which is consistent with the (011) spacing observed by Perrault et al. (1975) from XRD patterns. Regions of open channel defects (labeled O) in this crystal are rhombus- and parallelogram-like in shape and vary in cross-sectional area from approximately 16–75 nm². These defects allow for additional adsorption than the structure tunnels would generally allow and also affect the chemical composition (From Krekeler and Guggenheim, 2008, reproduced with the kind permission of Elsevier Press).

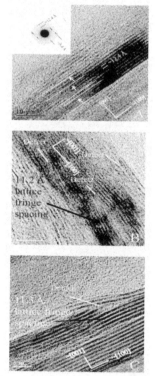

FIGURE 1.10   (A) Yofortierite TEM image along the [010] direction. This image shows stacking disorder of polysomes. Inset SAED pattern shows streaking along the *c* axis. (B) TEM image (approximately along [010] direction) of palygorskite from the Hawthorne Formation (southeast United States) showing a planar-angular defect near the crystal center. (C) TEM image of palygorskite fiber from the Hawthorne Formation along the [010] direction showing a planar-angular defect at a palygorskite fiber edge. (From Krekeler and Guggenheim, 2008, reproduced with the kind permission of Elsevier Press).

1.5.5.1 POLYSOME-WIDTH DISORDER

Krekeler and Guggenheim (2008) showed by TEM that variable widths occur in palygorskite, yofortierite (Mn-rich palygorskite, Figure 1.9a), and sepiolite (Figure 1.9b). For example, Figure 1.9b shows that a sepiolite from Helsinki has regions where quadruple-chain width (21.7 Å) polysomes were observed. Krekeler et al. (2005) also found polysome widths of triple (tetrahedral) chains (14.8 Å), and 5-chain (24.5 Å) polysomes in palygorskite. Such variations in polysome widths, especially if common in most occurrences, may impact industrial mineral performance, including nanoparticle substrates. For example, the accommodation of organic molecules of corresponding size would be expected for polysomes of variable widths, with larger molecules being accommodated in larger defect interstices. Therefore, sorptive properties are probably not limited to molecules fitting only in an ideal sepiolite-palygorskite structure (e.g., Maya Blue) where the polysomes are two pyroxene-like chains wide. If future research shows that defects affect sorption properties significantly, then industrial applications such as pesticide carriers, filtering, cleaning products, etc., may be enhanced.

1.5.5.2 OPEN CHANNEL DEFECTS

Open channel defects (or OCDs, Krekeler and Guggenheim, 2008) are defects consisting of the omission of polysomes in a palygorskite-sepiolite fiber, and they may form as isolated occurrences in single fibers (Figure 1.9c) or as dense groups. The cross-sectional area of OCDs may be as small as ~3.9nm^2 for a single omission to >75nm^2 for multiple polysome omissions, as found in a study on yofortierite. The cause of OCD formation is most likely crystallization rate and possibly the low temperatures of crystallization, but other factors may be important, such as changes in direction of crystal growth (i.e., possible local elemental concentration gradients) where (011) faces coalesce during crystallization to produce large channels of several tens of nanometers wide (Krekeler and Guggenheim, 2008). Anomalously high H_2O contents are known for palygorskite-sepiolite minerals and it may be the presence or absence of OCDs that affects H_2O content in palygorskite and sepiolite. For example, Jones and Galán (1988) reported that differential thermal analysis (DTA) techniques show variations commonly of 2–7 wt. % H_2O from palygorskite and sepiolite.

The monitoring of organocation exchange is primarily based on infrared spectroscopy, DTA, and other related techniques (e.g., Ruiz-Hitzky, 2001). Large organic molecules, such as cationic dyes and aromatic hydrocarbons, apparently exchange in the channels of palygorskite and sepiolite (e.g., Ruiz-Hitzky, 2001; Jones and Galán, 1988; Serna and Fernandez-Alvarez, 1974). However, the issues of steric constraints and how large molecules fit into the channels and replace cations and H_2O in the channels remain. Exchange requires a seemingly ordered transfer along the linear geometry of the confined channels along the [100] direction, with fiber lengths of several hundred to thousands of Ångströms (Jones and Galán, 1988). However, OCD structures in palygorskite and sepiolite may partly explain the absorption of large molecules in some samples, but not in others. Steric effects are reduced, mobility increases, and the kinetics of exchange may be better explained if OCD structures are present.

1.5.5.3   STACKING ERRORS AND PLANAR DEFECTS

Stacking errors are 180° rotations of polysomes with respect to each other along the *c* axis, and these errors may be observed along the [010] of fibers often as cross fringes in the TEM (Figure 10). Planar defects are characterized by 2° to 5° offsets normal to [001]. Common lengths of planar defects are 75–125 nm along the [100] direction, and displacements are generally 1–4 lattice fringes. These defects are most common in fibers that are 50–100 nm in width. [Portions of this section were updated from Guggenheim and Krekeler, 2011, and reprinted with the kind permission of the Elsevier Books]

ACKNOWLEDGMENTS

Thanks are extended to M.F. Brigatti and J. Churchman for reviewing the manuscript.

KEYWORDS

- **Chrysotile**
- **Halloysite**
- **Mesoporous structures**
- **Nanotubular phyllosilicates**
- **Sepiolite**
- **Smectite**

REFERENCES

1. Abidin, Z.; Matsue, N.; and Henmi, T.; Validity of the new method for imogolite synthesis and its genetic implication. In International studies in Environmental chemistry–Environmental research in Asia. Eds. Obayashi, Y.; Isobe, T.; Subramanian, A.; Suzuki, S.; Tanabe, S.; Terrapub: **2009**, 331–341.

2. Álvarez, A.; Santarén, J.; Esteban-Cubillo, A.; and Aparicio, P. Current industrial applications of palygorskite and sepiolite. In: Developments in Clay Science, Vol. 3; Eds. Galán, E.; Singer, A.; Elsevier: **2011**; 281–298.
3. Andréani, M.; Grauby, O.; Baronnet, A.; and Muñoz, M; *Eur. J. Min.* **2008**, *20*, 159–171.
4. Artioli, G.; and Galli, E.; *Mater. Sci. Forum.* **1994**, 166–169, 647–652.
5. Artioli, G.; Galli, E.; Burattini, E.; Cappuccio, G.; Simeoni, S.; and Neues J. B. *Min. Mh.* **1994**, 271–229.
6. Bailey, S. W.; *Clays. Clay. Min.* **1969**, 17, 355–371.
7. Bailey, S. W.; Structures of layer silicates. In: Crystal structures of clay minerals and their X-ray identification; Eds. Brindley, G. W.; Brown, G.; Monograph 5, Mineralogical Society: London; **1980**, 1–124.
8. Bailey, S. W.; Halloysite–A critical assessment. In: Proceedings of the 9th International Clay Conference, **1989**; Eds. Farmer, V. C.; Tardy, Y.; *Sci. Géol. Mém*: 86, **1990**; Strasbourg.
9. Baker, L. L.; and Strawn, D. G.; *Phys. Chem. Min;* **2012**, *39*, 675–684.
10. Baronnet, A. Equilibrium and kintetic processes for polytype and polysome generation. In: EMU Notes in Mineralogy 1, Modular aspects of minerals; Ed. Merlino, S.; European Mineralogical Union: Eötvös University Press: Budapest **1997**, 119–152.
11. Baronnet, A.; Andréani, M.; Grauby, O.; Devouard, B.; Nitsche, S.; and Chauanson, D.; *Am. Min.* **2007**, *92*, 687–690.
12. Bates, T. F.; Hildebrand, F. A.; and Swineford, A.; *Am. Min.* **1950**, *35*, 463–484.
13. Bishop, J. L.; Rampe, E. B.; Bish, D. L.; Abidin, Z.; Baker, L. L.; Matsue, N.; and Henmi, T.; *Clays Clay Min.* **2013**, *61*, 57–74.
14. Blaauw, C.; Stroink, G.; Leiper, W.; and Zentilli, M.; *Can. Min.* **1979**, *17*, 713–717.
15. Bobos, I.; Duplay, J.; Rocha, J.; and Gomes, C.; *Clay. Clay. Min.* **2001**, *49*, 586–607.
16. Brindley, G. W.; and Robinson, K.; *Min. Mag.* **1948**, *28*, 393–406.
17. Cámara, F.; Garvie, L. A. J.; Devouard, B.; Groy, T. L.; and Buseck, P. R.; *Am. Min.* **2002**, 87, 1458–1463.
18. Chernosky, J. V.; Jr.; *Am. Min.* **1975**, *60*, 200–208.
19. Chiari, G.; Giustetto, R.; and Ricchiardi, G.; *Eur. J. Min.* **2003**, *15*, 21–33.
20. Childs, C. W.; Parfitt, R. L.; and Newman, R. H.; *Clay. Min* **1990**, 25, 329–341.
21. Chrisholm, J. E.; *Can. Min.* **1992**, *30*, 61–73.
22. Christ, C. L.; Hathaway, J. C.; Hostetler, P. B.; and Shepard, A. O.; *Am. Min.* **1969**, *54*, 198–205.
23. Chryssikos, G. D.; Gionis, V.; Kacandes, G. H.; Stathopoulou, E. T.; Suárez, M.; and García-Romero, E.; *Am. Min.* **2009**, *94*, 200–203.
24. Chukanov, N. V.; Britvin, S. N.; Blass, G.; Belakovskiy, D. I.; and Van, K. V.; *Eur. J. Min.* **2012**, *24*, 171–179.
25. Chukhrov, F. V.; and Zvyagin, B. B.; Halloysite, a crystallochemically and mineralogically distinct species. In: Proceedings of the International Clay Conference, **1966**. Eds. Heller, L.; and Weiss, A.; Israel Program for Scientific Translation: Jerusalem, Israel, 1966.
26. Churchman, G. J.; *Clays. Clay. Min.* **1990**, *38*, 591–599.
27. Churchman, G. J.; Aldridge, L. P.; and Carr, R. M.; *Clays. Clay. Min.* **1972**, *20*, 241–246.
28. Churchman, G. J.; and Carr, R. M.; *Am. Min.* **1972**, *57*, 914–923.
29. Churchman, G. J.; and Carr, R. M.; *Clays. Clay. Min.* **1975**, *23*, 382–388.
30. Churchman, G. J.; and Gilkes, R. J.; *Clays. Min.* **1989**, *24*, 579–590.
31. Churchman, G. J.; Whitton, J. S.; Claridge, C. G. C.; and Theng, B. K. G.; *Clays. Clay. Min.* **1984**, *32*, 241–248.
32. Costanzo, P. M.; Giese, R. F., Jr; and Lipsicas, M.; *Clays. Clay. Min.* **1984**, 32, 419–428.
33. Costanzo, P. M.; Giese, R. F., Jr; Lipsicas, M.; and Straley, C.; *Nature* (London) **1982**, *296*, 549–551.

34. Costanzo, P. M.; and Giese, R. F., Jr.; *Clays. Clay. Min.* **1985**, *33*, 415–423.
35. Cradwick, P. D. G.; Farmer, V. C.; Russell, J. D.; Masson, C. R.; Wada, K.; and Yoshinaga, N.; *Nat. Phys. Sci.* **1972**, *240*, 187–189.
36. Cressey, B. A.; *Can. Min.* **1979**, *17*, 741–756.
37. Cressey, G.; Cressey, B. A.; and Wicks, F. J.; *Min. Mag.* **2008**, *72*, 1229–1242.
38. Cressey, G.; Cressey, B. A.; Wicks, F. J.; and Yada, K.; *Min. Mag.* **2010**, *74*, 29–37.
39. Cressey, B. A.; and Zussman, J.; *Can. Min.* **1976**, *14*, 307–313.
40. Dixon, J. B.; and McKee, T. R.; *Clays. Clay. Min.* **1974**, *22*, 127–137.
41. Dódony, I.; and Buseck, P. R.; *Intl. Geol. Rev.* **2004**, *46*, 507–527.
42. Drits, V. A.; and Aleksandrova, V. A.; *Zap. Vses. Min. Obsh.* **1966**, *95*, 551–560 (in Russian).
43. Drits, V. A.; and Sokolova, G. V.; *Soviet. Phys. Crys.* **1971**, *16*, 183–185.
44. Duarte, H. A.; Lourenço, M. P.; Heine, T.; and Guimarães, L.; Clay mineral nanotubes: stability, structure, and properties. In: Stoichiometry and Materials Science-When Numbers Matter. Ed. Innocenti, A. InTech: Croatia **2012**; 3–24.
45. Eggleton, R. A.; Taylor, G.; and Walker, P.; High cation exchange capacity kaolinite revisited. In Program and Abstracts, Australian Clay Mineral Society 12th Biennial Conference, 1991: Ballarat, Australia, **1991**; p 9.
46. Eggleton, R. A.; and Tilley, D. B.; *Clay. Clay. Min.* **1998**, *46*, 400–413.
47. Evans, B. W.; *Internat. Geol. Rev.* **2004**, *46*, 479–506.
48. Fahey, J. J.; Ross, M.; and Axelrod, J. M.; *Am. Min.* **1960**, *45*, 270–281.
49. Falini, G.; Foresti, E.; Gazzano, M.; Gualtieri, A. F.; Leoni, M.; Lesci, I. G.; and Roveri, N. *Chem. Eur. J.* **2004**, *10*, 3043–3049.
50. Farmer, V. C.; Adams, M. J.; Fraser, A. R.; and Palmieri, F. *Clay Min.* **1983**, *18*, 459–472.
51. Farmer, V. C.; and Fraser, A. R.; Synthetic imogolite, a tubular hydroxyaluminium silicate. In: International Clay Conference, Oxford. Mortland, M. M.; Farmer, V. C.; Eds.; Elsevier: Amsterdam **1979**, 547–553.
52. Faust, G. T.; Fahey, J. J.; Mason, B.; and Dwornik, E. J.; *Science.* **1969**, *165*, 59–60.
53. Ferraris, G.; Khomyakov, A. P.; Belluso, E.; and Soboleva, S. V.; *Eur. J. Min.* **1998**, *10*, 865–874.
54. Foresti, E.; Hocella, M. F.; Komishi, H.; Lesci, I. G.; Madden, A. S.; Roveri, N.; and Xu, H.; *Adv. Funct. Mater.* **2005a**, *15*, 1009–1016.
55. Foresti, E.; Hocella, M. F.; Lesci, I. G.; Roveri, N.; and Xu, H.; Morphological and chemical/physical characterization of Fe, Ti, and Al doped synthetic chrysotile nancrystals. In: 32nd International Geological Congress, Florence, **2005b**, (20–28 August 2004),
56. Galán, E.; *Clay. Min.* **1996**, *31*, 443–453.
57. Galán, E.; and Carretero, I.; *Clays. Clay. Min.* **1999**, *47*, 399–409.
58. García-Romero, E.; and Suárez, M.; *Clays. Clay. Min.* **2010**, *58*, 1–20.
59. Giese, R. F.; Jr.; Kaolin minerals: Structures and stabilities. In: Hydrous Phyllosilicates (Exclusive of micas). Ed. S.W. Bailey, Reviews in Mineralogy, 19, Mineralogical Society of America: Washington, D. C. **1988**, 29–66.
60. Giustetto, R.; and Chiari, G.; *Eur. J. Min.* **2004**, 16, 521–532.
61. Grauby, O.; and Baronnet, A.; Synthesis of polyhedral serpentine. XIV International Clay Conference, Italy, **2009**, Abstract Volume 2, **2009**, p 466.
62. Grim, R. E.; *Clay. Min.* New York: McGraw-Hill, **1953**, 384 p.
63. Guggenheim, S.; An overview of order/disorder in hydrous phyllosilicates. In: EMU Notes in Mineralogy: Layered Mineral Structures and their Application in Advanced Technologies, Eds. Brigatti, M. F.; and Mottana, A. **2011**, 73–121.
64. Guggenheim, S.; and Eggleton, R. A.; Crystal chemistry, classification, and identification of modulated layer silicates. In: Hydrous phyllosilicates (exclusive of the micas). Ed. Bailey, S.

W. Reviews in Mineralogy, 20, Washington, D. C.: Mineralogical Society of America: **1988**, 675–725.
65. Guggenheim, S.; and Krekeler, M. P. S.; The structures and microtextures of the palygorskite-sepiolite group. In Developments in Clay Science, 3, Eds. Galán, E. and; Singer, A.; Elsevier: **2011**, 3–32.
66. Guimarães-Enyashin, A. N.; Seifert, G.; and Duarte, H. A.; *J. Phys. Chem. C.* **2010**, *114*, 11358–11363.
67. Gustafsson, J. P.; *Clays. Clay. Min.* **2001**, *49*, 73–80.
68. Hawthorne, F. C.; Abdu, Y. A.; Tait, K. T.; and Back, M. E.; *Can. Min.* **2013**, *51*, 243–251.
69. He, H.; Barr, T. L.; and Klinowski, J.; *Clay. Min.* **1995**, 30, 201-209.
70. Heller-Kalai, L.; and Rozenson, I.; *Clay. Clay. Min.* **1981**, *29*, 226–232.
71. Hendricks, S. B.; and Jefferson, M. E.; *Am. Min.* **1938**, *23*, 863–875.
72. Henmi, T.; and Wada, K.; *Am. Min.* **1976**, 61, 379–390.
73. Honjo, G.; Kitamura, N.; and Mihama, K.; *Clay. Min. Bull.* **1954**, *2*, 133–141.
74. Honjo, G.; and Mihama, K.; *Acta. Cryst.* **1954**, 7, 511–513.
75. Jagodzinski, H.; Kunze, G.; and Neues J. B. *Min. Mh.* **1954**, 137–150.
76. Jancar, B.; and Suvorov, D.; *Nanotechnology.* **2006**, *17*, 25–29.
77. Janik, L. J.; and Keeling, J. L.; *Clay. Min.* **1993**, *28*, 365–378.
78. Jasmund, K.; and Sylla, H. M.; *Naturwissen.* **1970**, 45, 494–495.
79. Jones, B. F.; and Galán, E; Sepiolite and palygorskite. In: Hydrous phyllosilicates (exclusive of the micas), Bailey, S. W.; Ed., Reviews in Mineralogy, 19, Washington, D. C.: Mineralogical Society of America: **1988**, pp 631–674.
80. Joussein, E.; Petit, S.; Churchman, J.; Theng, B.; Roghi, D.; and Delvaux, B. *Clays. Clay. Min.* **2005**, *40*, 383–426.
81. Kadir, S.; Bas, H.; and Karakas, Z.; *Can. Min.* **2002**, *44*, 1091–1102.
82. Kerr, P. F.; Reference clay minerals. *Am. Petr. Inst. Res. Proj.* **1951**, *49*, Columbia University, N.Y.
83. Kirkman, J. H.; *Clay. Min.* **1977**, *12*, 199–216.
84. Kitagawa, Y.; *Am. Min.* **1971**, *56*, 465–475.
85. Kitagawa, Y.; *Clay. Sci.* **1973**, 4, 151–154.
86. Kogure, T.; Mori, K.; Drits, V. A.; and Takai, Y.; *Am. Min.* **2013**, *98*, 1008–1016.
87. Kogure, T.; Mori, K.; Kimura, Y.; and Takai, Y.; *Am. Min.* 2011, *96*, 1776–1780.
88. Kohyama, N.; Fukushima, K.; and Fukami, A. *Clay. Clay. Min.* **1978**, *26*, 25–40.
89. Korytkova, E. N.; Maslov, A. V.; Pivovarova, L. N.; Polegotchenkova, V. Yu; Povinich, V. F.; and Gusarov, V. V.; *Inorg. Mater.* **2005**, *41*, 743–749.
90. Krekeler, M. P. S.; Guggenheim, S.; *Appl. Clay Sci.* **2008**, *39*, 98–105.
91. Krekeler, M. P. S.; Guggenheim, S.; Rakovan, J;. *Clays. Clay Min.* **2004**, *52*, 263–274.
92. Krekeler, M. P. S.; Hammerly, E.; and Rakovan, J.; Guggenheim, S.; *Clays. Clay. Min.* **2005**, *53*, 94–101.
93. Krstanovic, I.; Pavlovic, S.; *Am. Min.* **1967**, *52*, 871–876.
94. Kunze, G. W.; and Bradley, W. F.; *Clays. Clay. Min.* **1964**, *12*, 523–527.
95. Kuroda, Y.; Ito, K.; Itabashi, K.; and Kuroda, K.; *Langmuir* 2011, *27*(5), 2028–2035.
96. Lafay, R.; Montes-Hernandez, G.; Janots, E.; Chiriac, R.; Findling, N.; and Toche, F.; *Chem. Eur. J.* **2013**, 19, 5417–5424.
97. Leoni, M.; Gualtieri, A. F.; and Roveri, N.; *J. Appl. Cryst.* **2004**, *37*, 166–173.
98. Levard, C. et al; J.-Y.; *J. Am. Chem. Soc.* **2008**, *130*, 5862–5863.
99. Lipsicas, M.; Straley, C.; Costanzo, P. M.; Giese, R. F. Jr.; and J. Coll; *Interf. Sci.* **1985**, *107*, 221–230.
100. Logar, M.; and Mellini, M. Structural refinement of polygonal serpentine. XIV International Clay Conference, Italy, **2009**, Abstract Volume 2, **2009**, p 469.

101. Lvov, Y. M.; Shchukin, D. G.; Mohwald, H.; and Price, R. R.; *Nano* **2008**, *2*, 814–820.
102. Ma, C.; and Eggleton, R. A.; *Clay. Clay Min.* **1999**, *47*, 181–191.
103. Ma, W.; Yah, W. O.; Otsuka, H.; and Takahara, A.; *J. Mater. Chem.* **2012**, *22*, 11887–11892.
104. Maillet, P. et al.; *J. Am. Chem. Soc. Comm.* **2010**, *132*, 1208–1209.
105. Martin-Vivaldi, J. L.; and Cano-Ruiz, J.; *Clays. Clay. Min.* **1955**, *4*, 173–176.
106. Martinez, E.; and Comer, J. J.; *Am. Min.* **1964**, *49*, 153–157.
107. Mellini, M.; and Zanazzi, P. F.; *Am. Min.* **1987**, *72*, 943–948.
108. Mellini, M.; *Am. Min.* **1982**, 67, 587–598.
109. Mellini, M.; Cressey, G.; Wicks, F. J.; and Cressey, B. A.; *Min. Mag.* **2010**, *74*, 277–284.
110. Nagasawa, K.; and Noro, H.; *Clay. Sci.* **1987**a, 6, 261–268.
111. Nagasawa, K.; and Noro, H.; *Chem. Geol.* **1987**b, 60, 145–149.
112. Newman, A. C. D.; and Brown, G.; The chemical constitution of clays. In: Chemistry of Clays Clay Minerals, Newman, A. C. D.; Ed. Mineralogical Society (of Great Britain): London, England, **1987**, pp. 1–128.
113. Newman, R. H.; Childs, C. W.; and Churchman, G. J.; *Clay. Min.* **1994**, 305–312.
114. Noll, W.; Kircher, H.; and Sybertz, W. Beitr. *Min. Petrogr.* **1960**, *7*, 232–241.
115. Noro, H.; *Clay. Min.* **1986**, 21, 401–415.
116. O'Hanley, D. S.; and Dyar, M. D.; *Can. Min.* **1998**, *36*, 727–739.
117. Olivieri, R.; *Accd. Nazion. Sci. Lett. Arti Modena* **1961**, Series VI, III, 44–59.
118. Ookawa, M.; Synthesis and characterization of Fe-imogolite as an oxidation catalyst. In: Clay minerals in nature–Their Characterization, Modification and Application, Eds. Valaškova, M.; Martynkova, G. S.; Intech: **2012**.
119. Pabst, S.; Zack, T.; Savov, I. P.; Ludwig, T.; Rost, D.; and Vicenzi, E. P.; *Am. Min.* **2011**, *96*, 1112–1119.
120. Parfitt, R. L.; and Henmi, T;. *Clays. Clay. Min.* 1980, *28*, 285–294.
121. Parfitt, R. L.; Furkert, R. J.; and Henmi, T. *Clays. Clay. Min.* **1980**, *28*, 328–334.
122. Perbost, R.; Amouric, M.; and Olives, J.; *Clays. Clay. Min.* **2003**, *51*, 430–438.
123. Perrault, G.; Harvey, Y.; and Pertsowsky, R.; *Can. Min.* **1975**, 13, 68–74.
124. Pluth, J. J. et al.; *Proc. Nat. Acad. Sci, USA.* **1997**, 94, 12263–12267.
125. Post, J. E.; and Bish, D. L.; Rietveld refinement of crystal structures using powder X-ray diffraction data. In: Modern powder diffraction, Eds. Bish, D. L.; Post, J. E.; Reviews in Mineralogy, 20, Mineralogical Society of America: Washington, D. C.; **1989**, 277–308.
126. Post, J. E.; Bish, D. L.; and Heaney, P. J.; *Am. Min.* **2007**, *92*, 91–97.
127. Post, J. E.; and Heaney, P. J. *Am. Min.* **2008**, *93*, 667–675.
128. Proenza, J. A.; Zaccarini, F.; Lewis, J.; Longo, F.; and Garuti, G.; *Can. Min.* **2007**, 45, 211–228.
129. Quantin, P.; Herbillon, A. J.; Janot, C.; and Stefferman, G. L.; *Clay. Min.* **1984**, 19, 629–643.
130. Range, K. J.; Range, A.; and Weiss, A.; Fire-clay type kaolinite or fire clay mineral? Experimental classification of kaolinite-halloysite minerals. In: Proceedings of the International Clay Conference, Tokyo, **1969**. v. 1, Ed. Heller, L.; Israel University Press: Jerusalem, **1969**, 3–13.
131. Rastsvetaeva, R. K.; Aksenov, S. M.; and Verin, I. A.; *Cryst. Rep. 2012*; 2012, *57*, 43–48.
132. Robertson, I. D. M.; and Eggleton, R. A.; *Clays. Clay. Min.* **1991**, *39*, 113–126.
133. Robertson, R. H. S.; *Clay. Min. Bull.* **1962**, 5, 41–43.
134. Roveri, N.; Falini, G.; Foresti, E.; Fracasso, G.; Lesci, I. G.; and Sabatino, P. J.; *Mat. Res.* **2006**, *21*, 2711–2725.
135. Rozenson, I.; Bauminger, E. R.; and Heller-Kallai, L.; *Am. Min.* **1979**, *64*, 893–901.
136. Ruiz-Hitzky, E.; *J. Mat. Chem.* **2001**, *11*, 86–91.
137. Saalfeld, H.; Wedde, M.; *Zeits. für Kristall.* **1974**, *139*, 129–135.
138. Santaren, J.; Sanz, J.; and Ruiz-Hitzky, E.; *Clays. Clay. Min.* **1990**, *38*, 63–68.
139. Serna, C.; and Fernadez-Alvarez, T. *Anal. Quimica* **1974**, *71*, 371–376.

140. Serna, C.; VanScoyoc, G. E.; and Ahlrichs, J. L.; *Am. Min.* **1977**, *62*, 784–792.
141. Shitov, V. A.; and Zvyagin, B. B.; *Soviet. Phys. Crystall.* **1966**, *10*, 711–716.
142. Singh, B. *Clays. Clay. Min.* **1996**, *44*, 191–196.
143. Singh, B.; Gilkes, R. J.; *Clays. Clay. Min.* **1992**, *40*, 212–229.
144. Singh, B.; Mackinnon, I. D. R.; *Clays. Clay. Min.* **1996**, *44*, 825–834.
145. Stathopoulou, E.T. et al.; *Eur. J. Min.* **2011**, *23*, 567–576.
146. Suárez, M.; and García-Romero, E.; Advances in the crystal chemistry of sepiolite and palygorskite. In: Developments in Clay Science. V. 3, Galán, E.; Singer, A. Eds. Elsevier: **2011**, 33–65.
147. Tazaki, K.; Analytical electron microscopic studies of halloysite formation processes– Morphology and composition of halloysite. In: Proceedings of the International Clay Conference, Bologna and Pavia, **1981**, Eds. van Olphen, H.; Veniale, F.; Elsevier: Amsterdam, **1981**, 573–584.
148. Toman, K.; and Frueh, A. J.; Jr.; *Acta Cryst.* **1968**, *A24*, 374–379.
149. Tomura, S.; Shibasaki, Y.; Mizuta, H.; and Kitamura, M.; *Clays. Clay. Min.* **1985**, *33*, 200–206.
150. Veblen, D. R.; and Buseck, P. R.; *Science* **1979**, 206, 1398–1400.
151. Veerabadran, N. G.; Price, R. R.; and Lvov, Y. M.; *Nano.* **2007**, *2*, 115–120.
152. Viti, C.; and Mellini, M.; *Eur. J. Min.* **1997**, 9, 585–596.
153. Wada, K.; Allophane and omogolite. In: Minerals in Soil Environments, Eds. Dixon, J. B.; Weed, S. B.; Madison: Soil Science Society of America, WI. **1977**, 603–638.
154. Wada, S.-I.; *Clays. Clay. Min.* **1987**, *35*, 379–384.
155. Wada, S. I.; Mizota, C.; *Clays. Clay. Min.* **1982**, *30*, 315–317.
156. Wada, S.-I.; Wada, K.; *Clays. Clay. Min.* **1982**, *30,* 123–128.
157. Walker, J. R.; Bish, D. L.; *Clays. Clay. Min.* **1992**, *40*, 319–322.
158. Weiss, A.; and Russow, J. Über das Einrollen von Kaolinkristallen zu Halloysitähnlichen Röhren und einen Unterschied zwischen Halloysit und röhrchenförmigem Kaolinit. In: Proceedings of the International Clay Conference: Stockholm, 2, **1963**, 69–74.
159. White, R. D.; Bavykin, D. V.; and Walsh, F. C.; J.; *Phys. Chem. C* **2012**, *116*, 8824–8833.
160. Whittaker, E. J. W.; *Acta Cryst.* **1953**, *6*, 747–748.
161. Whittaker, E. J. W.; *Acta Cryst.* **1956a**, *9*, 855–862.
162. Whittaker, E. J. W.; *Acta Cryst.* **1956b**, *9*, 862–864.
163. Whittaker, E. J. W.; *Acta Cryst.* **1956c**, *9*, 865–867.
164. Whittaker, E. J. W.; *Acta Cryst.* **1957**, *10*, 149.
165. Wicks, F. J.; *Can. Min.* **1986**, *24*, 775–788.
166. Wicks, F. J.; and Chatfield, E. J.; *Can. Min.* **2005**, *43*, 1993–2004.
167. Wicks, F. J.; and O'Hanley, D. S.; Serpentine minerals: Structures and petrology. In: Hydrous Phyllosilicates (Exclusive of micas). Bailey, S. W.; Ed., Reviews in Mineralogy, 19, Mineralogical Society of America: Washington, D. C.; **1988**, 91–167.
168. Wicks, F. J.; and Plant, A. G.; *Can. Min.* **1979**, *17*, 785–830.
169. Wicks, F. J.; and Whittaker, E. J. W.; *Can. Min.* **1975**, *13*, 227–243.
170. Xiangping, G.; Xiande, X.; Xiangbin, W.; Guchang, Z.; Jianqing, L.; Kenich, H.; and Jiwu, H.; *Eur. J. Min.* **2013**, *25*, 177–186.
171. Yada, K.; *Acta Cryst.* **1967**, 23, 704–707.
172. Yada, K.; *Acta Cryst.* **1971**, A27, 659–664.
173. Yada, K.; *Can. Min.* **1979**, 17, 679–691.
174. Yada, K.; and Tanji, T.; Direct observation of chrysotile at atomic resolution. In: Fourth International Conference on Asbestos, Torino, Italy, **1980**, 335–346.
175. Yah, W. O.; Yamamoto, K.; Jiravanichanun, N.; Otsuka, H.; and Takahara, A.; *Materials* **2010**, *3*, 1709–1745.

176. Yamnova, N. A.; Egorov-Tismenko, Yu. K.; and Khomyakov, A. P.; *Cryst. Rep.* **1996**, 41, 239–244 as translated from the Russian in: Kristall. **1996**, 41, 257–262.
177. Zhang, H.; et al. Analysis on crystal structure of 7Å halloysite. In: Advanced Materials, Part 3., Second International Conference on Advances in Materials and Manufacturing Processes (ICAMMP **2011**), Eds. Bu, J.; Jiang, Z.; Jiao, S.; **2012**, 2206–2214.
178. Zussman, J.; Brindley, G. W.; and Comer, J. J.; *Am. Min.* **1957**, *42*, 133–153.
179. Zvyagin, B. B.; Electron-diffraction analysis of clay mineral structures. Ed. Fairbridge, R. W, **1967**, Plenum Press: NY.

PART II
THE IDENTIFICATION AND NOMENCLATURE
OF NATURAL MINERAL NANOTUBES
(A HISTORICAL PERSPECTIVE)

CHAPTER 2

THE IDENTIFICATION AND NOMENCLATURE OF HALLOYSITE (A HISTORICAL PERSPECTIVE)

G. JOCK CHURCHMAN

CONTENTS

2.1   INTRODUCTION

Halloysite is a kaolin mineral with the same essential chemical composition as kaolinite, except for a commonly higher water content. Its higher water content led Berthier (1826) to distinguish halloysite as a separate mineral species from kaolinite. He named the mineral, first located near Liège in Belgium, in honor of its original discoverer, Omalius d'Halloy.

The provenance of the original sample of halloysite was karstic cavities within limestone that were affected by the acid weathering action of sulfides (Dupuis and Ertus, 1995). Halloysites have since been found worldwide. They have formed in a wide variety of geological environments, including weathered or hydrothermally altered volcanic rocks and ash, lateritic profiles, weathered granite, gabbro and dolorite, and also other rocks or deposits subjected to acid weathering (Joussein et al., 2005; Pasbakhsh et al., 2013; Keeling Chapter 5 of this book)

For a long time, halloysite has been of industrial interest mainly as a raw material for ceramics. Since about 2007, however, its frequent tubular morphology has led to an explosion of studies of it as a potentially valuable nanomaterial, particularly in polymer nanocomposites and as a carrier for active agents in, for example, medicine, agriculture, and environmental remediation (Du et al., 2010; Pasbakhsh et al., 2013). Nonetheless, historical analyses of the literature show that tubular morphology, which is the property that makes many halloysites potentially useful as a nanomaterial, cannot be seen as neither a necessary nor a sufficient characteristic of the mineral species (Churchman and Carr, 1975; Joussein et al., 2005).

2.2   IDENTIFICATION AND DEFINITION

Halloysites have inevitably been compared with kaolinites and the results of the application of successive newly developed instrumental analytical techniques have apparently brought new insights into its distinction from the kaolinite species, some of which have suggested alternative definitions of halloysite as a mineral species to that of a hydrated kaolinite.

2.2.1   X-RAY DIFFRACTION

In particular, early application of X-ray diffraction (XRD) to a number of halloysites gave rise to patterns which implied that halloysite was a highly disordered form of kaolinite (Ross and Kerr, 1934; Brindley and Robinson, 1946). This technique also located the additional water in halloysite over that in kaolinite within the interlayer region of the mineral (Hofmann et al., 1934; Mehmel, 1935). Halloysite with a fully hydrated interlayer gives a 001 peak for a spacing of 10 Å whereas the 001 peak for a kaolinite is 7 Å. However, when water is lost from the interlayer region by even mild heating—or by long-term dehydration in a dry atmosphere—its most distinctive peaks in XRD closely resembled those of a poorly ordered kaolinite. The loss of water from the interlayers of halloysite was found to be irreversible through simple addition of water and thus distinction between the two minerals was difficult by XRD alone.

2.2.2 EARLY APPLICATIONS OF OTHER INSTRUMENTAL TECHNIQUES

When halloysites were first viewed by electron microscopy (Alexander et al., 1943), they displayed a fibrous morphology that appeared to distinguish them sharply from kaolinites with their typical hexagonal platy particles. Other techniques such as differential thermal analyses (DTA), infrared (IR) spectra, and chemical reactivities were also applied to seek distinctive indicators for halloysite (Churchman and Carr, 1975). Generally, the results of these analyses reflected either the intercalation of water, in the present or the past, between aluminosilicate layers of halloysite, or a lower extent of structural order in halloysites than in kaolinites, or both. Thus, halloysites generally showed the distinctive kaolin endothermic peak in the 500–600°C region of DTA traces at lower temperatures than kaolinites and the peak was often considerably more asymmetric than for kaolinites (Bramao et al., 1952), suggesting weakly bound hydroxyls in its structure, possibly due to residual interlayer water molecules (Chukhrov and Zvyagin, 1966). A more disordered structure than for kaolinite was also suggested by more diffuse bands for hydroxyl stretching in IR spectra for halloysite than for kaolinite (Lyon and Tuddenham, 1960; Farmer and Russell, 1964).

2.2.3 CHEMICAL REACTIVITY

Many studies have found an apparently clear distinction between halloysites and kaolinites in their reactivities toward the formation of complexes with ionic or polar compounds. Halloysites are more reactive than kaolinites with a number of these compounds. They often form an intercalation complex with halloysites, but not at all with kaolinites (MacEwan, 1946; Wada, 1959). Some compounds, for example potassium acetate among ionic compounds (Andrew et al., 1960) and hydrazine as a polar organic compound (Weiss et al., 1963), can form intercalation complexes with at least some kaolinites as well as with halloysites. However, a common distinction between intercalation complexes of halloysites and those of kaolinites is that, upon addition of water, those with halloysite exchange the ionic or polar organic compounds with water and remain swollen in the c-direction, while those with kaolinite divest intercalated compounds without any uptake of water. The preconditioning of the interlayer association of halloysite with water in its natural state has enabled its easy retention of water in the interlayer region while there is a natural tendency for the layers of kaolinite to associate closely together without intervening water molecules.

This distinction was developed as the basis for early tests for halloysite that involved examination by XRD after addition of water following either potassium acetate (Wada, 1961) or hydrazine (Range et al., 1969). Halloysite remained expanded in each case although Range et al. (1969) also suggested adding glycerol after water (in the so-called hydrazine/water/glycerol (HWG) test) in the belief that some forms of disordered kaolinite also remained expanded with water alone after hydrazine. In contrast, Theng et al. (1984) noted no measurable extra effect of glycerol upon 42 samples of earth materials containing halloysite and kaolinite in different proportions. Nonetheless, Churchman and Theng (1984) found that halloysites from different locations reacted to differing extents with members of a series of amides that differed in chain length and extent of branching. Even so, all halloysites tested showed a complete intercalation of the simplest amides (formamide and n-methyl formamide) and the reaction was practically instantaneous. More studies were made with formamide as an inexpensive chemical that has a relatively low toxicity. Churchman et al. (1984) found that any kaolinites tested either showed no reaction with formamide or only a slow uptake into the interlayer, with no perceptible change being noted within ~4 h. This difference from kaolinite in its rate of reaction with formamide was proposed as a simple test for halloysite, especially when dehydrated. More recently, Hillier and Ryan (2002) have demonstrated that addition of ethylene glycol to a halloysite that is only slightly hydrated leads to changes in 001:002 peak intensity ratios that are diagnostic for halloysite in comparison to kaolinite without bringing about complete intercalation of the glycol.

2.2.4 MORPHOLOGY

Without a doubt, it is the peculiar and varied morphology of halloysites that has marked out their particle shape as the most curious aspect of the mineral species. Whereas the earliest applications of electron microscopy to halloysites revealed fibrous/tubular shapes for the mineral, as noted, and many halloysites worldwide occur in this form, halloysites can also occur in a wide variety of shapes. Joussein et al. (2005) has classified the halloysites described in >50 reports up to 2004 into ten different types, as summarized in Table 2.1.

TABLE 2.1 Summary of the ten different types of morphologies for halloysite that were identified by Joussein et al. (2005).

Morphological Type	Earliest Report	No. of Reports*
1. Tubular, long and thin, short and stubby tubes	1950	28
2. Pseudospherical and spheroidal	1969	23
3. Platy or tabular	1964	9

TABLE 2.1   *(Continued)*

Morphological Type	Earliest Report	No. of Reports*
4. Fiber	1965	2
5. Prismatic, rolled, crinkly, walnut-meat	1979	2
6. Cylindrical, disk	1977	3
7. Spherulitic, irregular lath with rolling edge	1977	1
8. Crumpled lamellar	1982	1
9. Lath, scroll	1976	2
10. Glomerular or "onion-like"	1969	1

[a]See Joussein et al. (2005), Table 2.1, for identification of reports

The recognition of different shapes for halloysite particles has continued beyond 2004 and not only further examples of some of these types, but also other morphological types for halloysites have been recognized.

Wyatt et al. (2010) found a few occurrences in buried tephra of a clay mineral occurring in platy particles stacked together in "books" that was positively identified as halloysite by the formamide test of Churchman et al. (1984) and which showed no 7-Å peaks for kaolinite. The arrangement of platy particles in book-like shapes has long been regarded as a typical morphology for kaolinite.

Overall, the ten types of shapes for halloysites that were recognized by Joussein et al. (2005) and are listed above, together with the book-like shapes of Wyatt et al. (2010), may be summarized into three or four major types:

I.   *Tubular:* includes Nos. 1, 4, and 5 of Joussein et al.'s (2005) groups;
II. *Platy:* includes Nos. 3 and 9 of Joussein et al.'s (2005) groups and also Wyatt et al.'s (2010) "books";
III. *Spheroidal:* includes Nos. 2, 6, 7, and 10 of Joussein et al.'s (2005) groups; and, perhaps also,
IV. *Prismatic:* No. 5 in Joussein et al. (2005). This apparently rarer type (Type IV) may be isolated from the other major types because some publications, for example, Chukrov and Zvyagin (1966) and Kogure et al. (2013), have identified it as a highly ordered form of the mineral. These authors identified the prismatic form only in dehydrated samples of halloysite and its origin has been ascribed by both Chukrov and Zvyagin (1966) and Kogure et al. (2013) to the alignment of groups of layers that results when the initially formed hydrated halloysite that is usually tubular loses its interlayer water upon dehydration. This form appears also in transmission electron micrographs (TEMs) in Dixon and McKee (1974) and also Churchman et al.

(1995). Examples of halloysites with predominantly tubular shapes (Type I), but different lengths and widths and also of a sample in which "blocky" platelike particles (Type II) dominate over other shapes in the halloysite assemblage, are shown in Figure 5.1 in Keeling (this volume).

Joussein et al.'s (2005) Group No. 7, comprising "spherulitic, irregular lath with rolling edge" morphologies, from Wilson and Tait (1977), describes an occurrence of halloysite with a range, or mixture of shapes. Many of the reports of occurrences of halloysite have revealed that this mineral may occur within the same sample in particles with a variety of shapes. Later examples of halloysites with mixed particle morphologies than those given by Joussein et al. (2005) include tubular and polygonal together (De Oliviera et al., 2007) and also spheroidal and tubular together (Ece and Schroeder, 2007; Cravero et al., 2012), while Wyatt et al. (2010) reported that halloysite occurred in the buried weathered tephra as tubes and spheres as well as plates within books. Several authors (Sudo and Yotsumoto, 1977; Jeong, 1998; Singer et al., 2004; and Papoulis et al., 2004) have proposed that one form of halloysite may transform to another with increasing extent of weathering. Often, a spheroidal form becomes transformed to a tubular form.

Many (see, e.g., Joussein et al., 2005) have found a relationship between impurities in the structure (usually Fe, but also, e.g., Ti – Singer et al., 2004) and particle size of halloysites, so that extraordinary long tubes as in Patch Clay have very little structural Fe. Furthermore, when structural Fe is very high, halloysite tends to give rise to platy forms (e.g., Bailey, 1990; Joussein et al., 2005). However, there seems to be no consistent correlation between spheroidal shapes and Fe contents of halloysites (Bailey, 1990; Adamo et al., 2001; Singer et al., 2004; Churchman and Lowe, 2012). Churchman et al. (2010), studying the formation of halloysite within saprolite from the same rock type (granite) in different locations in Hong Kong, found that crystal size, rather than shape, was a function of the concentration of impurities such as iron and manganese in the immediate environment, thought to be the environment of formation. Examples from their study are given in Figure 2.1 to illustrate the nature of the relationship between the contents of impurity Fe and/or Mn and the size of tubular particles in scanning electron micrographs (SEMs).

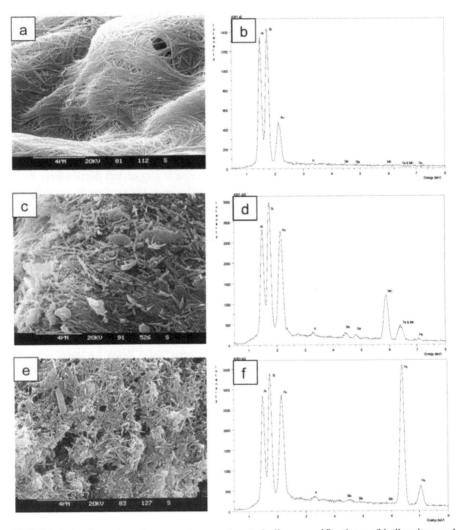

FIGURE 2.1 Scanning electron micrographs at similar magnifications of halloysites and also associated minerals were present within infill veins in weathered granite at different locations in Hong Kong. (a) and (b) respectively show SEM and EDX analysis of a white, mottled pink (Munsell color 7.5YR 8/4) vein; (c) and (d) respectively show SEM and EDX analysis of a very pale brown (Munsell color 10YR 8/4) vein; and (e) and (f) respectively show SEM and EDX analysis of a reddish yellow (Munsell color 5YR 6/8) vein containing red (2.5YR 5/8) streaks. Photos and analyses by Stuart McClure (Permission for publication has been granted by the Director of Civil Engineering and Development and the Head of the Geotechnical Engineering Office, Government of the Hong Kong Special Administrative Region, China).

2.2.5   RELATIONSHIPS BETWEEN HALLOYSITE AND KAOLINITE FROM THEIR OCCURRENCE

Halloysites and kaolinites quite often occur together in the same sample. In Hong Kong, Churchman et al. (2010) found that either pure halloysite, almost pure kaolinite, or a range of mixtures of halloysite and kaolinite had formed by weathering in granite or tuffaceous rocks. There are some kaolinite plates (confirmed by XRD analyses of intercalated samples) among the dominant tubular particles in Figure 2.1c and e, for instance. Many studies (see reviews by Churchman, 2000 and Churchman and Lowe, 2012) have inferred that halloysite tends to transform to kaolinite over time under weathering. This trend is suggested particularly by a general trend from halloysite at depth toward kaolinite at the surface in profiles on weathered undisturbed rocks or deposits of many types. It is consistent with a lower thermodynamic stability for halloysite than for kaolinite (e.g., see Lindsay (1979) and Churchman (2000)). The occurrence of these two types of kaolin minerals together has prompted several authors (e.g., Jeong (1998), Singer et al. (2004), and Papoulis et al. (2004)) to propose that kaolinite can form from preformed halloysite in weathering. Even so, only some of these studies have used intercalation tests to confirm that kaolinite has formed this way. Instead, some have relied upon morphology to a greater or lesser extent, with platy particles being regarded as sufficient evidence for the identification of kaolinite. This is not necessarily the case, as we have seen. In deep weathering profiles in Western Australia, platy particles were in greater concentrations than tubular particles in the uppermost parts of profiles when they were rare or absent in their lower parts (Churchman and Gilkes, 1989). These authors attributed this pattern with depth to the later formation of kaolinite alongside preformed halloysite within the same weathering system.

Churchman et al. (2010) concluded from their Hong Kong study that halloysite formed and persisted in constantly wet conditions and kaolinite formed when drying occurred, albeit even intermittently. This pattern supports Churchman's (2003) generalization that water is a necessary and sufficient formation for the formation of halloysite rather than kaolinite.

2.2.6   DEFINITION

Consequently, provenance and genetic studies bear out Churchman and Carr's (1975) definition of halloysite as "those minerals with a kaolin layer structure which either contain interlayer water in their natural state or for which there is unequivocal evidence of their formation by dehydration from kaolin minerals containing interlayer water" and their identification by intercalation methods rely upon this definition. Furthermore, their unusual and varied morphologies are often a consequence of their intercalation of two water molecules per unit cell at the time of their formation. As outlined elsewhere (see e.g., Joussein et al., (2005) and Churchman and Lowe (2012)), the water interleaved between the layers allows rolling to occur to correct mismatches between the dimensions of the tetrahedral and octahedral sheets rather than the inversion of silica tetrahedra in kaolinites, giving tubular shapes, unless sufficient impurity Fe is present to prevent rolling. Other shapes may originate from their particular origin, with spheroidal forms apparently resulting from recrystallization of products of the rapid dissolution of volcanic glass (Dixon and McKee, 1974; Bailey, 1990; Adamo et al., 2001; Singer et al., 2004; Churchman and Lowe, 2012).

2.3 NOMENCLATURE

2.3.1 HALLOYSITE VERSUS KAOLINITE

The first question of nomenclature that needs to be settled—that of halloysite versus kaolinite—might appear to be trivial. Even so, earlier studies (e.g., Honjo et al. (1954))—as was noted critically by Chukhrov and Zvyagin (1966)—insisted on referring to their halloysite samples as "tubular kaolinites." Early identification of halloysites as hydrated forms of kaolinites, or, when dehydrated, as poorly crystalline forms of kaolinite, are complicated by several indications that halloysites display some important structural differences from kaolinites, and that they may also display some important differences from kaolinites in chemical composition, at least while they remain wet. Their prime distinction from kaolinites is that of their incorporation of extra molecules of water in their structure. This distinguishing characteristic nonetheless begs the as-yet unanswered question of why they hold more water than kaolinite, with its apparently "identical" chemical composition. Bailey (1990) put forward the view that the aluminosilicate layer in halloysite is more highly (negatively) charged than that in kaolinite and attributed the difference to halloysites having a higher content of tetrahedral aluminum than kaolinites. Newman et al. (1994) tested this hypothesis and found it wanting insofar as ^{27}Al-NMR analyses indicated that a range of halloysites had similar contents of Al(IV) to standard kaolinites.

Churchman (2009) proposed instead that, while Bailey's suggestion of a higher charge for the aluminosilicate layers in halloysite than in kaolinite may explain the propensity for halloysite to incorporate polar water molecules in its interlayer, the charge may originate in a higher content of ferrous iron Fe^{2+} in halloysites than in kaolinites. A tendency toward the lower oxidation state for Fe would be expected from the formation of the mineral in a hydrous environment. It may also explain why soils formed on basalt which is underlain by calcareous material by weathering under a xeric (winter wet but summer dry) climate in the limited area encompassed by Mts Gambier and Schank (which are 10 km apart) in South Australia, gave rise to allophane and kaolinite in the main but no halloysite (Lowe et al., 1996; Lowe and Palmer, 2005; Takesako et al., 2010; Churchman and Lowe, in preparation).

Halloysite is the dominant clay mineral in soils on volcanic materials under a similar mean annual precipitation (MAP) in North Island, New Zealand (Parfitt et al., 1984). The MAP at Mt Gambier is ~700 mm while that of soils found to have halloysite dominance in clay fractions in North Island was 800–1,050 mm per year (Parfitt et al., 1984). In North Island, soils formed on volcanic materials under a higher MAP tended to give rise to allophane rather than halloysite. They did not give rise to kaolinite regardless of MAP. However, Parfitt et al. (1983, 1984) found that the key factor controlling the formation was the annual throughflow of water, rather than its receipt by precipitation. High values for throughflow led to allophane, and low values to halloysite. The soils at Mts Gambier and Schank have a rate of flow of water (~280 mm year ^{1}) through their upper horizons (Lowe and Palmer, 2005) similar to the upper threshold for halloysite formation in North Island soils (Parfitt et al., 1984), and there is considerably less flow of water through the lower horizons of the South Australian volcanogenic soils. It remains curious as to why some kaolinite, but no halloysite, forms in these latter soils.

The North Island soils are subjected to a more even distribution of precipitation over the year than those at Mts Gambier and Schank. It may be that the seasonal drying that is experienced by these soils during hot summers in comparison with their North Island counterparts may have limited the supply of ferrous ions sufficiently at the time of clay mineral synthesis so that a more ferric kaolinite formed rather than halloysite. Parfitt (1990) observed that kaolinite—and also smectite—may form in tephra weathered under climatic regimes with long, dry spells, including ustic regimes, which are normally dry for >90 days. Nonetheless, halloysite has been found to form under xeric climate regimes elsewhere. Both Southard and Southard (1989) and Takahashi et al. (1993) found halloysite to be a prominent clay mineral in soils formed from volcanic materials in the xeric moisture regime of Northern California.

Nonetheless, the material underlying the volcanic eruptive material at Mts Gambier and Schank is calcareous. Generally, soils within each profile have a high pH (between 6.1 and 8.1 in 0.01M $CaCl_2$ and between 6.4 and 8.7 in water). While ferrous ions are favored over ferric ions in wet conditions (where pE values are negative), they also become unstable in relation to solid phases such as siderite $FeCO_3$ (in a CO_2-rich system) (Sposito, 1994) and ferrihydrite $Fe_5HO_8.4H_2O$ (in a CO_2-poor system) as pH rises (Sposito, 2008). In the former, assuming realistic values for the concentration of Fe^{2+} and CO_2 partial pressure, ferrous ions are hardly present in solution in the pH 6–7 range and above. We may therefore conclude that they are hardly present in the suite of 9 soils examined by Lowe et al. (1996) in the Mts Gambier and Schank area (and another two volcanogenic soils examined by Takesako et al. (2010) in the same area). The lack of halloysite in these soils is consistent with a requirement for ferrous ions for the formation of halloysite. It is also consistent with the common occurrence of halloysite under acid conditions. The ease of oxidation of ferrous ions to ferric ions under oxidizing conditions including drying prior to analysis may have led to this requirement for halloysites being overlooked. In apparent contradiction, however, Silber et al. (1994) claimed to find halloysite in low concentrations in weathered tuffs at exceptionally high pHs in Israel, but the evidence given for halloysite is based on neither morphology nor chemical reactivity, and hence is somewhat equivocal.

Several pieces of evidence over a number of years have suggested that not only the chemical composition, but also the structure of halloysites may differ from that of kaolinites. It has been observed, mainly using electron diffraction, that their layer stacking can show two-layer periodicity (see Guggenheim, Chapter 1 of this book). In this regard, at least, they resemble the rarer, highly ordered kaolin mineral, dickite. Furthermore, with one of the features which distinguish halloysites from most kaolinites being the lack of any XRD peaks attributable to hkl reflections in the former, many authors (Sanchez Camazano and Gonzalez Garcia, 1966; Jacobs and Sterckx, 1970; Barrios et al., 1977, Anton and Rouxhet, 1977; Churchman, 2009) have noted a remarkable appearance of new hkl peaks in XRD patterns from complexes of halloysite with highly polar compounds. Generally, it was assumed that the interaction led to a reordering of halloysite toward kaolinite, consistent with the belief that the former was a hydrated, and hence, disordered form of the latter.

Churchman and Theng (unpub. data) found that up to three extra sharp peaks suggesting hkl reflections appeared for spacings of near 4.1 Å in XRD patterns for complexes of halloysites with formamide (10.4 Å c-spacing), n-methyl formamide (10.9 Å), acetamide (11.0 Å), and some other amides. A similar number of peaks in this region appeared for complexes of halloysites with pyridine (11.9 Å). One sharp peak, for a spacing of 4.05 Å, was retained in the pattern for the halloysite after first pyridine had been removed by washing in water, and then water had been removed by heating, suggesting that some ordering was retained after removal of the intercalate. Even more peaks appeared in this region for complexes of halloysite with dimethylsulfoxide DMSO (11.3 Å). Furthermore, complexes of DMSO could be formed with all of halloysite, kaolinite (standard low and high crystalline forms), and also dickite. All gave the same c-spacing and also definite peaks at 4.32, 4.20, 4.13, and 4.10 Å in common, but a peak at 3.83 Å appeared in the patterns for dickite and halloysite, but not those for either of the kaolinites. It appears that the ordering induced by complex formation was more similar to that in dickite than in kaolinite, consistent with suggestions (Honjo et al., 1954; Chukhrov and Zvyagin, 1966; Bailey, 1990; Kogure et al., 2013) that halloysite, like dickite, may have two-layer periodicity in its structure. At the least, the results point to an inherent difference between the structure of halloysite and that of kaolinite, in support of Chukhrov and Zvyagin's contention to this effect in 1966, which was supported by Kogure et al. (2013), at least for some (prismatic) halloysites. Churchman and Theng's results will be presented in detail and discussed further in another publication.

Even if much evidence from its provenance and also from its reordering upon intercalation points to possible chemical and structural differences from kaolinite, there have been occurrences of minerals showing some characteristics of both halloysites and kaolinites. These include tubular kaolin minerals found high within a profile in Australia dominated by halloysite at depth but which did not expand with formamide, although they remained expanded after Range et al.'s (1969) HWG test (Churchman and Gilkes, 1989). They also include a proportion of the kaolin minerals within soils throughout New Zealand which similarly remained expanded following HWG treatment but did not expand with formamide, although their particles were sometimes platy rather than curled (Churchman, 1990). Churchman (1990) confirmed Churchman and Theng's (1984) conclusion that extent of expansion with formamide is diminished when halloysite is heated, even to 40°C. Similarly, Hellier and Ryan (2002) showed that the test for halloysite using ethylene glycol was ineffective when water is completely driven out at 300°C. Janik and Keeling (1993), using statistical methods to relate IR spectra to the appearance of tubular shapes in electron micrographs, confirmed Churchman and Gilkes' (1989) finding that this morphological form of halloysite can persist under prolonged drying in Australia that has prevented entry of formamide to their interlayers. Furthermore, although the formamide intercalation test appears to have largely withstood the test of time worldwide (Christidis, 2013; Lagaly et al., 2013; Środoń, 2013), Joussein et al. (2007) found that some halloysites from soils showed a strong reduction in expansion upon immediate treatment with formamide after they had been left to dehydrate upon storage from their fully hydrated state. These samples also showed a high concentration of typical shapes for halloysite among their particles. These particular samples were also characterized by elevated cation exchange capacity (CEC) values, a high selectivity for K^+ ions, and a high selectivity for Cs^+ ions, suggesting that they incorporated some interstratified expandable 2:1 layers. Takahashi et al. (1993, 2001) had noted that halloysite with similar characteristics had also reacted only incompletely with formamide. These instances and others (Joussein et al., 2007) indicate both that halloysites may be highly variable in many aspects and also that there are probably no definitive tests to distinguish all halloysites from all kaolinites. Indications of the presence of halloysite therefore need to conform to its definition (Section 2.2.6) for their validation.

2.3.2   HYDRATED VERSUS DEHYDRATED HALLOYSITE

Early work on halloysite drew a sharp distinction between the hydrated form, with a 10 Å d-spacing and the dehydrated form, with a spacing of 7 Å (e.g., Churchman and Carr, 1975). The hydrated form was given a number of names: "halloysite" (Mehmel, 1935), "hydrated halloysite" (Hendricks, 1938), and "endellite" (Alexander et al., 1943). The dehydrated form was commonly known to early workers as "metahalloysite," although Alexander et al. (1943) called it "halloysite." However, a number of studies (MacEwan, 1947; Brindley and Goodyear, 1948; Harrison and Greenberg, 1962; Hughes, 1966; Churchman and Carr, 1972; Churchman et al., 1972; Lowe, 1986) came to the conclusion that halloysite could adopt a continuous series of hydration states, from 2 to 0 molecules of H_2O per $Al_2Si_2(OH)_5$ aluminosilicate layer, showing d-spacings over the whole range from 10 to 7 Å, although generally as a broad spread of reflections over this range with concentrations in intensity toward either 10 or 7 Å. Churchman et al. (1972) interpreted their behavior on dehydration as a type of interstratification of the two end-member types in which the end-member types tend to show some segregation by type. Most commonly, the nomenclature used to denote the two end-members is "halloysite −10 Å" and "halloysite − 7 Å," but it should be recognized that both fully hydrated (particularly) and the fully dehydrated forms of halloysite are rare, due respectively to the ease and the irreversibility of loss of interlayer water from halloysite in the one case, and its tendency to sequester residual interlayer water even to temperatures above 100°C (Churchman, 1970; Churchman and Carr, 1972), in the other.

2.3.3 NOMENCLATURE FOR USE OF HALLOYSITE AS NANOTUBES

Clearly, some halloysites are unsuitable for use in polymer nanocomposites or as nanocontainers. These two general classes of their use as nanomaterials both require them to belong to either the Type I (tubular), and probably also Type IV (prismatic), major types described in Section 2.2.4 above. Those in Type II (platy) and Type III (spheroidal) are unlikely to confer any particular advantage over other clays (e.g., smectites) as components of nanocomposites with polymers whereas tubular halloysite shows superior properties to smectites for these applications (Pasbakhsh et al., 2010; 2013). Nontubular halloysites offer no useful properties for use as nanocontainers (e.g., Pasbakhsh et al., 2013). According to Pasbakhsh et al. (2013), halloysites which are tubular and have a high dispersibility are most useful as nanofillers for polymer nanocomposites. Those which have central lumen within their tubular particles that are cylindrical and hollow and which offer easy access to the molecules they are to carry and supply are the most suitable for use as nanocontainers and also as nanotemplates for reactions (Pasbakhsh et al., 2013). It has become common for halloysites proposed for use in these types of applications to be known as "halloysite nanotubes" (HNTs) and this is an appropriate nomenclature for all tubular (including prismatic and tubular) forms of halloysite. A nomenclature that recognizes their shape is the most suitable for the selection of halloysites for applications as nanomaterials.

2.4 CONCLUSIONS

Halloysites have the same essential chemical composition as kaolinites but can have higher water contents. Early XRD analyses first characterized halloysite as a highly disordered form of kaolinite and also located its additional water within the interlayer region. Early electron microscopic analyses identified a fibrous morphology and other instrumental analyses gave results that reflected their past or present intercalation of water. Compared with kaolinites, they generally show greater reactivities with ionic or polar compounds and the displacement of these by water leads to the reentry of water to the interlayers of halloysites but not kaolinites.

Halloysite particles display three or four major types of shapes—tubular, spheroidal, platy, and possibly also, prismatic. More than one type can occur together in a sample. Concentrations of impurities such as Fe and Mn in the formation environment largely govern particle size and Fe and Ti, at least, can become incorporated into the structure, leading to the unrolling of tubes. Halloysite forms and persists as long as the environment remains wet. Churchman and Carr's (1975) definition of halloysite as kaolin minerals which contain interlayer water or which have resulted from the dehydration of kaolin minerals with interlayer water is confirmed by subsequent provenance and genetic studies.

Bailey's (1990) proposal for the interlayer water in halloysite arising from a higher-layer charge than that of kaolinite is seen as valid but his explanation of its origin in a greater Al(IV) content for halloysites than kaolinites has been invalidated; it is proposed instead that halloysites may incorporate more Fe^{2+} than kaolinites at the time of their formation. This follows from its absence in favor of kaolinite in soils with high pH, outside the stability field for Fe^{2+} in relevant pE –pH diagrams. Reordering of a three-dimensional structure occurs upon intercalation of polar organic molecules leading to a structure that resembles dickite more closely than kaolinite, consistent with the frequent attribution of a dickite-like two-layer structure for halloysite. Hence halloysite cannot be seen as simply "tubular kaolinite" although some minerals show characteristics of both halloysite and kaolinite, particularly when sustained dehydration has occurred. There does not appear to be any one test which is able to distinguish between all halloysites and all kaolinites.

Halloysites may hold any fraction of water between 0 and 2 molecules per $Al_2Si_2(OH)_5$ aluminosilicate layer and this is often reflected in a broad spread of reflections spanning between the end-member peaks at 7 and 10 Å. Generally, the fully hydrated form is known as "halloysite - 10 Å" and the fully dehydrated form as "halloysite - 7Å," but both end-member forms are rare due to the easy loss of most water, the difficulty of losing all the water, and the irreversibility of the process. Tubular forms of halloysite are eminently suitable for nanofillers in polymer nanocomposites or else for nanocontainers or nanotemplates for reactions, but other forms of halloysite are unsuitable for these niche uses. Therefore, the adoption of "halloysite nanotubes" or HNTs as a nomenclature for tubular forms (and usually also prismatic forms of the mineral) is appropriate.

ACKNOWLEDGMENTS

I am grateful to David J. Lowe of the University of Waikato, New Zealand, for reading the manuscript critically prior to its submission.

KEYWORDS

- Chemical reactivity
- Formamide
- Halloysite nanotube
- Kaolinite
- Morphology
- Nanocomposites
- Nomenclature of halloysite

REFERENCES

1. Adamo, P.; Violante P.; and Wilson. M. J.; *Geoderma.* **2001**, *99*, 295–316.
2. Alexander, L. T.; Faust, G. T.; Hendricks, S. B.; Insley, H.; and McMurdie, H. F.; *Am. Min.* **1943**, *28,* 1–18.
3. Andrew, R. W.; Jackson, M. L.; and Wada, K.; *Soil. Sci. Soc. Am. Proc.* **1960**, *24*, 422–424.
4. Anton, O.; and Rouxhet, P. G.; *Clays. Clay. Min.* **1977**, *25*, 259–263.
5. Bailey, S. W.; **1990**. *Sci. Géol.* **1990**, *86*, 89–98.
6. Barrios, J.; Plançon, A.; Cruz, M. I.; and Tchoubar, C.; *Clays. Clay. Min.* **1977**, *25*, 422–429.
7. Berthier, P.; *Annales de Chimie et de Physique.* **1826**. 32, 332–335.
8. Bramao, L.; Cady, J. G.; Hendricks, S. B.; and Swerdlow, W. L.; *Soil. Sci.* **1952**, *73*, 273–287.
9. Brindley, G. W.; and Goodyear, J.; *Min. Mag.* **1948**; *28*, 407–422.
10. Brindley, G. W.; and Robinson, K.; *Trans. Faraday. Soc.* **1946**, *42B*, 198–205.
11. Christidis, G. E.; Assessment of industrial clays. In: Handbook of Clay Science. 2nd edition. Part B. Techniques and Applications. Eds. Bergaya, F.; and Lagaly, G.; Amsterdam: Elsevier, **2013**, 425–450.
12. Chukhrov, F. V., and Zvyagin, B. B.; Halloysite, a crystallochemically and mineralogically distinct species. In: Proceedings of the International Clay Conference, Jerusalem,1966; Eds. Heller, L.; Weiss, A., Israel program for scientific Translation, Jerusalem, **1966**; Volume 1, 11–25.
13. Churchman, G. J.; Interlayer water in halloysite. Ph.D. Thesis, University of Otago, **1970**.
14. Churchman, G. J.; *Clays. Clay. Min.* **1990**, 38, 591–599.
15. Churchman, G. J.; The alteration and formation of soil minerals by weathering. In: Handbook of Soil Science; Ed. Sumner, M. E. Boca Raton, Florida: CRC Press; **2000**, F3–F76.
16. Churchman, G. J.; Euroclay. Modena, Italy, Abstracts; **2003**, 66–67.
17. Churchman, J.; Halloysite: Are we there yet? XIV International Clay Conference Italy, **2009**, Book of Abstracts; Volume 1, p 340.
18. Churchman, G. J.; Davy, T. J.; Aylmore, L. A. G.; Gilkes, R. J.; and Self, P. G.; *Clay. Min.* **1995**, *30*: 89–98.
19. Churchman, G. J.; Aldridge, L. P.; and Carr, R. M.; *Clays. Clay. Min.* **1972**, *20*, 241–246.
20. Churchman, G. J.; and Carr, R. M.; *Am. Min.* **1972**. 57, 914–923.
21. Churchman, G. J.; and Carr, R. M.; *Clays. Clay. Min.* **1973**, *21*, 423–424.
22. Churchman, G. J.; and Carr, R. M.; *Clays. Clay. Min.* **1975**, *23*, 382–388.
23. Churchman, G. J.; and Gilkes, R. J. *Clay. Min.* **1989**, *24*, 579–590.

24. Churchman, G. J.; and Lowe, D. J.; Alteration, formation and occurrence of minerals in soils. In: Handbook of Soil Sciences. Properties and Processes, 2nd edition: Eds. Huang, P.M.; Li, Y.; Sumner , M.E.; Boca Raton, Florida: CRC Press, **2012**, 20.1–20.72.
25. Churchman, G. J.; Pontifex, I. R.; and McClure, S. G.; *Clay. Clay. Min.* **2010**, *58*, 220–237.
26. Churchman, G. J.; and Theng, B. K. G.; *Clay. Min.* **1984**, *19*, 161–175.
27. Churchman, G. J.; Whitton, J. S.; Claridge, G. G. C.; Theng, B. K. G.; *Clay. Clay. Min.* **1984**, *32*, 241–248.
28. Cravero, F.; Maiza, P. J.; and Marfil, S. A.; *Clay Min.* **2012**, *47*, 329–340.
29. De Oliviera, M. T. G.; Furtado, S. M. A.; Formoso, M. L. L.; Eggleton, R. A.; and Dani, N.; *Anais da Academia de Ciências.* **2007**, *79*, 665–681.
30. Dixon J. B.; and McKee T. R.; *Clays. Clay. Min.* **1974**, *22*, 127–137.
31. Du, M.; Guo, B.; and Jia, D.; *Polym. Int.* **2010**, *59*, 574–582.
32. Dupuis, C.; and Ertus, R.; The Karstic Origin of the Type Halloysite in Belgium. In: Proceedings of the 10th International Clay Conference **1993**; Eds. Churchman, G. J.; Fitzpatrick, R. W.; and Eggleton, R. A.; Melbourne: CSIRO Publishing, **1995**, 362–366.
33. Ece, Ö. I.; and Schroeder, P. A.; *Clays. Clay. Min.* **2007**, *55*, 18–35.
34. Farmer, V. C.; and Russell, J. D.; *Spectrochimica Acta.* **1964**, *20*, 1149–1173.
35. Harrison, J. L.; and Greenberg, S. S.; *Clays. Clay. Min.* **1962**, *9*, 374–377.
36. Harvey, C. C.; Halloysite for high quality ceramics. In: Industrial Clays, 2nd edition: Ed. Kendall, T.; London: Metal Bulletin pcl. **1996**, 71–73.
37. Hendricks, S. B.; *Am.Min.* **1938**, *23*, 295–301.
38. Hillier, S.; and Ryan, P. C.; *Clay. Min.* **2002**, *37*, 487–496.
39. Hofmann, U.; Endell, K.; Wilm, D.; *Angewandte Chemie.* **1934**, *47*, 539–547.
40. Honjo, G.; Kitamura, N.; and Mihama, K.; *Clay. Min. Bull.* **1954**, *2*, 133–141.
41. Hughes, I. R.; *New Zealand J. Sci.* **1966**, *9,* 103–113.
42. Jacobs, H.; and Sterckx, M.; Contribution à l'étude de l'intercalation du dimethylsulfoxyde dans le reseau de la kaolinite. In: Proceedings Reunion Hispano-Belga de Minerals de la Arcilla, Madrid **1970**; 154–160.
43. Janik, L. J.; and Keeling, J. L.; *Clay. Min.* **1993**, *28*, 365–378.
44. Jeong G. Y.; *Clay. Clay. Min.* **1998**, *46*, 270–279.
45. Joussein, E.; Petit, S.; Churchman, G. J.; Theng, B.; Righi, D.; and Delvaux, B.; *Clay. Min.* **2005**, *40*, 383–426.
46. Joussein, E.; Petit, S.; and Delvaux, B.; *App. Clay. Sci.* **2007**, *35*, 17–24.
47. Keeling, J. L.; The Mineralogy, Geology and Occurrence of Halloysite *(this volume)*
48. Kogure, T.; Mori, K.; Drits, V.A.; and Takai, Y.; *Am. Min.* **2013**, *98*, 1008–1016.
49. Lagaly, G., Ogawa, M., and Dékány, I.; Clay mineral-organic interactions. In: Handbook of clay science. 2nd edition. Part A. Fundamentals. Eds. Bergaya, F; Lagaly, G.; Amsterdam: Elsevier, **2013**, 437–505.
50. Levis, S. R.; and Deasy, P. B.; *Int. J. Pharm.* **2002**, *243*, 125–134.
51. Lindsay, W. L.; *Chemical Equilibria in Soils.* New York: Wiley-Interscience, New York.
52. Lowe, D. J.; Controls on the rates of weathering and clay mineral genesis in airfall tephras: a review and New Zealand case study. In: Rates of Chemical Weathering of Rocks and Minerals Eds. Colman, S. M.; Dethier, D. P. Academic Press, Orlando, **1986**; 265–330.
53. Lowe, D. J.; Churchman, D. J.; Merry. G. J.; Fitzpatrick, R. H.; Sheard, R. W.; and Hudnall, M. J. Australian and New Zealand National Soils Conference **1996**. Vol. 2, Oral Papers, 153–154.
54. Lowe, D. J.; and Palmer, D. J.; Andisols of New Zealand and Australia. *J. Integr. Field. Sci.* **2005**, *2*, 39–65.
55. Lyon, R. J. P.; and Tuddenham, W. M.; *Nature.* **1960**, *185*, 835–836.
56. MacEwan , D. M. C.; *Nature.* **1946**, *157*, 159.

57. MacEwan, D. M. C.; *Min. Mag.* **1947**, *28*, 36–44.
58. Mehmel, M.; *Zeitschrift Kristallographie.* **1935**, *90*, 35–43.
59. Newman, R. H.; Childs, C. W.; and Churchman, G. J.; *Clay. Min.* **1994**, *29*, 305–312.
60. Norrish, K.; An unusual fibrous halloysite. In: Proceedings of the 10th International Clay Conference 1993. Eds. Churchman, G. J.; Fitzpatrick, R. W.; Eggleton, R. A.; CSIRO Publishing, Melbourne, **1995**, 275–284.
61. Papoulis, D.; Tsolis-Katagas, P.; and Katagas, C.; *Clays. Clay. Min.* **2004**, *52*, 275–286.
62. Parfitt, R. L.; Soils formed in tephra in different climatic conditions. Transactions 14th Congress, International Society of Soil Science, Kyoto, 7, 134–139.
63. Parfitt, R. L.; Saigusa, M.; and Cowie, J. D.; *Soil. Sci.* **1984**, 138, 360–364.
64. Parfitt, R. L.; Russell, M; and Orbell, G. E.; *Geoderma.* **1983**, *29*, 41–57.
65. Pasbakhsh, P.; Churchman, G. J.; and Keeling, J. L.; *Appl. Clay. Sci.* **2013**, *74*, 47–57.
66. Pasbakhsh, P.; Ismail, H.; Fauzi, M. N. A.; and Bakar, A. A.; *Appl. Clay. Sci.* **2010**, *48*, 405–413.
67. Range, K. J.; Range, A.; and Weiss, A.; Fire-clay type kaolinite or fire-clay mineral? Experimental classification of kaolinite-halloysite minerals. In: Proceedings of the International Clay Conference Tokyo 1969, Ed. Heller, L. Jerusalem: Israel University Press, **1969**; Volume 1, 3–13.
68. Ross, C. S.; and Kerr, P. F.; U.S. Geological Survey Professional Paper **1934**, 185-G, 135–148.
69. Sanchez Camazano, M.; and Gonzalez Garcia, S.; *Anales de Edafología y Agrobiología.* **1966**. *25*, 9–25
70. Silber, A.; Bar-Yosef, B.; Singer, A.; and Chen, Y.; *Geoderma.* **1994**, *63*, 123–144.
71. Singer, A.; Zarei, M.; Lange, F. M.; and Stahr. K.; *Geoderma* **2004**; *12*, 279–295.
72. Southard, S. B.; and Southard, R. J.; *Soil. Sci. Soc. Am. J.* **1989**, *53*, 1784–1791.
73. Sposito, G.; Chemical Equilibria and Kinetics in Soils. Oxford University Press, New York, **1994**.
74. Sposito, G. The Chemistry of Soils. Second edition. Oxford University Press, New York, **2008**.
75. Środoń, J. Identification and quantitative analysis of clay minerals. In: Handbook of Clay Science. 2nd edition. Part B. Techniques and Applications. Eds. Bergaya, F; Lagaly, G. Amsterdam: Elsevier, **2013**, 25–49.
76. Sudo, T.; Yotsumoto, H. *Clays. Clay. Min.* **1977**, *25*, 155–159.
77. Takahashi, T.; Dahlgren, R.; and van Susteren, P.; *Geoderma.* **1993**, *59*, 131–150.
78. Takahashi, T.; Dahlgren, R. A.; Theng, B. K. G.; Whitton, J. S.; and Soma, M.; *Soil. Sci. Soc. Am. J.* **2001**, *65*, 516–526.
79. Takesako, H.; Lowe, D. J.; Churchman, G. J.; and Chittleborough, D.; Holocene volcanic soils in the Mt. Gambier region, South Australia. In: Proceedings of the 19th World Congress of Soil Science Brisbane, **2010**, Symposium 1.3.1, 47–50. Published at http://www.iuss.org.
80. Theng, B. K. G.; Churchman, G. J.; Whitton, J. S.; and Claridge, G. G. C.; *Clays. Clay. Min.* **1984**, *32*, 249–258.
81. Wada, K.; *Am. Min.* **1959**, 44, 153–164.
82. Wada, K.; *Am. Min.* **1961**; 53, 334–339.
83. Weiss, A.; Thielepape, W.; Goring, G.; Ritter, W.; Schafer, H.; Kaolinit-Einlagerungs-Verbindungen. In: Proceedings of international clay conference, Stockholm 1963, T.R.P. Ed. Graff-Petersen, London: Pergamon Press, **1963**; Volume 1, 287–305.
84. Wilson, M. J.; and Tait, J. M.; *Clay. Min.* **1977**, 12, 59–76.
85. Wyatt, J.; Lowe, D. J.; Moon, V. G.; Churchman, G. J.; Extended Abstracts, 21st Australian Clay Minerals Society Conference, Brisbane **2010**, 39–42.

CHAPTER 3

THE IDENTIFICATION AND NOMENCLATURE OF SEPIOLITE AND PALYGORSKITE (A HISTORICAL PERSPECTIVE)

EMILIO GALÁN and PATRICIA APARICIO

CONTENTS

Sepiolite and palygorskite clay minerals have received very less attention than other clay minerals. This neglect partly reflects the fact that these clay minerals are less common than the layer phyllosilicates (Singer and Galán, 1984). Despite this, however, these minerals have been known for centuries because of their many and useful properties. Their unusual crystal structure is mainly responsible for their unique physicochemical properties and important characteristics related to surface area, porosity, dehydration, and sorption active centers.

In the emerging field of nanotechnology, mostly focused on carbon- and inorganic-based nanomaterials (carbon nanotubes, graphene, transition metal nanotubes, and nanowires), systems containing aluminosilicates minerals (i.e., clay minerals) can also form nanostructure-layered materials and nanotubes with remarkable geometric properties. Among these minerals, sepiolite and palygorskite show particularly suitable structures and features for the preparation of nanotubes.

3.1   NOMENCLATURE: A HISTORICAL PERSPECTIVE

Sepiolite has been used for hundreds of years in the manufacture of pipes (Turkey, Hungary, and Germany). The sepiolite from Vallecas (Spain) was used for the ceramic paste of the Buen Retiro, founded by King Carlos III of Spain in 1760, until 1808 (Martín Vivaldi and Robertson, 1971). Also sepiolite was used as a building material (Villanova, 1875).

Another interesting old use of sepiolite was the *Tierra del Vino* (Wine Earth), a mixture of sepiolite and calcite (50/50%) with minor portions of smectite, palygorskite, and quartz. This material was formerly used to clear and purify wine in South Spain (sherry wine) (Galán and Ferrero, 1982)

Palygorskite was also used for ceramics by the Pre-Columbian Indians from Yucatán Pensinsula (Mexico) and they called the palygorskite clay *"Sac lu'um"* (Maya for "White Earth"), nearly 800 years ago. Sacalum (a Spanish corruption of Sac lu'um), a village in the northern peninsula of Yucatan, has served as a source of this clays for over 800 years (Van Olphen, 1966; Isphording, 1984)

Maya Blue was an artificial pigment fabricated by the Maya in the early first millennium A.D. from indigotin mixed with palygorskite of Yucatan. The pigment was used in pre-hispanic times from the southern Maya region to most of Mesoamerican cultures (Mexico, Guatemala, Nicaragua, Belize). It was rediscovered in 1931 by Merwin (Merwin, 1931), but the name "Maya Blue" was coined by Gettens and Stout (1942). This pigment is of a great archeological importance for tracing Maya trade routes. For a detailed review of these pigments and their composition see Sanchez del Rio et al. (2011).

A mineral, probably sepiolite, was formerly named as Keffekil Tartatorum by Cronsted in 1758 (Cronsted, 1758). Werner (1788) was apparently the first to use the name *meerschaum* in allusion to the lightness and color of the material. Sepiolite is referred as *myrsen* and *meerschaum* by Kirwan (1794), and later Haüy (1801) called it *eçume de mer*, as a magnesium and siliceous carbonate. But the name sepiolite was given by Ernst Friedrich Glocker, in 1847 (Glocker, 1847), from the Greek *cuttle fish,* the bone of which is light and porous like the mineral. Different varieties of sepiolite earlier described by Fersman (1908, 1913) and later by Efremov (1939), have since been discarded. Nevertheless other varieties have been accepted: Fe-sepiolite, corresponding to the obsolete names on *gunnbjarnite* (Boggild, 1951), *xylotile* (Hermann, 1845), and *ferrisepiolite* (Strunz, 1957); *Ni sepiolite* (Caillere, 1936), now officially named *falcondoite* (Springer, 1976); *Na sepiolite* described by Fahey et al, (1960) and named *loughlinite*; *Al sepiolite* (Roger et al., 1956); etc.

The name palygorskite comes from the locality "in der Palygorischen Distanz" of the second mine of the Popovka River, Urals, Russia (Mitchel, 1979). The name was first used by Von Ssaftschenkow (1862). According to Ovcharenko and Kukovsky (1984), "When mountain-leather deposits were prospected in the Palygorsk Division Mine near Popovka River, Perm Province, Russia, this mineral was assumed to represent a variety of asbestos." Nevertheless, the mineral has been known since Teofrastus' time, ca. 314 B.C. (Robertson, 1963). Two other names were used for this mineral: *pilolite* (Heddle, 1879) and *lassalite* (Friedel, 1901). Later, Fersman (1913) applied the name palygorskite to a family of fibrous hydrous silicates forming an isomorphous series between two end-members: an aluminum form, called *paramontmorillonite* and magnesium form, *pilolite*. But these names were never accepted and used.

"Attapulgite" represents the same mineral and it was first applied by Lapparent (1936) to the clay that he found in fuller's earth from Attapulgus, Georgia (USA); Quincy, Florida (USA); and Mormoiron (France). Although *"attapulgite"* still appears in the literature, the name "palygorskite" should be used preferably because it is older and has priority (Bailey et al., 1971). However, the name *attapulgite* is so well entrenched in commercial circles that it continues to be used.

Two minerals, *yofortierite* (Perrault et al. 1975) and *tuperssuatsiaite* (Karup-Möller and Petersen, 1984), have been accepted as belonging to the palygorskite group, the first as a manganese analog of palygorskite and the latter as Na-Fe-palygorskite. More recently a new palygorskite-like mineral, *raite*, has been described which contains Na, Mn, and Ti (Pluth et al., 1997).

Presently, after the corresponding studies of all the varieties cited, the palygorskite-sepiolite mineral group consists of palygorskite, sepiolite, falcondoite (Springer, 1976), intersilite (Khomyakov, 1995; Yamnova et al., 1996), kalifersite (Ferraris et al., 1998), loughlinite (Fahey and Axelrod, 1948; Fahey et al., 1960), raite (Khomyakov, 1995; Pluth et al.,1997), tuperssuatsiaite (Karup-Möller and Petersen, 1984; Cámara et al.,2002), and yofortierite (Perrault et al.,1975).

3.2   IDENTIFICATION OF SEPIOLITE-PALYGORSKITE MINERALS

3.2.1   MORPHOLOGY

Sepiolite and palygorskite are easily identified by their fibrous macro- and micro-morphologies, showing a fibrous habit with channels running parallel to the fiber length (Figure 3.1). They are included as members of the "modulated" phyllosilicates (Guggenheim and Eggleton, 1988), with the octahedral sheet being the modulated component. As lamellar morphologies have never been observed, it must be assumed that sepiolite, palygorskite, and other less frequent minerals of the group occur with only a fibrous habit. Most of the occurrences of the so-called *moun tain leather, mountain wood,* and *cardboard paper* historically described have been identified as palygorskite and some as sepiolite, but many others were chrysotiles (Brauner and Preisinger, 1956).

Textural analyses have shown that sepiolite and palygorskite fibers vary considerably in shape and size from one deposit to another, and even among samples from the same deposit (Nolan et al., 1991; Torres-Ruiz et al., 1992). Normally, the elongated particles vary in length from about 1–10 µm and are approximately 0.01 µm in width. Average fiber length ranges from 1 to 2 µm, with only a very small proportion (<1%) of >5 µm fibers (Bellman et al., 1997). Although it is common to find fiber lengths over 20 µm in hydrothermal sepiolite, in sedimentary deposits, which are usually exploited commercially, 95 percent of the particles are less than 1.5 µm in length (López-Galindo and Sánchez-Navas, 1989).

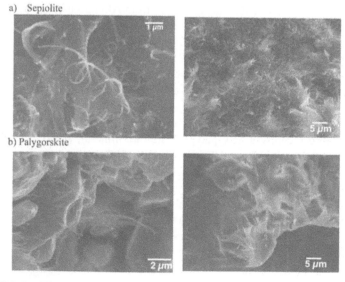

FIGURE 3.1   SEM microphotographs of sepiolite from Vicalvaro (Spain) and palygorskite from Lebrija and Ciudad Real (Spain).

Recently García Romero and Suarez (2013) presented a detailed study of the microtextural features of wide-ranging deposits of extremely pure sepolite and palygorskite, and indicated that each deposit has its own characteristic signatures that vary greatly from one locality to another, explaining why each deposit has different physical and chemical properties. Sepiolite and palygorskite can consist of fibers with different sizes, curls, or types of aggregation. Several morphological fiber classifications have been made according to length, the width–length ratio (W/L), or curliness. They also proposed some terms to describe the width of fibers: lath (the smallest units that can be observed, the true unit crystal), rod (several laths in a crystallographical arrangement), and bundle (several rods parallel to the c-axis).

Palygorskite–sepiolite minerals were investigated using transmission electron microscopy (TEM) to describe defects within individual fibers (Krekeler and Guggenheim, 2008). The defects observed by these authors would create unusually wide channels (open channel defect: OCD), enhancing the possibility to absorb large organic molecular and anomalously high contents of H_2O in some samples, and may explain variations in a number of properties of these minerals and chemical composition.

3.2.2 CHEMICAL ANALYSIS AND CRYSTAL-CHEMISTRY FORMULAS

From a historical point of view, the chemical analysis of sepiolite (in the form of a *"meerschaum"* pipe from Turkey) was first attempted in the second half of the eighteenth century (before 1784) by Johann Christian Wiegleb (Martín Vivaldi and Robertson, 1971). In 1794, Martin Heinrich Klaproth made a chemical analysis of a sepiolite from Eskisehir, Turkey (Klaproth, 1794). The chemical composition of the two pure dehydrated minerals calculated from the theoretical formulae as mass is 69.1 percent SiO_2 and 30.9 percent MgO, for sepiolite; and 72.47 percent SiO_2, 15.37 percent Al_2O_3, and 12.5 percent MgO, for palygorskite.

Published analytical data mostly refer to bulk samples except for a few studies (e.g., Yalçın and Bozkaya, 1995 and 2004). As such, they are affected by both crystal-chemical variations and admixed contaminants (other clay minerals and associated minerals). The most frequent admixtures in sepiolite and palygorskite are smectite, illite, chlorite, quartz, feldspars, carbonates, zeolites, iron (hydr)oxides, and silica gels. In the past decades, the use of TEM has enabled analyses (by Analytical Electron Microscopy, AEM) of very small particles, thus providing more accurate data from the individual particles and avoiding the influence of impurities.

Galán and Carretero (1999) reviewed the literature on chemical analyses, including bulk chemical analyses, and EDX analyses of selected individual particles and pure samples. Their assessment indicates that sepiolite can be a true trioctahedral mineral with eight octahedral positions filled by Mg^{2+} and negligible structural substitutions. A very pure (near end-member) specimen is close to the theoretical formula of $Mg_8Si_{12}O_{30}(OH)_4(OH_2)_4(H_2O)_8$. Palygorskite is intermediate between di and trioctahedral, with a theoretical formula of $(Mg_2R_2^{3+}{}_1)$ $(Si_{8-x}$ $Al_x)O_{20}(OH)_2(OH_2)_4R^{2+}{}_{x/2}(H_2O)_4$, (Galan and Carretero, 1999; Yalçın and Bozkaya, 2011). The octahedral sheet contains mainly Mg^{2+}, Al^{3+}, and Fe^{3+}, with an R^{2+}/R^{3+} ratio close to 1, and shows four of the five structural positions occupied.

Recently, García-Romero and Suárez (2010) and Suárez and García-Romero (2011) compared and studied a great number of sepiolites and palygorskites to obtain a consistent composition of these minerals. According to this review, sepiolite can be classified into two types: sepiolite and Al-sepiolite. Al-sepiolites are those that present more than 10 percent of octahedral positions vacant and more than 0.5 ^{VI}Al atoms. Palygorskite classifies into (i) ideal palygorskite, with an octahedral composition near to the ideal palygorskite $(Mg_2 Al_2)$; (ii) common palygorskite, where the ^{VI}Al content is less than in the ideal formula and thus the Mg content is higher, and the number of octahedral cations is close to 4; (iii) magnesic palygorskite, the most trioctahedral extreme; and (iv) aluminic-palygorskite, which is defined by a total number of octahedral cations (p.h.u.c.) <4, with Mg <2, and consequently $(Al+Fe^{3+})$ >Mg.

As magnesic palygorskite and aluminic sepiolite can show very similar chemical compositions, these authors propose that there is no compositional gap between the two minerals. Palygorskite can be so rich in Mg and sepiolite so rich in Al that it is possible to affirm that a continuous composition series exists and all the intermediate compositions between the two extreme, corresponding to the two pure minerals, can be found.

3.2.3 IDENTIFICATION BY XRD, DTA-TG, AND IR

Sepiolite is identified by a strong 110 (or 011, depending on crystal orientation) X-ray reflection at 1.2 nm, whereas the corresponding value for palygorskite by a reflection at 1.05 nm (Figure 3.2). The diffraction patterns are usually unaffected by ethylene glycol (EG) solvation and by mild heat treatment. Nevertheless some "expandability" has been described with EG and dimethylsulfoxide solvation for sepiolite and palygorskite, corresponding to about 0.01–0.03 nm for the main peak (see Jones and Galán (1988). This behavior of some sepiolite or palygorskite with polar molecules suggests a mechanism by which the chain structure may be weakened or ruptured.

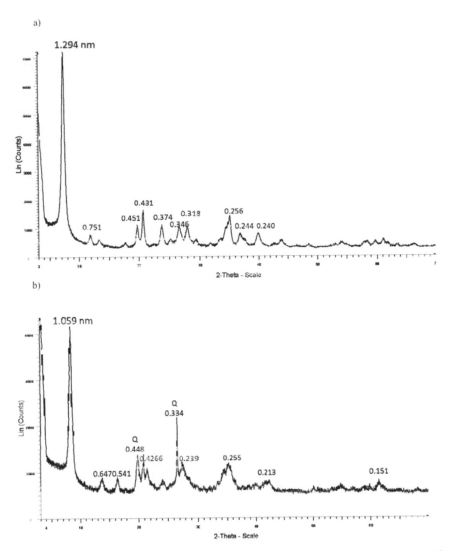

FIGURE 3.2 XRD patterns of (a) sepiolite from Vicálvaro (Spain) and (b) palygorskite from Guanshan (China).

Many impurities of other clay minerals (smectites, illite, mixed-layered clay minerals) usually occur, which makes the identification of sepiolite or palygorskite difficult. Hence a profile-fitting peak decomposition program, part of MacDiff 4.1.2 by Petschick (2000), can be used to assess the main representative peaks and impurities, as well as their crystal order. Using this fitting procedure the reflection 110 shown in Figure 3.3 could be a resultant of two different peaks which contribute to both sepiolite and palygorskite, probably related to the occurrence of different crystal structural order (Figure 3.3 air dried patterns). A small expandability of sepiolite and palygorskite can also be detected (Figure 3.3 showing EG patterns).

Giustetto et al. (2011) used the synchrotron X-ray powder diffraction pattern to study the sepiolite structure, based on both molecular mechanical simulation and Rietveld structure refinement. The refined structure of the studied sepiolite does not significantly differ from the starting model (Post et al., 2007), also obtained by synchrotron X-ray powder diffraction. The sepiolite structure showed the omission of TOT ribbons (polysomes), which is a common defect in palygorskite/sepiolite minerals, as earlier observed by Krekeler and Guggeheim (2008).

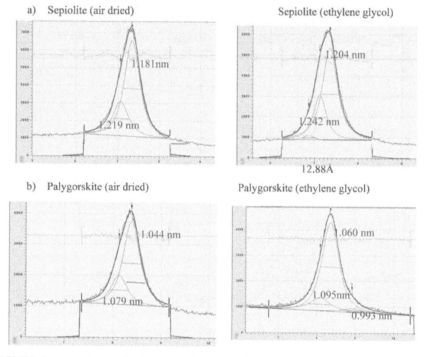

FIGURE 3.3 Fitting of the 110 reflection for sepiolite and palygorskite before and after EG solvation. In both samples the 110 peaks are decomposed into another two, indicating the possible presence of different structural order grades. Also sepiolite and palygorskite swell slightly after EG solvation.

These minerals contain four water molecules (zeolitic water) in the channels, and four others are bound to the octahedral edge inside the channels. For sepiolite, the differential thermal analysis (DTA) curve can be divided into three parts: (i) The low-temperature region (<300°C), where the mineral loses water adsorbed on outer surfaces and zeolitic water (peak at 120–150 °C). (ii) The central region (300–

600°C), where two endothermic peaks occur at about 350 and 500–550°C. The first endothermic peak is ascribed to the loss of the first two water molecules coordinated to the inner octahedral edge, causing rotation of alternate ribbons and particle folding (Nagata et al., 1974; Serna et al., 1975; Van Scoyoc et al., 1979); and the second central endotherm peak is due to the loss of the other two edge-coordinated water molecules that are "trapped" inside the collapsed channels (Perez-Rodriguez and Galán, 1994). (iii) The high-temperature region (>600°C), where an endothermic effect (at about 800°C) is immediately followed by an exothermic maximum, represents the dehydroxylation of the structure and the formation of clinoenstatite. Frost et al. (2009) indicated that the number of steps in the thermal analysis of sepiolites are greater compared to previous published results, when using dynamic high resolution differential thermogravimetry (DTG) and Controlled Rate Thermal Analysis (CRTA) techniques. Two dehydration steps for sepiolite were around 70–90°C and 210–320°C.

For palygorskite, the DTA curve can be divided into three parts: (i) The low-temperature region <200°C), where the mineral loses adsorbed water, both in the channels (zeolitic) and on the outer surface of the mineral (peak at 120–150°C). (ii) The central region (300–600°C), where two endothermic peaks occur at about 200–350 and 350–550°C. The first endothermic peak is ascribed to the loss of the first half of the coordination water giving rise to semi-anhydrous palygorskite and the second central endotherm peak is due to the dehydroxilation of the rest of coordination water and of the hydroxyls or structural water. (iii) The progressive weight loss at high temperature (>650°C) corresponds to the residual OH− of the silanol groups (Suarez et al., 1995), but at higher temperature (800–950°C) an exothermic peak, which is not represented in Figure 3.5b, occurs due to the pyroxene crystallization (Haden and Schwint, 1967).

The structural changes that occur on heating are also evidenced by a decrease in the intensity of the principal XRD reflections. For instance, in sepiolite the reflections at 1.2, 0.45, 0.38, and 0.34 nm decrease when the mineral is heated at 250°C for 1 h, while new reflections appear at 1.04 and 0.82 nm. Figure 3.4a shows XRD patterns using a heating camera difractometer. At 300°C, the reflection at 1.2 nm decreases in intensity but the other reflections are not affected. At 400°C, the reflections at 1.2, 0.78, 0.64 and 0.45 nm disappear, while new reflections appear at 1.00, 0.803, and 0.438 nm. These changes have been related to the disappearance of orthorhombic sepiolite and the formation of anhydrous sepiolite (monoclinic) (Valentin et al., 2007). At 800°C, sepiolite is practically X-ray amorphous. In palygorskite, the intensity of the reflections at 1.05, 0.45, and 0.323 nm decrease on heating, and new reflections appear at 0.92 and 0.47 nm (Hayashi et al., 1969). Heating to 300°C completely eliminates the 1.05 nm reflection and, according to our data, the 0.92 nm reflection is also eliminated (Figure 3.4b) and shifts to 0.81 nm with less intensity at 600°C. At 700°C, palygorskite is practically X-ray amorphous. Figure 3.4b shows XRD patterns using a heating camera difractometer, confirming the Hayashi et al. (1969) description.

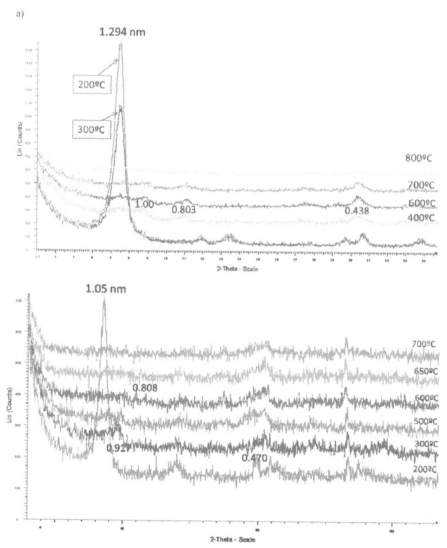

FIGURE 3.4   Evolution of XRD patterns with temperature of (a) sepiolite from Vicálvaro (Spain) and (b) palygorskite from Guanshan (China).

Sepiolite and palygorskite infrared (IR) spectra are complex, with absorption bands in stretching and bending regions of water molecules. As both minerals contain water in the channels of their structures, they display bands originating in bound molecular water directly coordinated to Al or Mg ions at the edges of 2:1 ribbon layers (3,580, 3,737, 1,650 cm^{-1} in palygorskite, and 3,618, 3,565, 1,627 cm^{-1} in sepiolite) and unbound zeolitic water molecules within the channels (3,448, 3,400, 1,672 cm^{-1}

for palygorskite, and 3,410, 3,245, 1,657 cm $^{-1}$ for sepiolite) (Russell, 1987, Serna et al., 1977) (Figure 3.5).

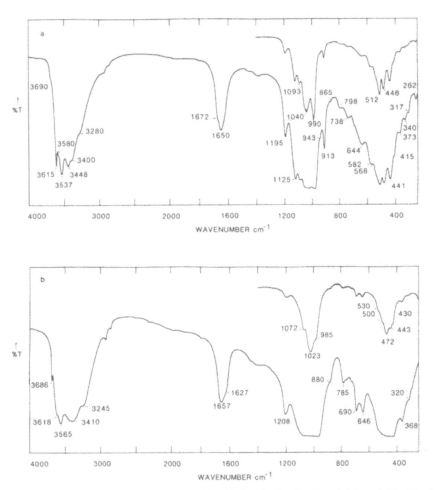

FIGURE 3.5 IR spectra of (a) palygorskite (Cabrach, Scotland) and (b) sepiolite (Spain) (From Russell, 1987).

The diagnostic bands Si-O-Si (980–1,030 cm 1) of alternating ribbons are characteristic of the two fibrous minerals. Palygorskite shows an Al-Al-OH vibration at 3,616 cm ^{1} and Al-Mg-OH at 3,550 cm ^{1} and sepiolite shows an Mg$_3$-OH vibration at 3,680 cm ^{1}.

The influence of chemical composition in the position and intensity of the absorption bands observed in the Fourier transform infrared spectra (FTIR) spectra of palygorskite has been studied by Suarez and García Romero (2006). The decrease of the intensity of the most characteristic peaks in the spectra of palygorskite is clearly related to the increase in Mg content.

IR studies combined with powdered diffraction EM and TA have provided insight into the nature of the water in sepiolite and palygorskite, as well as the structural changes that occur after heating/dehydration (Hayashi et al., 1969; Serna et al., 1975, 1977; Mifsud et al., 1978; Van Scoyoc et al., 1979; Blanco et al., 1988).

3.3.3 MAIN PROPERTIES OF INDUSTRIAL INTEREST AND PRINCIPAL APPLICATIONS

In Table 3.1 the most important physicochemical properties of these minerals are summarized. These characteristics clearly make them different from other phyllosilicates and give them significant economic value.

On the basis of these properties, the wide range of industrial applications of sepiolite and palygorskite can be classified as sorptive, rheological, and catalytic.

Thermal or acid treatments produce important structural changes which modify the properties of these clays, particularly the surface area and porosity which mainly affect their sorptive properties, as summarized in Table 3.2 (Álvarez et al., 2011).

These clays show a high density of silanol groups (-SiOH) on the external surface in the edges of the structural polysomes, which explains their marked hydrophilicity. In contrast with bentonitic (smectite-rich) clays, these fibrous clays are nonswelling and their structure is stable even in systems with high salt concentrations. Palygorskite and sepiolite particles do not flocculate because the elongated crystals hinder settling (Murray, 2007). These minerals retain colloidal properties when dried at moderate temperatures, while higher temperature drying enhances useful absorptive and other properties. An historical and future perspective of the applications of palygorskite-sepiolite has been outlined by Álvarez et al. (2011).

Combination at the nanometric scale with both microfibrous silicates and carbonaceous materials, such as graphene-like compounds and carbon nanotubes, offers the possibility to develop conductive materials of interest for diverse electrochemical applications.

TABLE 3.1 Properties of palygorskite and sepiolite (Galán, 1987; Murray, 2007).

Particle Shape	Needle-Like	
Length (µm)	0.2–2.0	
Width (nm)	10–30	
Thickness (nm)	5–10	
Channels dimensions (nm)	Palygorskite:	0.37 × 0.64
	Sepiolite:	0.37 × 1.06
Specific surface area SSA, BET N$_2$ (m²/g) (high)	Palygorskite:	150
	Sepiolite:	320
Specific gravity (g/cm³)		
Mohs' hardness	2.0–2.3	
Cation Exchange Capacity (mEq/100 g) (moderate)	<25meq/100g	
Sorptivity		
Melting point (°C)	High	
Brookfield viscosity of a suspension at 6% in water at 5 rpm (cP)	1.550	
	Palygorskite: 10.000–12.000	
	Sepiolite: 30.000–50.000	

TABLE 3.2 Summary of effect of thermal and acid treatment on sepiolite and palygorskite (Álvarez et al., 2011).

		Change Produced in the Structure	Effect on SSA, BET N$_2$ (m²/g)	Effect on Porosity
Thermal treatment	200°C	Loss of absorbed water on the external surface and zeolitic water in the intercrystalline channels	Maximum values	No modification
	300 400°C	Rotation of alternate ribbons and particle folding due the loss of coordinated water to the octahedral magnesium	Decrease from 300 to 125 in sepiolite and 150–125 in palygorskite	Loss of microporosity
	500°C	Intercrystalline channels collapse folding the structure due the loss of coordinated water, which are strongly bound		Micropores disappear
Acid treatment		Leaching of the magnesium cations and production of amorphous material	Sepiolite 549 Palygorskite 286	The micropore volume increases from 16 to 20%
				More surface acid centers

KEYWORDS

- **Thermal and acid treatment**
- **Clay minerals**
- **Fourier transform infrared spectra**
- **Graphene**
- **Heating camera difractometer**
- **Sepiolite**
- **Palygorskite**

REFERENCES

1. Álvarez, A.; Santaren, J.; and Aparicio, P.; Current industrial applications of palygorskite and sepiolite. In: Developments in Palygorskite-Sepiolite Research A New Outlook on these Nanomaterials, Developments in Clay Science, Vol. 3: Eds. Galán E., Singer A., Elsevier, **2011**, 281–298.
2. Bailey, S. W.; Brindley, G. W.; Johns, W. D.; Martin, R. T.; and Ross, M.; *Clays. Clay. Min.* **1971**, *19*, 132–133.
3. Bellmann, B.; Muhle, H.; and Ernst, H.; *Environ. Health. Perspect.* **1997**, *105* (Suppl. 5), 1049–1052.
4. Blanco, C.; Herrero, J.; Mendioroz, S.; and Pajares, J. A.; *Clays. Clay. Min.* **1988**, *36*, 364–368.
5. Boggild, O. B.; *Gronland.* **1951**, *142*, 3–11.
6. Brauner, K.; and Preisinger; A.; Tschermak's Mineralogische und Petrographische Mitteilungen, **1956**, *6*, 120–140.
7. Caillere, S.; *Bull. Soc. Fr. Min.* **1936**, *59*, 353–374.
8. Cámara, F.; Garvie, L. A. J.; Devouard, B.; Groy, T. L.; and Buseck, P. R.; *Am. Min.* **2002**, *87*, 1458–1463.
9. Cronsted, A. A.; System of Mineralogy. Ed. Dana J. D. John Wiley and Sons, **1758**, 696 pp.
10. Efremov, N. E.; *Dokl. Akad. Nauk.* **1939**, *24*, 2877–298.
11. Fahey, J. J.; and Axelrod, J. M. *Am. Min.* **1948**, *33*, 195.
12. Fahey, J. J.; Ross, M.; and Axelrod, J. M.; *Am. Min.* **1960**, *45*, 270–281.
13. Ferraris, G.; Khomyakov, A. P.; Belluso, E.; and Soboleva, S. V.; *Eur. J. Min.* **1998**, *10*, 865–874.
14. Fersman, A.; *Mém. Acad. Sci.* St. Petetersburg, **1913**, *32*, 321–430.
15. Fersman, A. E.; *Bull. Acad. Sci.*, **1908**, *2*, 255–274.
16. Friedel, G. *Bull.*; *Soc. Franc. Min.* **1901**, *24*, 12–14.
17. Frost, R.L., Kristóf, J., and Horváth, E.; *J. Therm. Anal. Calorim*, **2009**, *98*,423–428
18. Galán, E.; and Carretero, M. I.; *Clays. Clay. Min.*, **1999**, *47*, 399–409.
19. Galán E.; Ferrero. *Clays. Clay. Min.*, **1982**, *30*, 191–199.
20. Galán, E.; Industrial applications of sepiolite from Vallecas-Vicalvaro, Spain. A review. In: Eds. Schultz, L. G., Van Holphen, H., Mumpton, F. A. In: Proceedings of the International Clay Conference, Denver, **1985**. The Clay Minerals Society, Bloomington, Indiana, **1987**, 400–404.
21. García-Romero, E.; and Suárez, M.; *Clays. Clay. Min.* **2010**, *58*, 1–20.

22. García Romero, E; and Suarez, M.; *App. Clay Sci.* **2013**, *86*, 129–144.
23. Gettens, R. J.; and Stout, G. L.; Painting Materials, A Short Encyclopaedia. New York :Van Nostrand, **1942**.
24. Giustetto, R.; Levy, D.; Wahyudi, O.; Ricchiardi, G.; and Vitillo, J. G.; *Eur. J. Min.*, **2011**, *23*, 449–466.
25. Glocker, E. F.; Synopsis, Halle. In: A system of Mineralogy. Ed. Dana J. D. London:Truner and Co, **1847**, 456.
26. Guggenheim, S.; and Eggleton, R. A.; Crystal chemistry, classification, and identification of modulated layer silicates. In: Hydrous Phyllosilicates (Exclusive of the Micas). Reviews in Mineralogy, vol. 20. Bailey, S. W. Ed., Mineralogical Society of America, Washington, D. C., **1988**, 675–725.
27. Haden, W. L.; and Schwint, L. A; *Ind. Eng. Chem.*, **1967**, *59*, 59–69.
28. aüy, R. J.; Traite die Mineralogie. Paris vol. 4, Culture et Civilisation, Bruxelles. **1801**.
29. Hayashi, H.; Otsuka, R.; and Imai; N. *Am. Min.* **1969**, 53, 1613–1624.
30. Heddle, M. F. *Min. Mag*, **1879**, *2*, 206–219.
31. Hermann, R. J.; *Prakt. Chem.*, **1845**, *34*, 177–181
32. Jones, B. F.; and Galán, E.; Palygorskite–sepiolite. In: Hydrous Phyllosilicates (Exclusive of Micas). Reviews in Mineralogy. Bailey S. W. Ed., Chapter 16, vol. 19. BookCrafters, Inc., Chelsea, Michigan, **1988**, 631–674.
33. Karup-Möller, S.; Petersen, O. V.; *Am. Min*, **1984**, *70*, 1332.
34. Khomayok, A. P.; Mineralogy of hyperagpaitic alkaline rocks. Oxford: Clarendon Press, **1995**.
35. Kirwan, R.; Elements of Mineralogy. 2nd edition, London Elmsly, **1794**, 144–145.
36. Klaproth, M. H.; Beobachtunge und entdeckungenan der natiurkunde, Berlin, **1974**, *5*, 149.
37. Krekeler, M. P. S., Guggenheim, S.; *App. Clay Sci.*, **2008**, *39*, 98–105.
38. Lapparent De, *J. Compt. Rend.* **1936**, 202, 1728–1731.
39. López Galindo, A.; and Sánchez Navas, A.; *Bol. Soc. Esp. Min.* **1989**, *12*, 375–384.
40. Martin Vivaldi J. L.; and Robertson R. H. S.; Palygorskite and sepiolite (the hormites). In: The Electron-Optical Investigation of Clays, vol. 41. Gard, J. A. Ed., London: The Mineralogical Society, **1971**, 255–257.
41. Merwin, H. E.; Chemical analysis of pigments. In: Temple of the Warriors at Chitchen-Itza, Yucatan. Morris, E. H., Charlot, J., Morris, A. A. Eds., Carnegie Institution of Washington, Washington, D.C, **1931**, 355–356.
42. Mifsud, A.; Rauterau, M.; and Fornes, V.; *Clays. Clay. Min.*, **1978**, *13*, 367–374.
43. Mitchel, R. S.; Minerals Names: What do they mean? Van NostrandReuinhold. New York, **1979**, 229.
44. Murray, H. H.; Applied Clay Mineralogy: Occurrences, Properties and Applications of Kaolins, Bentonites, Palygorskite-Sepiolite, and Common Clays. Amesterdam: Elsevier, **2007**, 179 pp.
45. Nagata, H.; Shimoda S.; and Sudo T.; *Clays. Clay. Min.* **1974**, *22*, 285–293.
46. Nolan, R. P.; Langer, A. M.; and Herson, G. B.; *Br. J. Ind. Med.*, **1991**, *48*, 463–475.
47. Ovchaenko, F. D.; and Kukovsky, Y. E. G.; Palygorskite and sepiolite deposits in the URSS and their uses. Playgorskite-Sepiolite. Occurrences, genesis and uses. Developments in Sedimentology 37. Eds. Singer A., Galán, E. **1984**, 233–242.
48. Pérez Rodríguez, J. L.; and Galán, E. J.; *Ther. Anal.* **1994**, *42*, 131–141.
49. Perrault, G.; Harvey, Y.; and Pertsowsky, R.; *Can. Min.*, **1975**, *13*, 68–74.
50. Petschick, R.; MacDiff 4.1.2. Powder diffraction software. Available from the author at http://www.geol.uni-erlangen.de/html/software/ Macdiff html. **2000**.
51. Pluth, J. J.; Smith, J. V.; Pushcharovsky, D. Y.; Semenov, E. I.; Bram, A.; and Riekel, C., et al. Third-generation synchrotron x-ray diffraction of a 6-mm crystal of raite, ~ $Na_3Mn_3Ti0.25$-$Si_8O_{20}(OH)_2.10H_2O$, opens up new chemistry and physics of low-temperature minerals. *Proc. Natl. Acad. Sci. USA 94*, **1997**, 12263–12267.

52. Post, J. E., Bish, D. L.; and Heaney, P. J. *Am. Min*, **2007**, *92*, 91–97.
53. Robertson, R. H. S.; *Class. Rev. New. Ser*. **1963**, *13*, 132.
54. Roger, L. E.; Quirk, J.; and Norrish, K. J.; *Soil. Sci*. **1956**, *7*, 177–184.
55. Russell, J. D.; Infrared method. In: A Handbook of Determinative Methods in Clay Mineralogy. Chapter 4 Blackie and Son. Ed. Glasgow, M.J. Wilson, **1987**, 99–132.
56. Sanchez del Río, M.; Domenech, A.; Domenech-Carbo, M. T.; Vázquez Agredos-Pascual, M. L; Suárez, M.; and García Romero, M.; The Maya blue pigment. In: Developments in Palygorskite-Sepiolite Research A New Outlook on these Nanomaterials, Developments in Clay Science, Vol. 3, Galán, E. Singer, A., Eds., Elsevier; **2011**, 453–481.
57. Serna, C.; Ahlrichs, J. L.; and Serratosa, J. M.; *Clay. Clay. Min*. **1975**, *23*,452–457.
58. Serna, C.; Van Scoyoc, G. E.; and Ahlrichs, J. L.; *Am. Min*. **1977**, *62*, 784–792.
59. Singer, A.; and Galán, E.; Palygorskite–Sepiolite. Occurrences, Genesis and Uses. Developments in Sedimentology, vol. 37. Amsterdam: Elsevier, **1984**.
60. Springer, G.; Falcondoite, nickel analogue of sepiolite. *Can. Min*. **1976**, *14,* 407–409.
61. Strunz, H. MineralogischeTabellen, AkademischeVerlagsgesellschaft. **1957**, Leipzig.
62. Suarez, M.; Flores, L. V.; and Martín Pozas, J. M. *Clay. Min*. **1995**, *30*, 261–266.
63. Suarez, M.; and García Romero M.; Advances in the Crystal Chemistry of Sepiolite and Palygorskite. In: Developments in Palygorskite-Sepiolite Research A New Outlook on these Nanomaterials, Developments in Clay Science, Vol. 3, Galán, E. Singer, A., Eds., Elsevier. **2011**, 33–65.
64. Suarez, M.; and García Romero, E.; *App. Clay Sci*. **2006**, 31, 154–163.
65. Torres-Ruiz, J.; López-Galindo, A.; González-López, J. M.; and Delgado, A.; Spanish fibrous clays: an approach to their geochemistry and micromorphology. In: Electron Microscopy, Materials Sciences, Vol. 2., López-Galindo, A., Rodríguez-García, M.I. Eds., Secretariado de Publicaciones, Universidad de Granada, **1992**, pp. 595–596.
66. Valentín, J. L.; López Manchado, M. A.; Rodríguez, A.; Posadas, P.; and Ibarra, L.; *App. Clay Sci*., **2007**, *36*, 245–255.
67. Van Olphen, H.; *Science*, **1966**, 154, 645–646.
68. Van Scoyoc, E. G.; Serna, C., and Ahlrichs, J. L.; *Am. Min*., **1979**, *64*, 215–223.
69. Villanova, J.; *Anal. Soc. Esp. Hist. Nat., IV Acta*, **1875**.
70. Von Ssaftschenkow, T. V.; Palygorskit. Verhandlungen der Russisch Kaiserlichen Gesellschaftfu¨r Mineralogie, Sankt Petersburg, **1862**, 102–104.
71. Werner, A. G.; Definition of the species, Bergm. J., p. 377. In: A System of Mineralogy London: Dana J. D. Truner and Co, **1788**, p 456.
72. Yalçın, H., and Bozkaya, Ö.; Sepiolite palygorskite from the Hekimhan region (Turkey). *Clays. Clay. Min*. **1995**, *43*, 705–717.
73. Yalçın, H.; andBozkaya, Ö.; Ultramafic rock hosted vein sepiolite occurrences in the Ankara ophiolitic **mélange, Central Anatolia, Turkey.** *Clays. Clay. Min*. **2004**, *52*, 227–239.
74. Yalçın, H., and Bozkaya, Ö.; Sepiolite Palygorskite Occurrences in Turkey. In: Developments in Palygorskite-Sepiolite Research: A New Outlook on these Nanomaterials. Eds. E. Galan and A. Singer, A., Elsevier, Amsterdam, The Netherlands, Developments in Clay Science 3, 520 pp, ISBN-13: 978-0-444-53607-5, **2011**, 175–200.
75. Yamnova, N. A.; Egorov-Tismenko, Yu. K.; and Khomyakov, A. P.; Crystal structure of a new natural (Na,Mn,Ti)-phyllosilicate. *Crystallogr. Rep. 41*, 239–244 as translated from the Russian in: Kristallografiya, **1996**, *41*, 257–262.

CHAPTER 4

THE IDENTIFICATION AND NOMENCLATURE OF CHRYSOTILE (A HISTORICAL PERSPECTIVE)

SAVERIO FIORE and F. JAVIER HUERTAS

CONTENTS

86

The serpentine-mineral group is constituted by three main minerals: antigorite, lizardite, and chrysotile. They are hydrous magnesium phyllosilicates having the approximate formula $Mg_3Si_2O_5(OH)_4$ (Bailey, 1980). The basic structural unit is an Mg-rich octahedral sheet tightly linked on one side to a single tetrahedral silicate sheet, the so-called T-O layer, whose stacking produces the three-dimensional structure. The crystal morphologies range from planar (lizardite) to alternating waves (antigorite) to cylindrical rolls (chrysotile). For a detailed description of the crystal structure of serpentine minerals, see Guggenheim (see Chapter 1 of this book by Guggenheim).

Chrysotile, originally called *amianthus*, was named in 1834 by F. von Kobell from the Greek, in allusion to its golden fibers (Gaines et al., 1997). Its crystal structure has been a challenge and it has not been completely solved. Pauling (1930) suggested a curved structure to solve the misfit between T and O sheets in serpentine. Warren and Bragg (1930) proposed a double Si-O chain structure resembling that of fibrous amphiboles. It was Warren and Hering (1941) who first proposed a layered structure to account for X-ray data and mineral composition.

Transmission electron microscopy (TEM) has been essential to unravel the morphology and structure of chrysotile fibrils. First TEM images of chrysotile fibers were obtained by Kühn (1941) in ferruginous bodies. Ruska (1943) published TEM images of chrysotile powder and electron-diffraction pattern rings. The tubular morphology of chrysotile was not recognized until Bates et al. (1950) and Noll and Kircher (1951) published electron micrographs showing cylindrical and apparently hollow fibers. Later, Yada (1967, 1971), using high-resolution TEM, confirmed that the tubes are formed by concentric cylinders or scrolls of serpentine layers.

The identification of chrysotile is closely related to the concern for asbestos. There exist two different approaches to the problem: (i) to identify chrysotile (or other asbestos) fibers in filters from air or water sampling and (ii) identification of chrysotile in man-made or in natural materials, as for example serpentinites. In the first case, the interest is focused on identifying and distinguishing fibers that belong to chrysotile or fibrous amphiboles minerals with chemical, physical (optical), and structural differences. In serpentinites, the objective is to identify the presence and abundance of minerals and polytypes for environmental, mineralogical, or petrographic studies. In other bulk materials where chrysotile is not the main mineral, its identification is possible but what is almost impossible is a quantitative measure of the chrysotile content.

Health professionals use monitoring airborne fiber concentrations as an important tool for exposure assessment and environmental monitoring. National and international health and environmental protection agencies have proposed a number of methods to determine and count fibers in filters and bulk samples (e.g., Perkins and Harvey, 1993; NIOSH, 1994; WHO, 1997). They include protocols based on light and electron microscopy, as well as X-ray diffraction (XRD) or infrared spectroscopy. Phase-contrast optical microscopy (PCOM) has been the most commonly used method for fiber counting in filters (NIOSH, 1994; WHO, 1997). It requires a trained analyst (Harper et al., 2009). However, PCOM does not allow for differentiating the type of fibers. A detailed description of these methods in asbestos is beyond the scope of this section.

Specific identification of chrysotile requires polarized light microscopy, electron microscopy with microanalysis, or XRD (NIOSH, 1994). Chrysotile optical properties permit distinction from fibrous amphiboles in thin sections. A trained analyst can provide a rough estimation of the percentage of serpentine minerals. However, the very similar properties of serpentine minerals and the fine grain size of most serpentinites, with frequent intergrown textures, make it impossible to completely describe chrysotile in samples using only optical techniques. Cross-fiber asbestos can usually be identified with confidence as chrysotile. However, the occurrence of fibrous/asbestiform antigorite in cross-fiber veins (Groppo and Compagnoni, 2007; Keeling et al., 2008; Fitzgerald et al., 2010) may produce false-positive identification.

The identification of serpentine minerals is not possible without diffraction data (Bailey, 1988) (Fig.1). Chrysotile is a mineral that is difficult to study, and requires the use of several techniques and sample preparation, particularly if chrysotile is a minor component of the bulk material. However, the identification of chrysotile as minor component is beyond the scope of this book. Chrysotile structure and morphology introduce a component of low crystallinity that produces broad reflections, compared with lizardite and antigorite. Furthermore, polytypism introduces changes in the XRD patterns. The mismatch derived from the larger dimension of the Mg-octahedral sheet compared with the tetrahedral sheet is partially relieved in chrysotile by curling and buckling of the T-O layer, with the O sheet outside. The layer stacking produces several cylindrical polytypes (see Chapter 1 of this book by Guggenheim), some of them observed and others theoretical (Wicks and Whittaker, 1977; Bailey, 1980). Tentative identification of polytypes can be done in pure chrysotile specimens, as well as for chrysotile in serpentine-rich rocks. Three different XRD patterns were observed for chrysotile fibers (Whittaker and Zussman, 1956). The criteria to identify the chrysotile polytypes were reported by Whittaker and Zussman (1956), and later updated by Bailey (1980) for modern powder-diffraction patterns. Identification is based on diagnostic reflections, on the relative intensity of some reflections, and the absence of others depending on mineral or polytype (Figure 4.1). Clinochrysotile ($2M_{cl}$ and $1M_{cl}$) is the major component of silky chrysotile asbestos, occurring as small fibers in more massive or splintery specimens. The XRD patterns consist of sharp $h0l$ and diffuse $hk0$ reflections. Fiber directions is along the 5.3 Å repeated axis, normally a for a layer silicate (Whittaker, 1956a). The other lattice parameters, b and c, respectively, are curved with the rolling layers and are radial. Orthochrysotile ($2Or_{cl}$) occurs in small amounts in fibrous and splintery specimens. The fiber axis is coincident with the X crystallographic axis (Whittaker, 1956b). The two-layer character is more patent than in clinochrysotile, resulting in more intense $h0l$ with $l=2n$ reflections. Parachrysotile is a minor constituent in chrysotile specimens with a 9.2 Å axis (b) parallel to the fiber axis (Whittaker, 1956c). Compared to other chrysotile structures, layers are randomly displaced around the cylinder walls, and the fiber direction is Y instead of X. In the XRD pattern, the strong $h0l$ reflections in phyllosilicates are absent and the weak $0kl$ reflections are present only for k=6n (Bailey, 1980).

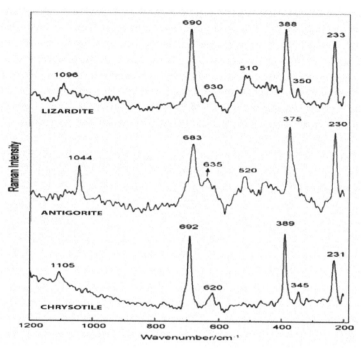

FIGURE 4.1 Simulated X-ray diffraction patterns of the serpentine polymorphs.

The use of electron microscopy coupled with microanalytical techniques (energy-dispersive X-ray spectroscopy, EDS) has progressively advanced as a technique, in preference to optical microscopy. Scanning electron microscopy (SEM) and transmission electron microscopy (TEM) are time-consuming techniques and also not often representative of the bulk sample, but allow an indisputable identification of chrysotile from morphology and composition. Some agencies have used SEM) and/or TEM as complementary or as alternatives to optical microscopy (Perkins and Harvey, 1993; NIOSH, 1994; USEPA, 2009). Their application to filters from air or water sampling is relatively simple. SEM can be applied to pieces of bulk materials and polished sections, providing morphological identification of chrysotile, chemical analysis, and textural information. TEM samples need further preparation as bulk material should be disaggregated and ground to be transferred to a grid, or ion-milled in a metal ring. Small angle electron diffraction (SAED) on oriented particles permits polytype identification and chemical analysis of single nanometric fibers.

Vibration spectroscopy has also been proposed for chrysotile identification in asbestos (Lewis et al., 1996). Initially bulk samples were used diluted in pressed pellets for transmission (Farmer, 1974; Heller-Kalai et al., 1975; Kraincva et al., 1982; Madejova, 2003) or in powder for reflectance (Madejova, 2003; Bishop et al., 2008) spectra, identifying chrysotile by characteristics vibrations of the OH, SiO, and MgOH groups in infrared (Farmer, 1974; Anbalagan et al., 2010). The Fourier Transform Infrared (FTIR) microscope allows analysis of micro-scopic samples in single crystals or polished sections, with a beam size as small as 10 µm. Raman spectroscopy is also applied to chrysotile analysis (Bard et al. 1997; Kloprogge et al., 1999, Wang et al., 2002). Despite the similarities in structure and composition, serpentine minerals can be identified by the analysis of the MgOH vibration bands (Rinaudo and Gastaldi, 2003) (Figure 4.2). Raman spectra are more complex, depending on the excitation frequency and fluorescence phenomena that may render useless a part of the spectra (Rinaudo et al., 2003; Petry et al., 2006). Nevertheless, Raman absorption bands are narrower than IR bands and beam size can be focused to 1 µm. Micrometric samples can be analyzed in polished sections or as single crystals. Micro-Raman has been successfully used for identification of serpentine minerals in ultramafic rocks (Groppo et al., 2006) or particles in histological sections (Rinaudo et al., 2009). Recently, Raman spectrometers have been coupled with SEM (Petry et al., 2006); although beam size does not go below 1 µm, SEM provides a better spatial and chemical (EDS) resolution, improving substantially the capacity of analysis of the SEM-Raman with respect to a Raman microscope. Experimental spectra can be compared with reference samples in databases (Downs, 2006) or laboratory collections.

Thermal analysis is a successful technique applied to phyllosilicates identification, based on their dehydroxylation events (e.g., Mackenzie, 1970). Serpentine minerals exhibit characteristic temperatures of the dehydroxylation event that can be used to discriminate one another, particularly in natural massive samples, where different members may be mixed (Viti, 2010; Viti et al., 2011). Further research may be necessary to apply thermal analysis to bulk asbestos-containing material, although it is not expected to find interferences with amphibole asbestos thermal events (Glasser, 1970).

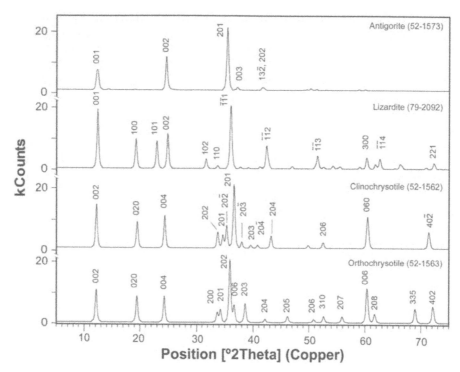

FIGURE 4.2 FT-Raman spectra for chrysotile, antigorite, and lizardite. The minerals are easily distinguishable on the presence/absence and position of some frequencies (Modified after Rinaudo et al. 2003).

We can conclude that identification of chrysotile in natural or man-made samples is a complex task that requires simultaneous use of several techniques. Probably XRD is the most appropriate if the sample can be ground. Depending on the nature of the sample, additional determinations may be done using microscopic, spectroscopic, and thermal analyses. If the final aim is the quantitative estimation of the content of chrysotile, the task is generally more complex. Reliable quantitative estimation of chrysotile content of bulk samples is still a challenge.

KEYWORDS

- **Crystal morphology**
- **Crystal structure**
- **Lizardite**
- **Raman microscope**
- **Serpentine minerals**
- **X-ray diffraction**

REFERENCES

1. Anbalagan, G.; Sivakumar, G.; Prabakaran, A. R.; and Gunasekaran, S.; *Vibrat. Spectros.* **2010**, *52*, 122–127.
2. Bard. D.; Yarwood, J.; and Tylee, B. *J. Raman. Spectros.* **1997**, *28*, 803–809.
3. Bates, T. F.; Sand, L. B.; and Mink, J. F.; *Science* .**1950**, *3*, 512–514.
4. Bailey, S. W. In: Crystal structures of clay minerals and X-ray identification, Eds. Brindley, G. W., Brown, G. Monograph 5; Mineralogical Society: London, UK, **1980**; 1–123.
5. Bailey, S. W.; *Clay. Clay. Min.* **1988**, *36*, 193–213.
6. Bishop, J. L.; Lane, M. D.; Dyar, M. D.; and Brown, A. J.; *Clay. Min.* **2008**, *43*, 35–54.
7. Downs, R. T.; Program and Abstracts, 19th General Meeting of the International Mineralogical Association, Kobe, Japan, July 23–28, **2006**; 03–13.
8. Fitzgerald, J. D.; Eggleton, R. A.; and Keeling, J. L.; *Eur. J. Min.* **2010**, *22*, 525–533.
9. Farmer, V. C.; The Infrared Spectra of Minerals; Monograph 4; Mineralogical Society: London, UK, **1974**.
10. Gaines, R. V.; Skinner, H. C. W.; Foord, E. R.; Mason, B.; and Rosenzweig, A.; Dana's New Mineralogy, 8th edition; New York: John Wiley & Sons, Inc, **1997**.
11. Glasser F. P.; In: Differential Thermal Analysis, Ed. Mackenzie, R. C. London, UK: Academic Press, **1970**; 575–608.
12. Groppo, C.; Compagnoni, R. *Periodico. Di. Mineralogia.* **2007**, *76*, 169–181.
13. Groppo, C.; Rinaudo, C.; Cairo, S.; and Gastaldi, D.; Compagnoni, R.; *Eur. J. Min.* **2006**, *18*, 319–329.
14. Heller-Kalai, L.; Yariv, S.; and Gross, S.; *Min. Mag.* **1975**, *40*, 197–200.
15. Harper, M.; Slaven, J. E.; and Pang, T. W. S.; *J. Environ. Monitor.* **2009**, *11*, 434–438.
16. Keeling, J. L.; Raven, M. D.; Self, P. G.; and Eggleton, R. A.; In: Conference proceedings, 9th International Congress for Applied Mineralogy, Brisbane, Australia, September 8–18, **2008**; AusIMM: Melbourne, Australia, **2008**; 329–336.
17. Kloprogge, T. J.; Frost, R. L.; and Rintoul, L.; *Physl. Chem. Chem. Phys.* **1999**, *1*, 2559–2564.
18. Kraineva, É. P.; Petrov, V. L.; and Polupanova, T. I.; *J. Appl. Spectros.* **1982**, *37*, 1157–1159.
19. Kühn, J.; *Archiv für Gewerbepathologie und Gewerbehygiene.* **1941**, *10*, 473–485.
20. Lewis, I. R.; Chaffin, N. C.; Gunter, M.; and Griffiths, P. R.; Spectrochimica Acta Part A. **1996**, *52*, 315–328.
21. Madejová, J.; *Vibrat. Spectros.* **2003**, *31*, 1–10.
22. Mackenzie, R. C.; In: Differential Thermal Analysis, Ed. Mackenzie, R. C., London: Academic Press, UK, **1970**; 498–537.

23. NIOSH. Manual of Analytical Methods (NMAM), 4th edition. – 3rd Supplement. DHHS (NIOSH) Publication No. 2003–154. http://www.emedco.info/nmam/nmampub html
24. Noll, W.; and Kircher, H.; *Neues. Jahrbuch. Mineralogie. Monat.* **1951**, *10*, 219–40.
25. Pauling, L.; *Proc. Natl. Acad. Sci, Washington* **1930**, *16*, 578–582.
26. Perkins, R. L.; and Harvey, B. W.; Test Method for the determination of Asbestos in bulk building materials. U.S. Environmental Protection Agency, EPA/600/R-93/116, **2003**.
27. Petry, R.; Mastalerz, R.; Zahn, S.; Mayerhöfer, T.G.; Völksch, G.; Viereck-Götte, L.; Krejer-Hartmann, B.; Holz, L.; Lankers, M.; and Poop, J. *Chem. Phys. Chem.* **2006**, *7*, 414–420.
28. Rinaudo, C.; and Gastaldi, G.; *Canad. Min.* **2003**, *41*, 883–890.
29. Rinaudo, C.; Allegrina, M.; Fornero, E.; Musa, M.; Croce, A.; and Bellis, D.; *J. Raman. Spec tros.* **2009**, *41*, 27–32.
30. Ruska, H.; *Archiv für Gewerbepathologie und Gewerbehygiene* **1943**, *11*, 575–578.
31. USEPA.; Standardized Analytical Methods for Environmental Restoration Following Homeland Security Events, Revision 5.0. US Environmental Protection Agency, EPA/600/R-04/126E September 29, **2009**.
32. Viti, C.; *Am. Min.* **2010**, *95*, 631–638.
33. Viti, C.; Giacobbe, C.; Gualtieri, A. F.; *Am. Min.* **2011**, *96*, 1003–1011.
34. Wang, A.; Freeman, J.; and Kuebler, K. E.; Spectroscopic characterization of phyllosilicates. In: Lunar and Planetary Science Conference XXXIII, Huston, TX, March 11–15, **2002** [CD-Rom]; Lunar and Planetary Science Conference: Huston, TX, **2002**; Abstract 1374.
35. Warren, B. E.; and Bragg, W. L.; *Zeitschrift für Kristallographie* **1930**, *76*, 201–210.
36. Warren, B. E.; and Hering, K. W.; *Phys. Rev.* **1941**, *59*, 925.
37. Whittaker, E. J. W.; *Acta. Crystallogr.* **1956a**, *9*, 855–862.
38. Whittaker, E. J. W.; *Acta. Crystallogr.* **1956b**, *9*, 863–864.
39. Whittaker, E. J. W.; *Acta. Crystallogr.* **1956c**, *9*, 865–867.
40. Whittaker, E. J. W.; Zussman, J.; *Min. Magaz.* **1956**, *31*, 107–126.
41. Wicks, F. J.; and Whittaker, E. J. W.; *Canad. Min.* **1977**, *15*, 459–488.
42. WHO.; Determination of Airborne Fibre Number Concentrations. A recommended method, by phase-contrast optical microscopy (membrane filter method); World Health Organization: Geneva, **1997**.
43. Yada, K.; *Acta. Crystallogr.* **1967**, *23*, 704–707.
44. Yada, K.; *Acta. Crystallogr.* **1971**, *A27*, 659–664.

PART III
MINERALOGY, GEOLOGY AND OCCURRENCE OF NATURAL MINERAL NANOTUBES

CHAPTER 5

THE MINERALOGY, GEOLOGY AND OCCURRENCES OF HALLOYSITE

JOHN L. KEELING

CONTENTS

5.1 INTRODUCTION

Halloysite was named by Berthier in recognition of Belgium geologist J.J. d'Omalius d'Halloy (Daltry and Deliens, 1993) who, in the early nineteenth century, collected samples of waxy, white clay at Angleur, Liége, which Berthier (1826) later analyzed and described. While the type locality for the Angleur halloysite has long been lost, Dupuis and Ertus (1995) assert that original descriptions clearly identify the clay as having been recovered from karstic cavities within Dinantian (Lower Carboniferous) limestone in association with Pb and Zn sulfides, and the products of sulfide weathering. Similar occurrences have been described from various buried karst on the Entre-Sambre-et-Meuse plateau in southern Belgium, and also at Aïn Khamoudain, Central Tunisia, and the Djebel Debbagh region in northeastern Algeria (Perruchot et al., 1997; dePutter, 2002; Dupuis et al., 2003; Renac and Assassi, 2009; Bruyère et al., 2010). The formation of relatively pure masses of halloysite close to where the fluids have interacted with carbonate rocks is common to these and other important halloysite occurrences, as described below. Substantial volumes of halloysite-rich clays also result from weathering or hydrothermal alteration of aluminum silicates in other geological settings, the most significant being the alteration, under neutral to acidic conditions, of Cenozoic volcanic deposits.

Why halloysite forms in preference to kaolinite—which is thermodynamically favored and much more abundant—is most likely due to local environmental factors that assist reaction kinetics to have greater influence on the atomic arrangement of the mineral phase precipitated from solution and its preservation (e.g., de Ligny and Navrotsky, 1999; Jeong, 2000; Ziegler et al., 2003; Kleber et al., 2007). Halloysite is a metastable phase of kaolin, with lower activation energy of nucleation than that required for kaolinite crystallization (Steefel and Van Cappellen, 1990). Halloysite would be expected to be the first kaolin mineral to precipitate from solution. The likelihood is increased, where preexisting silicates undergo rapid dissolution and where few suitable mineral substrates are available for nucleation of kaolinite crystallites. Such a situation arises where silicates interact with warm and/ or corrosive (acidic) fluids or where less stable silicates, such as Ca-plagioclase, are exposed to weathering. A high nucleation rate of halloysite can rapidly reduce fluid saturation levels to the point where any kaolinite precursors are resorbed and the predominance of halloysite is enhanced (Fritz and Noguera, 2009). If saturation levels were maintained, then kaolinite would ultimately predominate through recrystallization. Requirements for halloysite formation and preservation might be summarized, therefore, to include the following conditions: high water saturation, Al and Si in solution, a chemical or thermal gradient that modifies solubility leading to rapid nucleation of crystallites, and successive influx of Al and Si to accumulate crystals deposited from either solution or colloidal gel.

5.2 HALLOYSITE MORPHOLOGY

Aspects of halloysite structure and identification are described in earlier chapters and a review of halloysite mineralogy is included in Joussein et al. (2005). The following focuses on variation in halloysite morphology. In specialist nanotechnology applications, morphology of tubular forms of halloysite in particular, may be a key attribute in achieving or maximizing the desired function. Consequently, assessment of halloysite deposits requires analyses that provide information on halloysite morphology, in addition to halloysite content and chemistry.

Tubular morphology is widely associated with halloysite, but spheroidal forms are equally common and may predominate in some settings, particularly, but not exclusively, with the alteration of glassy volcanic deposits. More rarely, platy or tabular forms are observed (e.g., Carson and Kunze, 1970; Wada and Mizota, 1982; Noro, 1986). Wyatt (2009) described halloysite, from Tauriko near Tauranga in New Zealand, that crystallized as coarse vermicular stacks of platelets, ranging from 1 to 20 μm across and 1.5–50 μm in length. These formed alongside halloysite tubes and spheres. Whereas X-ray diffraction and chemistry may confirm high halloysite content, electron microscopy combined with data on surface area and pore size distribution and shape are generally required to characterize morphology.

Halloysite crystallizes with interlayer water, as halloysite 10Å, but the interlayer water is readily and irreversibly lost either naturally, or during sample storage, handling, and processing. The loss of interlayer water may alter halloysite morphology, for example by collapsing thin tubes to form ribbons or laths, or development of slit-shaped pores in thick tubes by uneven dehydration of clay layers making up the tube walls. In the description of halloysites, given below, no distinction is made with regard to the hydration state of clays from particular deposits; this information is often inconsistently reported. For individual halloysite sources, the choice of process method, particularly conditions of drying, may affect halloysite morphology and is a factor that may need to be accounted for and modified, as necessary.

Halloysite tubes show wide variation in shape and size, and also in the diameter of the central cylindrical pore or lumen (Figure 5.1). Tubes vary from long and thin, >30 μm length by 0.03 μm diameter (Norrish, 1995), to short and thick, 0.2 μm length by 0.1 μm diameter (Pasbakhsh et al., 2013), with maximum diameter reported of ~0.5μm for long thick tubes (de Oliveira et al., 2007). A tubular kaolin from Piedade, São Paulo, Brazil, is described by de Souza Santos et al. (1965) as several centimeters in length and between 0.03 μm and 0.15 μm in width. The clay occurred as infill in 1–3 cm-wide cracks in weathered porphyritic granite and dry clay fragments had a distinctly fibrous appearance. This source may represent the longest halloysite fibers recorded.

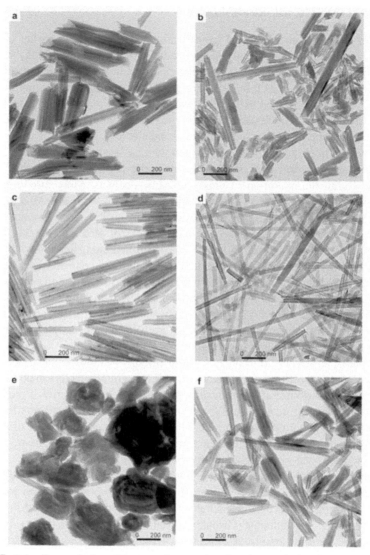

FIGURE 5.1 Transmission electron micrographs showing variation in morphology of halloysite from various geological settings. (a) Matauri Bay, New Zealand— thick tubular forms from weathered alkaline rhyolite, (b) Dragon Mine, Utah—long and short tubular forms showing a range of diameters, from hydrothermal replacement of dolomite, (c) Camel Lake, South Australia—regular tubes from acidic groundwater alteration of probable illite-smectite clays, (d) Patch Nickel mine, Western Australia—regular long thin tubes from weathered ultramafic rock, (e) Te Puke, New Zealand—blocky forms and minor thin tubes from soil developed on weathered alkaline rhyolite, (f) Jarrahdale, Western Australia—thin tubes of variable diameter from deeply weathered dolerite.

Tube cross sections are generally circular or oval, but may also be polyhedral. Churchman et al. (1995) measured the size distribution of pores for various halloysites and showed that wall thickness can influence the central pore diameter. Thick tubes were often folded or more tightly rolled to give a very small lumen, whereas tubes with thinner walls commonly had a large central pore. Packets of layers forming thick tube walls are more likely to act as rigid blocks during formation, or subsequent dehydration. This can result in slit-shaped pores where parting occurs between layers or wedge-shaped pores at points where layers are bent, resulting in flat surfaces and polygonal cross-section (e.g., Dixon and McKee, 1974). Thin halloysite tubes can collapse on drying to form ribbons or laths. Singh (1996) argued that curved layers in hydrated halloysite with a radius of 9.8 nm (~20 nm diameter) have the ideal curvature to accommodate the mismatch in dimensions between the larger silica tetrahedral sheet and smaller Al-OH octahedral sheet. The diameter is close to that observed for halloysite from the "Patch" nickel mine in Western Australia (Figure5.1d), where regular, long, thin halloysite tubes typically have a wall thickness of around 5–6 nm (5–6 layers),with a central pore diameter calculated at 18.7 nm (Norrish, 1995). For layers with a radius greater than 9.8 nm, the misfit between the sheets must be accommodated also by some rotation of silicon atoms in the tetrahedral layer (Singh, 1996).

5.3   GEOLOGICAL SETTING OF HALLOYSITE DEPOSITS

Halloysite is most commonly an alteration product resulting from weathering or hydrothermal activity. The conditions for formation, outlined in the introduction, are similar to that for kaolinite and the two polytypes often occur together. Halloysite, however, is much less stable under conditions of mechanical erosion, transport, and deposition, and, unlike kaolinite, secondary sedimentary deposits are not common. More often halloysite in sediments is authigenic, and the result of groundwater interaction with silicate minerals or with glassy volcanic ash deposits within the sediment.

5.3.1   WEATHERING ENVIRONMENT

The process of kaolinization by weathering is favored during warm and wet conditions, with annual precipitation >1,000 mm, pH between 6 and 4, and fluids that maintain K^+, Na^+, Ca^{2+}, and Mg^{2+} in solution and aid in their removal, along with some leaching of SiO_2 (Galán, 2006). In this environment, halloysite is preferentially formed in situations where reaction kinetics promote rapid nucleation of halloysite crystallites, and the conditions for crystallization and growth of kaolinite are restricted. These situations can arise where the silicates being altered are most susceptible to weathering, such as volcanic glass or calcic feldspar (Jeong and Kim, 1993), where fluid ion saturations fluctuate in response to cyclic changes yet the site remains wet, or where groundwater is highly acidic. Good drainage is necessary to remove cations, but highly drained sites promote formation of gibbsite (Churchman et al., 2010), and sites that are subject to wetting and drying favor formation of kaolinite, including transformation of earlier formed halloysite to kaolinite (Churchman and Gilkes, 1989, Inoue et al., 2012). Weathering of glassy volcanic rocks can give rise to extensive halloysite deposits, as is the case in North Island, New Zealand. Halloysites formed in this environment of high permeability and unstable silicates can show wide variation in size and shape, in part influenced by precursor allophane (e.g., Nagasawa and Miyazaki, 1975) (Figure5.1a,e). For other rock types, local environmental conditions of weathering probably have greater influence on the morphology of the halloysite than the type of silicate mineral being altered (e.g., McCrea et al., 1990; Churchman et al., 2010).

Sulfide oxidation and groundwater acidification during weathering was responsible for halloysite formation at the original "type" locality in karstic limestone on the Entre-Sambre-et-Meuse plateau in Belgium (Dupuis and Ertus, 1995). Here, meteoric waters passed through reduced sandy sediments, comprising organic matter and pyrite with interbedded, lignitic clays that fill in and cover large karstic features in the underlying Carboniferous limestone (Figure 5.2). The pyrite was progressively oxidized, which acidified the groundwater leading to enhanced dissolution of silicate minerals in the sediments and clayey fill in solution cavities within the limestone. The pH gradient established at the interface with the limestone reduced the solubility of Al which reacted with Si in solution to precipitate halloysite. Soluble ions, including excess Si, were transported further into the porous limestone, where silica was precipitated along the fluid pathways (Dupuis et al., 2003). Groundwater acidification leading to halloysite precipitation in karst occurs at other localities, including the Carlsbad Cavern in New Mexico, one of the world's largest natural limestone chambers and the result of enhanced excavation by sulfuric acid-bearing waters. Here smectite clay, infilling small solution cavities, was altered to halloysite and alunite (Polyak and Güven 1996). While this style of halloysite deposit is generally too small and variable for economic extraction, large deposits occasionally develop. Examples include Karst 46 in the Djebel Debbagh region of Algeria (Renac and Assassi, 2009) and at Camel lake, near Maralinga in South Australia (Keeling et al., 2010). The Camel lake site is a desert playa where erosion has exposed a thick layer of halloysite that formed beneath Miocene limestone by acid groundwater discharged from underlying lignitic sands. Halloysite from this site is unusual in that the tubular forms are very regular (Figure 5.1c) with a consistent central pore that produces a high surface area of 76 m^2/g (Pasbakhsh et al., 2013).

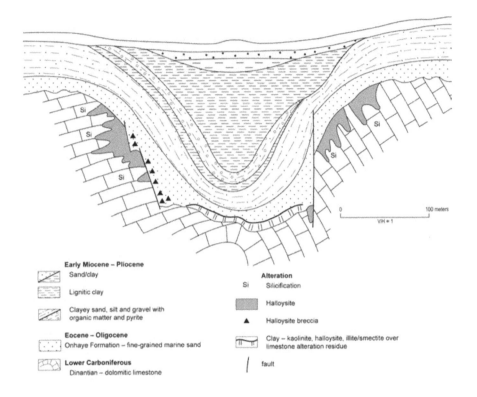

FIGURE 5.2 Halloysite in karstic limestone on the Entre-Sambre-et-Meuse plateau, Belgium. Generalised section modified from Dupuis et al. (2003) and incorporating aspects of the d'OnhayeWeillen karst shown in Perruchot et al. (1997). Approximate scale only.

5.3.2 HYDROTHERMAL ENVIRONMENT

Circulating hydrothermal fluids from igneous activity accelerates rock alteration which can lead to halloysite deposits. This generally occurs near the surface where geothermal, sulfur-bearing fluids become acidic, and fluid temperatures drop below 100°C. The pattern of alteration is usually controlled by fault structures, and where these intersect permeable or reactive lithologies. Zoned alteration is often apparent and reflects changes in fluid chemistry and temperature, relative to distance from the main fluid conduits. At the Kohdachi kaolin deposit in Miyoshi, Japan, Kitagawa and Köster (1991) described zoning in altered granite of (i) halloysite with weak silicification, (ii) halloysite-kaolinite, and (iii) kaolinite, which together with extensive clay veining were taken as indicative of hydrothermal activity. Kyne et al. (2013) recognized supergene and hypogene halloysite alteration in a porphyry-epithermal system at Cerro la Mina, Mexico. This was described as part of the argillic alteration assemblage of a high-sulfidation epithermal mineral system in brecciated, trachyandesite volcaniclastic rocks. Hypogene halloysite extended to depths of at least 800 m. Where warm, acidic hydrothermal fluids interact with dolomite or limestone, substantial deposits of halloysite may result. This is the situation at the Dragon Mine in Utah and with smaller high-grade halloysite deposits on the Biga Peninsula in Turkey; both are described below.

5.4   DESCRIPTION OF SELECTED HALLOYSITE DEPOSITS

Despite the widespread occurrence of halloysite, large deposits, from which high purity halloysite can be economically extracted, are comparatively rare. These include deposits with high-grade zones of dominantly halloysite, and lower grade deposits where halloysite can be readily separated to give a high purity product. Relatively pure halloysite typically occurs as narrow lenses or pockets in altered rock and requires selective mining and sorting to produce a high-grade product. More commonly, halloysite is associated with fine-grained kaolinite, silica, or other fine-grained mineral contaminants, and is difficult to separate using beneficiation methods that rely primarily on wet size classification. Whereas halloysite is a significant component of some bulk kaolinitic clays used extensively by the heavy clay and ceramics industries, the production of high-grade halloysite from large deposits, for specialist industrial use, has, to the present time, been restricted to two regions. These are in Northland, New Zealand, from weathered rhyolite, and in the Tintic district of Utah, United States, from hydrothermal clay masses replacing dolomite. Smaller or lower grade deposits occur in many countries including Japan, Korea, China, Thailand, Indonesia, Australia, South America, and Europe. Often these have a history of use in local clay blends, especially for specialist ceramic applications. An example is halloysite deposits on the Biga Peninsula in Turkey which provide small quantities of specialist clay used in the ceramic industry as a suspension agent in glaze preparations and as an additive to modify viscosity to improve slip-casting characteristics of ceramic raw material blends used in sanitary ware manufacture (Clarke, 2008).

5.4.1   NORTHLAND, NEW ZEALAND

The Kerikeri-Matauri Bay area of Northland, North Island, New Zealand, ~240 km north of Auckland, has, since the late 1940s, been a source of halloysite clays for industry. In 1969, a wet process plant was installed at the Matauri Bay deposit to reduce the silica content of the halloysite clay product in order to access higher-value markets (Townsend and Marsters, 2002). Since the mid 1970s, this has been the dominant continuous supplier of halloysite worldwide, averaging around 20,000 tons per annum (tpa). The operation was acquired in 2000 by Imerys, from former owner New Zealand China Clays Ltd. Imerys Tableware NZ Ltd currently mine raw clay from open pit mines developed in the Matauri Bay and nearby Mahimahi rhyolite domes (Figure 5.3). The raw clay contains around 50% halloysite, 50% silica minerals (quartz, cristobalite, tridymite, amorphous silica), and minor feldspar. The halloysite is separated using wet process methods. As-mined clay is blended from stockpiles and wet ground in a pan mill to <5 mm. This is followed by high solids sand grinding and initial classification using hydro cyclones and settling boxes. Centrifuges are used to reduce the levels of fine-grained silica and to achieve separation of clay with median particle size 0.3 μm and 96 percent <2 μm (Luke, 1997, Wilson, 2004a). The beneficiated clay is thickened, then dewatered by filter pressing and the filter cake is either extruded at 37 percent moisture content or shredded and dried to produce granules and powders with 3–4 percent moisture content (Townsend and Marsters, 2002). Halloysite product from the Matauri Bay plant has high natural whiteness due to low Fe_2O_3 and TiO_2 contents of <0.30% and <0.1 percent, respectively. Almost all the product is exported, primarily for use in quality tableware ceramics, porcelain, and bone china, with around 15 percent for technical ceramics, principally synthetic zeolite-base molecular sieves and in the manufacture of honeycomb catalyst supports (Christie et al., 2000; Clarke, 2008).

The Northland halloysite deposits are all within alkaline rhyolite domes that were emplaced during the youngest phase of volcanism, which commenced in the late Miocene and continued into the late Pleistocene (Figure 5.3). The rhyolites are a minor component of the voluminous basaltic Kerikeri Volcanic Group, and are most probably fractionation products of the basalt (Smith et al. 1993). At the Matauri Bay deposit, rhyolite intruded an earlier basalt flow and is in turn partly overlain by thin silty clay and peat, and a younger basalt flow dated at 4.0 ± 0.7 Ma (Brathwaite et al., 2012) (Figure 5.2). The upper portion of the rhyolite is highly altered to a halloysite-rich clay layer, between 10 and 30 m thick. Harvey et al. (1990) considered the alteration was the result of combined hydrothermal and weathering activity, but Brathwaite et al. (2012) argued that weathering was the primary agent of alteration. Their conclusions were based on the predictable pattern of chemical changes down the altered profile and oxygen and hydrogen isotopes of the <2 μm clay fraction that showed the clay crystallized at temperatures in equilibrium with meteoric water. They concluded that under a subtropical weathering environment, primary sanadine and plagioclase phenocrysts, along with the ground mass of volcanic glass, were altered to halloysite to give the clay-rich zone, with halloysite and silica minerals in approximately equal proportions.

Beneficiated halloysite from the Matauri Bay plant typically contains some fine-grained silica, estimated by Pasbakhsh et al. (2013) at ~11 percent (Table 5.1). This can be reduced with further processing to give a product with ~97 percent halloysite (see sample MB-NN, Table 5.1). The halloysite has a comparatively low surface area of between 22 and 29 m^2/g and is characterized by tubular forms, 0.1–3 μm in length, showing wide variation in dimensions and morphology, including a high proportion of thick tubes, and minor sub spherical and platy forms (Yuan et al., 2008; Pasbakhsh et al., 2013) (Figure 5.1a).

TABLE 5.1 Chemistry and mineralogy of selected beneficiated halloysites.

Halloysite Sample	SiO$_2$ (%)	TiO$_2$ (%)	Al$_2$O$_3$ (%)	Fe$_2$O$_3$ (%)	MnO (%)	MgO (%)	CaO (%)	Na$_2$O (%)	K$_2$O (%)	P$_2$O$_5$ (%)	SO$_3$ (%)	Sum (%)	LOI (%)	Halloysite (%)	Accessory minerals (%)
MB	51.41	0.08	34.15	0.30	0.01	0.08	0.00	0.12	<0.01	0.06	0.06	86.27	13.73	87	Quartz/cristobalite (11), anatase (0.08)
*MB-NN	45.82	0.12	38.21	0.48	0.01	0.17	0.01	0.12	0.01	0.32	0.02	85.30	14.70	97	Quartz/cristobalite (1), anatase (0.12)
DG	43.50	0.02	38.88	0.33	0.001	0.12	0.26	0.07	0.07	0.83	0.26	84.34	15.7	84	Kaolinite (8), quartz (3), gibbsite (3), alunite/woodhousite (1)
PATCH	45.64	0.08	36.86	0.65	0.02	0.21	0.80	0.07	0.02	0.01	0.22	84.82	15.18	96	Quartz (2), Fe-oxides (0.5), anatase (0.08)
CLA	44.96	0.15	37.57	1.21	0.01	0.19	0.28	0.09	0.31	0.01	0.63	85.47	14.53	95	Quartz (1), alunite (3), Fe-oxides (1), anatase (0.15)
TP	44.63	0.36	36.44	3.38	0.02	0.09	0.18	0.10	0.09	0.02	0.02	85.34	14.66	98	Quartz/cristobalite/amorphous silica (1.5), anatase (0.36)

Source: Source: Pasbakhsh et al. (2013)—*MB-NN reprocessed Matauri Bay halloysite supplied by NaturalNano Inc. and analyzed by Pasbakhsh (unpublished data).
Matauri Bay, New Zealand (MB),Dragon Mine Utah, USA (DG),Patch Nickel Mine, Western Australia (PATCH), Camel lake, South Australia(CLA),–Te Puke, New Zealand(TP).

In addition to the Matuari Bay and Mahimahi deposits, halloysite is present in other rhyolite domes in the district. The largest is the Maungaparerua dome covering 140 ha. Clay alteration is mostly within the weathered zone between 8 and 30 m thick, averaging 15 m. The weathered rhyolite is a mixture of~50% quartz and ~50 percent halloysite, kaolinite, and allophane with minor plagioclase feldspar (Murray et al., 1977). Iron oxide and titania contents in the clay zone are generally higher than those at the Matauri Bay and Mahimahi deposits (Townsend et al., 2006).

5.4.2 TINTIC DISTRICT, UTAH—DRAGON MINE

Late magmatic fluids associated with intrusion of the "monzonite porphyry of Silver City," during Oligocene time, were responsible for numerous silver and base metal vein and limestone replacement deposits in the Tintic Mining District of Utah—and for extensive patches of argillic alteration, including halloysite (Cook, 1957). Large masses of halloysite were encountered while mining ironstone flux at the Dragon Mine, ~4 km south of the township of Eureka and ~96 km south of Salt Lake City. From 1949 to 1976, halloysite was extracted by open cut and underground methods and processed by Filtrol Corporation, a subsidiary of The Anaconda Co, for use as a petroleum-cracking catalyst (Wilson, 2004a). Around 1 million tons of halloysite were mined. Operations ceased following development of synthetic zeolite catalysts that replaced halloysite (Wilson, 2004a). The clay masses at the Dragon Mine are possibly the largest deposits of high-grade halloysite yet encountered. Substantial resources remain, and developments are currently underway by Applied Minerals Inc. to bring the mine back into production. Smaller deposits are known in the district, including the North Star halloysite deposit, 0.7 km north of the Dragon Mine.

Halloysite and iron oxide alteration at the Dragon Mine (Figure 5.4) formed by irregular replacement of Cambrian Ajax Dolomite by hydrothermal fluids channeled along the Dragon Fissure Zone, a 150 m-wide, north-northeast trending fault structure. Latite flows from widespread volcanic activity during late Eocene—early Oligocene time (~34 to 33 Ma) partly covered the area of Ajax Dolomite prior to the dolomite being intruded by "monzonite porphyry of Silver City" at around 32.7 Ma (Pampeyan, 1989). The extent of the halloysite and iron oxide replacement is postulated by Kildale and Thomas (1957) to result from intrusion-driven fluids within the Dragon fissure being "trapped" beneath an overhanging "roof" of igneous rocks allowing for extensive reaction between the fluids and the host dolomite. Original bedding traces in dolomite replaced by clay are still evident in the pattern of unreplaced chert nodules (Morris, 1964). Near the surface, the halloysite deposit forms two irregular pipes, 60–80 m wide, separated by a mass of mixed iron oxide and clay, localized along the main Dragon fissure (Figure5.4). The pipes plunge steeply to around 100 m at which point, bedding in the dolomite becomes a significant control on the pattern of alteration (Kildale and Thomas, 1957). Fine-grained pyrite is present with high-grade halloysite. Other alteration minerals include iron and manganese oxides, kaolinite, illite/smectite, alunite, and silica minerals, and minor gibbsite and nontronite.

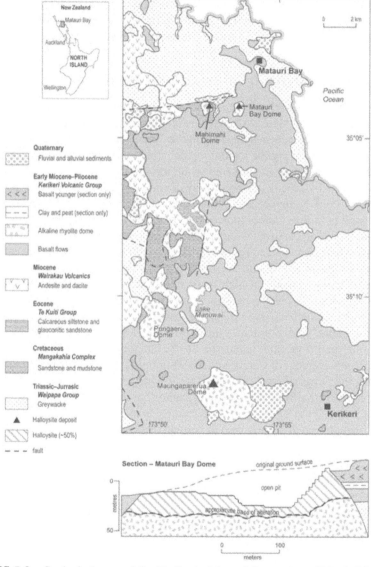

FIGURE 5.3   Geological map of the Kerikeri—Matauri Bay region of North Island New Zealand (Derived from Web Map (GNS Science, 2013)) showing rhyolite domes and halloysite deposits, with sketch west-east section through the Matauri Bay halloysite deposit (After Brathwaite et al. (2012)).

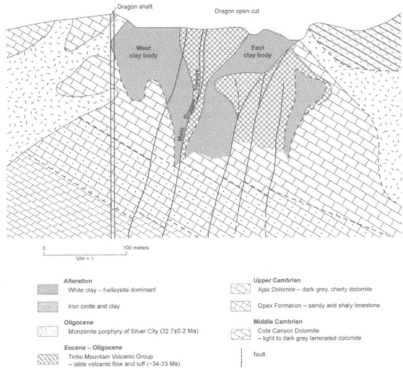

FIGURE 5.4 Geological section, approximately west-east, through the Dragon Mine, Tintic mining district, Utah, showing distribution of halloysite ore zones—redrawn from Kildale and Thomas (1957).

During 2009–2010, drilling was undertaken below the original Dragon Mine open pit and established a measured resource of ~500,000 t of clay averaging 65 percent halloysite (Wilson, 2011). Kaolinite (ave. 18%) and illite-smectite (ave.11%) were also identified as components of the clay fraction. Additional resources of clay were determined for the "Western area" and within waste stockpiles. The clay fraction (<5 μm) from these areas had lower halloysite content ranging from 4 to 42 percent halloysite, together with variable amounts of kaolinite and illite-smectite. The waste stockpiles also contained high amounts of iron oxides—hematite, goethite, and ferrihydrite (Wilson, 2011). Beneficiated halloysite product from the Dragon Mine, supplied by Applied Minerals in 2012 and analyzed by Pasbakhsh et al. (2013), contained ~84 percent halloysite and 8 percent kaolinite with the remainder, quartz, gibbsite, and traces of aluminum sulfate/phosphate minerals (Table 5.1). The halloysite was dominantly tubular with a mixture of short and long tubes, up to ~1.5 μm length and 0.15 μm diameter (Figure 5.1b). The sample had a high pore volume of ~30 percent and relatively high surface area of ~57 m^2/g (Pasbakhsh et al., 2013).

5.4.3 BIGA PENINSULA, TURKEY

Hydrothermal alteration of Neogene alkaline, volcanic lava and pyroclastic deposits by geothermal fluids moving up fracture zones on the Biga Peninsula in northwestern Turkey has resulted in significant kaolin deposits, with localized zones of high-grade halloysite (Laçin and Yeniyol, 2006; Sayin, 2007; Ece et al., 2013). These are part of more extensive clay resources from altered volcanic deposits that have, over the centuries, sustained the development of local clay-based industries such that Turkey is now a major ceramics producer, third biggest in Europe and sixth largest in the world (Yöruköğlu and Delibaş, 2012). Ece et al. (2008) reported six halloysite deposits in the Yenice-Gönen-Balya district on the Biga Peninsula, two of which were worked out and the remaining four being actively mined at the time by the ceramics industry. Reserves of halloysite in the region were stated to be ~50,000 tons (Ece et al. 2008). Production of halloysite in Turkey, at the time, was estimated at 3,000 tpa (Clarke, 2008), mostly by Esan Eczacibaşi, the mining division of the country's leading ceramics manufacturer.

The location of halloysite deposits on the Biga Peninsula was controlled by a combination of minor faults and proximity to Permian limestone blocks entrained within either Triassic Karakaya Complex or overlying Miocene calc-alkaline volcanic rocks. The geological setting, shown in Figure 5.5, is adapted from a model by Laçin and Yeniyol (2006) for the Soğucak deposit, ~3 km north of the village of Sogucak in Çanakkale Province; the adaptions reflect studies by Ece and Schroeder (2007) and Ece et al. (2008) and their genetic models for the Turplu halloysite and alunite deposits, in nearby Balikesir Province. Sulphur-bearing geothermal fluids, ascending during Pleistocene time along the reactivated main North Anatolian Fault, flowed up minor faults and spread out into zones of high permeability associated with porous Miocene andesitic tuffs and pyroclastics. Here the geothermal fluids, containing gasses H_2S and SO_2, mixed with oxygenated meteoric water to form H_2SO_4. The acidified waters intensified clay alteration of the volcanics closest to the faults. Where the fluids encountered limestone blocks, either at the unconformity with Karakaya Complex or entrained within the Miocene volcanics, the interaction established a geochemical gradient that favored precipitation of halloysite (Figure 5.5). High permeability associated with karstic features in the limestone was also a factor in maintaining fluid circulation and continuous leaching of elements, except for Al, which reacted with Si in solution to form high-grade halloysite deposits (Ece et al., 2008).

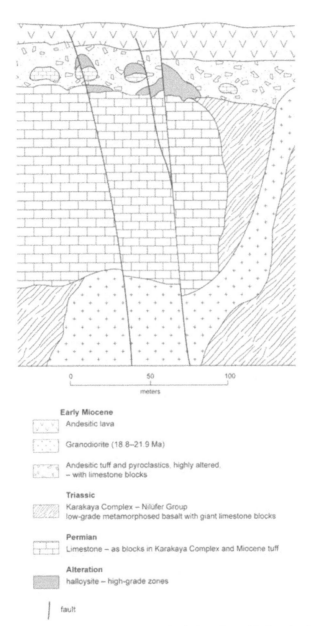

Early Miocene

Andesitic lava

Granodiorite (18.8–21.9 Ma)

Andesitic tuff and pyroclastics, highly altered.
– with limestone blocks

Triassic

Karakaya Complex – Nilüfer Group
low-grade metamorphosed basalt with giant limestone blocks

Permian

Limestone – as blocks in Karakaya Complex and Miocene tuff

Alteration

halloysite – high-grade zones

fault

FIGURE 5.5 Sketch section showing the geological setting of halloysite deposits on the Biga Peninsula, Turkey. Section based on the Soğucak deposit described by Laçin and Yeniyol (2006) and the model for low-temperature hydrothermal alteration at the Turplu deposit proposed by Ece et al. (2008). Approximate scale only.

Laçin and Yeniyol (2006) described the Soğucakhalloysite deposit as being 45–50 m wide, 125–130 m long, and 12–15 m thick; associated minerals included kaolinite, smectite, illite, alunite, pyrite, and iron and manganese oxides. Electron microscopy by Saklar et al. (2012) of halloysite from the high-grade Taban deposit, on the western border of Balikesir region, ~3.7 km southeast of the town of Tabanköy, recorded tubular particles up to 5 μm long, but more typically between 0.1 and 1.0 μm in length and 0.04–0.05 μm in diameter. Chemical analysis of commercial halloysite product from the district, reported in Saklar et al. (2012), were SiO_2 46.22%, Al_2O_3 37.30%, Fe_2O_3 1.02% and TiO_2 0.21%. Acid leach experiments on a sample from the Taban deposit showed that much of the iron oxide was leachable, and was present mostly as goethite (Saklar et al., 2012).

5.4.4   OTHER DEPOSITS

Halloysite clays have a long history of use, mainly in whiteware ceramics and clay-based refractories, particularly in Japan and Korea. Most contain variable content of kaolinite and also minor amounts of fine-grained silica minerals, illite, smectite, and iron oxides. Sand and silt-sized contaminants can be removed by wet process but selective mining and blending is commonly used to manage grade variation due to the presence of other fine-grained contaminants.

In Japan, halloysite clays are widespread, associated with weathering or hydrothermal alteration of Pliocene andesite lavas, andesitic breccias and Pleistocene pyroclastic deposits. These have been described by Nagasawa (1978, 1992). The area east-northeast of Nagoya in central Japan is an important source of kaolin clay, from Pliocene sediments with intercalated altered volcanic tuff, and from weathered granitic basement of Cretaceous age. Sedimentary *Kibushi* clay, a kaolinitic ball clay containing organic matter, and *Gaerome* clay, a kaolinitic sand, both contain minor amounts of halloysite. Post-depositional alteration of intercalated pyroclastic tuff deposits has produced a dominantly halloysite or mixed halloysite/kaolinite clay. Halloysite in the sediments is typically in the form of short tubes, <0.5μm in length. Altered, interbedded pyroclastics contain mostly spheroidal halloysite, together with platy forms and short tubes, which appear to have evolved from spheroidal halloysite (Nagasawa and Miyazaki, 1975). Kaolin in the weathered granite is a mixture of kaolinite and short and long halloysite tubes, which can exceed 5 μm in length. Hydrothermal kaolin with high halloysite content is mined at various sites, including the Omura and Okuchi mines on Kyushu, Japan's southernmost main island, and at the Joshin mine in the northern part of the Gumma Prefecture in central Japan (Harvey, 1996).

The Sancheong district of South Korea has a long history of kaolin production, including clays with high halloysite content. These formed by weathering of late Paleoproterozoic anorthosites intruded into gneissic rocks over an area of~150 km² (Jeong, 1998). The anorthosites are predominantly medium to coarse-grained plagioclase feldspar, of labradorite composition, which was extensively altered by weathering under humid, temperate climatic conditions. The resulting clay is a mixture of kaolinite and halloysite, which precipitated from solution and progressively filled microfractures formed by plagioclase dissolution. The halloysite is predominantly as tubes 1–12 μm long and 0.1–0.3 μm diameter; spheroidal forms are a minor component only (Khan and Kim, 1991). Zones of high halloysite content are concentrated in areas of more fractured rock and at sites of increased porosity arising from dissolution of feldspar grains (Jeong and Kim, 1993; Jeong, 1998). Vermiculite, illite, chlorite-vermiculite, and iron oxides are present as minor alteration mineral phases (Jeong and Kim, 1992).

China is today a major kaolin producer, supplying mostly internal markets for paper-manufacture and the ceramics industry. Halloysite is present with kaolinite in residual kaolins extracted from the Longyan area in Fujian Province, and is common in kaolinized Jurassic volcanics mined in the Suzhou district in Jiangsu Province. Several small deposits in Guangxi region have been mined specifically for halloysite (Wilson, 2004b). These include the Dafang, Qingxi, Sunyi, and Shijin areas where halloysite is selectively extracted from patchy zones of hydrothermal alteration in volcanic rocks, adjacent to limestone. Wilson (2004b) estimated production at ~2,000 tpa, including halloysite with very low iron oxide and titania contents and, in the case of Dafang halloysite, predominantly as thin tubular forms 0.2 to 4 μm in length.

Halloysite is a major component of kaolin product derived from hydrothermally-altered granite in Ranong Province in Thailand, an important source of kaolin used in the ceramics industry and as filler in paper. Ranong halloysite is tubular and reportedly up to 20 μm in length, which gives the clay excellent casting properties for sanitaryware ceramics manufacture (Wilson, 2013). In the Thung Yai district, pinkish clay lenses of dominantly halloysite clay were reported by Bordeepong et al. (2011). These occur within 6-8 m-thick kaolinitic clays that underlie sedimentary ball clay resources on the flood plain of the Tapi River in southern Thailand. The halloysite is tubular, ranging from 0.5 to 4 μm in length, 0.08 to 0.2μm in diameter, and has high iron oxide and titania contents of 2.27 percent and 2.72 percent, respectively.

Minor halloysite production in Poland has come from the Dunino deposit near Legnica in Lower Selesia. The clay is a mixture of tubular, spheroidal, and platy forms, with some kaolinite, and is the weathering product of basalt (Komusiński et al., 1981; Matusik, 2009). Contaminants include hematite, iron hydroxides and siderite.

Mixed kaolinite and halloysite is typical of kaolin mined from small deposits in weathered pegmatite in the Minas Gerais district of Brazil, from weathered granitic basement in the São Paulo region, and weathered anorthosite at Encruzilhada, east of Porto Alegre (Wilson et al., 2006). The morphology of halloysites from these deposits is generally tubular and their characteristics are described in de Souza Santos (1993 and references therein). Halloysite mined in Argentina at Mamil Choique and Buitrera in the southwest of the province of Rio Negro is reported by Cravero et al. (2012) as a late Paleogene weathering product of ignimbrites and pyroclastics of rhyolitic composition. Mamil Choique halloysite is dominantly spheroidal with minor kaolinite; Buitrera halloysite is mostly as short tubes that appear to crystallize on the surface of partly dissolved volcanic glass (Cravero et al., 2012).

Despite the small number of identified high-grade deposits, halloysite is widely distributed, and the global resource base for halloysite could likely be expanded to meet substantial growth in demand. This would certainly be the case if demand was from higher-value markets that would justify higher costs for mining and processing. Such expansion would be anticipated both through the discovery of new deposits and through the adoption of more elaborate process methods to separate halloysite occurring within lower-grade sources.

ACKNOWLEDGMENTS

Electron micrographs of halloysite samples were taken by Pooria Pasbakhsh and Peter Self in collaboration with the author. Figures were compiled or redrafted by Zöe French, DMITRE. Comments by reviewer and the book editors on an early draft improved the manuscript.

KEYWORDS

- **Hydrothermal alteration**
- **Kaolinization**
- **Karstic**
- **Mineralogy**
- **Reaction kinetics**
- **Rhyolite domes**
- **Slit-shaped pores**

REFERENCES

1. Berthier, P.; *Ann. Chim. Phys.* **1826**, *32,* 332–335.
2. Bordeepong, S.; Bhongsuwan, D.; Pungrassam, T.; Bhongsuwan, T.; and Songklanakarin.; *J. Sci. Tech.* **2011**, *33,* 599–607.
3. Brathwaite, R. L.; Christie, A. B.; Faure, K.; Townsend, M. G.; and Terlesk, S.; *Min. Deposita.* **2012**, *47,* 897–910.
4. Bruyère, D et al.; *J. Afr. Earth. Sci.* **2010**, *57,* 70–78.
5. Carson, C. D.; and Kunze, G. W.; *Soil. Sci. Soc. Am. Pro.* **1970**, *34,* 538–540.
6. Christie, T.; Thompson, B.; and Braithwaite, B.; *Min. Comm. Rep Clays*, New Zealand Institute of Geological and Nuclear Sciences, Lower Hutt, New Zealand **2000**.

7. Churchman, G.J.; Davy, T.J.; Aylmore, L.A.G.; Gilkes, R.J.; and Self, P.G.; *Clay Min.* **1995,** 30, 89–98
8. Churchman, G.J.; and Gilkes, R.J.; *Clay Min.* **1989,** 24, 579–590.
9. Churchman, G.J.; Pontifex, I.R.; and McClure, S.G.; *Clays Clay Min.* **2010,** 58, 220–237.
10. Clarke, G.; *Ind. Min.*, March **2008,** 58–59.
11. Cook, D.R.; Ore Deposits of the Main Tintic Mining District. In Geology of the East Tintic Mountains and Ore Deposits of the Tintic Mining Districts; Ed. Cook, D.R. Utah Geological Society, Guidebook to the Geology of Utah **1957,** 57–79.
12. Cravero, F.; Maiza, P.J.; and Marfil, S.A.; *Clay. Min.* **2012,** *47*, 329–340.
13. Daltry, V. D. C.; and Deliens, M.; *Ann. Soc. Geol. Belgique* **1993,** *116*, 15–28.
14. de Ligny, D.; and Navrotsky, A.; *Am. Min.* **1999,** *84*, 506–516.
15. de Oliveira, M. T. G.; Furtado, S. M. A.; Formoso, M. L. L.; Eggleton, R. A.; and Dani, N. *An. Acad. Bras. Cienc.*; **2007,** *79*, 665–681.
16. de Putter, T.; André, L.; Bernard, A.; Dupuis, C.; Jedwab, J.; Nicaise, D., and Perruchot, A.; *Appl. Geochem.*; **2002,** 17, 1313–1328.
17. de Souza Santos, P.; *Clay Min.* **1993,** *28*, 539–553.
18. de Souza Santos, P.; and Brindley, G. W.; de Souza Santos, H.; *Am. Min.* **1965,** *50*, 619–628.
19. Dixon, J.B.; and McKee, T.R.; *Clays. Clay. Min.* **1974,** *22*, 127–137.
20. Dupuis, C.; and Ertus, R.; The Karstic Origin of the Type Halloysite in Belgium. In: Clays Control the Environment – Proceeding of the 10th International Clay Conference **1993**; Eds. Churchman, G. J.; Fitzpatrick, R. W.; and Eggleton, R. A.; CSIRO Publishing: Melbourne, **1995**; pp 262–366.
21. Dupuis, C.; Nicaise, D.; DePutter, T.; Perruchot, A.; and Demaret, M.; Roche, E.; *Geol. Fra.* **2003,** *1*, 27–31.
22. Ece, Ö. I.; Ekinci, B.; Schroeder, P. A.; Crowe, D.; and Esenli, F. J.; *Volcanol. Geotherm. Res.* **2013,** *255*, 57-78.
23. Ece, Ö.I.; and Schroeder, P.A.; *Clays Clay Min.* **2007,** *55*, 18–35.
24. Ece, Ö. I.; Schroeder, P. A., Smilley, M. J.; and Wampler, J. M.; *Clay. Min.* **2008,** 43, 281–315.
25. Fritz, B.; and Noguera, C.; *Rev. Min. Geochem.* **2009,** *70*, 371–410.
26. Galán, E.; Genesis of clay minerals. In: Handbook of clay science. Eds. Bergaya, F.; Theng, B. K. G.; Lagaly, G.; Developments in Clay Science **2006,** Vol. 1, 1129–1162, Elsevier Ltd.
27. GNS Science; New Zealand Geology Web Map. http://data.gns.cri nz/geology/. Accessed June 12, **2013.**
28. Harvey, C.C.; Halloysite for High Quality Ceramics. In Industrial Clays, 2nd ed.; Kendall, T. Ed.; Metal Bulletin Plc: London, **1996;** 71–73.
29. Harvey, C.C.; Townsend, M.G.; and Evans, R.B.; In: Proceeding. AusIMM Annual Conference; **1990,** 229–238.
30. Inoue, A.; Utada, M.; and Hatta, T.; *Clay. Min.* **2012,** 47, 373–390.
31. Jeong, G. J.; *Clays Clay Min.* **1998,** *46*, 509–520.
32. Jeong, G. J.; *Clays Clay. Min.* **2000,** 48, 196–203.
33. Jeong, G. J.; and Kim, S. J.; Kaolinites in the Sancheong Kaolin, Korea: Their Textures, Chemistry and Origins. In: Clay minerals, their natural resources and uses; Nagasawa, K. Ed.; Proceedings, 29th International Geological Congress Workshop WB-1 **1992**; 129–135.
34. Jeong, G. J.; and Kim, S. J.; *Clays. Clay. Min.* **1993,** *41*, 56–65.
35. Joussein, E.; Petit, S.; Churchman, J.; Theng, B.; Righi, D.; and Delvaux, B.; *Clay. Min.* **2005,** 40, 383–426.
36. Keeling, J. L.; Self, P. G.; and Raven, M. D.; *MESA J.* **2010,** *59*, 9–13.
37. Khan, A. M.; and Kim, S. J.; *Geol. Bull., Uni. Peshawar* **1991,** *24*, 63–70.

38. Kildale, M. B.; and Thomas, R. C.; Geology of the Halloysite Deposit at the Dragon Mine. In: Geology of the East Tintic Mountains and Ore Deposits of the Tintic Mining Districts; Ed. Cook, D. R. Utah Geological Society, Guidebook to the Geology of Utah **1957**; 94–96.

39. Kitagawa, R.; and Köster, H. M.; *Clay. Min.* **1991**, *26*, 61–79.

40. Kleber, M.; Schwendenmann, L.; Weldkamp, E.; Rößner, J; and Jahn, G.; *Geoderma* **2007**, *138*, 1–11.

41. Komusiński, J.; Stoch, L.; and Dubiel, S. M.; *Clay. Clay Min.* **1981**, *29*, 23–30.

42. Kyne, R.; Hollings, P.; Jansen, N. H.; and Cooke, D. R.; *Econ. Geol.* **2013**, *108*, 1147–1161.

43. Laçin, D.; and Yeniyol, M.; *Istanbul* Üniversity, *Yerbilimleri* J.; **2006**, 19, 27–41.

44. Luke, K. A.; Geology and Extraction of the Northland Halloysite Deposits. New Zealand Minerals & Mining Conference Proceedings; Ministry of Commerce: Wellington, NZ, **1997**; 193–198.

45. Matusik, J.; Gaweł, A.; Bielańska, E; Osuch, W.; and Bahranowski, K.; *Clays. Clay. Min.* **2009**, *57*, 452–464.

46. McCrea, A. F.; Anand, R. R.; and Gilkes, R. J.; *Geoderma* **1990**, *47*, 33–57.

47. Morris, H. T.; Geology of the Eureaka Quadrangle Utah and Juab Counties, Utah. *Geol. Surv. Bull.* 1142-K, 1964, United States Government Printing Office, Washington.

48. Murray, H. H.; Harvey, C.; and Smith, J. M.; *Clays Clay Min.* **1977**, 25, 1–5.

49. Nagasawa, K.; Weathering of Volcanic Ash and Other Pyroclastic Materials. In: Clays and clay minerals of Japan; Eds. Sudo, T. and Shimoda, S.; Developments in Sedimentology **1978**, *26*, 105–123.

50. Nagasawa, K.; Geology and Mineralogy of kaolinitic Clay Deposits Around Nagoya. In: Clay minerals, their natural resources and uses; Nagasawa, K. Ed.; In: Proceedings, 29th International Geological Congress Workshop WB-1 1992; 1–15.

51. Nagasawa, K.; Miyazaki, S.; Mineralogical properties of halloysite related to its genesis. In: Proceedings international clay conference **1975**, 257–265.

52. Noro, H.; *Clay. Min.* **1986**, *21*, 401–415.

53. Norrish, K.; An unusual fibrous halloysite. In: Clays control the environment – Proceedings 10th International Clay Conference **1993**; Eds. Churchman, G. J.; Fitzpatrick, R.W.; and Eggleton, R. A.; CSIRO Publishing: Melbourne, **1995**; 275–284.

54. Pampeyan, E. H.; Geological Map of the Lynndyl 30- by 60- Minute Quadrangle, West-Central Utah **1989**, United States Geological Survey, Denver, US.

55. Pasbakhsh, P.; Churchman, G. J.; and Keeling, J. L.; *Appl. Clay. Sci.* **2013**, *74*, 47–57.

56. Perruchot, A.; Dupuis, C.; Brouard, E.; Nicaise, D.; and Ertus, R.; *Clay. Min.* **1997**, *32*, 271–287.

57. Polyak, V.J.; and Güven, N.; *Clays. Clay. Min.* **1996**, *44*, 843–850.

58. Renac, C.; and Assassi, F.; *Sediment. Geol.* **2009**, *207*, 140–153.

59. Saklar, S; Ağrili, H.; Zimitoğlu, O; Başara, B; and Kaan, U.; The Characterisation Studies of the Northwest Anatolian Halloysites/Kaolinites. Mineral Research and Exploration Bulletin (Turkey) **2012**, *145*, 48–61.

60. Sayin, A. L.; Origin of Kaolin Deposits: Evidence From the Hisarick (Emet-Kutahya) Deposits, West Turkey. *Turkish J. Earth. Sci.* **2007**, 16, 77–96.

61. Singh, B.; *Clays. Clay Min.* **1996**, *44*, 191–196.

62. Smith, I. E. M.; Okada, T.; Itaya, T.; and Black, P. M.; *N. Z. J. Geol. Geophys.* **1993**, *36*, 385–393.

63. Steefel, C. I.; and Van Cappellen, P.; *Geochim. Cosmochim. Acta* **1990**, *54*, 2657–2677.

64. Townsend, M. G.; Luke, K. A.; and Evans, R. B.; Recent Developments in the Exploration and Uses of Halloysite Clay Deposits in Northland. In Geology and Exploration of New Zealand Mineral Deposits; Eds. Christie, A.B.; Brathwaite, R. L.; AusIMM Mono., **2006**, 25; 59–64.

65. Townsend, M. G.; and Marsters, S.; Northland Halloysite—Past, Present and Future. In: Proceedings of the AusIMM **2002** Annual Conference, publication series 6/02, Australasian Institute of Mining and Metallurgy: Melbourne, **2002**; 139–142.
66. Wada, S. I.; and Mizota, C.; *Clay. Clay. Min.* **1982**, *30*, 315–317.
67. Wilson, I. R.; *Ind. Min.*, November **2004a**, 54–61.
68. Wilson, I. R.; *Clay. Min.* **2004b**, 39, 1–15.
69. Wilson, I. R.; Progress Report Including Resource Statement – Dragon Mine, Eureka, Utah, USA. Applied Minerals Inc., April 13, **2011**. *http://appliedminerals.com* (accessed August 15, **2013**).
70. Wilson, I. R.; *Ind. Min.* March **2013**, 23–31.
71. Wilson, I. R.; de Souza Santos, H.; de Souza Santos, P.; *Clay. Min.* **2006**, *41*, 697–716.
72. Wyatt, J.; Sensitivity and clay mineralogy of weathered Tephra-derived Soil Materials in the Tauranga Region. M.Sc. Thesis University of Waikato, **2009**.
73. Yörüköğlu, A; and Delibaş, O.; Mineral Potential of Turkey. Mining Turkey March **2012**, vol. 2, no. 2, 18–23.
74. Yuan, P.; Southon, P. D.; Liu, Z.; Green, M. E. R.; Hook, J. M.; Antill, S. J.; and Kepert, C. J. J.; *Phys. Chem. C.* **2008**, *112*, 15742–15751.
75. Ziegler, K.; Hsieh, J. C. C.; Chadwick, O. A.; Kelly, E. F.; and Hendricks, D. M.; Savin, S. M.; *Chem. Geol.* **2003**, *202*, 461–478.

CHAPTER 6

THE MINERALOGY, GEOLOGY, AND MAIN OCCURRENCES OF SEPIOLITE AND PALYGORSKITE CLAYS

EMILIO GALÁN and MANUEL POZO

CONTENTS

6.1 INTRODUCTION

Palygorskite and sepiolite are relatively rare minerals in nature; however, their origins have been studied frequently owing to the unique conditions required for their formation and stability, and the need to find new commercial deposits. Moreover, the properties are origin dependent because there are close relationships between origin, crystal structure, chemical composition, beneficiation treatments, and industrial utilization, as occurs with other clay minerals (Murray, 2007). Knowledge of clay mineral genesis from studies of existing deposits also contributes to improve exploration programs and to the selective mining and processing of clay-rich materials (Harvey and Murray, 1997).

The annual tonnage of palygorskite and sepiolite is estimated to be 1,300,000 and 850,000 tons, respectively. By far, the largest producer of palygorskite is United States, which accounted for about 75 percent of the world's production. Spain is the largest producer of sepiolite and accounted for about 95 percent of the world's annual production. The total world production of all fuller's earth clays including palygorskite, sepiolite, and calcium bentonites is estimated to be in excess of 3.3 million tons (Murray et al., 2011).

6.2 OCCURRENCES OF SEPIOLITE AND PALYGORSKITE: MINERALOGY AND GENESIS

The possible environments for occurrences of palygorskite and sepiolite range from soils to marine and lacustrine deposits, hydrothermal veins in serpentinite, diorite and dolostone, and the weathering of volcanic rocks, although detrital palygorskite also occurs frequently in some environments (e.g., soils and deep sea) (Galán and Pozo, 2011; Thiry and Pletsch, 2011; Yalçin and Bozkaya, 2011).

Sepiolite and palygorskite are two excellent examples of authigenic clays, formed by neoformation or transformation[1]. Neoformation and transformation, either in lacustrine or perimarine environments, are the two major mechanisms that control the occurrence and distribution of sepiolite and palygorskite (Galán and Pozo, 2011).

With respect to their geotectonic setting, neoformed palygorskite and sepiolite are predominant in shallow shelf basins on passive margins or intraplate basins. Continental rift basins, continental "sag" basins, or some perimarine basins tend to contain significant deposits of these minerals, commonly formed in saline or hypersaline conditions (Merriman, 2005). Palygorskite of detrital origin may come from the continent, as in the case of deep-sea Atlantic palygorskite close to Morocco,

---

[1]Neoformation is commonly the crystallization of a new mineral structure from simple or complex ions, in which there is no inheritance of a pre-existing mineral structure. Transformation is the formation of a new mineral in which part or all of the preexisting structure is inherited.

which was transported by southwesterly winds from near shore shallow African basins (Pletsch et al., 1996). But marine palygorskite can also be derived from smectite by diagenetic transformation (Lopez Galindo et al., 1996), sometimes associated with marine phosphorites (Chahi et al., 1999). The large volumes of palygorskite in the Mid-Cretaceous (Albian) deep-sea deposits of the Central Atlantic formed in situ in the deep sea (Thiry and Pletsch, 2011). Inheritance from neritic, perimarine, or continental formations was unlikely because the potential source areas are generally palygorskite-free. Palygorskite developed due to the contact with hypersaline and warm brines in the bottom waters of the Atlantic Ocean during the Mid-Cretaceous by transformation or aggradation (or early diagenesis) of inherited clay minerals (mainly smectite) at the seafloor.

Also, the formation of Lower Eocene palygorskite in the deep sea correlates with the distribution of warm sub-Antarctic bottom water during the Early Eocene Climatic Optimum. Elevated temperature (at least 8°C warmer than at present), alkalinity, and ion concentrations were all favorable for the formation of palygorskite on the seafloor. Authigenic deep-sea palygorskite was a vestige of the Cretaceous Paleogene greenhouse period and has no modern analog (Thiry and Pletsch, 2011).

In soils of arid regions, palygorskite and sepiolite are common neoformed minerals (Singer, 1979). Traces of palygorskite in some arid soils may be inherited (Shadfan and Dixon, 1984; Mackenzie et al., 1984), but most palygorskite in soils developed authigenically on calcite and smectite in caliche (Yalçin and Bozkaya, 2011) or was precipitated by evaporation from soil solutions (Hojati and Khademi, 2011), forming in many cases duricrusts (palycretes) (Rodas et al., 1994).

Sepiolite can be formed by the diagenetic transformation of magnesite at pH 10.5–11.5 in silica-rich lake waters (Ece, 1998). However, most of the palygorskite and sepiolite deposits in the world are authigenic, formed by neoformation from solution, amorphous silica gels, or by transformation from smectite, illite, or chlorite, commonly in lacustrine and perimarine environments (Galán and Pozo, 2011; Murray et al., 2011).

Jones and Galán (1988) tabulated the favorable environmental conditions for palygorskite and sepiolite formation as compared to trioctahedral smectite (Table 6.1). As expected, with high values of Al, Mg, and Si activity, palygorskite is favored over sepiolite (Figure 6.1). However, besides the presence of Al-rich clay phases (e.g., smectites), short-term variations in chemistry related to changes in environmental conditions such as evaporation, rainfall, freshwater inflow, etc. affect the formation of palygorskite and sepiolite (Garcia-Romero et al., 2007).

TABLE 6.1 Environmental Conditions of Formation of Mg- Clays (Modified After Jones and Galán, 1988).

+++ favored. ++ less favored. – not favored		PALYGOR-SKITE	SEPIOLITE	Mg-SMEC-TITE
CHEMISTRY				
pH and alkalinity	moderate pH <8.5	+++	++	
	intermediate pH = 8–9.5	++	+++	++
	high pH >9.5			+++
Major constituents ratios	high Mg+Si/Al	++	+++	
	high Mg+Fe/Si			+++
Sediment-water pCO$_2$	high			+++
	low	+++	+++	
Alkali salinity	high			+++
	intermediate	++	+++	++
	moderate	+++	++	
ENVIRONMENT				
Pedogenic calcrete, or alluvial siliciclastic or akosic matrix		+++	++	
	carbonate or mafic matrix	++	+++	
Closed basin lacustrine groundwater input dominant			+++	++
	surface runoff dominant	?		+++
	hypersaline			+++
Marine	lagoon or tidal	+++	++	
	deep sea	++		++

Palygorskite originates in many cases from the transformation of smectite via a dissolution–precipitation process (Galán and Ferrero, 1982; Jones and Galán, 1988, Suarez et al., 1994; Lopez Galindo et al., 1996; Sánchez and Galán, 1995; and Chen et al., 2004). In some shallow restricted basins with evaporative conditions, fibrous clay minerals, Mg-rich smectites (saponite, stevensite), and kerolite, form as authigenic minerals (Figure 6.2), usually together with sulfate and carbonate minerals (Pozo and Casas, 1999). In a typical playa lake sequence, detrital illite, kaolinite, and smectite dominated at the lake edge; palygorskite with Al^{3+}- smectite occurred near the shore; while sepiolite and other Mg^{2+}-clay minerals were abundant toward the center. Sepiolite, stevensite, and kerolite equilibria depended on salinity (Na^+ concentration), pH, and Mg^{2+} activity (Jones, 1986). High salinity favored stevensite formation, while sepiolite and kerolite were formed in less saline conditions. High Si/Mg ratios favored sepiolite formation. In this environment, at lower pH, smectite can also be transformed into palygorskite (Sanchez and Galán, 1995), whereas at slightly higher pH, sepiolite, amorphous silica, and palygorskite can precipitate. Sepiolite is favored over palygorskite at higher pH in silica-poor solutions (Birsoy, 2002).

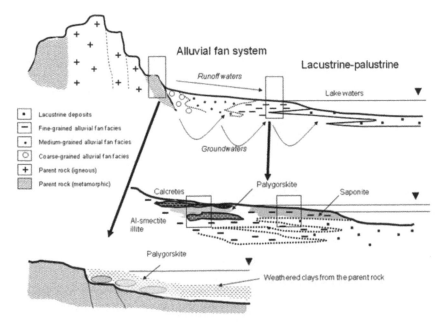

FIGURE 6.1 Sketch of palygorskite deposit formation in continental sedimentary environments. The transformation of Al-bearing clays to palygorskite can take place close to the parent rock supplying Mg^{2+}, $Si(OH)_4$, and Al-rich minerals but also related to marginal alluvial–palustrine–lacustrine environments (After Galán and Pozo, 2011).

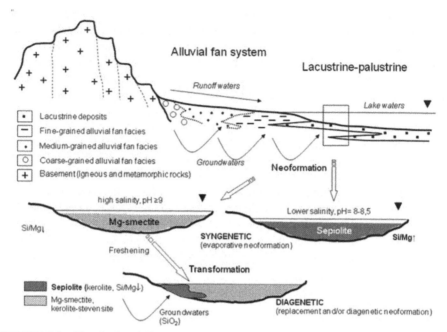

FIGURE 6.2 Sketch of sepiolite deposit formation in continental sedimentary environments. Salinity, pH, and Mg/Si ratio play an important role in the syngenetic formation of sepiolite, kerolite, or stevensite. A drastic change in salinity–pH (freshening) during early diagenesis of Mg- clays favors the intrasedimentary formation of sepiolite (After Galán and Pozo, 2011).

The stability of palygorskite and sepiolite is mainly a function of pH. The transformation of palygorskite to smectite has been suggested by Golden and Dixon (1990). Their work indicated that palygorskite readily converts to smectite above 100°C, although the reaction was sluggish at room temperature (22°C). They showed that at pH conditions near 12, the palygorskite to smectite transformation occurred over a period of several months. The transformation of palygorskite, from the Meigs Member of the Hawthorne Formation, in Southern Georgia, USA, to smectite was analyzed in detail using atomic force microscopy (AFM) and transmission electron microscopy (TEM) techniques by Krekeler et al. (2005). AFM analysis indicated that palygorskite fibers in this horizon were commonly altered to particles with platy morphology, which were interpreted as smectite.

This transformation to smectite also accounts for the very low abundance of palygorskite in Mesozoic and older sediments. An implication of the transformation is that palygorskite deposits may have existed in abundance in the Mesozoic and perhaps even in older sedimentary systems. However, Tertiary sedimentary rocks have different proportions of palygorskite. Orogenic activities, which resulted in the Tethys Sea being cut off during the late Cretaceous period, gave rise to the development of shallow saline lakes during the Tertiary that were chemically favorable for the formation of fibrous clay minerals. In an evaporative environment, these conditions promoted the formation of gypsum and resulted in an increase in the Mg/Ca ratio that account for the authigenic formation of a large amount of palygorskite, particularly in Neogene sediments. The positive correlation between the occurrence of palygorskite/sepiolite and gypsum and carbonates in sediments supports this hypothesis (Yalçin and Bozkaya, 2011; Hojati and Khademi, 2011). The geochemistry of the post-Tethys Sea environment, which was significantly affected by climatic conditions and orogenic events during the Tertiary, controlled the formation of palygorskite and sepiolite. The present-day arid to hyperarid environments prevailing in many areas have caused the preservation of these minerals (Singer, 1984). The deep-sea environment of the Tethys Ocean (geological formations older than Cretaceous) does not seem to be the most suitable for the formation of palygorskite and sepiolite.

Source materials for the neoformation of sepiolite and palygorskite can vary, depending on the regional geology (Figure 6.3). Available source of solute magnesium and silica were carbonate and pyroclastic materials for sepiolite formation in the Southwest Nevada deposits (Jones, 1983). For the Lebrija palygorskite (Seville, Spain), Mg was supplied from surrounding dolomites and diorites, and Si from dissolution of diatomites (Galán and Ferrero, 1982). This last source for Si was also reported by Fleischer (1972) in the Santa Cruz Basin of California for sepiolite. Also Patterson (1974) suggested the same origin for the silica in the palygorskite of Attapulgus, Georgia. Palygorskite from Warren Quarry, Leicestershire, England, was also formed when groundwater penetrated cracks in diorite rocks, causing Si and Mg to be released (Tien, 1973). Chloritic slate alteration was suggested for the Torrejón (Spain) palygorskite deposit (Galán and Castillo, 1984). Weathering of acidic igneous rocks was the origin of Si, Al, and Mg for most sepiolite and Mg-smectites found in the Madrid Basin (Galán and Pozo, 2011). High concentration of Mg and Si released from olivine basalts at the foot of nearby Mt. Kilimanjaro were the source for sepiolite, kerolite, and stevensite in Amboseli, Kenya (Hay and Stoessel, 1984).

Typical mineral associations for these fibrous clay minerals include illite, dioctahedral smectites and minor kaolinite and chlorite as inherited clay minerals, trioctahedral smectites (saponite and stevensite), cristobalite, sulfates and chlorides (gypsum, halite, thenardite) and carbonates (dolomite in particular), and sometimes zeolites also, as neoformed minerals, and detrital quartz (Galán, 1979).

Loughlinite (Na-rich sepiolite) occurs in veins in dolomitic oil shale replacing shortite, northupite, and searlesite (Fahey et al., 1960) in the Green River Formation, Wyoming, USA. Tank (1972) suggested that loughlinite (or early sepiolite) was forming from dissolved Mg and Na with Si derived from devitrification of volcanic rocks, and it is restricted to the saline facies of the Wilkins Peak Member. Loughlinite has been also described in Eskisehir, Turkey (Kadir et al., 2002) in a Neogene lacustrine environment of variable depth and salinity, controlled by synsedimentary step-faulting in arid and semiarid climatic conditions. Under these conditions, volcanic glass and dolomite release Si and Mg, and when an intense evaporation caused the dominance of Mg with Na and K, the formation of loughlinite was favored.

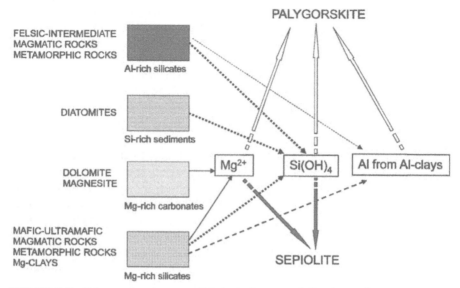

FIGURE 6.3   Scheme showing some lithological sources delivering Mg^{2+} and Si $(OH)_4$ in the sedimentary environment. The presence of inherited Al-rich clays favored palygorskite over sepiolite formation (After Galán and Pozo, 2011).

6.3   MAIN DEPOSITS OF SEPIOLITE AND PALYGORSKITE: LOCATION AND GEOLOGY

The largest presently mined deposits of these minerals are located in Spain for sepiolite and in United States and China for palygorskite. Other mined deposits of sepiolite are in United States, Somalia, and Turkey. For, palygorskite are also mined deposits in Senegal, Greece, Ukraine, and Russia. Minor occurrences, sometimes exploited, are located in India, Australia, and Guatemala (Figure 6.4).

The sepiolite deposits of the Madrid Basin, the most economically important in the world, were mainly formed by direct precipitation in saline lacustrine-palustrine environments (Figure 6.5) of Miocene age with other Mg^{2+}-rich trioctahedral clay minerals (Galán and Pozo, 2011). The reserves are over 20 Mt, but some authors have reported up to 45 Mt (Sánchez Rodriguez et al., 1995), which is 70 percent of the world deposits. The annual production is over 600,000 tons. The sepiolite beds are composed mainly of sepiolite (>80%), smectites (15%), calcite and dolomite (2%), quartz (<2%) and feldspars (<1%).

Sepiolite deposits: 1 Amargosa Deposit, Nevada (USA); 6. Vallecas-Vicálvaro-Yunclillos Deposit and Mara deposits (Spain); 9. Eskisehir deposit (Turkey); 10. El-Bur (Somalia).

Palygorskite deposits: 2. Meigs-Attapulgus-Quincy District, South Georgia-North Florida (USA); 3. Guatemala; 4. Theis and Nianming deposit (Senegal); 5 Torrejón and Bercimuel deposits (Spain); 7. Ventzia, Grevena (Greece); 8. Cherkassy (Ukraine); 11. Andhra Pradesh and Gujarat (India); 12. Guanshan deposit, Anhui (China); 13. Ipswich (Queensland) and Lake Nerrarnyne (Australia).

FIGURE 6.4   Map showing the location of main deposits for sepiolite and palygorskite (After Murray et al. 2011).

Sepiolite has been identified in the sedimentary filling of grabens in the south-western US Basin and Range Physiographic Province. The most interesting and exploitable deposit is located in the Amargosa Desert along the California–Nevada border of late Pliocene to Pleistocene age (Miles, 2011). In this area, there are large deposits of Mg-rich smectites (saponite) and kerolite/stevensite (Khoury et al., 1982, Eberl et al., 1982). Based on textural evidence, Khoury et al. (1982) indicates that the sepiolite could have been formed not only as the result of direct precipitation but also from kerolite/stevensite dissolution. A secondary solution, for instance, groundwater with lower pH and sodium activity than the waters in which the kerolite/stevensite precipitated, may have been the origin of secondary sepiolite. Amargosa sepiolite is presently mined with an annual production of more than 50,000 metric tons.

Ece and Coban (1994) have described the presence of sepiolite in beds and as nodules (meerschaum) in the Eskisehir Basin (Turkey). The bedded sepiolite exhibits different colors (from white to black) ranging from pure sepiolite to sepiolitic dolomite. The deposit is lacustrine and dates from the Miocene, filling a graben of extensional origin (rifting). The sepiolite was formed by direct precipitation in the saline-alkaline waters of the lake, with supersaturation of silica. The environment was alkaline-saline in arid to semiarid climatic conditions with possible wet intervals owing to seasonal fluctuations. The total reserves for meerschaum sepiolite are about 17,000 metric tons (DPT, 2001) with an annual production of up to 40 tons.

The estimate for bedded sepiolite and dolomite-rich sepiolite reserves is around 1–2 million metric tons. The production is about 50,000 metric tons per year.

The Meigs-Attapulgus-Quincy District (Georgia-Florida) is a clay-rich region chiefly of palygorskite (Weaver, 1984). Its importance justifies that this district supplied 66 percent of the production of fuller's earth in the United States. The estimated resources of this area are over 3,000,000 tons. The palygorskite deposits are located in the Gulf Trough-Apalachicola embayment, a northeast-southwest trending linear structure in Southern Georgia and the eastern panhandle of Florida. According to Weaver and Beck (1977), the palygorskite formed mainly by montmorillonite alteration in shallow waters of a perimarine-lagoon environment. The paleoenvironmental interpretation has been confirmed in later, more detailed studies (Krekeler et al., 2004). Salinity, pH, and temperature have been critical factors in the formation of the fibrous clay mineral. The mineral content of the palygorskite beds consists mostly of palygorskite with minor quantities of smectite, sepiolite, kaolinite, quartz, and dolomite. Trace amounts of clinoptilolite occur as small euhedral crystals in association with opal-CT (Zhou, 1996). Annual production is in the order of 1,000,000 metric tons.

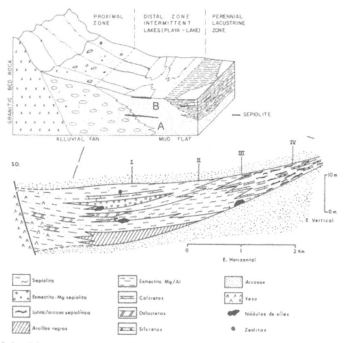

FIGURE 6.5 Diagram showing the Madrid Basin sedimentary environments during Neogene and representative cross-section of the Vicalvaro sepiolite deposit. The main sepiolite beds are labeled A and B. Roman numbers indicate boreholes position (Adapted from Galán and Castillo, 1984 and Sánchez Rodríguez et al., 1995).

The Chinese palygorskite (Guanshan deposit), the second most important deposit, originated from weathering products of basaltic rock that were deposited in an alkaline basin rich in Si and Mg, where they were transformed to palygorskite from smectite (Zhou and Murray, 2011). The palygorskite content is higher than 50 percent (commonly more than 90%). The reserves are about 22 million metric tons.

The Senegal deposits of palygorskite are located in the so-called Basin of Senegal—Mauritania (Africa), where during the Early Eocene formed in an epicontinental marine environment (Wirth, 1968). The exploitable deposits, which range from 2 to 6m in thickness, are located in the western zone of the country, and near to the coast, principally in the areas of Theis (NE of Dakar) and Nianming (in Southern Senegal). In Theis, the palygorskite overlies an aluminum phosphate deposit which is also mined. The major mineral present is palygorskite with minor amount of quartz, dolomite, chert, and sepiolite. The coexistence of Mg-rich palygorskite and Al-rich sepiolite has been observed by García Romero et al. (2007), suggesting a possible epitactic growth. Production is higher than 200,000 metric tons per year.

Another large palygorskite deposit was discovered in Western Macedonia, Greece (Kastritis et al., 2003). The Ventzia Basin is northeast of the city of Grevena. The basin represents a small section of a much larger continental basin that developed during the late Pliocene/Early Pleistocene time. The palygorskite-rich beds contain concentrations ranging from 60 to 95 percent. The reserves of palygorskite are estimated to be about 6 million tons.

The Ukrainian palygorskite deposit is situated along the borders of the Cherkassy and Kiev regions and is located in the central part of the Ukrainian crystalline massif. The age of the palygorskite beds is Tertiary and Quaternary. Fursa (1958) believes that the palygorskite resulted in the replacement of amphiboles in a relatively high alkaline medium. The estimated reserves are 10 million metric tons (Heivilin and Murray, 1994).

6.4 CONCLUDING REMARKS

Deposits of sepiolite and palygorskite clays of economical interest have complex geological origins. One deposit is not like another in mineralogy and geological setting, but genetic integrative models can establish useful patterns for exploration purposes, or assist in identifying the geological processes responsible for their unique properties. The examples cited here present the most important patterns to base further exploration.

KEYWORDS

- Atomic force microscopy (AFM)
- Clay mineral genesis
- Fibrous clay mineral
- Madrid Basin
- Neoformation
- Perimarine
- Stevensite formation
- Transmission electron microscopy (TEM)

REFERENCES

1. Birsoy, R.; *Clays. Clay. Min.* **2002**, *50*, 736–745.
2. Chahi, A.; Duringer, P.; Ais, M.; Bouabdelli, M.; Gauthier-Lafaye, F.; and Fritz, B.; *J. Sed. Res.* **1999**, *69*, 1123.
3. Chen, T. Xu; H. Lu; A. Xu; X., Peng, S.; and Yue, S.; *Sci China. Ser. D. Earth Sci.* **2004**, *47*, 985–994.
4. DPT Mining; Republic of Turkey, State Plan Organization, 8th Five-Year Development Plan, Ankara, **2001**, 1–32.
5. Ece, Ö. I.; *Clay. Clay. Min.* **1998**, *46*, 436–445.
6. Ece, Ö. I.; and Çoban, F.; *Clay. Clay Min.*, **1994**, *42*, 81–92.
7. Eberl, D. D.; Jones, B. F.; and Khoury, H. N.; *Clay. Clay. Min.* **1982**, *30*, 321–326.
8. Fahey, J. J.; Ross, M.; and Axelrod, J. M.; *Am. Min.* **1960**, *45*, 270–281.
9. Fleischer, P.; *Am. Min.* **1972**, *57*, 903–913.
10. Fursa, A. E. In: Sbornik Bentonitouye Glini Ukrainy, Ñor. 5, Kiev, Izd, AN Ukr SSR, **1958**.
11. Galán, E.; and Castillo, A.; Sepiolite-Palygorskite in Spanish Tertiary Basins: Genetical Patterns in Continental Environments. In: Palygorskite-Sepiolite. Occurrences, Genesis and Uses. Eds. Singer, A., Galán, E. Developments in Sedimentology, 37, Elsevier, **1984**: 87–124.
12. Galán, E.; and Ferrero, A.; *Clay. Clay. Min.* **1982**, *30*: 191–199.
13. Galán, E.; and Pozo, M.; Palygorskite and sepiolite deposits in continental environments. Description, genetic patterns and sedimentary settings. In: Developments in Palygorskite-Sepiolite Research. A new outlook of these nanomaterials. Eds. Galán, E., Singer, A.; Development in Clay Science, Elsevier vol. 3, **2011**, 125–173. Elsevier.
14. Galán, E.; The fibrous clay minerals in Spain. In: 8th Conference Clay Mineral. Ed. J. Konta; Petrol. Teplice. University. Carolinae, Prague, **1979**, 239–249.
15. García-Romero, E.; Suárez, M.; Santarén, J.; and Alvarez, A.; *Clays. Clay. Min.* **2007**, *55*, 606–617.
16. Golden, D. C.; and Dixon, J. B.; *Clays. Clay. Min.* **1990**, *38*, 401–408.
17. Harvey, C. C.; and Murray, H. H.; *Appl. Clay. Sci.* 1997, 11 285–310.
18. Hay, R. L.; and Stoessel, R. K.; Sepiolite of the Amboseli Basin of Kenia: a new interpretation. In: Palygorskite-Sepiolite: Occurrences, Genesis and Uses. Eds. Singer, A., Galán, E.; Developments in Sedimentology, Elsevier, **1984**, 37: 125–136.
19. Heivilin, F. G.; and Murray, H. H.; Hormites: palygorskite (attapulgite) and sepiolite. In: Industrial Minerals & Rocks; Griffith, J. Ed., Littleton (Colorado), USA, **1994**; 249–254.

20. Hojati, S.; and Khademi, H.; Genesis and distribution of palygorskite in Iranian soils and sediments. In: Developments in palygorskite-sepiolite research. A new outlook of these nano-materials., Eds. Galán, E., Singer, A., Development in Clay Science, vol. 3, Elsevier, **2011**, 201–218.
21. Jones, B. F.; *Sci. Geol. Min.* **1983**, *72*, 81–92.
22. Jones, B. F.; Clay mineral diagenesis in lacustrine sediments. U.S. Geological Survey Bulletin, **1986**, 1578: 291–300.
23. Jones, B. F.; and Galán, E.; Sepiolite and palygorskite. In: Hydrous Phyllosilicates. Reviews in Mineralogy; Ed. Bailey, S. W.; Mineralogy Society of America, **1988**, 19, 631–674.
24. Kadir, S.; Baş, H.; and Karakaş, Z.; *The. Can. Min.* **2002**, *40*, 1091–1102.
25. Kastritis, D.; Mposkos, E.; Gionis, V.; and Kacandes, G.; The palygorskite and Mg-Fe-smectite clay deposits of the Ventzia Basin, western Macedonia, Greece. In: Mineral Exploration and Sustainable Development. Eds. Eliopoulos et al.; Proceedings of the 7th SGA Meeting. Rotterdam: Mill Press, **2003**.
26. Khoury, H. M.; Eberl, D. D.; and Jones, B. F. *Clays. Clay. Min.* **1982**, *30*, 327–336.
27. Krekeler, M. P. S.; Guggenheim, S.; and Rakovan, J. *Clays. Clay. Min.* **2004**, *52*, 263–274.
28. Krekeler, M. P. S.; Hammerley, E.; Rakovan, J.; and Guggenheim, S.; *Clays. Clay. Min.*, **2005**, *53*, 1, 92–99.
29. López-Galindo, A.; Ben Aboud, A.; Fenoll, P.; and Casas, J.; *Clay. Min.* **1996**, *31*, 33–44.
30. Mackenzie, R. C.; Wilson, M. J.; and Mashhdy, A. S.; Origin of palygorskite in some soils of the Arabian Peninsula. In Palygorskite-Sepiolite. Occurrences, genesis and uses. Eds. Singer, A., Galán, E.; Developments in Sedimentology, 37, Elsevier, **1984**, 177–186.
31. Merriman, R. J.; *Eur. J. Min.* **2005**, *17*, 7–20.
32. Miles, W. J. Amargosa Sepiolite and Saponite: Geology, Mineralogy and Markets. In Developments in Palygorskite-Sepiolite Research. A new outlook of these nanomaterials. Eds. Galán E., Singer A.; Development in Clay Science, vol.3, Elsevier, **2011**, 265–277.
33. Murray, H. H.; Pozo, M.; and Galán, E.; An introduction to palygorskite and sepiolite deposits-location, geology and uses. In Developments in Palygorskite-Sepiolite Research. A new outlook of these nanomaterials. Galán E., Singer A. Eds.; Development in Clay Science, vol. 3, Elsevier, **2011**, 85–99.
34. Murray, H. H.; Applied clay mineralogy. *Dev. Clay. Sci. 2*, **2007**, 59–61
35. Patterson, S. H.; Fuller 's Earth and Other Industrial Mineral Resources of the Meigs-Attapulgus-Quincy District, Georgia and Florida, Prof. Paper 828, U.S. Geological Survey, **1974**, 45 p.
36. Pletsch, T.; Daoudi, L.; Chamley, H.; Deconinck.; and Charroud, M.; *Clay. Min.* **1996**, *31*, 403–416.
37. Pozo, M.; and Casas, J. C.; *Clay. Min.* **1999**, *34*, 395–418.
38. Rodas, M.; Luque, F. J.; Mas, R.; and Garzón, M. G.; *Clay. Min.* **1994**, *29*, 273–285.
39. Sánchez, C.; and Galán, E.; *Clay. Min.* **1995**, *30*, 225–238.
40. Sánchez Rodríguez, A; et al. Libro Blanco de la Minería de la Comunidad de Madrid. *Inst. Tecnológ. Geomin. España Comunidad de Madrid*, **1995**, 286 pp. + 2 maps. Madrid, Spain.
41. Shadfan, H.; and Dixon, J. B.; Occurrence of palygorskite in the soils and rocks of the Jordan Valley. In: Palygorskite-Sepiolite. Occurrences, Genesis and Uses. Eds. Singer, A.; and Galán, E.; Developments in Sedimentology, 37, Elsevier, **1984**, 187–198.
42. Singer, A.; Palygorskite in sediments: Detrital, diagenetic or neoformed–A critical review. Sonderdruck aus der Geologischen Rundschau Band, **1979**, *68*, 996–1008.
43. Singer, A.; Pedogenic palygorskite in the arid environment. In: Palygorskite-Sepiolite. Occurrences, Genesis and Uses. Eds. Singer, A., Galán, E., Developments in Sedimentology, 37, Elsevier, **1984**: 169–177.

44. Suárez, M.; Robert, M.; Elsass, F., and Martín Pozas, J. M.; *Clay. Min.* **1994**, *29*: 255–264.
45. Tank, R. W. *Clay. Min.* **1972**, *9*, 297–308.
46. Thiry, M.; and Pletsch, T.; Palygorskite Clays in Marine Sediments: Records of Extreme Climate. In: Developments in Palygorskite-Sepiolite Research. A new outlook of these nanomaterials. Eds. E. Galán and Singer, A. Development in Clay Science, vol. 3, Elsevier, **2011**, 102–124.
47. Tien, P. L. *Clay Min.* **1973**, *10*, 27–34.
48. Weaver, C. E.; Origin and geologic implications of the palygorskite deposits of SE United States. In: Palygorskite-Sepiolite: Occurrences, Genesis and Uses. Eds. Singer, A.; and Galán, E. Developments in Sedimentology, Elsevier, **1984**, *37*, 39–58.
49. Weaver, C. E.; and Beck, K. C.; Miocene of the S.E. United States: A model for chemical sedimentation in a peri–marine environment. *Sedimen. Geol.*, **1977**, *17*, 1–234.
50. Wirth, L.; Attapulgites du Senegal Ocidental, Rapport No. 26, Laboratoire de Geologie, Faculté des Sciences, Universite de Dakar, **1968**, 55 pp.
51. Yaçin, H.; and Bozkaya, Ö.; Sepiolite-Palygorskite Occurrences in Turkey. In Developments in Palygorskite-Sepiolite Research. A new outlook of these nanomaterials., Eds. Galán, E., Singer, A. Development in Clay Science, vol. 3, Elsevier, **2011**, 175–200.
52. Zhou, H.; Industrial Clay Mineralogy of Palygorskite from Guanshan, Anhui Province, P.R. China, Ph.D. Thesis, Indiana University, **1996**, 196 pp.
53. Zhou, H.; and Murray, H. H.; Overview of Chinese Palygorskite Clay Resources-Their Geology, Mineralogy, Depositional Environment, Applications and Processing. In: Developments in Palygorskite-Sepiolite Research. A new outlook of these nanomaterials. Eds. Galán, E., Singer, A.; Development in Clay Science, vol.3, Elsevier, **2011**, 239–263.

CHAPTER 7

THE MINERALOGY, GEOLOGY, AND MAIN OCCURRENCES OF CHRYSOTILE

SAVERIO FIORE and F. JAVIER HUERTAS

CONTENTS

7.1 INTRODUCTION

Chrysotile has been described as a "wonder" mineral because of its long, ultrathin, durable, flexible, and sometimes woven fibers. This peculiar crystal morphology, together with its chemical composition and crystal structure, determine unique thermal, electrical, and mechanical properties that have enabled chrysotile to be used in a number of industrial products or applications, from plastics to cement, from friction materials to vinyl tiles, and so on (e.g., Ross, 1981; Virta and Mann, 1994; Ross and Virta, 2001). Chrysotile fibers are classified, for commercial use, into fiber length groups, each one with its own subgroups (grade), with the longest fibers assigned to Group 0 and the shortest to Group 7. Each group has a specific use, the longer has been used in textiles, the shorter as a filler in various materials (cement, plastic, vinyl, etc.)

The use of chrysotile for manufacturing industrial products began in the early decades of the twentieth century and its story is reflected in the worldwide consumption of asbestos that has increased from about 175 kilotonne (in the 1920) up to 4.8 megatonne (in the 1980) (Virta, 2006). Production of chrysotile asbestos in 2012 was around 2 kilotonne a year with the main producers being Russia (50%), China (22%), Brazil (15%), Kazakhstan (12%), (U.S. Geological Survey, 2013). USA and Canada were once among the main producers but now the mineral extraction is banned there, as well as in many European countries, because all forms of asbestos have been judged to be carcinogenic by the World Health Organization (WHO, 1988) and International Agency for Research on Cancer (IARC, 1977, 1987). However, it is widely recognized that chrysotile is less dangerous than amphiboles and that its low inhalation does not represent a risk for health although a prolonged exposure can produce health diseases (e.g., Sporn and Roggli, 2004; Bernstein et al., 2013 and references therein). Experimental evidence shows that chrysotile can be degraded by acid fluids such as those present in the lung (e.g., Rozalen et al., 2014). This implies that the genotoxic and carcinogenic potentials of this serpentine mineral are surely enhanced by the presence of amphiboles, and more frequently tremolite asbestos. Chrysotile free from other potentially dangerous minerals, or its analogous synthetic material, may be looked at with interest for new applications as nanotubes.

7.2 MINERALOGY AND GEOLOGY

Chrysotile, as well as the other two minerals forming the serpentine group–antigorite and lizardite-, is the principal mineralogical components of serpentinites, fine grained ultramafic rocks that are exposed in almost all continents, including Antarctica. These hydrous magnesium silicate phases generally form through hydrothermal alteration of ultrabasic rock-forming minerals, such as olivine and ortho/clinopyroxenes. The reaction that leads to the formation of chrysotile, antigorite, and lizardite may be conventionally described as follows:

olivine + water + oxygen + silica(aq) -> serpentine + brucite + magnetite (+talc)

Chrysotile and lizardite crystallize in a similar interval of temperature and pressure ranging from surficial conditions to more than 400°C was documented by synthesis experiments (Evans, 2004) that also proved that lizardite is more stable at low temperatures. Antigorite (+brucite) is more stable at temperatures higher than 300°C thus implying that chrysotile is never the most stable of the three polymorphs. Viti and Mellini (1997) proposed the sequence chrysotile-polygonal serpentine-lizardite as an example of metastable phases due to kinetic effects during crystallization. More recently, experimental and field observations led to the conclusion that competition in crystallization between chrysotile and lizardite was most likely controlled by nucleation and growth rates and by the degree of supersaturation of the circulating fluids, rather than by temperature and pressure (Evans, 2004). Further discussion and information on the conditions governing the formation of serpentine minerals in serpentinites is given in Evans et al. (2013).

Ultramafic lithologies in serpentinite belts are quite varied, reflecting differences in precursor rocks and tectonic settings. Host environments for chrysotile mineralization include variably massive and schistose harzburgite, dunite, wehrlite, peridotite, lherzolite, and pyroxenite. Serpentinites associated with pillow lava and chert (the Steinmann's trinity) are generically referred to as ophiolites. Occurrences of chrysotile are possible in other lithologies, such as dolostones and carbonatites, given the appropriate physicochemical conditions, that is, a high magnesium content in host rocks, and suitable structural and thermal conditions.

Deposits of chrysotile are found in three geological settings (Ross, 1984): Type I – alpine type ultramafic rocks (including ophiolites); Type II—stratiform ultramafic intrusions, and Type III—serpentinized limestone. Type I deposits probably account for more than 85 percent of the asbestos mined and are found in Canada (Quebec) and Russia (Urals). Type II deposits are those in South Africa, Swaziland, and Zimbabwe. Type 3 are found in USA (Arizona) and South Africa (Transvaal).

Serpentinites occur as metric (or kilometric) lenses or bodies contained in ancient continental belts as well as large massifs mainly located in suture zones associated with the closure of paleo-oceans. Much more common are serpentinites occurring in submarine environments since 20–25 percent of the oceanic floor is composed of mantle-derived ultramafic rocks (Cannat et al., 2010). On the seafloor, serpentinites are associated with mid-ocean spreading ridges, subduction zones, and tectonically active areas involving pieces of the mantle. Active mud volcanoes and seamounts of serpentines were observed in the shallow forearc region of the Mariana-Izu-Bonin system (Fryer et al, 1985).

Chrysotile in serpentinites may occur as local transformation of primary olivine and pyroxenes. However, chrysotile frequently precipitates from circulating hot fluids that fill cracks, shear planes and voids originated by tectonic deformations and volumetric expansions usually, but not always, associated with the transformation of primary minerals. The result is a network of irregular white veins having variable thickness (from a few hundred micrometers up to some centimeters) and length (from few millimeters to meters). Chrysotile may grow in veins and seams as *cross fibers* (fiber axes at right angles to the seam or vein), *slip fibers* (fiber axes parallel to the seam) or *mass fibers* (fiber parallel or divergent), but it also occurs in serrate veinlets formed by replacement (Evans, 2004). Formation of fibers by the crack seal mechanism (Ramsay, 1980) is described by Andreani et al. (2004).

Formation of veins usually occurs at different times and under distinct ambient conditions and their detailed study (orientation, thickness, and composition of filling minerals) may provide significant information on the physicochemical conditions existing during their formation (shear planes, deformations, folds, faults, and fluid chemical composition) useful for geological reconstruction at local and regional scale.

Depending on the genetic environment, a number of minerals may be associated with chrysotile. In deposits from ultramafic rocks, iron is ubiquitous and may form magnetite with an unusual fibrous morphology (Perry, 1930; Gahlan et al., 2006). Brucite is also common and it can be found as plates or fibers (nemalite). Altered micas and feldspars, amphiboles, talc, calcite, zeolites, and many other mineral phases, in variable concentrations, have been reported in a number of studies (e.g., Addison and Davies, 1990; Langer and Nolan, 1994; Gibbs, 1998). Single fibers of chrysotile have a white color and a silky appearance (it is also called "white asbestos"), but when they form aggregates, their color may be very pale green or yellow or gray depending on the presence of other mineral phases and impurities (e.g., magnetite, lizardite, brucite, chlorite, talc, pyroaurite, amorphous phases) or alteration products (Rozalen and Huertas, 2013). The fibers have variable lengths, from a few micrometers to several centimeters; bundles of fibers 20–30 cm long can be observed in many museums around the world.

Chrysotile occurring in most serpentinites and exploited deposits contain several volume percent of this mineral. It was intensively mined some decades ago; production of chrysotile asbestos between 1931 and 1999 was 145–158 million tons (Ross and Virta, 2001). The most important exploitable chrysotile outcrops are located in Quebec, Canada; western Alps, Italy; northern India; Turkey; California, USA; Oman; Philippines; Portugal; the central and southern Ural Mountains of Russia-Kazakhstan; South Africa; Swaziland; Rhodesia; China; Brazil; Zimbabwe; Greece; the Troodos Massif, Cyprus; New Caledonia; Corsica. A compilation of continental outcrops of serpentinites is given in Coleman (1977).

Nowadays mining and use of chrysotile, being an asbestos mineral, is banned in many countries. Nevertheless, it is likely that in the future exploitation of chrysotile-containing rocks could be reconsidered in an environmental friendly way since intensely weathered serpentinites are the main source of nickel and cobalt (Berger et al., 2011).

7.3   OCCURRENCES

As stated above, the most significant chrysotile asbestos deposits are found in a few countries: Russia, China, Brazil, South Africa, USA, and Canada. These last two countries have banned the extraction of asbestos. Russia, Brazil, and China are important current sources of chrysotile and descriptions of deposits in these countries may be useful for currently available sources of chrysotile.

Russia. The most important chrysotile asbestos deposits in Russia occur in gabbros, peridotites, and pyroxenites associated with ophiolites of the Ural Mountains (Petrov and Znamensky, 1981; Spadea and Scarrow, 2000). The Uralasbest Mine is the world largest reserves of chrysotile asbestos and extracts nearly half a million tons annually.

China. China has 467 asbestos deposits, the vast majority (95%) of which contains chrysotile asbestos (Lu, 1998). Eleven of the 14 major commercial deposits are found in serpentinized ultramafic rocks, while the others are in dolomite-hosted and occur as groups of tabular and lenticular ore bodies lying along the bedding plane. Tremolite is associated with chrysotile from every deposit, but anthophyllite is found only in dolomite-hosted chrysotile (from Tossavainen et al., 2001).

Brazil. The most important Brazilian deposit of asbestos occurs within serpentinites of the Cana Brava Complex, an anorogenic stratiform mafic-ultramafic body (Oliveira et al., 2004). Chrysotile occurs in extensional fractures and in veinlets and it is very pure. The fibers present in the Cana Brava mine have variable lengths ranging from Grade 3 to Grade 7 (from Bernstein et al., 2006). Brazil is currently the world's third largest chrysotile producer.

South Africa. There are many deposits of chrysotile in South Africa; the most important are hosted in ultramafic intrusions within the Swartkoppies Formation (Hart, 1988; McCulloch, 2003). Chrysotile production peaked in 1989 at about 115 kilotonnes and was 6,000 kilotonnes in 2003 (from Virta 2006, p. 9).

United States. In USA chrysotile is hosted in both serpentinites and carbonates. Californian serpentinites consists of a large amount of highly sheared and pulverized rock fragments and powders, as well as boulders of partially altered serpentine-rich rock. The average New Idria rock contains 5–15 volume percent of chrysotile (Mumpton and Thompson, 1975), but rock in the area mined by the King City Asbestos Corporation contains up to 60 percent short-fibers.

The largest serpentinized carbonate chrysotile deposits are in Arizona. They formed by contact metamorphism with diabase intrusions (Sliver, 1978). The chrysotile grows as veins and masses (Lindberg, 1989) and it contains about half as much iron as Canadian (Quebec) chrysotile, important for certain electrical applications (Pierce and Garcia, 1983). The individual chrysotile veins vary from microscopic in size to a maximum of 36 cm thick, with most less than 5 cm thick (Wilson, 1928; Stewart, 1955, 1956).

7.4 FINAL REMARKS

The recent discoveries of seafloor serpentinites and the unique mechanical and physical properties of chrysotile have given rise to a renewed scientific interest for serpentine minerals, more specifically for chrysotile, that includes earth and environmental sciences (e.g., Guillot and Hattori, 2013; Evans et al., 2013; Power et al., 2013), but also geomicrobiology (e.g., Schulte et al. 2006; McCollom and Seewald, 2013) and materials science (Lourenço et al., 2012; Korytkovaa et al. 2013).

ACKNOWLEDGMENTS

The authors are indebted to John Keeling for comments, suggestions, and reviews. The editorial assistance of Pooria Pasbakhsh and Jock Churchman is greatly appreciated.

KEYWORDS

- **Cross-fibers**
- **Geomicrobiology**
- **Serpentinites**
- **Slip-fibers**
- **Tremolite**
- **Ultramafic**
- **White asbestos**

REFERENCES

1. Addison, J.; and Davies, S. T.; *Ann. Occup. Hyg.* **1990**, *34*, 159–175.
2. Andreani, M.; Baronnet, A.; Boullier, A-M.; and Gratier, J-P.; *Eur. J. Min.* **2004**, *16*, 585–595.
3. Berger, V. I.; Singer, D. A.; Bliss, J. D.; and Moring, B. C.; *Ni Co Laterite Deposits of the World Database and Grade and Tonnage Models.* USGS Open-File Report, 2011–1058, **2011**; 26 pp.
4. Bernstein, D.;et al.; *Crit. Rev. Toxicol.* **2013**, *43*, 154–183.
5. Bernstein, D. M.; Rogers, R.; Smith, P.; and Chevalier, J.; *Inhal. Toxicol.* **2006**, *18*, 313–32.
6. Cannat, M.; Fontaine, F.; and Escartín, J.; In: *Diversity of hydrothermal systems on slow spreading ocean ridges*; Eds. Rona P. A., Devey, C. W., Dyment J., and Murton B. J.; Geophysical Monograph 188; American Geophysical Union: Washington D.C., USA, **2010**; 241–264.
7. Coleman, R. G;. Ophiolites: Ancient Oceanic Lithosphere? Berlin: Springer-Verlag; **1977**; 229 pp.
8. Coleman, R. G.; Field guide book to New Idria area, California, In: 14th General Meeting, International Mineralogical Association, July, **1986**; Stanford University: Stanford, California, **1986**; 36 pp.
9. Cressey, G.; Cressey, B.; Wicks, F. J.; and Yada, K.; *Min. Magaz.* **2010**, *74*, 29–37.

10. Evans, B. W. *Int. Geol. Rev.* **2004**, *46*, 479–506.
11. Evans, B. W. Hattori, K.; and Baronnet, A. *Elements.* **2013**, *9*, 99–106.
12. Fryer, P.; Ambos, E. L.; and Hussong, D. M. *Geology.* **1985**, *13*, 774–777.
13. Gahlan, H. A.; Aria, S.; Ahmed, A. H.; Ishida, Y.; Abdel-Aziz, Y. M.; and Rahimi, A.; *J. Afr. Earth. Sci.* **2006**, *47*, 318–330.
14. Gibbs, G. In: *Chrysotile Asbestos*; Environmental Health Criteria 203; World Health Organization: Geneva, **1998**; 10–22.
15. Glen, R. A.; and Butt, B. C.; *Econ. Geol.* **1981**, *76*, 1153–1169.
16. Guillot, S.; Hattori, K.; *Elements.* **2013**, *9*, 85–98.
17. Hart, H. P.; *J. South. Afr. Inst. Min. Metal.* **1988**, *88*, 185–198.
18. Hora, Z. D.; Ultramafic-hosted chrysotile asbestos. In: British Columbia Geological Survey – Geological Fieldwork; British Columbia Ministry of Employment and Investment, Paper 1998-1; 1997; 24K-1to 24K-4.
19. IARC.; *Asbestos.* IARC Monographs on the Evaluation of Carcinogenic Risk of Chemicals to Humans 14; International Agency for Research on Cancer: Lyon, France, **1977**.
20. IARC.; In: Overall evaluations of carcinogenicity: An updating of IARC monographs volumes 1 to 42. IARC Monographs on the Evaluation of Carcinogenic Risk of Chemicals to Humans, Supplement 7. International Agency for Research on Cancer: Lyon, France, **1987**; 106–116.
21. Karkanas, P; *Ore. Geol. Rev.* **1995**, *10*, 19–29.
22. Korytkova, E. N.; Semyashkina, M. P.; Maslennikova, T. P.; Pivovarova, L. N.; Al'myashev, V. I.; and Ugolkov V. L.; *Glass. Phys. Chem.* **2013**, *39*, 294–300.
23. Lindberg, P. A.; In: Geologic Evolution of Arizona. Eds. Jenney, J. P., Reynolds, S. J.; Digest 17; Arizona Tucson: Geological Society; **1989**; 187–210.
24. Lourenço, M. P.; de Oliveira, C.; Oliveira, A. F.; Guimarães, L.; Duarte, H. A.; *J. Phys. Chem C.* **2012**, *116*, 9405 9411.
25. Lu, W. Chinese Industria Minerals. Surrey, UK: Industrial Minerals Information Ltd; **1998**.
26. McCollom, T. M.; and Seewald, J. S.; *Elements.* **2013**, *9*, 129–134.
27. McCulloch, J.; *Int. J. Occup. Environ. Health.* **2003**, *9*, 230–235.
28. Mumpton, F. A.; and Thompson, C. S.; *Clays Clay Min.* **1975**, *23*, 131–143.
29. O'Hanley, D. S.; *Geology.* **2012**, *20*, 705–708.
30. O'Hanley, D. S.; and Offler, R.; *Canad. Min.* **1992**, *30*, 1113–1126.
31. O'Hanley, D. S.; *Canad. J. Earth. Sci.* **1987**, *24*, 1–9.
32. O'Hanley, D. S.; *Econ. Geol.* **1988**, *83*, 256–265.
33. O'Hanley, D. S.; *Canad. Min.* **1991**, *29*, 21–35.
34. O'Hanley, D. S.; *Serpentinites Records of Tectonic and Petrological History.* New York, N.Y: Oxford University Press, **1996**.
35. Oliveira, M. C. B.; Coutinho, J. M. V.; Bagatin, E.; and Kitamura, S. In: *Applied Mineralogy: Developments in Science and Technology 200.* In: Proceedings of the VII International Council for Applied Mineralogy, *Águas de Lindoia, Brazil, September 19–24*, 2004; Pecchio, M.; Andrade, F. R. D.; D›Agostino, L. Z.; Kahn, H.; Sant'agostino, L. M.; Tassinari, M. M. M. L. Eds.; ICAM, Sao Paulo, Brazil, 2004; 1, 443–445.
36. Perry, E. L.; *Am. J. Sci.* **1930**, *20*, 177–179
37. Petrov, V.P.; and Znamensky, V.S. In: *Geology of Asbestos Deposits*; Riordon, P. H., Ed.; Society of Mining, Metallurgical, and Petroleum Engineers, Inc.: New York, N.Y., **1981**; 45–52.
38. Pierce, H. W.; and Garcia, M. M. *Asbestos – towards a new perspective*; Fieldnotes 13; Tucson, AZ: Arizona Bureau of Geology and Mineral Technology, **1983**, 13, 1–8.

39. Power, I. M.; and Wilson, S. A.; Dipple, G. M. *Elements* **2013**, *9*, 115–121.
40. Ramsay, J. G. *Nature.* **1980**, *284*, 135–139.
41. Ross, M.; In: *Definitions for Asbestos and Other Health related Silicates.* Ed. Levadie, B.; Philadelphia : American Society for Testing Materials: **1984**; 51–104.
42. Ross, M. In: *Amphiboles and other Hydrous Pyriboles   Mineralogy.* Ed. Veblen, D.R.; *Rev Mineral* 9A; Mineralogical Society of America: Chelsea, Michigan, **1981**; 279–324.
43. Ross, M.; and Nolan, R. P.; In: *Ophiolite concept and the evolution of geological thought*; Eds. Dilek, Y., Newcomb, S.; Special Paper 373; Geological Society of America: Boulder, CO, **2003**; 447–470.
44. Ross, M.; and Virta, R. L.; In: *The health effects of chrysotile asbestos—Contribution of Science to Risk Management Decisions.* Eds. Nolan, R.P., Langer, A.M., Ross, M., Wicks, F.J., Martin, R.F.; The Canadian Mineralogist Special Publication 5; Ottawa, Canada: Mineralogical Association of Canada: **2001**; 79–88.
45. Rozalen, M.; and Huertas, F. J.; *Chem. Geol.* **2013**, *352*, 134–142
46. Rozalen, M.; Ramos, M. E.; Fiore, S.; Gervilla, F.; and Huertas, F. J.; *Am. Min.* **2104**, in press.
47. Schulte, M.; Blake, D.; Hoehler, T.; and McCollom, T.; *Astrobiology.* **2006**, *6*, 364–376.
48. Silver, L. T.; In: *Land of Cochise (Southeastern Arizona)*; Eds. Callender, J. F., Wilt, J., Clemons, R. E., James, H. L.; Fall Field Conference Guidebook 29; Socorro, NM: New Mexico Geological Society, **1978**; 157–163.
49. Spadea, P.; and Scarrow, J. H.; Geological Society of America Special Paper **2000**, *349*, 461–472.
50. Sporn, T. A.; and Roggli V. L.; In: *Pathology of Asbestos Associated Diseases*, 2nd edition; Eds. Roggli, V. L., Oury, T. D., Sporn, T. A.; New York, USA: Springer-Verlag, **2004**; 104–168.
51. Stewart, L. A.; *Chrysotile asbestos Deposits of Arizona*; U.S. Bureau of Mines Information Circular 7706; US Department of Interior: Washington DC, **1955**.
52. Stewart, L. A.; *Chrysotile asbestos deposits of Arizona (Supplement to Information Cir cular 7706)*; U.S. Bureau of Mines Information Circular 7745; US Department of Interior: Washington DC, **1956**.
53. Tossavainen, A.; Kotilainen, M.; Takahashi, K.; Pan, G.; and Vanhala, E.; *Ann. Occup. Hyg.* **2001**, *45*, 145–152.
54. U.S. Geological Survey; *Min. Comm. Summ. 2013*; Reston, VA: U.S. Geological Survey, **2013**.
55. Virta, R. L.; *Worldwide Asbestos Supply and Consumption Trends from 1900 through 2003*; Circular 1298; U.S. Reston, VA: Geological Survey: 2006. http://pubs.usgs.gov/circ/2006/1298/c1298.pdf
56. Virta, R. L.; and Mann, E. L.; In: *Industrial Minerals and Rocks*; Ed. Carr, D. D.; Littleton, Colorado: Society for Mining, Metallurgy and Exploration, Inc. **1994**; 97–124.
57. Virta, R. L.; In: *Industrial Minerals and Rocks*, 7th ed.; Eds. Kogel, J.E., Trivedi, N.C., Barker, J.M., Krukowski, S.T., Society for Mining, Metallurgy, and Exploration, Inc.: Littleton, Colorado, **2006**; 195–217.
58. Viti, C.; and Mellini, M.; *Eur. J. Min.* **1997**, *9*, 585–596.
59. WHO. *Man made mineral fibres*; Environmental Health Criteria 77; World Health Organization: Geneve, **1988**.
60. Wilson, E.D. *Asbestos Deposits of Arizona*; Arizona Bureau of Mines Bulletin 126; University of Arizona: Tucson, USA, **1928**.

PART IV
APPLICATIONS OF NATURAL MINERAL NANOTUBES IN NANOCOMPOSITES

CHAPTER 8

POLYMER NANOCOMPOSITES REINFORCED BY HALLOYSITE NANOTUBES: A REVIEW

VAHDAT VAHEDI and POORIA PASBAKHSH

CONTENTS

8.1 INTRODUCTION

A review of the recent studies on polymer halloysite nanocomposites is presented in this chapter. Halloysite nanotubes (HNTs) are naturally occurring clay minerals which predominantly exist in nanotubular form. Their unique structure and properties have drawn interest to use them for new high-tech applications including polymer based nanocomposites. Effects of halloysite on a wide range of polymer matrixes including thermoplastics, rubbers, and thermosets have been studied in this context. Different aspects of preparation process, morphology and microstructure, key mechanical properties, and structure-property relationship of each polymer/halloysite nanocomposites have been discussed. In general, HNTs have easier dispersion and preparation process compared to other nanofillers such as carbon nanotubes (CNTs) and montmorrilonite (MMT); they cause considerable improvement in mechanical and thermal properties of polymer matrixes; and they have reasonable price and availability, which makes them promising nanofiller for polymer industry.

Incorporation of fillers into polymeric materials in order to improve their properties and/or lowering the cost has been studied since 1850 (Seymour, 1990). Conventional fillers used in industry usually have particles of micron size (Tsou and Waddell, 2002), while with the developments in the area of nanotechnology, researchers managed to produce and incorporate nanofillers into polymers to prepare nanocomposites. From the first time that the Toyota research group reported an exceptional increase in mechanical properties of polyamides (PA) by incorporation of nanoclays, various nanofillers with different shapes and morphologies such as layered silicates (Montmorillonite) (Sinha Ray and Okamoto, 2003; Pavlidou and Papaspyrides, 2008), CNTs (Breuer and Sundararaj, 2004), and nanosilica (Zou et al., 2008) have been used to reinforce polymer matrixes. Remarkable increases in strength, modulus, heat resistance, flammability, and barrier properties at low filler loading were advantages that attracted great interest in this field both in industry and academia. Despite these advantages, high cost, sophisticated synthesis and modification procedures and complicated mixing processes with polymers are drawbacks that restricted the usage of the nanofillers (Sinha Ray and Okamoto, 2003; Breuer and Sundararaj, 2004; Crosby and Lee, 2007; Hussain, 2006; Pavlidou and Papaspyrides, 2008; Sengupta et al., 2007).

Recently, HNTs have become a matter of interest for researchers in the field of nanotechnology. Halloysites are natural aluminosilicate which are formed in nanotubular shapes. There are abundant deposits of HNTs in all over the world such as USA, New Zealand, Turkey, Australia, and China. Historically, halloysites have been used to produce the highest-quality tableware in the manufacturing of bone china, fine china, and porcelain products (Wilson, 2004; Joussein et al., 2005), but unique properties of halloysites have drawn attention to find new potential applications.

HNTs can be used in high-tech applications such as nanocontainers (Shchukin et al., 2008; Shchukin and Möhwald, 2007), sustained release (Qi et al., 2010;

Levis and Deasy, 2003), cosmetics (Suh et al., 2011), nanoreactors (Shchukin et al., 2005; Tierrablanca et al., 2010), and catalyst support (Wang et al., 2011; Zatta et al., 2011; Liu and Zhao, 2009). HNTs can be coated with metallic and other substances to achieve a wide variety of electrical, chemical, and physical properties, ideal for use in electronic fabrication, radiation absorbing materials (RAM), and high-tech ceramic composite applications (Lvov et al., 2008; Baral et al., 1993; Cavallaro et al., 2011; Liu et al., 2011c; Cao et al., 2009; Zhang and Yang, 2012). Halloysites can be used as molecular sieves in applications such as separation of liquids and gaseous mixtures, water purification in refining industries, and the remediation of acid mine drainage (Lvov et al., 2008; Li et al., 2011; Liu et al., 2011c). Halloysites can act as controlled- or sustained-released capsules, and can be filled with additives such as paints and sealants, lubricants, herbicides, pest repellents, household, food and personal products, cosmetics, pharmaceuticals, and other agents that could benefit from a controlled release (Lvov et al., 2008; Levis and Deasy, 2002; Shi et al., 2011; Lvov and Price, 2008; Suh et al., 2011; Shchukin et al., 2008; Wei et al., 2012). Halloysite can also be used as filler in polymer composite industry to reinforce various plastics and rubbers (Lecouvet et al., 2011b; Tang et al., 2012; Ismail and Shaari, 2010).

Recent studies showed that incorporation of HNTs can cause a remarkable increase in mechanical and thermal performance of polymeric matrixes such as Polypropylene (PP) (Liu et al., 2008a, b, 2009a, b, c, 2010 Ning et al., 2007; Du et al., 2009; Mingliang Du, 2008; Du et al., 2006b; Prashantha et al., 2011), PA (Handge et al., 2010; Guo et al., 2009b; Hedicke-Höchstötter et al., 2009; Guo et al., 2009c; Lecouvet et al., 2011a), Ethylene propylene diene monomer (EPDM) (Ismail and Shaari, 2010; Ismail et al., 2008), Styrene-butadiene rubber (SBR) (Du et al., 2008; Jia et al., 2009a; Zhixin et al., 2010; Guo et al., 2008; Guo et al., 2009a), Epoxy (Tang et al., 2011; Ye et al., 2011a; Deng et al., 2009; Ye et al., 2007; Ye et al., 2011b; Liu et al., 2007a, c) etc. Two major deposits, Imerys' Matauri Bay openpit operations in Northland (New Zealand) and the Dragon Mine in Utah (USA), currently owned by Applied Minerals Inc., are the source of halloysites for the researchers and industries and recently they have been used by Naturalnano Inc. as a new type of additive to prepare the first masterbatch of PP/HNTs nanocomposites. In this chapter, the preparation of polymer/HNTs nanocomposites and the effects of the interesting nanotubular mineral on the properties of different types of polymer composites are reviewed and discussed.

8.2 POLYMER/HNTS NANOCOMPOSITES

One of the easiest methods to achieve a nanostructured polymeric composite is to incorporate nano-sized fillers into the polymer matrix. However in this matter, the key challenge is to achieve a well-dispersed structure of nanofillers in the polymeric matrix. The dispersion of fillers in polymers generally depends on the structure of filler, structure of the polymer, and the mixing method. Different methods have been used for making polymeric nanocomposites including melt mixing, solution blending, and in situ polymerization. Well-dispersed nanofillers provide much higher surface to volume ratios, compared to conventional fillers. So incorporation of small amounts of nano-sizes particles can enhance the properties of polymer matrixes dramatically. Recently, HNTs were found to show good performance as a new kind of nanofiller for reinforcing of the polymers due to their unique features such as:

- HNTs are naturally occurring, readily obtainable minerals. They are much cheaper in contrast with the other nano-sized fillers such as MMTs, CNTs, and nanosilica. Comparison of HNTs' price with the other fillers and polymers is given in Table 8.1.
- HNTs do not need exfoliation and can be easily processes due to their discrete nanotubular shape. The tubular structure of HNTs resembles that of multiwalled CNTs. Therefore, HNTs may have the potential to be used as a cheap alternative to the expensive CNTs. The aspect ratio of halloysite can vary from 8 to 50 (Pasbakhsh et al., 2013). This aspect ratio gives a large amount of filler–polymer interactions and makes HNTs a potential competitor to silicate fillers.
- HNTs have simple modification process compared to other nanoparticles due to their unique tubular shape and crystalline structure. They can be functionalized on both internal and external surfaces to increase their performance.
- A large space inside the HNTs' lumen makes it possible to encapsulate the chemical agents for variety of applications like self-healing, sustained release, and drug delivery.
- HNTs are nontoxic, environment-friendly, and biocompatible. Regarding exposure and environmental risk issues associated with nanomaterials, HNTs have competitive advantages over other nanomaterials such as CNTs (Shi et al., 2011; Vergaro et al., 2010).

TABLE 8.1   The comparison between the price and availability of fillers.

Type of Filler	Price ($ per Kg)
HNTs	2–20
MMT	35–200
CNT[a]	400–1,000
Carbon Black	1.1
Silica	2.2–4.2

TABLE 8.1   *(Continued)*

Type of Filler	Price ($ per Kg)
Type of Filler	Price ($ per Kg)
Calcium carbonate	0.15–13.64
PP	1.49–2.35
PE	1.38–2.42
PVC	1.21–2.59
PA	1.0–7.0
NR	1–5
SBR	1.5–1.98

[a]CNT is available in grams; however, other fillers are available in thousands of tons

Effects of HNTs on properties of thermoplastics, thermosets and elastomers—such as PP, PA, epoxy, natural rubber (NR), SBR, EPDM, and so on—was studied by various researchers (Ning et al., 2007; Hedicke-Höchstötter et al., 2009; Ismail et al., 2008; Lei et al., 2011; Ye et al., 2007). Melt mixing method was mainly used to prepare polymer/HNTs nanocomposites, while in some cases such as polylactic acid (PLA) and polyvinyl alcohol (PVA), polystyrene etc., solution casting and in situ polymerization were also successfully utilized (Lin et al., 2011; Liu et al., 2007b). Studies showed that HNTs could be dispersed very well in different types of polymer matrixes even in their unmodified state; although at high loading HNTs, agglomerates may form. Due to this uniform dispersion HNTs can effectively enhance the physical and mechanical properties of polymers. Effects of HNTs on mechanical properties of different polymers are shown in Table 8.2. HNTs could establish various morphologies in different polymer matrixes (Table 8.3 and Figure 8.1), which is influential on the overall properties of final nanocomposites and have been studied by various researchers.

Natural Mineral Nanotubes

TABLE 8.2 Effects of HNTs on mechanical properties of different polymer nanocomposites.

Matrix/HNTs Content (phr)	Modifier/Content (phr)	Tensile Modulus	Changes in properties (%)					Reference
			Tensile Strength	Elongation at Break	Flexural Modulus	Flexural Strength	Impact Strength	
PP/30	–	–	–3	–	+43	+10 2	–27	Liu et al (2008b)
PP/30	BBT/10	–	+16 8	–	+127 4	+46 9	–112	Liu et al (2008b)
PP/30	–	–	+3	–	+21	+71	–17.9	Liu et al (2009c)
PP/30	CBS/3	–	+21	–	+128 1	+44 4	–31	Liu et al (2009c)
PP/30	–	–	–1	–	+41	+31 2	–28.2	Liu et al (2008a)
PP/30	EPB /3	–	+5 3	–	+109 1	+62 2	–34.7	Liu et al (2008a)
PP/10	–	–	–1.6	–	+13 9	+41	+13 0	Du et al (2006b)
PP/10	PP	–	+44	–	+73	+32 17	+33 9	Du et al (2006b)
PP/6	–	+31	+28 6	–22	+38	+16 5	+66 8	Prashantha et al (2011)
PP/6	QM	+39 8	+35 7	–20.5	+47 6	+19 4	+76 7	Prashantha et al (2011)
PE/40	–	–	+37 7	–	+217 5	+132 9	–78.8	Jia et al (2009b)
PE/40	Graft PE/ 5	–	+55 1	–	+315	+175 6	–53.2	Jia et al (2009b)
PA/2	–	+26 3	+12 2	+88 2	–	–	–	Hedicke-Höchstötter et al (2009)
PA/5	–	+31 4	+14 4	–2.8	–	–	–	Hedicke-Höchstötter et al (2009)
PA/5	MAPTS	–	+15 5	–	+33 3	+21 5	+11 9	Guo et al (2009c)
PA/10	MAPTS	–	+20 7	–	+56 0	+28 5	+6 7	Guo et al (2009c)
PVC/10	–	–	+41	–	+40 2	+9 4	+48	Liu et al (2011a)
PVC/5	PMMA	–	+10	–	+42	+12 7	+146 1	Liu et al (2011b)
Epoxy/2 3 wt %	–	–	–	–	+3 6	+10	+400	Ye et al (2007)
Epoxy/10	–	+3 1	+4	–6.8	+17 7	+8 3	+50 (K_IC)	Deng et al (2008)
Epoxy/10 Ball mill mixed	–	+5 5	+6 6	0 0	–	–	+50 (K_IC)	Deng et al (2009)
Epoxy/10 Ball mill mixed	Potassium acetate (PA)	+11 1	+4 6	–2.3	–	–	+51(K_IC)	Deng et al (2009)

TABLE 8.2 (Continued)

Matrix/HNTs Content (phr)	Modifier/Content (phr)	Changes in properties (%)						Reference
		Tensile Modulus	Tensile Strength	Elongation at Break	Flexural Modulus	Flexural Strength	Impact Strength	
Epoxy/10 Ball mill mixed	N-(β-aminoethyl)-ɤ-aminopropyl-trimethoxysilane	+7 6	+5 2	+2 3	–	–	+40 2(K_{IC})	Deng et al (2009)
Epoxy/10 Ball mill mixed	CTAC	+8 9	-22.5	-59.1	–	–	+32 6(K_{IC})	Deng et al (2009)
Epoxy/10 Ball mill mixed	PPA	+14 8	0 0	+0 6	–	–	+78 3(K_{IC})	Targ et al (2011)
NR/10	-	+37 5(M300%)	+28 5	-15.1	–	–	-	Rooj et al (2010)
NR /10	Si69	+94 6(M300%)	+7 1	-29.8	–	–	-	Rooj et al (2010)
EPDM/100	-	+179 4(M100%)	+874 3	+305	–	–	-	Ismail et al (2009)
EPDM/100	MA-gr-EPDM compatibilizer/20	+333 6(M100%)	+763 6	+133 7	–	–	–	Pasbakhsh et al (2009a)
EPDM/30	ɤ-MPS	+72 9(M100%)	+264 4	+59 6	–	–	–	Pasbakhsh et al (2010)
SBR/40	SA/12	+218(M300%)	+311 4	+15 7	–	–	–	Guo et al (2009a)
SBR/40	MAA/12	+190 4(M300%)	+265 9	+42 2	–	–	–	Guo et al (2008)
SBR/40	RH complex/2 4	+282 7(M300%)	+250	-0.03	–	–	–	Jia et al (2009a)
SBR/40	MimMP/4	+115 7(M300%)	+70 3	-11.9	–	–	–	Lei et al (2011)
SBR/40	BMimMS/3	+115 7(M300%)	+109 3	-7.3	–	–	–	Lei et al (2011)
xSBR/30	-	+256 8(M100%)	+53	-25.8	–	–	–	Du et al (2008)
PLGA/1 wt % (Electrospun nano-fiberous mat)	-	+22 9%	+57	+19 4	–	–	–	Qi et al (2010)
PGS/20	-	+85	+129	+106	–	–	–	Chen et al (2011)

TABLE 8.3   Different morphologies of HNTs nanocomposites.

Morphology	Polymer Matrix	Effect	Micrographs
Uniform dispersion, Separate tubes at low HNTs loading (<30 phr)	PP, PVA, Epoxy, NR, SBR, EPDM, etc.	Increased mechanical properties	Figure 8.1a
HNTs agglomerates (>30 phr)	PP, PVA, Epoxy, NR, SBR, EPDM, etc.-at high loading	Lower mechanical properties compared to separately dispersed HNTs	Figure 8.1b
HNTs rich/polymer rich regions at low and high modified-HNTs laoding	Epoxy, EPDM	Increased fracture toughness via crack arrest and deflection	Figure 8.1c
Zigzag network at high HNTs loading (>50 phr)	EPDM	Filler networking increased mechanical properties	Figure 8.1d
Fibril like	PP (BBT, EPB Modifier)	Increased crystallinity and mechanical properties	Figure 8.1e

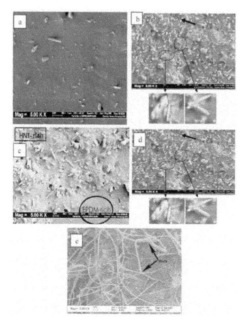

FIGURE 8.1   Different morphologies of HNTs nanocomposites
(a) Separate nanotubes (Pasbakhsh et al., 2010); (b) HNTs agglomerates (Guo et al., 2008); (c) HNTs rich/Polymer rich (Pasbakhsh et al., 2009a); (d) Zigzag network (Ismail et al., 2008); (e) Fibril like BBT modified HNTs (Liu et al., 2008b).

8.2.1   THERMOPLASTIC/HNTS NANOCOMPOSITES

8.2.1.1   POLYPROPYLENE (PP)/HNTS NANOCOMPOSITES

Different researchers (Du et al., 2009; Liu et al., 2008b, 2009c, b, 2010; Ning et al., 2007; Du et al., 2006b) studied the crystallization behavior of PP/HNT composites prepared by melt blending. Ning et al. (2007) reported that HNTs served as nucleation agents in the PP matrix and accelerated its crystallization, resulting in higher crystallization temperatures. However, HNTs could increase the activation energy of crystallization due to the hindrance of HNTs to the motion of PP chains (Ning et al., 2007). HNTs showed dual nucleating ability for α- and β-PP crystals, due to their unique surface characteristics (Liu et al., 2009a) and the percentage of β-crystals was increased by their incorporation. However, they did not significantly increase the total crystallinity of PP. Surface modification of HNTs and preparation method also influenced the formation of β-crystals and total crystallinity of PP/HNTs nanocomposite. For example, HNTs modified with 2,5-bis (2-benzoxazolyl) thiophene (BBT) has remarkably increased crystallinity (16.6% increase in crystallinity at 3 phr BBT and 30 phr HNTs), which was attributed to the formation of ribbonlike or wirelike phase (fibrils) in the PP/HNTs/BBT nanocomposites (Figure 8.1e) (Liu et al., 2008b). Benzothiazole sulphide (CBS) compatibilized HNTs showed lower β-nucleating ability, because CBS changed the surface properties of HNTs. Du et al. (2006b) reported that PP grafted HNTs substantially lowered the crystallinity of PP. β-crystals were not observed in PP/HNTs nanocomposites prepared by a solution casting method, while total crystallinity increased, as reported by Liu et al. (2010).

Ning et al. (2007) did not observe much enhancement on mechanical properties of PP/HNTs nanocomposites. However, improvement of mechanical properties was reported by other researchers (Liu et al., 2008a, b, 2009c; Du et al., 2006b; Prashantha et al., 2011). A substantial increase reported in the tensile strength, flexural strength, and flexural modulus of PP/HNTs nanocomposites with reduction in their impact strength (Liu et al., 2008a, b, 2009c). It has been shown by researchers that mechanical properties can be further improved by modification of HNTs due to better dispersion and stronger interfacial bonding between PP and HNTs. Prashantha et al. (2011) used masterbatches of unmodified HNTs and quaternary ammonium-salt-treated HNTs (QM-HNTs) (from NaturalNano) in PP and reported an increase in storage modulus (E'), tensile and flexural properties, and decrease in elongation at break with increasing halloysite content; while this reinforcement effect was more pronounced in case of modified HNTs. Furthermore, they observed an increase in impact properties of PP/HNTs nanocomposites (up to 77% in case of 6% QM-HNTs), due to HNTs pullout and breakage (and bridging in case of QM-HNTs). Du et al. (2006b) also reported an increase in impact strength of PP by incorporation of PP-grafted HNTs. However, grafting of PP on HNTs deteriorated their dispersion in PP matrix and resulted in microscale-sized clusters. This discrepancy in impact results may be attributed to the fact that different amounts and types of HNTs were used in these researches.

Incorporation of HNTs in PP increased the thermal stability (60°C increase in $T_{5\%}$ weight loss) and flame retardancy of the PP/HNTs nanocomposites (Du et al., 2006a). Acting as barriers for heat and mass transport, the presence of iron in HNTs and most importantly the entrapment of the decomposition products in the lumen of HNTs were reasons given for these improvements.

8.2.1.2   POLYETHYLENE (PE)/HNTS NANOCOMPOSITES

Jia et al. (2009b) investigated the reinforcement and flame retardancy effects of HNTs on linear low-density polyethylene (LLDPE). Tensile strength, flexural strength, and flexural modulus of LLDPE/HNTs nanocomposites gradually increased with HNT's content; but the impact strength of the respected nanocomposites decreased rapidly (up to a tenth at 60 wt % of HNTs). As well-dispersed structure of HNTs in LLDPE matrix was hardly achieved, maleic anhydride, methyl methacrylate, and butyl-acrylate-grafted PE was used as the interfacial modifier resulting in more uniform dispersion of HNTs particles in LLDPE matrix with better interfacial bonding and improvement in the strength, modulus, impact, thermal stability, and flame retardancy of LLDPE/HNTs nanocomposites.

8.2.1.3   POLYAMIDE (PA)/HNTS NANOCOMPOSITES

Studies of Ulrich et al. (2010), Guo et al. (2009b) and Lecouvet et al. (2011a) showed that HNTs could act as nucleating agents in the PA matrix, and increased the formation of the γ-crystals, especially for the low molar mass PA6. Incorporation of HNTs increased the crystallization temperature, although it lowered the crystallinity. Interestingly, the crystallinity of the PA6/HNTs nanocomposites increased with cooling rate, and the proportion of γ-crystals in the total crystallinity grew instantly as cooling rate and HNTs content increased. This crystallization behavior of the PA6/HNTs nanocomposites was ascribed to the multiple roles of HNTs in the crystallization: First, HNTs served as a nucleating agent, resulting in a higher initial crystallization temperature (T_0); second, there was an enthalpic interaction between the PA6 and the HNTs, resulting in restricted mobility of PA6 chains; and finally and more importantly, HNTs favored the formation of the gamma phase crystals (Guo et al., 2009b). The formation of an effective arrangement of this crystalline phase in the presence of halloysite could possibly cause the reinforcement effect at low filler loading (Handge et al., 2010).

It has been reported by various researchers (Handge et al., 2010; Guo et al., 2009c) that the tensile strength, flexural strength, and flexural modulus of the PA6 composites increased with incorporation of HNTs. This reinforcing effect was much pronounced at low filler loading for the low molar mass PA6. The modulus of elasticity and the storage modulus linearly increased with concentration of halloysite (Lecouvet et al., 2011a; Hedicke-Höchstötter et al., 2009), while the reinforcement effect of HNTs was very dominant above the glass transition temperature of PA6. The incorporation of halloysite did not lower, and even enhanced, the toughness of the PA6 matrix.

Silane modification of the surface of HNTs improved their dispersion and the interaction between HNTs and the PA matrix, resulting in further improvement of mechanical properties. Strong interfacial interactions caused an effective load transfer from the PA matrix to HNTs and resulted in significant increase in HDT (48°C at 10 phr modified HNTs) (Guo et al., 2009c).

Rheological studies showed that the flow properties of PA6 were only moderately altered by the addition of HNTs and even at a filler concentration of 30 wt. % no indication of a filler network was observed. Hence PA6/halloysite composites can still be processed under similar conditions as neat PA6 (Handge et al., 2010).

Marney et al. (2008) reported that HNTs retarded burning of PA, without stopping its burning, and more than doubled the total burning time by formation of a thermal insulation barrier at the surface of the composite and fuel dilution effect during the burning.

8.2.1.4 POLYVINYL CHLORIDE (PVC)/HNTS NANOCOMPOSITES

Mondragon et al. (2009) claimed that incorporation of unmodified HNTs into rigid PVC did not effectively improve mechanical strength and impact properties and reduced its stiffness. However, their incorporation significantly increased elongation at break due to the orientation of HNTs in PVC matrix (Mondragon et al., 2009). On the contrary, Liu et al. (2011a) reported that loading of HNTs steadily increased the tensile strength, flexural strength and flexural modulus of PVC/HNTs nanocomposites; while the impact strength of the PVC/HNTs nanocomposites raised with HNTs loading to a maximum of 48 percent at 10 phr and then decreased steadily to −15 percent at 40 phr. PMMA-gr-HNTs also showed the same trend in reinforcement of PVC (Liu et al., 2011b), while grafting PMMA on HNTs increased the reinforcing effect of HNTs as a result of stronger interfacial bonding. The tensile strength, flexural strength and flexural modulus of PVC/PMMA-gr-HNTs nanocomposites were increased with increasing PMMA-gr-HNTs content up to 10, 12.7 and 42 percent, respectively, at 5 phr PMMA-grafted HNTs. The notched izod impact strength of nanocomposites increased up to nearly double compared to pristine PVC at 3 phr. The improvement of strength and modulus was attributed to the efficient load transfer from PVC to HNTs as a result of strong interfacial bonding between HNTs and PVC. Good dispersion of HNTs, stabilization of cracks by the bridging, debonding, and pull-out of HNTs nanotubes, and most importantly plastic shear deformation of PVC matrix around the HNTs particles contributed to improve the toughness of PVC. However, aggregation of HNTs at high loading deteriorated the impact properties.

Moreover, incorporation of HNTs and PMMA-gr-HNTs slightly increased the stiffness, glass transition temperatures (T_g), and thermal stability of PVC. These enhancements in physical and thermal properties were attributed to the good dispersion of HNTs and strong interfacial interaction that reduced mobility of PVC chains, and the hindrance effect of HNTs serving as a barrier for heat and volatile material (Liu et al., 2011a, b).

8.2.1.5   POLYVINYL ALCOHOL (PVA)/HNTS NANOCOMPOSITES

Liu et al. (2007b) prepared PVA/HNTs nanocomposite films via solution casting and coagulating of PVA/HNTs aqueous solutions. The size of HNTs' particles in PVA/HNTs solutions was significantly lower than that in the pure HNTs solution and practically independent of HNTs content. More uniform dispersion of HNTs was achieved by using coagulation method since well-dispersed HNTs could be separated by the help of a PVA during the coagulation process. However, in the casting procedure, the gradual evaporation of water increased the viscosity and caused PVA chains to precipitate. Consequently, the interaction between water and HNTs as well as interaction between PVA and HNTs decreased and resulted in reaggregation of HNTs.

HNTs had a nucleating effect on PVA, causing an increase in the crystallization temperatures and broadened the range of the crystallization temperature. However, overloading of HNTs (53 wt. %) depressed the increase in crystallization temperature due to aggregation of overloaded HNTs. T_g of PVA/HNTs nanocomposites decreased slightly with HNTs content due to increase in the free volume, as HNTs had the length of microscale and the polymer chains could not wrap the nanotubes in three dimensions.

HNTs substantially restricted the decomposition of PVA backbone, but they could not improve decomposition of the side chain, which was attributed to the ability of PVA to supply oxygen by itself.

8.2.1.6   POLYSTYRENE (PS)/HNTS NANOCOMPOSITES

Various methods have been utilized to prepare PS/HNTs nanocomposites (Shamsi et al., 2010; Lin et al., 2011). Shamsi et al. (2010) prepared PS/HNTs nanocomposites by solution blending of HNTs and PS in benzene followed by freeze and vacuum drying of the mixture. Lin et al. (2011) made PS/HNTs nanocomposites by in situ polymerization of styrene in aqueous emulsion of HNTs and styrene using sodium dodecyl sulfate (SDS) as surfactant. Lin et al. reported very uniform dispersion of individually separated HNTs in PS matrix, whereas Shamsi et al. could not study the dispersion of HNTs in their nanocomposites. SDS served as dispersing agent for HNTs in water and caused a very well dispersed suspension of HNTs in water, which was stable for a few days due to formation of a bilayer SDS coat on the surface of halloysite (Figure 8.2a). This bilayer coat of SDS impeded the adsorption of styrene onto the surface of HNTs during polymerization. Consequently, a mixture of HNTs surrounded by polystyrene nanosphere (Figure 8.2b) was produced by the in situ emulsion polymerization method.

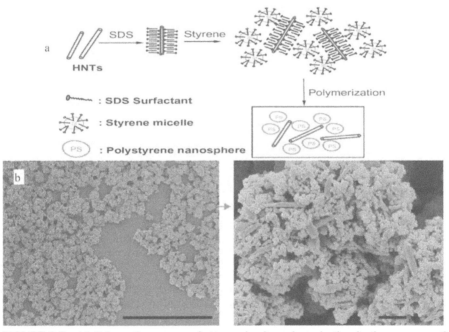

FIGURE 8.2   (a) Schematic process of styrene in situ polymerization in the presence of HNTs and SDS (b) SEM micrographs of the PS/HNTs nanocomposites (Lin et al., 2011).

Incorporation of HNTs considerably increased the impact strength of PS up to 300 percent at 5 wt % HNTs loading. However, overloading of HNTs deteriorated the impact strength (Lin et al., 2011). Moreover, the storage modulus of PS increased with addition of HNTs (Lin et al., 2011; Shamsi et al., 2010). Massive crazing at the HNTs' tips, nanotube bridging and pull out, debonding cavitation at the HNTs–PS interface due to weak interaction, and localized shear yielding of PS was identified as toughening mechanisms in PS/HNTs nanocomposites (Lin et al., 2011). Dimple-like structure was observed in SEM micrographs of fracture surface of PS/

HNTs nanocomposites which was attributed to massive crazing and plastic deformation of PS around HNTs (Lin et al., 2011).

Thermal stability of PS was improved by incorporation of HNTs. Loading of 5 and 10 percent HNTs in PS increased the temperature of 5 percent weight loss ($T_{5\%}$ weight loss) of PS by 50 and 25°C, respectively (Lin et al., 2011). This enhancement in thermal stability was mainly attributed to entrapment of decomposition products in the HNTs' lumen. Glass transition temperature (T_g) of PS decreased by addition of HNTs (Lin et al., 2011; Shamsi et al., 2010), which was attributed to weak interactions between the PS and HNTs resulting in higher free volume and chain mobility in PS matrix.

In order to make HNTs more compatible with PS, Shamsi et al. coated halloysite by PS using plasma polymerization of styrene in the presence of HNTs. Although the plasma coating process did not result in a covalent bond between HNTs and PS, the mechanical properties of PS/plasma-modified HNT nanocomposites were slightly higher than those of unmodified HNTs, due to improved interfacial interaction (Shamsi et al., 2010).

8.2.1.7 POLYVINYLIDENE FLUORIDE (PVDF)/HNTS NANOCOMPOSITES

Tang et al. (2013) prepared PVDF/HNTs nanocomposites with uniform dispersion of HNTs, even at high loading, by solution blending of HNTs and PVDF in N-Methylpyrrolidone (NMP) solvent. They used deionized water as antisolvent for the PVDF/HNTs mixture in NMP which caused fast precipitation and formation of PVDF/HNTs film in 3–5 min. Then these nanocomposite PVDF/HNTs films were compression molded into sheets by hot press.

HNTs affected the crystallization behavior of PVDF and increased the crystallization temperature of PVDF due to their nucleating ability. However, overloading of HNTs reduced the crystallization temperature. Enthalpy of crystallization (ΔH_c) was significantly decreased by incorporation of HNTs in PVDF. Moreover, HNTs promoted the formation of γ–phase crystals. The amount of γ–phase crystals increased to a maximum at 10 phr HNTs loading and then decreased by higher loading (Tang et al., 2013).

Dynamic mechanical tests showed that the storage modulus of PVDF did not change at low HNTs loading (2, 5 phr), while it noticeably increased at high HNTs loading (10, 20 phr). Loading with HNTs did not significantly affect the T_g of PVDF (−30°C). However, the relaxation peak at about 25°C, correlated to the α_2 transition of the amorphous region at the surfaces of the crystals, passed a maximum at 2 phr and then dropped at higher loadings. The relaxation peak at about 90°C (α_1 transition), correlated to the liberation of PVDF chains within the crystals, shifted to a lower temperature with HNTs loading and almost disappeared at 20 phr HNTs.

These observations were attributed to the effect of halloysite on the crystalline structures of PVDF (Tang et al., 2013).

Tensile modulus of PVDF/HNTs nanocomposites increased with HNTs loading up to 56 percent at 20 phr HNTs. The tensile strength is slightly reduced with HNTs loading which was attributed to the weak interfacial interaction between HNTs and PVDF. The characteristic yielding and long plastic deformation behavior of PVDF was preserved at low HNTs loading, while high HNTs loading considerably reduced the cold drawing of PVDF/HNTs nanocomposites. SEM micrographs revealed that the fracture mechanisms of PVDF changed by addition of HNTs. Fractured surface of PVDF showed a brittle fracture near the core of sample—where the fracture started and propagated outwards—and a microductile fracture which happened at edges (Figure 8.3). The same phenomena was observed at low concentrations of HNTs, while a lot of voids appeared in brittle core of samples and the microductile surface of the edges became rougher due to the change of local stress distribution as a result of the presence of HNTs (Figure 8.4). At high concentrations of HNTs, brittle fracture started at the edges of the sample and propagated inward to a fibrillation zone where severely deformed fibrils of PVDF were observed (Figure 8.5). This transition of fracture mechanism was attributed to change in stress field and reduction of PVDF ligament thickness which was caused by incorporation of HNTs (Tang et al., 2013).

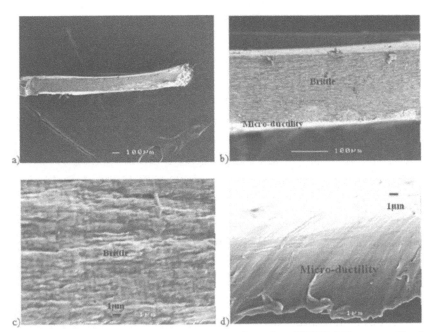

FIGURE 8.3 SEM micrograph of the fracture surface of neat PVDF: (a) low magnification; (b) high magnification; (c) brittle; (d) microductility (Tang et al., 2013).

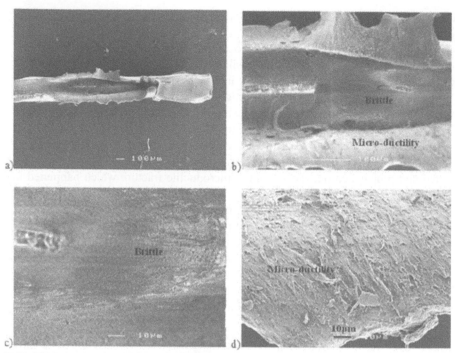

FIGURE 8.4   SEM micrograph of the fracture surface of PVDF/HNTs nanocomposites (10 phr HNT): (a) low magnification; (b) high magnification; (c) brittle; (d) microductility (Tang et al., 2013).

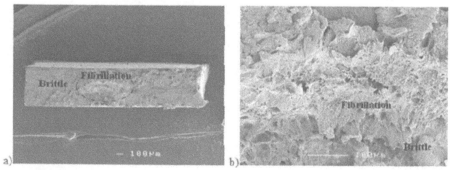

FIGURE 8.5   SEM micrograph of the fracture surface of PVDF/HNTs nanocomposites (20 phr HNT): (a) low magnification; (b) high magnification in the fibrillation area (Tang et al., 2013).

8.2.2 RUBBER/HNTS NANOCOMPOSITES

8.2.2.1 NATURAL RUBBER (NR)/HNTS NANOCOMPOSITES

Properties of natural rubber (NR)/HNTs nanocomposites were studied by Rooj et al. (2010). Their studies showed that incorporation of HNTs into the NR affects the curing characteristics of NR composites. Their finding indicated that the scorch time of sulfur curing systems increased due to the adsorption of vulcanization additives such as accelerators onto HNTs. However, the silane (Si69) modified HNTs caused shorter scorch times due to generation of some additional sulfur during mixing besides decrease in additives adsorption. Curing time (t_{90}) was also increased in the case of NR/silane-modified HNTs nanocomposites, since the presence of silane modifier prolonged the cross-linking reactions. Moreover, addition of HNTs into the NR matrix raised the maximum torque of NR/HNTs nanocomposites compared to the NR gum which was more pronounced in case of the silane-modified HNTs.

HNTs increased the tensile strength and modulus of NR/HNTs nanocomposites, but decreased their elongation at break. Reinforcing effects of HNTs in NR matrix were higher compared to silica fillers, especially in case of silane-modified HNTs. TEM observations showed that the silane-modified HNTs were very well dispersed in the NR due to the strong interaction between HNTs and NR matrix caused by silane coupling agent (Figure 8.6) (Rooj et al., 2010).

HNTs had prominent effects on the thermal stability of the NR/HNTs nanocomposites. $T_{5\%}$ weight loss of NR vulcanizate increased 64°C and $T_{85\%}$ weight loss increased by 70 and 79°C by loading 10 phr of unmodified and silane-modified HNTs, respectively. This large amount of increase in degradation temperature was attributed to interfacial and intertubular interactions between NR and HNTs and entrapment of degradation products of NR inside the lumen of the HNTs (Rooj et al., 2010).

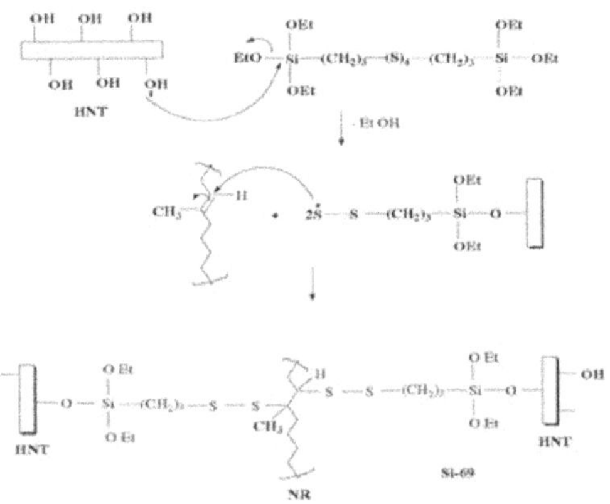

FIGURE 8.6   Possible mechanism of interaction between halloysite nanotubes and natural rubber with bis (triethoxysilylpropyl)-tetrasulphide as a coupling agent (Rooj et al., 2010).

8.2.2.2   ETHYLENE PROPYLENE DIENE MONOMER (EPDM)/HNTS NANOCOMPOSITES

Pasbakhsh et al. (2010) and Ismail et al. (2008) reported using HNTs to reinforce EPDM by applying three different procedures including traditional mechanical mixing, melt blending, and solution casting (unpublished results). In the reported studies, HNTs showed a very high potential to disperse uniformly inside the EPDM even at high loading and unmodified form. Beside the ability of EPDM to accept high amounts of filler compared to the other polymers, HNTs formed face-to-edge and edge-to-edge structures especially at high loading (zigzag structures) due to the low amount of OH groups at the external surface of the tubes. TEM microstructures of EPDM/HNTs nanocomposites revealed that HNTs were well dispersed in EPDM matrix with a three-dimensional structure even at 100 phr loading and this is the reason for the simultaneous enhancement of strength, modulus, and ductility (E_b), particularly at high filler loading (>30 phr). The tensile strength, modulus (M_{100}), and elongation at break (E_b) of EPDM/HNTs nanocomposites at 100 phr HNTs loading were 874.3, 179.4, and 305 percent higher than those of the unfilled EPDM, respectively.

HNTs were used as a co-filler with silica and calcium carbonate in EPDM composites (Pasbakhsh et al., 2009b). The tensile properties of EPDM/CaCO$_3$/HNTs composites increased with increasing HNTs loading. However, tensile strength of EPDM/Silica/HNTs composites showed an optimum value at 100/25/5 phr of the EPDM/Silica/HNT. The partial replacement of HNTs with silica led to a better dispersion of silica inside the EPDM and resulted in higher tensile strength and E_b. The storage modulus of the nanocomposites increased and the Tanδ (at 50°C), damping, and curing time decreased with replacement of silica or CaCO$_3$ by HNTs.

Functionalization of HNTs by γ-methacryloxypropyl trimethoxysilane (MPS) improved HNTs dispersion and reduced their aggregation in EPDM matrix. The tensile strength and modulus of EPDM/HNTs nanocomposites was increased by modification of HNTs, while their elongation at break (E_b) reduced slightly. The intertubular and interfacial bonding between the MPS-modified HNTs and EPDM and the better dispersion of the HNTs within the EPDM matrix contributed to increase the tensile strength of EPDM-modified HNTs nanocomposites compared to unmodified ones which consequently reduced their ductility (Pasbakhsh et al., 2010).

Compatibilization of the EPDM/HNTs nanocomposites by maleic anhydride grafted EPDM (MAH-g-EPDM) (20 phr) also resulted in the increment of the tensile strength and modulus and reduction of E_b, due to the improvement in the interfacial interactions between HNTs and EPDM. However, at high HNTs loading, the tensile strength is reduced, because of the poor dispersion of the HNTs and formation of two different phases of EPDM-rich and HNT-rich areas (Figure 8.7) (Pasbakhsh et al., 2009a).

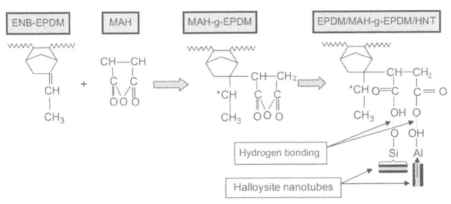

FIGURE 8.7 Possible mechanism of interactions between MAH-g-EPDM and halloysite nanotubes (Pasbakhsh et al., 2009a).

8.2.2.3 STYRENE-BUTADIENE RUBBER (SBR)/HNTS NANOCOMPOSITES

Nanocomposites of HNTs and SBR were studied by Guo et al. (2010a, b, 2008, 2009a), Lei et al. (2011), and Jia et al. (2009a). Different agents such as unsaturated acids (methacrylic acid (MAA) or sorbic acid (SA)), complex of resorcinol and hexamethylenetetramine (RH), and ionic liquids (ILs) have been used to improve interfacial interactions between HNTs and SBR. Reactions between modifying agents, HNTs, and SBR resulted in tightly held together components and formed a strong composite structure with uniform dispersion of HNTs (Figure 8.8). The mechanical properties of SBR/HNTs nanocomposites increased with HNTs and modifier content. However, overloading modifiers deteriorated tensile strength and modulus. In optimal condition, four- to fivefold increase in tensile and tear strength was achieved compared to samples without a modifier. The significantly enhanced mechanical properties were attributed to the strong interfacial bonding between HNTs and the rubber matrix, as a result of grafting/complexation mechanism, and significantly improved dispersion of HNTs by virtue of the interactions between HNTs and modifiers (Guo et al., 2008, 2009a, 2010a; Jia et al., 2009a).

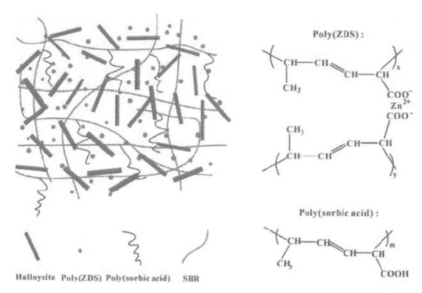

FIGURE 8.8   Schematic illustration of the composite structure in SA compatibilized SBR/HNTs nanocomposites (Guo et al., 2010a).

Epoxidized Natural Rubber (ENR) was used as a macromolecular coupling agent to improve the compatibility between HNTs and SBR (Zhixin et al., 2010). With incorporation of ENR into SBR/HNTs nanocomposites, the dispersion and orientation of HNTs improved and the agglomerates of HNTs remarkably reduced. The tensile modulus, tensile strength, tear strength, and shore A hardness of SBR/HNTs/ENR nanocomposites increased with increasing ENR content up to 3–4 phr, whereas the elongation at break and permanent set decreased. Moreover, the values of Tanδ of the composite at 0°C is increased, while the Tanδ values at 60°C decreased which benefits the improvement of rolling resistance and wet skid resistance of SBR/HNTs composites simultaneously.

8.2.2.4   CARBOXYLATED STYRENE BUTADIENE RUBBER (XSBR)/ HNTS NANOCOMPOSITES

Well-dispersed xSBR/HNTs nanocomposites was prepared by Du et al. (2008) using a co-coagulation process of the aqueous suspension of xSBR/HNTs latex. The sulfur vulcanization of xSBR/HNTs compounds was promoted at high HNTs loading, while the vulcanization was delayed at low HNTs loading. At low HNTs content, adsorption of the accelerators onto the surface of HNTs delayed curing behavior, but when the HNTs content was high enough, formation of hydrogen bonds between HNTs and carboxylic groups of xSBR resulted in accelerated vulcanization.

The incorporation of HNTs increased tensile strength, modulus, and hardness of the xSBR significantly. The modulus of xSBR/HNTs nanocomposites increased consistently by HNTs loading and is almost doubled at 5 phr HNTs. Incorporations of 2 phr and 5 phr HNTs led to increment of 7° and 15° in hardness, respectively, which was quite remarkable compared to silicate or carbon black fillers. Significant reinforcing effects of HNTs was attributed to strong interfacial interaction between the nanotubes and the matrix due to the hydrogen bonding (Figure 8.9), and uniformly dispersed nanotubes with high L/D ratio.

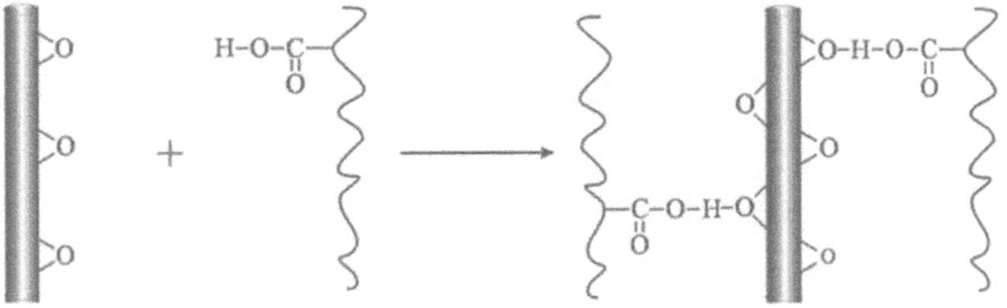

FIGURE 8.9   Schematic of physical cross-link via hydrogen bonding between xSBR and HNTs (Du et al., 2008).

8.2.3   THERMOSET RESIN/HNTS NANOCOMPOSITES

8.2.3.1   EPOXY/HNTS NANOCOMPOSITES

Effects of incorporation of HNTs on properties of epoxy resins have been studied by different researchers (Deng et al., 2009; Ye et al., 2007; Liu et al., 2007a, c). Although HNTs did not significantly affect the modulus and strength of epoxy resin, they had remarkable effect on impact properties of epoxies. Ye et al. (2007) reported unusual 400 percent increase in charpy impact strength of epoxy at 2.3 percent HNTs loading, without sacrificing the flexural modulus and strength. Deng et al. (2008) used a compact tension (CT) method to study the impact properties of epoxy/HNTs nanocomposites, and reported a noticeable increase in K_{IC} (critical stress intensity factor, up to 50%) and G_{IC} (critical strain energy rate, up to 127%). A two-phase structure, a HNT-rich particle phase, and an epoxy-rich matrix phase (Figure 8.10) was found in epoxy/HNTs nanocomposites. The toughening mechanisms responsible for the unusual increase in impact strength were identified as massive microcracking, nanotubes bridging/pull-out/breaking, and crack deflection by HNT-rich phase; whereas the former two were proposed as the dominant mechanisms that dissipate extensive amount of impact energy. Although HNT-rich particle phases are supposed to have an important role in toughening of epoxy/HNTs nanocomposites via deflection and stopping the running crack, it was believed that any approach which can effectively reduce HNTs' cluster size or fully isolate HNTs and enhance the adhesion between the clay and epoxy—including optimization of mixing method or chemical treatments—could enhance mechanical properties of the epoxy/HNTs nanocomposites.

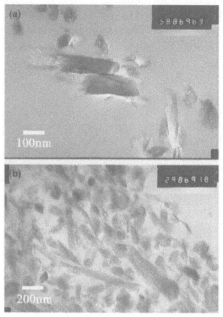

FIGURE 8.10 TEM micrographs of epoxy nanocomposites with 2.3 wt.% halloysite in: (a) the epoxy-rich region and (b) the HNT-rich region (Ye et al., 2007).

Deng et al. (2008) used a ball mill homogenization method to reduce the size of HNTs clusters in the epoxy matrix, which led to enhancements of strength, modulus, and glass transition temperature (T_g). Different chemicals—including potassium acetate (CH_3COOK), silane (N-(β-aminoethyl)-γ-aminopropyl trimethoxysilane), cetyl trimethyl ammonium chloride (CTAC, $[CH_3(CH_2)_{15}N(Cl)(CH_3)_3]$) (Deng et al., 2009), and phenylphosphonic acid (PPA) (Tang et al., 2011)—were used for treatment of HNTs to achieve better dispersion in epoxies. CH_3COOK treatment reduced the size of the particle clusters, but silane, and particularly CTAC, increased the tendency of agglomeration due to the increased surface reactivity and led to premature failure under tensile stresses. PPA (Tang et al., 2011) could effectively intercalate and unroll HNTs and changed the morphology of HNTs from nanotubes to nanoplatelets, depending on PPA treatment time (Figure 8.11). Unrolled and intercalated HNTs with PPA had better dispersion in the epoxy compared to untreated HNTs and significantly increased the fracture toughness (78.3% higher K_{1C} for the composite containing 100 h PPA-treated HNT), without sacrificing other properties such as strength, modulus, and thermal stability. Liu et al. (2007a) used cyanate ester as the hardener and prepared epoxy/HNTs nanocomposites with very uniform dispersion of individually separated HNTs, although aggregated HNTs coexisted at relatively higher HNTs loading (12 wt.%). Cyanate ester could be co-cured with epoxy resin and was reactive toward the silanols and aluminols of the HNTs by formation of the iminocarbonate linkage (Figure 8.12). The modulus of these epoxy/HNTs nanocomposites in the glassy and rubbery state was significantly higher (58.6% and 121.7%, respectively) compared to neat epoxy, and their coefficient of thermal expansion (CTE) was substantially lower (19.6% lower in the temperature range of 25–100°C and 21.8% lower in the temperature range of 100–160°C) that made them promising material for printed circuit boards (PCBs) in the electronics industry. Liu et al. (2007c) used γ-glycidoxypropyltrimethoxy silane to modify the surface of HNTs, but this modification did not improve the properties of cyanate ester–cured epoxy/HNTs nanocomposites, due to changes in the interactions between HNTs, cyanate ester, and epoxy matrix.

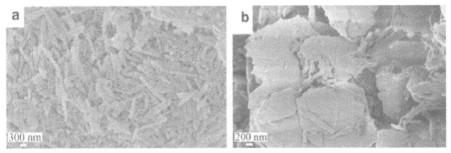

FIGURE 8.11   SEM images of (a) as-received and (b) 100 h-PPA-treated halloysite (Tang et al., 2012).

Incorporation of HNTs into carbonfiber–reinforced epoxy composites caused 20 percent enhancement in the interlaminar shear strength (ILSS) at 5 wt.% HNTs, and 37 percent enhancement in the delamination fracture resistance (G_{IIC}) at 2 wt.% HNT, with little or no effect on the flexural properties (Ye et al., 2011b). Prevention of the growth of microcracks by HNTs bridging, and deflection and arrest of the cracks by the HNT-rich area during the translaminar crack propagation contributed to the enhancement of ILSS, while the enhanced toughness of the epoxy matrix, due to incorporation of HNTs and strengthened interfacial adhesion between the carbon fibers and the epoxy matrix due to an unclear mechanism brought in by the HNTs, was the most important reason for the increased G_{IIC} (Ye et al., 2011a).

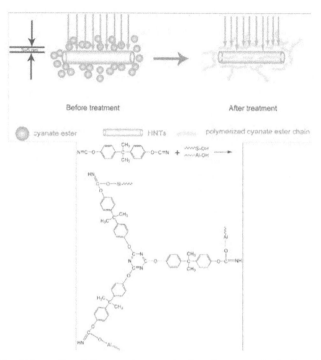

FIGURE 8.12   Schematic of the reaction mechanism between the hydroxyl groups and cyanate ester (Liu et al., 2007a).

8.2.3.2   UNSATURATEDPPOLYESTER RESIN (UPR)/HNTS NANOCOMPOSITES

Garea et al. (2011) used HNTs to improve the properties of unsaturated polyester resin (UPR). They modified HNTs with two coupling agents, triethoxyvinylsilane (TEVS) and γ-aminopropyltriethoxysilane (APTES), which resulted in grafting of vinyl and amine groups on the surface of HNTs. Incorporation of unmodified HNTs did not change the T_g of UPR, whereas vinyl-HNTs increased and amine-HNTs decreased the T_g. $T_{5\%}$ weight loss of UPR increased from 295°C to 314°C with incorporation of 1 wt.% vinyl-HNTs, while it decreased to 292°C in the case of amine-HNTs. The higher T_g and thermal stability especially at low concentration of vinyl-HNTs was attributed to the strong interfacial bonding between vinyl-HNTs and UPR due to the involvement of the grafted vinyl groups into the cross-linking process of UPR. Amine-HNTs had low compatibility with the UPR matrix, so they behaved more like a plasticizer and reduced T_g and thermal stability of the final composites. At higher concentrations, decrease in T_g and thermal stability was observed in all samples possibly due to the agglomeration of HNTs.

8.2.4   OTHER POLYMERS

Qi et al. (2010) prepared a new system for effective loading and sustained release of drugs by incorporation of HNTs into poly lactic-co-glycolic acid (PLGA) electrospun nanofibers. Loaded HNTs with a model drug (tetracycline hydrochloride, TCH) was mixed with PLGA solution and then the mixture was electrospun to form composite drug-loaded nanofibers (Figure 8.13). HNTs coaxially aligned along and embedded within the nanofibers, and did not change the morphology of the PLGA nanofibers. The electropsun PLGA/HNTs nanofibers had a smooth fiber surface with a diameter much smaller than the length of the HNTs. The addition of a small quantity of HNTs (1 wt % relative to PLGA) significantly improved the mechanical properties of the PLGA fibrous mat, while it did not significantly change its porosity and density. The breaking strength, Young's modulus, and failure strain of PLGA/HNTs nanocomposite fibrous mats increased by 57%, 22.9%, and 19.4%, respectively, which was attributed to the to the alignment of HNTs in the PLGA fibers and efficient load transfer from the PLGA to the HNTs. Incorporation of HNTs could effectively weaken the initial burst release of drug in PLGA fibers. They had slightly slower release rates than HNTs powder. However, after 15 days, the release rate is quite low and it was not clear yet whether the remaining drug would be able to be released and how long it would take. HNTs had no significant influence on the cell proliferation and viability of PLGA, and PLGA/HNTs nanocomposite scaffolds showed an excellent biocompatibility. This excellent biocompatibility and improved mechanical properties made PLGA/HNTs nanocomposites a promising material for the applications in tissue engineering and pharmaceutical sciences.

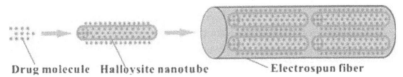

Drug molecule   Halloysite nanotube   Electrospun fiber

FIGURE 8.13   Schematic illustration of drug-loaded electrospun composite nanofibers (Qi et al., 2010).

Chen et al. (2011) incorporated HNTs into Poly (glycerol sebacate) (PGS) to improve their mechanical properties. To prepare PGS/HNTs nanocomposites, a prepolymer of PGS was synthesized by incomplete polycondensation of 1:1 molar ratio of the glycerol and the sebacic acid at 120°C for 24 h under nitrogen gas. This prepolymer melted at 50°C, mixed with HNTs by magnetic stirring and cast into sheets. The cast sheets were then postcured under vacuum at 120°C for further 3 days to complete their cross-linking process. HNTs had uniform distribution in the PGS matrix. Existence of SiO_2 on outer layers of HNTs resulted in an acidic environment, causing a lower cross-link density in the PGS matrix and higher swelling at low HNTs loading. However, at high HNTs loading, the swelling percentage is reduced due to the interfacial bonding of PGS chain and HNTs (physical cross-links, bound rubber structure) and formation of the tubular filler network. Young's modulus, tensile strength and strain at break (ε_{max}) increased up to 85%, 129% and 106%, respectively, in the nanocomposites containing 20 wt.% halloysite. The hysteresis of PGS/HNTs composites slightly increased and their resilience dropped from 96 percent in pure PGS to ~90 percent. The extra energy dissipation of PGS viscoelastic matrix was attributed to the chain slippage or breaking, the motion of filler particles, and reduction in the cross-link density of the PGS network. The degradation rate of PGS in tissue culture medium was reduced at low HNTs loading (3–5 wt.%), although higher loading of HNTs increased the degradation rate. This was attributed to two competing effects of HNTs: First, the acidic effect of HNTs and reduced cross-link density of PGS network which accelerates the degradation rate of material. Second, the bound rubber layer which effectively impeded the water attack and the hydrolysis degradation rate. About 1–5 wt.% PGS/halloysite nanocomposites had satisfactory biocompatibility, while in those with 10–20 wt.% significant cytotoxicity was observed, possibly due to the release of acidic degradation by-products from the low cross-linked PGS matrix.

8.3   CONCLUSION

Studies have shown that HNTs can be effectively used as reinforcing nanofillers in different polymer matrixes. Compared to other nanofillers such as CNT and MMT, they are cheap, abundant, nontoxic, and easy to process which gives them great potential to be used in various polymeric nanocomposites.

HNTs have good processability and various methods could be used to prepare their polymer nanocomposites. Well-dispersed structures of HNTs were possible to achieve by direct melt blending without any pretreatment. Solution casting, cocoagulation, and in situ polymerization were also successfully used to prepare HNT nanocomposites.

HNTs affect the crystallization behavior of crystalline thermoplastics and increase the γ-crystal structure. They usually improved the tensile and flexural strength of thermoplastics, while their effect on impact properties is controversial. In terms of toughness it has been found by researchers that HNTs cannot improve the impact properties of ductile polymers due to matrix shear yielding. However, improvement in impact properties of such polymers (PP) was reported by others.

HNTs considerably improved the impact properties of rigid thermoset polymers without decreasing their stiffness and strength. The toughening mechanism was identified as dissipation of impact energy by bridging/pull-out/breaking of nanotube and crack deflection. This toughening mechanism is a great contribution for improving impact properties of brittle polymers as no matrix plastic deformation is needed.

KEYWORDS

- **Biocompatibility**
- **Crystallinity**
- **Glass transition temperature**
- **Hydrogen bonding**
- **Montmorillonite (MMT)**
- **Nanofillers**
- **Natural rubber**
- **Polymer nanocomposites**

REFERENCES

1. Baral, S.; Brandow, S; and Gaber, B. P.; *Chem. Mater.* **1993**. *5(9)*, 1227–1232.
2. Breuer, O.; and Sundararaj, U.; *Poly. Compos.* **2004**. *25(6)*, 630–645.
3. Cao, J; Hu, X.; and Jiang, D.; e-*J. Surf. Sci. Nanotechnol.*, *7*, **2009**. 813–815.
4. Cavallaro, G.; Donato, D. I; Lazzara, G; and Milioto, S.; *J. Phys. Chem.* C. **2011**, *115(42)*, 20491–20498.
5. Chen, Q. Z.; Liang, S. L.; Wang J; and Simon, G. P.; *J. Mech. Behav. Biomed. Mater.* **2011**. *4(8)*, 1805–1818.
6. Crosby, A. J.; and Lee J. Y.; *Poly. Rev.* **2007**. *47(2)*, 217–229.
7. Deng, S.; Zhang. J.; and Ye. L.; *Compos. Sci. Technol.* **2009**. *69(14)*, 2497–2505.
8. Deng, S.; Zhang. J.; Ye. L.; and Wu. J.l *Polymer.* **2008**. *49(23)*, 5119–5127.
9. Du, M.; Guo, B.; and Jia, D.; *Eur. Poly. J. 42(6)*, 1362–1369.
10. Du, M.; Guo. B.; Lei, Y.; Liu M.; and Jia, D.; *Polymer.* **2008**, *49(22)*, 4871–4876.
11. Du, M.; Guo, B.; Liu, M.; and Jia, D.; *Polym. J.* **2006b**. *38(11)*, 1198–1204.
12. Du, M.; Guo, B.; Wan, J.; Zou, Q.; and Jia, D.; **2009**. *J. Poly. Res. 17(1)*, 109–118.
13. Gârea S. A.; Ghebaur, A.; Constantin, F.; and Iovu, H.; *Poly. Plast. Technol. Eng.* **2011** *50(11)*, 1096–1102.
14. Guo, B.; Chen, F.; Lei, Y.; and Jia, D.; **2010a**. *J. Macromol. Sci*, Part B *49(1)*, 111–121.
15. Guo, B.; Chen, F.; Lei, Y.; Liu, X.; Wan, J.; and Jia, D.; *Appl. Surf. Sci.* **2009a**. *255(16)*:7329–7336.
16. Guo. B.; Lei. Y.; Chen, F.; Liu. X.; Du. M.; and Jia. D.; *Appl. Surf. Sci.* **2008**. *255(5)*, 2715–2722.
17. Guo, B.; Liu, X.; Zhou W. Y.; Lei Y.; and Jia, D. J.; *Macromol. Sci. Part. B.* **2010b** *49(5)*, 1029–1043.
18. Guo, B.; Zou, Q.; Lei, Y.; Du, M.; Liu, M.; and Jia, D.; *Thermochimica Acta.* **2009b**. *484(1–2)*:48–56.
19. Guo, B.; Zou, Q.; Lei, Y.; and Jia, D.; *Poly. J.* **2009c**. *4(10)*, 835–842.
20. Handge, U. A.; Hedicke-Höchstötter K.; and Altstädt, V.; *Polymer.* **2010**. *51(12)*, 2690–2699.
21. Hedicke-Höchstötter. K.; Lim, G. T.; and Altstädt, V.; **2009**. *Compos. Sci. Technol. 69(3–4)*, 330–334.
22. Hussain, F. J.; *Compos. Mater.* **2006**, *40(17)*, 1511–1575.
23. Ismail, H.; Pasbakhsh, P.; Ahmad Fauzi M. N.; and Abu Bakar, A.; **2009** *Poly Plas. Technol. Eng. 48(3)*, 313–323.

24. Ismail, H.; Pasbakhsh, P.; Fauzi, M.; and Abubakar, A.; *Poly. Test.* **2008,** *27*(7), 841–850.
25. Ismail, H.; and Shaari, S. M.;. *Poly Test.* **2010.** *29*(7), 872–878.
26. Jia, Z.-X.; Luo, Y.-f.; Yang, S.-y.; Guo, B.-c.; Du, M.-l.; and Jia D.-M.; *Chin J Poly Sci.* **2009a.** *27*(6), 857–864.
27. Jia, Z.; Luo, Y.; Guo, B.; Yang, B.; Du, M.; and Jia, D.; **2009b**. *Polym. Plast. Technol. Eng.* *48*(6), 607–613.
28. Joussein, E.; Petit, S.; Churchman, J.; Theng, B.; Righi, D.; and Delvaux, B.; *Clay. Min.* **2005,** *40*(4), 383–426.
29. Lecouvet, B.; Gutierrez, J. G.; Sclavons M,. and Bailly, C.; *Poly. Degrad. Stab.* **2011a.** *96*(2), 226–235.
30. Lecouvet, B.; Sclavons, M.; Bourbigot, S; Devaux, J; and Bailly, C.; *Polymer.*; **2011b.** *52*(19), 4284–4295.
31. Lei, Y.; Tang, Z.; Zhu, L.; Guo, B.; and Jia, D.; *Polymer.* **2008.** *52*(5):1337–1344.
32. Levis, S. R.; and Deasy, P. B.; *Int. J. Phar.* **2002.** *243*(1–2), 125–134.
33. Levis, S. R.; and Deasy, P. B.; *Int. J. Pharm.* **2003.** *253*(1–2), 145–157.
34. Li, R.; Hu, Z.; Zhang, S.; Li, Z.; and Chang, X.; *Int. J. Environ. Anal. Chem.* **2011**, 1–13.
35. Lin, Y.; Ng, K. M.; Chan, C.-M.; Sun, G.; and Wu, J.; *J. Col. Interf. Sci. 358(2)*, **2011**. 423–429.
36. Liu, C.; Luo, Y.; Jia, Z.; Li, S.; Guo, B.; and Jia, D.; *J. Macromol. Sci. Part. B.* **2011a.** *51(5)*, 968–981.
37. Liu, C.; Luo, Y.; Jia, Z.; Zhong, B.; Li, S.; Guo, B.; and Jia, D.; *Exp. Polym. Lett.* **2011b.** *5*, 591–603.
38. Liu, M.; Guo, B.; Du, M.; Cai, X.; and Jia, D.; *Nanotechnology.* **2007a,** *18(45)*, (art. No. 455703).
39. Liu, M.; Guo, B.; Du, M.; Chen, F.; and Jia, D.; *Polymer.* **2009a,** *50(13)*, 3022–3030.
40. Liu, M.; Guo, B.; Du, M.; and Jia, D.; *Appl. Phys A.* **2007b,** *88(2)*, 391–395.
41. Liu, M.; Guo, B.; Du, M.; and Jia, D.; *Poly. J.* **2008a,** *40(11)*, 1087–1093.
42. Liu, M.; Guo, B.; Du, M.; Lei, Y.; and Jia, D.; *J. Poly Res.* **2007c.** *15(3)*, 205–212.
43. Liu, M.; Guo, B.; Du, M.; Zou, Q.; and Jia, D.; *J. Phys. D: Appl. Phys.* **2009b** *42(7)*, 075306.
44. Liu, M.; Guo, B.; Lei, Y.; Du, M.; and Jia, D.; *Appl. Surf. Sci.* **2009c.** *255(9)*, 4961–4969.
45. Liu, M.; Guo B.; Zou, Q.; Du, M.; and Jia, D.; *Nanotechnology.* **2008b.** *19(20)*, (art. No. 205709).
46. Liu, M.; Jia, Z.; Liu, F.; Jia, D.; and Guo, B.; *J. Coll. Interf. Sci.* **2010,** *350(1)*, 186–193.
47. Liu, P.; and Zhao, M.; *Appl. Surf. Sci.* **2009.** *255(7)*, 3989–3993.
48. Liu, R.; Fu, K.; Zhang, B.; Mei, D.; Zhang. H.; and Liu, J.; **2011c.** *J. Disper. Sci. Technol.* *33(5)*, 711–718.
49. Lvov, Y. M.; and Price, R. R.; Halloysite nanotubules, a Novel substrate for the controlled delivery of bioactive molecules. In: Bio-inorganic Hybrid Nanomaterials. Wiley-VCH Verlag GmbH & Co. KGaA, 2008, 419–441. doi:10.1002/9783527621446.ch14.
50. Lvov, Y. M.; Shchukin, D. G.; Möhwald, H.; and Price, R. R.; *ACS Nano,* **2008.** *2(5)*, 814–820.
51. Marney, D. C. O.; et al. Polymer Degradation and Stability, **2008,** *93(10)*, 1971–1978.
52. Mingliang, Du BG, Xiaojia Cai, Zhixin Jia, Mingxian Liu, and Demin Jia; *e Polymers,* **2008** 130.
53. Mondragon, M.; Roblero-Linares, Y.; Sanchez-Espindola, M.; and Zendejas-Leal, B.; In: Nanotech Conference & Expo 2009,Technical proceedings–Nanotechnology **2009**: Life Sciences, medicine, diagnostics, bio materials and composites, Nanotech conference and expo, BocaRaton, FL, USA., 2009.
54. Ning, N.-y.; Yin, Q.-j.; Luo, F.; Zhang, Q.; Du, R.; and Fu, Q.; *Polymer.* **2007.** *48(25)*, 7374–7384.
55. Pasbakhsh, P.; Churchman, G. J.; and Keeling, J. L.; *Appl. Clay. Sci.* **2013,** *74*, 47–57.

56. Pasbakhsh, P.; Ismail, H.; Fauzi, M. N. A.; and Bakar, A. A.; *Poly. Test.* **2009a**. *28(5),* 548–559.
57. Pasbakhsh, P.; Ismail, H.; Fauzi, M. N. A.; and Bakar, A. A.; *J. Appl. Poly. Sci.* **2009b**. *113(6),* 3910–3919.
58. Pasbakhsh, P.; Ismail, H.; Fauzi, M. N. A.; and Bakar, A. A.; *Appl. Clay Sci.* **2010**, *48(3),* 405–413.
59. Pavlidou, S.; and Papaspyrides, C. D. *Prog. Poly. Sci.* **2008**.; *33(12),* 1119–1198.
60. Prashantha, K.; Lacrampe, M.; and Krawczak, P. *Exp. Poly Lett.*; **2011** *5(4),* 295–307.
61. Qi, R; Guo R; Shen M; Cao X; Zhang L; Xu J; Yu J; and Shi, X.; **2010**. *J. Mater. Chem. 20*(47), 10622–10629.
62. Rooj, S.; Das, A.; Thakur, V.; Mahaling, R. N.; Bhowmick, A. K.; and Heinrich, G.; *Mater Des.* **2010** 31(4), 2151–2156.
63. Sengupta, R. et al.; *Polym. Eng. Sci.* **2007**. *47(11),* 1956–1974.
64. Seymour, R. B.; Polymer composites VSP, Utrecht, The Netherlands, **1990**.
65. Shamsi, M. H.; Luqman, M.; Basarir, F.; Kim, J.-S.; Yoon, T.-H.; and Geckeler, K. E.; *Poly. Int.* **2010**. *59(11),* 1492–1498.
66. Shchukin D. G.; Lamaka S. V.; Yasakau, K. A.; Zheludkevich, M. L.; Ferreira, M. G. S.; and Mohwald, H.; *J. Phys. Chem. C,* **2008**, *112(4)*, 958–964.
67. Shchukin, D. G.; and Möhwald, H. *Adv. Funct. Mater.* **2007**. *17(9),* 1451–1458.
68. Shchukin D. G.; Sukhorukov G. B.; Price R. R.; and Lvov Y. M.; *Small.* **2005**. *1(5),* 510–513.
69. Shi YF; Tian Z; Zhang Y; Shen HB; and Jia N. Q.; *Nanoscale. Res. Lett.* **2011**. *6(1)*, 608.
70. Sinha Ray, S.; and Okamoto, M., *Progr. Poly. Sci.* **2003**. *28(11),* 1539–1641.
71. Suh Y. J.; Kil, D. S.; Chung, K. S.; Abdullayev, E.; Lvov, Y. M.; and Mongayt, D,; *J. Nanosci. Nanotechnol.* **2011**. *11*(1), 661–665.
72. Tang, X. G.; Hou, M.; Zou, J.; and Truss, R;. *J. Appl. Poly. Sci.* **2013**. *128(1),* 869–878.
73. Tang, Y.; et al. *Compos. Part A. Appl. Sci. Manuf.,* **2011**. *42*(4), 345–354.
74. Tang, Y; Ye, L.; Deng, S.; Yang, C.; and Yuan, W.; *Mater. Des.* **2012**. *42(0),* 471–477.
75. Tierrablanca, E.; Romero-García, J.; Roman, P.; Cruz-Silva, R. *Appl. Catal. A. Gen.* **2010**. *381(1–2),* 267–273.
76. Tsou, A. H.; and Waddell, W. H.; Fillers. In: Eds. Mark, H. F.; *Encycl. Poly. Sci. Technol.* vol 10. John Wiley & Sons, Inc., **2002**, 1–20.
77. Vergaro, V.; et al. *Biomacromolecules.* **2010**. *11*(3), 820–826.
78. Wang, R.;et al. *Appl. Mater. Interf.* **2011**, *3(10),* 4154–4158.
79. Wei, W.; Abdullayev, E.; Hollister, A.; Mills, D.; and Lvov, Y. M.; *Macromol. Mater. Eng.* **2012**, *297(7),* 645–653.
80. Wilson, I. R.; *Clay. Min.* **2004**. *39*(1), 1–15.
81. Ye Y.; Chen, H,; Wu, J.; and Chan, C. M.; *Compos. Part B. Eng.* **2011a**, *42*(8), 2145–2150.
82. Ye, Y.; Chen, H.; Wu, J.; and Chan, C. M.; *Compos. Sci. Technol.* **2011b**. *71*(5), 717–723.
83. Ye, Y.; Chen, H.; Wu, J.; and Ye, L.; *Polymer.* **2007**. *48*(21), 6426–6433.
84. Zatta, L; Gardolinski, J. E. F. D. C.; and Wypych, F. *Appl. Clay. Sci.* **2011**, *51*(1–2), 165–169.
85. Zhang, Y.; and Yang, H.; *Appl. Clay. Sci.* **2012**, *56(0),* 97–102.
86. Zhixin, J.; Yuanfang, L.; Baochun, G.; Shuyan, Y.; Mingliang, D.; and Demin, J.; Styrene-butadiene rubber/halloysite nanotubes composites modified by epoxidized natural rubber. In: 3rd International Nanoelectronics Conference (INEC), Hong Kong, China, 3–8 Jan. **2010**, 1030–1031.
87. Zou, H.; Wu, S.; and Shen, J.; *Chem. Rev.* **2008**. *108*(9), 3893–3957.

PART V
BIOPOLYMER COMPOSITES OF NATURAL MINERAL NANOTUBES AND THEIR APPLICATIONS

CHAPTER 9

BIONANOCOMPOSITES OF SEPIOLITE AND PALYGORSKITE AND THEIR MEDICAL ISSUES

EDUARDO RUIZ-HITZKY, MARGARITA DARDER,
FRANCISCO M. FERNANDES, BERND WICKLEIN,
ANA C. S. ALCÂNTARA, and PILAR ARANDA

CONTENTS

9.1   SEPIOLITE AND PALYGORSKITE FIBROUS CLAYS IN BIOHYBRIDS PREPARATION

Among clay minerals sepiolite and palygorskite constitute a group of natural silicates of microfibrous morphology. Both silicates exhibit a crystalline structure consisting of T-O-T (T = tetrahedral; O=octahedral) structural blocks alternating with cavities (tunnels), along the fiber direction (c-axis) (Figure 9.1) (Bradley, 1940; Brauner and Pressinger, 1956; Santarén et al., 1990; Ruiz-Hitzky, 2001). The ideal unit cell formulae of sepiolite and palygorskite are $Si_{12}O_{30}Mg_8(OH)_4(H_2O)_4 \cdot 8H_2O$ and $Si_8O_{20}(Al_2\,Mg_2)(OH)_2\,(H_2O)_4 \cdot 4H_2O$, respectively (Suárez and García-Romero, 2011). Tunnel dimensions are 1.06×0.37 nm^2 and 0.64×0.37 nm^2 for sepiolite and palygorskite, respectively (Ruiz-Hitzky et al., 2011a). These structural micropores can be occupied only by small-size molecules such as water, methanol and acetone (Ruiz-Hitzky, 2001). Most interactions between the fibrous clays and organic or inorganic species involve the external surfaces of those silicates. The specific surface of both clays are high (ca. 300 m^2/g and 200 m^2/g, for sepiolites and palygorskites of different origin, respectively) being around a half of those values ascribed to the external surface (Alcântara et al., 2012). This external surface is covered by silanol ($\equiv$ Si-OH) groups regularly located at the edge of the T-O-T blocks due to the discontinuity of the silica sheets, and they can establish direct interactions with diverse organic species (Ruiz-Hitzky, 2001). These structural and textural features allow the preparation of diverse types of advanced nanostructured materials, essentially reliant on their ability to assemble inorganic species and nanoparticles (Ruiz-Hitzky and Aranda, 2014; González-Alfaro et al., 2011; Ruiz-Hitzky et al., 2012a), organic molecules (Ruiz-Hitzky et al., 2004, 2010a), as well as polymer and biopolymers (Ruiz-Hitzky et al., 2010b, 2011a, b, 2013a, b).

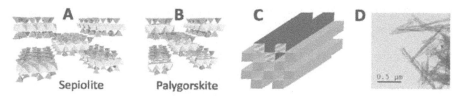

FIGURE 9.1   Schematic representation of structural and textural features of the fibrous clay minerals: (A) sepiolite structure; (B) palygorskite structure; (C) ideal representation of the cross-section of sepiolite fiber; (D) TEM image of sepiolite from Vicálvaro, Spain (Reprinted from Ruiz-Hitzky et al. (2013b), Copyright (2013), with permission from Elsevier).

Probably the first clay-based hybrid material was the so-called Maya Blue, consisting of a natural pigment (indigo) included into the intracrystalline cavities of palygorskite. A perfect fitting of these dye molecules within the structural tunnels of the silicate impedes their degradation by light, weathering and microorganisms by allowing the preservation of the initial vivid color for centuries (Ruiz-Hitzky et al., 2011a). Another way to obtain very stable organic-inorganic hybrid materials based on the fibrous clays contains the grafting reactions of organosilanes through silanol groups on the external surfaces (Ruiz-Hitzky and Fripiat, 1976; Ruiz-Hitzky, 2004; Ruiz-Hitzky et al., 2004). These silanol groups can act as assembling points for many polar molecules and polymers, being the basis for the generation of a large variety of hybrid and biohybrid materials. Among these latter, the use of biopolymers is of particular relevance for preparing bionanocomposite materials (Darder et al., 2007; Ruiz-Hitzky et al., 2008; 2013b; in press). The interest in this class of nanostructured systems is related to the biocompatibility and biodegradability properties inherent to the biopolymers and to the apparent lack of toxicity shown by the most fibrous clay minerals. This chapter will introduce various highlights of bionanocomposites based on the fibrous clays, especially sepiolite, showing their potential toward diverse applications including biomedical aspects.

9.2 PROCESSING OF BIONANOCOMPOSITES

The nature of the different phases is used to prepare bionanocomposites is a fundamental aspect. It ensures biocompatibility while accounting for demanding structural properties, enhances biodegradability and often reshapes common materials toward greener, more eco-friendly ones. However, besides the nature of bionanocomposites, another key aspect that must be taken in consideration before fully disclosing their growing applications is their processing. As is well known, a material is the result of three distinct features: its composition, its processing and finally its application. From this point of view, processing goes beyond the technical step of shaping a given composition to obtain a usable object, it is rather a rational step linking a given composition to a specific application. Given the wide range of applications which are considered for sepiolite- and palygorskite-based bionanocomposites, their processing requires thus some degree of attention. In general, processing bionanocomposites can be achieved by solution casting (for the great majority of reported bionanocomposites), *in situ* polymerization and melt processing (Ojijo and Ray, 2013).

The most common applications for biomaterials require the material to be either in a hydrogel state (Giraud Guille et al., 2010) (*e.g.* for extracellular matrix tissue repair) or in the form of fibers (Boccaccini et al., 2003) (*e.g.* for surgical sutures), nonwoven felts (Roh et al., 2008) (*e.g.* for vascular grafts), films (Lee et al., 2001; Lu et al., 2004) (*e.g.* for corneal substitutes and aortic wall and leaflets) or foams (*e.g.* for hemostatic resorbable dressings).

Hydrogels based on biopolymers are readily prepared in the presence of the fibrous clays. In fact, the presence of a fibrous clay moiety in either physical or chemical biopolymer gels seems to be fully compatible with the conventional water-based procedures. This is quite intuitive since the fibrous clay crystals display abundant silanol groups along their edges, which account for their important hydrophilicity. Moreover, the viscoelastic properties of fibrous silicates in water can actually enhance the gel-forming ability of biopolymers. Many examples are available from the literature where sepiolite or palygorskite are used along with different biopolymers both for structural or functional bionanocomposites (Darder et al., 2006; Chivrac et al., 2010). It is also worth noting that many of the different final forms of bionanocomposites are actually mediated by a hydrogel step. Highly homogeneous gelatin-sepiolite hydrogels have been cast and subsequently dried to obtain bionanocomposite films (Fernandes et al., 2011). This strategy can be assumed to be widely applicable for the preparation of bionanocomposite films, sheets and membranes. Aerogels can also be obtained through a controlled removal of the solvent to achieve highly porous, low density materials. Although not exclusive, the main techniques to achieve such constructs are freeze-drying (Wicklein et al., 2013), and supercritical drying (Fujimoto et al., 2003).

It is quite extraordinary to note that no fiber processing techniques such as melt spinning or solution spinning have been described for biopolymers in the presence of fibrous clays. Some reports however mention that spinning standard polymers, such as polyamides (Tsai et al., 2012) or polyurethane (Bilotti et al., 2010), in the presence of fibrous clays is not only possible, but also leads to improved mechanical properties. Moreover, Tsai and coworkers (Tsai et al., 2012) have demonstrated that a low palygorskite content can actually increase the applied draw ratio in the fiber spinning process. Some preliminary reports are however available in electrospinning poly-lactic acid in the presence of sepiolite (Alongi and Poskovic, 2011). The obtained microfibers show clear improvements in both thermal and gas barrier properties. From our perspective, there is a wide terrain for developing highly efficient materials if these strategies are explored more systematically.

The last of the main processing techniques used in bionanocomposites based on fibrous clays concerns the final form of powder-like composites. Several works have used the adsorption of the biopolymer moiety onto the fibrous silicate in aqueous suspension to obtain precise mass ratios between the silicate and the biopolymer in the form of a powder. In fact, such a strategy has been applied to sepiolite-based bionanocomposites using chitosan (Darder et al., 2006), xanthan (Ruiz-Hitzky et al., 2009) and more recently continuous phospholipid biomimetic membranes (Wicklein et al., 2010).

In summary, processing bionanocomposites is clearly dominated by the water-related approaches, which is readily understood in view of the high hydrophilicity of both the clay fibers and hydrocolloids. Moreover, since most biopolymers degrade before attaining a molten state, there is a clear incompatibility between the commonly applied techniques for thermoplastic-based nanocomposites and the thermal stability of many biopolymers.

9.3   GREEN NANOCOMPOSITES BASED ON SEPIOLITE AND PALYGORSKITE

Reduction of packaging waste was a strong motivation for replacing plastic packaging materials by biodegradable materials from renewable sources during the past decades. Bioplastics include the plastic materials made from biological/renewable resources or those degraded by the action of micro-organisms/biological activity or both. Thus, a bioplastic can be bio-based and/or biodegradable, that is a biopolymer derived from nature and/or a polymer that can return to nature (Vert et al., 2012). In this sense, much research is devoted to bioplastics derived from agropolymers (eg. starch, cellulose and proteins), bacterial sources such as polycaprolactone (PCL), polylactic acid (PLA), and polyhydroxyalkanoates biodegradable polymers. Since most of these biopolymer films have a high hydrophilic character and do not present good enough mechanical properties for their potential use as a bioplastic, their assembly to natural or synthetic clays nanofillers is a usual alternative for overcoming these drawbacks.

Although less studied as nanofillers compared to layered clays, sepiolite and palygorskite microfibrous clays can be employed as additives in bio-based matrices originating bionanocomposite materials (Ruiz-Hitzky et al., 2013b; in press). These fibrous clays show some characteristics that make them very attractive such as their large aspect ratio, high specific surface area and the presence of silanol groups at their external surface, which can act as interaction sites with other species. The incorporation of microfibrous clays normally causes a reinforcement of the polymer matrix, which derives mainly from the uniform dispersion of the filler particles inside the matrix and the strong interfacial interactions between both components. In this context, the first and important goal in the use of bionanocomposite material as bioplastics is related to their excellent mechanical properties. This reinforcement effect was reported by Darder et al. (2006) in self-supported chitosan-sepiolite bionanocomposite films prepared in acidic media conditions. The good integration of sepiolite fibrils within the biopolymer structure, evidenced by SEM images of the bionanocomposite films (Figure 9.2), confers an improvement in the mechanical properties of these materials about three times higher in comparison to the biopolymer alone. This phenomenon can be explained by the strong interaction through intermolecular hydrogen bonding, which is supported by spectroscopy techniques (FTIR and ^{13}C NMR). These bionanocomposite films also exposed interesting results as membranes for N_2 separation, being a promising material in membrane processes (Martínez-Frías, 2008). The combination with fibrous clays was extended to other polysaccharides such as alginate, pectin, starch, xanthan and carboxymethylcellulose (CMC) and hydroxypropylmethylcellulose (HPMC) cellulose derivatives (Alcântara et al., 2014).

In this case, bionanocomposite films prepared by the incorporation of sepiolite and palygorskite as reinforcing agents in the biopolymer matrixes not only exhibited an increase in the elasticity modulus compared to the pristine polysaccharide films, but also showed remarkable resistance to water and barrier properties to the passage of UV light, making them very attractive for their use as reinforced bioplastics or coatings for the food sector. Another strategy involving polysaccharides as a biopolymer matrix was the prior modification of the sepiolite surface with cationic starch in order to improve the compatibility of the clay particles with the starch matrix (Chivrac et al., 2010). In this case, the bionanocomposite films that incorporate organosepiolite in starch showed better compatibility and mechanical properties than starch films loaded with pure sepiolite, montmorillonite and montmorillonite modified with cationic starch.

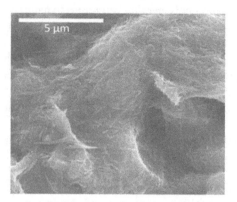

FIGURE 9.2   SEM image of chitosan-sepiolite bionanocomposite film.

The good compatibility evidenced between the polysaccharides and microfibrous clays was also verified in proteins. In this context, the assembly of gelatin, a structural protein derived from collagen and sepiolite was studied by Fernandes et al. (2009, 2011) in the preparation of homogeneous gelatin-sepiolite bionanocomposite films, obtaining an increase of the elastic modulus of up to 250 percent for the bionano-composites with 50 percent (w/w) of clay loading. Although an absolute knowledge of the interactions between both components could not be established, the important role of the sepiolite fibrils in the structural properties of these films is evident, which may be related to the adequate environment provided by the sepiolite for the crystal-lization of the protein, reflected in the mechanical properties of the resulting bion-anocomposites. In the context of gelatin-based bionanocomposite films, the effect was evaluated of the incorporation of sepiolite in gelatin-egg white films contain-ing clove essential oil, as an active compound against pathogenic and spoilage mi-croorganisms of foods (Gimenez et al., 2012). Besides providing good mechanical properties, the presence of the fibrous silicate in this bioactive packaging allowed a gradual release of clove oils components, favoring antioxidant and antimicrobial activities in this film, as a potential material for the food area.

Within of the scope of proteins, reinforced wheat gluten (WG) films were pre-pared by thermal compression-mold using palygorskite fibrous clay as filler (Yuan et al., 2010). In this case, SEM and TEM images of the transparent WG/palygorskite films revealed that the clay particles are evenly dispersed in the WG matrix, reflected in a significant improvement of the mechanical properties of these materials. Furthermore, these bionanocomposites showed attractive results for biodegradability, where only 20 wt % of the WG films loaded with 7 percent (w/w) of palygorskite remained after 20 days burial in the soil. On the other hand, taking advantage of the possibility of modification of the clay surface, a new bio-organoclay material based on the assembly of the hydrophobic zein protein to sepiolite and palygorskite fibrous clays, was synthesized in order to reduce the hydrophilic character of these clays, as is the case of conventional organoclays based on long-chain alkylammonium cations (Alcantara et al., 2010, 2012). The improved hydrophobic properties of zein-fibrous clay biohybrids were evaluated as biofillers in hydrophilic biopolymer matrices, such as those of alginate polysaccharide. Such bionanocomposite films showed a marked resistance to the passage of water molecules and improved barrier properties toward oxygen in comparison to pristine alginate films, with promising applications in food packaging and membrane processes.

In the search for ecological alternatives for plastics derived from renewable, biodegradable polymers such as PLA and PCL, which are produced by classical chemical routes using bio-based monomers, are widely employed as biopolymer matrices in the preparation of bioplastics for diverse applications. Several approaches to synthesis involving sepiolite and palygorskite fibrous clays as reinforcing agents of PLA and PCL films have been reported in literature (Duquesne et al., 2007; Liu and Chen, 2008); they include solution mixing and melt-blending processes. From this perspective, Fukushima et al. (2009) studied the influence of adding sepiolite in the biodegradation of both PCL and PLA bioplastics in compost (Fukushima et al., 2010). Although the biodegradation of PCL does not appear to be strongly influenced by the presence of clay, PLA films loaded with sepiolite showed significant changes in biodegradation compared to pure PLA films. In an earlier case, a possible effect of the fibers on polymer chains in preventing the enzymatic degradation of the bionanocomposite film is postulated. Furthermore, hydrolytic biodegradation experiments of PLA/sepiolite films carried out in a phosphate buffer (pH 7.0) showed a delay occurring in film degradation, while at 58°C the incorporation of sepiolite fibers does not seem to strongly influence the biodegradation of PLA, since an analogous decrease of the molecular weight for the bionanocomposite material and pristine polymer film is evidenced. Such behavior could be associated with an easy access of water molecules to the amorphous and semicrystalline regions, producing a hydrolytic chain scission in both materials.

Given the good adsorption properties of clays, as well as their abundance, low-cost, and nontoxicity, they are widely used in environmental applications such as the removal of pollutants (Gatica and Vidal, 2010), with special emphasis on the cleaning of drinking water (Srinivasan, 2011). The usefulness of organoclays in this type of application, which allow the removal of anionic pollutants and hydrophobic nonpolar species in addition to cationic compounds (Ruiz-Hitzky et al., 2010a) is also well-known. Within this context, the bionanocomposites resulting from the assembly of natural polymers and the fibrous clays can be also very valuable in the environmental field, showing in some cases a synergistic effect coming from the combination of adsorption and absorption properties of both types of components. This is the case of systems involving chitosan and sepiolite or palygorskite, in which the polysaccharide chains interact with the clay through hydrogen bonding interactions with the silanol groups present on its surface and also through electrostatic interactions, as the biopolymer is positively charged as a result of bearing amino protonated sites (Darder et al., 2006). Thus, it has been reported that the adsorption of Fe(III) and Cr(III) ions on the chitosan-clay bionanocomposites is significantly higher than on the separate components (Zou et al., 2011). These materials, as well as those based on alginate assembled to sepiolite or palygorskite, also showed their ability for the removal of Pb(II) and Cu(II) ions from aqueous solutions (Figure 9.3) (Alcântara et al., 2014). Similarly, chitosan-palygorskite materials were efficient for the uptake of U(VI) ions from aqueous solutions, with the absorption being a feasible, spontaneous, and endothermic process (Pang et al., 2010). In contrast, a bionanocomposite involving alginate, sepiolite and other components was also able to absorb U(VI) spontaneously, but the process was endothermic, and therefore was enhanced by increasing the temperature (Donat et al., 2009a). Another adsorbent of uranium ions was obtained by assembling the green macro marine algae *Ulva sp.* with sepiolite (Donat et al., 2009b). Although the nature of the Ulva *sp.*-sepiolite interaction has not been investigated, the reported study confirms that further desorption with HNO_3 leads to an almost complete recovery of U(VI) ions. A prior modification of the clay provides bionanocomposite materials with new functionalities, as in the case of palygorskite modified with magnetite to introduce magnetic properties (Mu et al., 2013). The further assembly of Fe_3O_4-palygorskite with polyelectrolytes like chitosan and cysteine-modified β-cyclodextrin, deposited on the surface of the fibers following a layer-by layer (LbL) process, results in bionanocomposites able to better adsorb traces of precious metal ions, particularly Ag^+ and Pd^{2+}.

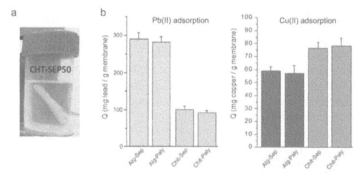

FIGURE 9.3   (a) Chitosan-sepiolite bionanocomposite with clay content of 50% wt. (CHT-SEP50) showing the enhanced stability in water. (b) Adsorption of Pb(II) and Cu(II) ions from aqueous solutions by different alginate (Alg) and chitosan (Chit) bionanocomposite membranes involving sepiolite (Sep) or palygorskite (Paly). (Alcântara et al., 2014).

Functional biopolymer-fibrous clay nanocomposites showing superabsorbent properties have been prepared based on a synthetic polymer, such as polyacrylic acid (PAA) or polyacrylamide (PAAm), grafted to the biopolymer chains. The disaggregation and dispersion at the nanometer scale of these fibrous silicates within the polymers is responsible for the enhanced properties of the resulting materials (Ruiz-Hitzky et al., 2011a). In fact, the addition of palygorskite in materials based on guar gum-g-PAA (Wang et al., 2008) or starch-g-PAAm (Li et al., 2005) gave rise to increased water absorbency and water absorption rates in the ternary systems. The water absorption properties can be also tuned by modifying the clay. For instance, Mg^{2+} ions can be filtered from the sepiolite structure by acid treatment and then the clay can be modified with metal salts like $AlCl_3$ (Xie et al., 2010). These multivalent ions can act as crosslinkers of the polymer matrix in chitosan-g-PAA/sepiolite systems, increasing the water absorption and the swelling rate. Usually, this type of material finds application in agriculture, but another original use of a CMC-based system involving sepiolite and a copolymer of PAA and PAAm, showing high moisture absorption capacity and a fast response to humidity changes, was the control of relative humidity in museums and similar places (Yang et al., 2011). Another interesting application of this type of nanocomposite is related to the removal of pollutants. For this purpose, the systems usually involve biopolymers bearing negatively charged sites, as in CMC-g-PAA/palygorskite bionanocomposites that allow the removal of heavy metal ions like Pb(II) (Liu et al., 2010), or biopolymers which shows complexing ability, such as chitosan-g-PAA in palygorskite-based superabsorbents able to adsorb Cu(II) (Wang et al., 2009) among other pollutants.

The removal of organophilic pollutants requires the modification of the fibrous clays with appropriate compounds, as for instance phospholipids that can self-organize as a supported lipid membrane on the clay surface (Wicklein et al., 2010). The organophilic environment afforded by the lipid membrane allows the retention of hydrophobic species such as mycotoxins (Wicklein et al., 2010). The lipid-modified sepiolite was more effective in the removal of aflatoxin B1 (AfB1) than commercial alkylammonium sepiolites, showing adsorption capacities of about 1440 mg/100 g.

The fibrous clays and their bionanocomposite derivatives can also act as matrices for the protection or the immobilization of biological species of environmental interest. Thus, chitosan-palygorskite materials were associated with diverse anionic dyes and applied as matrices for the photostabilization of microbial biocontrol agents (Cohen and Joseph, 2009). The phosphate-accumulating bacteria *Acinetobacter junii* was immobilized on the surface of sepiolite fibers by extracellular substances while retaining its bioactivity and the resulting material was able to remove phosphate from synthetic wastewater (Hrenovic et al., 2010). The assembly of polysaccharides like chitosan to the fibrous clays may contribute to an increase in their biocompatibility. In this sense, chitosan-sepiolite materials processed as macroporous foams were suitable for the accommodation of diverse algal cells (Ruiz-Hitzky et al., 2010b). The resulting systems could be exploited for the capture and transformation of CO_2 into O_2. It has been also reported that the presence of the biopolymer associated with sepiolite is useful to increase the flocculation of microalgal cells (*Microcystis aerugi nosa*) which present in freshwaters (Pan et al., 2006).

9.4 BIOMIMETIC INTERFACES

The biomimetic interfaces resembling those of cell membranes or mucosae can be built from different molecules with self-assembly ability, including phospholipids, polysaccharides, and natural and synthetic biosurfactants (Lu et al., 2007; Carnero Ruiz, 2009; Koutsopoulos et al., 2012). The advantages of biosurfactants besides their structural and functional variability are inherent biocompatibility, biodegradability, and nontoxicity (Pacwa-Plociniczak et al., 2011). Processes to obtain biomimetic interfaces are generally self-assembly-driven like the molecular self-assembly of dissolved molecules including supramolecular chemistry (Lehn, 2002; Ariga et al., 2008), self-assembly of preformed aggregates such as vesicles (Rapuano and Carmona-Ribeiro, 1997; Anderson et al., 2009), or the use of Langmuir-Blodgett technologies (Petty, 1996; Szabo et al., 2008).

The fibrous clays can act as excellent supports for these biomimetic interfaces with physicochemical and/or biological properties similar to their natural analogs (Ruiz-Hitzky et al., 2009; Wicklein et al., 2010, 2011, 2012). Such interfaces can serve as compatibilizers between the inorganic host and immobilized biological species. In this way, enzymes, proteins, viral particles, and nucleic acids could be adsorbed on fibrous clays and preserve their biological activity and be useful as active phases in biosensors, biocatalysts, vaccines, transfection agents, and more.

A recent example for the formation of biomimetic interfaces on sepiolite is the deposition of different lipid membrane structures on sepiolite (Wicklein et al., 2010, 2011). Here, the phospholipid molecules were adsorbed either from ethanol or as liposomes from aqueous media. Depending on the lipid concentration mono- or bilayer membranes can be obtained by more hydrophobic (monolayer) or more hydrophilic (bilayer) properties. NMR and IR investigations suggested hydrogen bonding between the lipid headgroup and the sepiolite silanol groups as the main interaction mechanisms (Wicklein et al., 2010). By subsequent adsorption of e.g. sugar-based surfactants (octyl-galactoside) onto the lipid monolayer, a hybrid bilayer membrane with sugar groups on the external surface could be prepared (Wicklein et al., 2011). Such mixed bilayers could be used as models for glycolipids or glycoproteins containing cell membranes such as erythrocytes. These sugar-based surfactant-containing interfaces are also interesting for their antifouling properties (Hederos et al., 2005; Fyrner et al., 2011) or their protein stabilization ability (Stubbs et al., 1976; Privé, 2007; Mukherjee et al., 2011).

The sepiolite-supported lipid interface was shown to be well-suited for the immobilization of functional biological species such as enzymes and proteins. In a comparative study the cytoplasmic enzyme urease and the membrane-associated enzyme cholesterol oxidase were adsorbed on phosphatidylcholine membranes and maintained their enzymatic activity over months. Contrarily, when the enzymes were adsorbed on the pristine sepiolite surface their catalytic activity was immediately lost due to strong ionic interactions with the clay surface (Figure 9.4). Thus the supported enzymes were successfully tested as the active phase in an urease biosensor and a cholesterol batch bioreactor (Wicklein et al., 2011). In addition, these materials show the advantage of using an eco-friendly modifier like PC, in comparison to analogous systems where the enzyme alkaline phosphatase was immobilized on sepiolite modified with a supported bilayer of cetyl trimethylammonium bromide (CTMAB) (Sedaghat et al., 2009). In a follow-up study, sepiolite-lipid-urease hybrids were incorporated into nanoarchitectured polyvinyl alcohol foams showing simultaneous enzymatic and conductive properties making these nanocomposite foams interesting for bioelectrocatalysis (Wicklein et al., 2013).

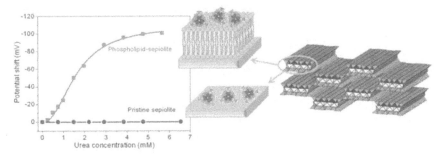

FIGURE 9.4   Illustration of a bilayer lipid membrane adsorbed on sepiolite with immobilized urease in comparison to urease directly adsorbed on the sepiolite surface. The biohybrid material was employed in a voltammetric biosensor for urea. The signal is a potential shift as result of a pH increase produced by enzymatic urea oxidation and plotted as function of urea concentration.

   Many functional proteins like receptors, transporters, or catalysts reside on the lipid membranes of cells, microorganisms, or viruses. The exploitation of their biological function is of great technological and medical interest; however, many proteins rapidly degenerate outside their natural environment which makes adequate immobilization imperative. One of these high-interest proteins is hemagglutinin (HA), a surface protein of the influenza virus. Its antigenic properties make this protein interesting for influenza subunit vaccine development. In a recent work, recombinant HA was immobilized on sepiolite-lipid biohybrids in order to mimic the viral surface of influenza A virions (Wicklein, 2011). It could be shown that HA supported on these lipid interfaces shows similar agglutination activity as HA present in the lipid envelop of the native virus particle. This strongly suggests that influenza vaccination doses formulated on the basis of hemagglutinin alone could be equally immunogenic as conventional whole virus vaccines.

Besides lipid-based membranes there exists a large variety of natural biointer-
faces constituted of polysaccharides and proteins such as capsular and extracellular polysaccharides (CPSs and EPSs) in bacterial (Esko, 1999) and fungal cell
walls (Latgé, 2007), secreted EPS for cell adhesion and protective films (Vu et al.,
2009; Bazaka et al., 2011), surface-layer (S-layer) proteins of prokaryotes (Sleytr
and Messner, 2009; Pum et al., 2013), or the glycosylated protein mucin that is
involved in respiratory epithelial lubrication and barrier interfaces, e.g. in mucous
membranes (Linden et al., 2008). Another biointerface that has lately received great
attention are the adhesive amino acid-catechol based surfaces of blue mussel (*Myti
lus edulis*) byssus threads (Lee et al., 2006, 2007).

Following the inspiration from mucous membranes Sandberg et al. (2009) describes the preparation of supports coated with a layer of mucin, which actively
interacts with host proteins adsorbed on this coating. Similarly, the assembling of
polysaccharides, mimicking the glycosylated structure of mucin, to solid supports
including clay minerals can also offer a biocompatible interface to accommodate
biological species and preserve their bioactivity. Thus, the surface of sepiolite fibers was modified with the anionic polysaccharide xanthan gum with the aim to
develop a biointerface that mimics features of mucous membranes in their interaction with virus particles (Ruiz-Hitzky et al., 2009). Xanthan gum consists of a
β-(1→4)-D-glucopyranose glucan backbone with side chains of -(3→1)-α-linked
D-mannopyranose-(2→1)-β-D-glucuronic acid-(4→1)-β-D-mannopyranose on alternating residues, about 40 percent of the terminal mannose residues being 4,6-py-
ruvated (Chaplin, 2013). Modification of the sepiolite fibers takes place by spontaneous assemblage of xanthan, driven by hydrogen bonding interactions between the
hydroxyl groups of the polysaccharide and the silanol groups located at the external
surface of the silicate. The biohybrid material provides water retention ability and
constitutes a favorable environment for biological species. It also bears the negative
charges coming from xanthan, which can interact electrostatically with the positively charged surface of the influenza viral particles. Thus, the resulting biointerface would mimic the interaction of natural mucosa with virions during infection
processes.

9.5   BIOMEDICAL APPLICATIONS

9.5.1   MEDICAL ISSUES OF SEPIOLITE AND PALYGORSKITE

Traditional uses of clay minerals include applications related to human healthcare
due to the unique properties of these silicates. They have been proposed as active
drugs by themselves or when used as additives or excipients in pharmacological
formulations. Antiseptic (antimicrobial and antifungal agents), skin protection,
anti-inflammatory, antidiarrheic, and other beneficial effects (Carretero and Pozo,
2009) have been attributed for centuries to different types of clays around the world.
In addition, physical properties of clays are convenient as drug diluents, carriers

and binders. Good quality smectites (bentonites, saponite, synthetic hectorites,.), kaolinite, as well as sepiolite and palygorskite have been traditionally used as pharmaceutical-grade clays due to their mechanical and rheological properties, acting as emulsifiers and stabilizers of drug formulations and also as suspending agents for hydrophobic drugs (Takahashi and Yamaguchi, 1991; Ito et al., 2001; Carretero and Pozo, 2009). Surface properties of clays mediated in adsorption-desorption processes allow a controlled release of drugs in interaction with the silicates. The most important effect is the retarded release of adsorbed pharmaceuticals, helping to modulate the drug release profile (Aguzzi et al., 2007). Excellent reviews by Pozo, Carretero, Viseras, and other researchers have been published on this topic including on uses of sepiolite and palygorskite (Aguzzi et al., 2007; Carretero and Pozo, 2009; Viseras et al., 2010).

The smectite clays can retain drugs mainly by adsorption mechanisms based on ion-exchange reactions and intercalation into the interlamellar space of these layer silicates (Choy et al., 2007; Joshi et al., 2009; Oh et al., 2009). This represents the encapsulation of drugs which are then protected and suitably released by different ways following their administration. Conversely, sepiolite and palygorskite offer a low cation-exchange capacity compared to the smectites but afford an elevated external surface that is covered by hydroxyls (silanol groups) which can be appropriated for drug immobilization. The surface properties of this type of clays have proved to be effective for adsorbing odor and exudates from wounds (Kose et al., 2005). Besides, the fibrous clay minerals can also interact directly with drugs of diverse nature such as β-blockers and 5-aminosalicylic acid and so can act as carriers for drug delivery (Aguzzi et al., 2007). Sepiolite and palygorskite also show possibilities from prior (bio) organic modification. Thus, bionanocomposites based on the fibrous clays, which show biocompatibility, nontoxicity, high water uptake or mucoadhesivity, can be promising candidates for biomedical applications as discussed below.

The use of clay minerals for biomedical applications must address toxicity issues in particular when involving fibrous clays. Toxicity of these latter silicates is controversial and apparently depends on their fiber length and crystallinity, which in turn is determined from their geological origin (Carretero and Pozo, 2007). It seems that the fibers of lengths below 5 micrometers do not cause serious harm in cells in mice as reported for *in vivo* experiments (Carretero and Pozo, 2007). Oscarson et al. (1986) have reported erythrocyte lysis by the action of sepiolite and palygorskite fibers. However, sepiolite is considered a nonhazardous mineral according to European directives (e.g. 67/548/EEC), being extensively used as a food additive in animal feed preparations. *In vitro* and *in vivo* tests, as well as epidemiological studies, point out that at least sepiolite from Taxus Basin deposits in Spain does not constitute a health risk (Denizeau et al., 1985; Santarén and Alvarez, 1994). Longer sepiolite fibers such as those from other deposits (e.g., in China and Finland) must be employed with caution due to potential hazardous effects in animal and humans.

9.5.2   FIBROUS CLAYS-BASED BIONANOCOMPOSITES AS DDS

Biopolymer-clay nanocomposites are being applied as carriers in drug delivery systems, profiting from their interesting properties such as swelling, film-forming ability, bioadhesion, and cell uptake (Viseras et al., 2008). They can afford controlled delivery of the drug, allowing the release of constant levels within the therapeutic dose, and thus avoiding overdosing or underdosing episodes during the treatment.

The modification of the clay surface with biopolymers can also be helpful to enhance the adsorption of nonionic or hydrophobic drugs, which nowadays represent a high percentage of used active principles. In addition, the assemblage of biopolymers can help to reduce the tendency of many pristine clays used as drug carriers to flocculate in physiological media (Luckham and Rossi, 1999).

Most of the bionanocomposites reported for drug delivery applications are based on layered silicates, but the possibility of using analogous materials involving fibrous clay minerals began to be also explored several years ago. In this sense, the hydrogel system based on sepiolite and the biocompatible polymer polyvinyl alcohol (PVA) was evaluated as a DDS for the controlled release of rifampicin, a drug of low solubility used in tuberculosis treatment (Viçosa et al., 2009). In this system, the fibrous clay does not act as a drug carrier, but is dispersed within the rifampicin-loaded PVA matrix in order to reduce swelling and thus slow the release rate of the drug.

Chitosan-palygorskite bionanocomposites were evaluated for oral delivery and controlled release of diclofenac sodium (DS), a nonsteroidal anti-inflammatory drug (Wang et al., 2011a). The addition of palygorskite is beneficial in the processing of these systems by means of spray-drying, as it contributes to an increase in the isoelectric point allowing uniform microspheres with a narrow size distribution to be obtained. In addition, its presence within the chitosan matrix helps to reduce the release of DS in gastric fluid and ensures the controlled release of DS at the pH conditions of the intestinal fluid.

The superabsorbent hydrogel materials (described in Section 9.3 of this chapter) are also useful in drug delivery applications. For instance, palygorskite–CMC–g-PAA hydrogels loaded with the drug DS were mixed with the polysaccharide alginate and processed as beads by means of ionic cross-linking of the alginate chains with Ca^{2+} ions (Figure 9.5a) (Wang et al., 2012). The slow release of the drug molecules is impeded in gastric fluid conditions but takes place as the material swells in simulated intestinal fluid (Figure 9.5b). The release rate is slower as the content in palygorskite increases due to the decrease in swelling.

As is the case for drug delivery, the fibrous clays-based systems can be also of interest for application as nonviral vectors for gene transfection. In spite of the negative charge of the silicate structure in sepiolite and palygorskite, DNA is able to adsorb on their surfaces through hydrogen-bonding interactions with the external silanol groups located on the edges of these silicate fibers. Ruiz-Hitzky et al. (2012b) showed that short-chain DNA molecules spontaneously adsorb on sepiolite and could find use as nonviral agents for gene-therapy. In fact, Yoshida and Sato (2009) investigated the transfer of exogenous plasmid DNA into bacterial cells like *E. coli* and other Gram-negative and Gram-positive bacteria using sepiolite fibers with diameters up to 50 nm.

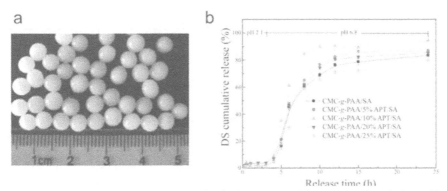

FIGURE 9.5   (a) Beads of the composite hydrogel comprising CMC-g-PAA/palygorskite (APT)/sodium alginate (SA) loaded with DS. (b) In vitro cumulative release profiles of DS from composite hydrogel beads with different palygorskite content in simulated gastric fluid (pH 2.1) and simulated intestinal fluid (pH 6.8) at 37°C (Reprinted from Wang et al. (2012). Copyright © 2011 Wiley Periodicals, Inc.).

9.5.3   VACCINES

Efficacious vaccines greatly rely on their stabilization on particulate carriers that can enable easy administration of immunogenic species, their protection from proteolysis, improved thermal storage performance, and provide an adjuvant effect of the solid vehicle (Heegaard et al., 2011). Based on these requirements clays have been recently recognized as potential alternatives to soft matter carriers like liposomes, micelles, virosomes, polymeric nanoparticles, polymersomes, emulsions, and dendrimers (Peek et al., 2008; Heegaard et al., 2011) or inorganic, mineral carriers such as Alum (Clapp et al., 2011), calcium phosphate, layered double hydroxides, or iron oxide nanoparticles materials (Xu et al., 2006). The reason for their attraction is the large protein adsorption capacity related to the large specific surface area of clays and the inherent biocompatibility of these minerals.

Using the pristine, layered clay bentonite, Patil et al. (2004) prepared a clay-adjuvated vaccine against hemorrhagic septicemia in calves which gave a longer-lasting immunity as compared to the alum-supported vaccine. However, it was observed in this study that adsorption of antigens on the pristine clay surface can alter their protein structure which resulted in decayed immunogenicity of the antigen and eventually, can compromise the vaccination efficacy (Ruiz-Hitzky et al., 2009; Rytwo et al., 2010). Therefore, Rytwo et al. (2010) explored the possibility of using clays with organo-interfaces (montmorillonite-berberine hybrids) in order to accommodate antigens while preventing adsorption-induced protein alterations. In this study, it could be shown that the organic modification of the clay permitted the co-adsorption of two antigenic proteins, heat-labile eteroxin and viral protein 2, without conformational protein alterations, as evidenced from infrared spectroscopy, while the secondary protein structure was changed on pristine clay; possibly through strong electrostatic interactions with the highly charged clay surfaces. The retained bioactivity of these organo-clay associated proteins enabled efficacious binding to cellular GM1 receptors, an important step for stimulation of an immune response to the infectious bursal disease virus.

The first attempt at employing biohybrids based on fibrous clay as adjuvants was conducted by Ruiz-Hitzky et al. (2009) who used the biopolymer xanthan for surface modification of sepiolite to design an influenza vaccine. Here, the governing idea was to provide a tissue-like environment for the adsorbed influenza virus similar to the nasal mucous membrane, the natural location of influenza virus entry into an organism. The formation of stable suspensions of this bionanocomposite offers the possibility for nasal, needle-free delivery of the vaccine, a requirement often demanded for its ease and low costs (Amorij et al., 2010). Influenza A virions adsorbed on sepiolite-xanthan in a dispersed manner and in large quantities, driven by electrostatic attraction between the negatively charged carboxylate sites of the polysaccharide and the cationic centers on the virus surface. Conversely, virus immobilized on the pristine sepiolite surface formed inhomogeneous agglomerates. Immunization of mice and posterior challenge with infectious influenza A virus demonstrated the high level of seroprotection imparted by this vaccine due the conservation of immunogenicity on sepiolite-xanthan. A further advantage of sepiolite-based vaccines originates from its microfibrous texture, which can help to elicit mucosal immune responses provoked from irritation of the nasal mucous membrane by the fibers. This can help to improve the efficacy of vaccination and thus, reduce the necessary dose for immunization.

Another important issue in vaccination is the thermostability of vaccines, which becomes critical in stockpiling of prepandemic vaccines and in discontinuous cold-chains, e.g. in low-income countries (Brandau et al., 2003). The stability against accidental freezing during transportation and storage can also compromise vaccine efficacy (ca. 30% of all vaccines are freeze-sensitive) (Chen and Kristensen, 2009). Both scenarios represent economic and healthcare threats in the multimillion dollar range (Chen and Kristensen, 2009). Therefore, the Ruiz-Hitzky group explored sepiolite-lipid hybrids as novel adjuvants in thermostable influenza A vaccines (Wicklein, 2011). Functional studies shows improved thermal stability at elevated temperatures up to 48°C together with enhanced resistance against lyophilization-induced antigen denaturation as often seen for Alum stabilized antigens (Maa et al., 2003; Clapp et al., 2011). This improvement in thermal stability was suggested to be related to the creation of a chemical microenvironment by the sepiolite-lipid biohybrid (Wicklein, 2011). It is known from many adjuvants, such as for instance

Alum, that they can create a chemical microenvironment with specific pH values, electrolytes, and ionic strength that can vary significantly from the bulk solution conditions (Wittayanukulluk et al., 2004; Clapp et al., 2011). Such microenvironments have been shown to influence the thermal stability of associated antigenic species (Wittayanukulluk et al., 2004). Therefore, it might be reasonable to attribute similar effects to the sepiolite-lipid microenvironment forming a kind of thermally protective scaffold for adsorbed influenza virions.

9.5.4 FOAMS AND SCAFFOLDS FOR TISSUE ENGINEERING

In the context of tissue engineering, bionanocomposite foams are often associated with trabecular bone replacement and healing. Such association is not deprived of meaning since bone is in fact a natural bionanocomposite porous solid. The porous character is more pronounced in trabecular bone, where porous volumes are estimated around 80 percent (Fratzl and Weinkamer, 2007). Regarding its composition, the extracellular matrix of bone is composed of mainly type I collagen which is reinforced by hydroxyapatite crystals in form of thin platelets. The third major component is water, accounting for around 10 wt % of the total mass. Tackling bone replacement and healing has thus been largely addressed by the preparation and introduction of temporary (scaffolds) or definitive (implants) bionanocomposite foams in bone defects.

It should also be noted that hard tissue replacement and healing has extremely important socioeconomic implications. Bone is the primary organ requiring surgical intervention with around one million cases per year (Salgado et al., 2004). The development of new, efficient and cost-effective procedures (and thus materials) to cope with such demand is therefore critical.

The field of bionanocomposite porous foams related to hard biological tissue applications has been mostly led forward by researchers working directly on the native bone composition (Kikuchi et al., 2004; Salgado et al., 2004; Zhu et al., 2006; Wang et al., 2011b) (i.e. collagen and hydroxyapatite). Following such approach, other biopolymers were slowly added to the initial formulation with interesting results (Zhang and Ma, 1999, 2004; Liao et al., 2004; Azami et al., 2006; Vidyarthi et al., 2007; Liu et al., 2009). Components such as fibrous clays have been considered to attain porous structures that partially reproduce the critical features of bone.

One of the significant examples consists on the use of a collagen matrix reinforced with sepiolite. Such work trend has been initiated by Perez-Castells et al. (1987) who have simply probed the adsorption of collagen onto sepiolite nanofibres. Although quite distant from the preparation of a foam or scaffold, such a physical-chemistry approach has established the high affinity between the most relevant protein in mammal extracellular matrix and sepiolite. This fundamental approach was followed by in vitro testing the effect of a pseudo-extracellular medium obtained from the sepiolite-collagen complex (Olmo et al., 1987). The dermal human fibroblasts showed similar cellular growth between control and sepiolite-containing samples. Additionally, these authors have demonstrated that collagenase activity was severely reduced when collagen was supported on sepiolite. The previously mentioned studies have led to the analysis of fibroblasts outgrowth direct from sepiolite collagen complexes (Lizarbe et al., 1987). According to these authors, the cellular behavior of fibroblasts was not significantly different from that exhibited toward a collagen control. The ensemble of these results is quite clearly followed by Herrera et al. (1995) by the implantation of collagen-sepiolite bionanocomposites in rat calvaria defect, a model for critical size defect (*i.e.* an orthotropic defect that will not naturally heal in the lifetime of an animal). Their results reveal that sepiolite-collagen composites did not induce inflammation or necrosis as could be predicted from the resemblance between sepiolite and asbestos nanofibres. While the composite materials inserted into bone defects did not enhance ossification in comparison with control, it is clear that the presence of a mechanically protective material can be most valuable during healing.

The high affinity between sepiolite and other proteins has also been probed in other systems. Fernandes et al. (2011) have addressed the preparation of gelatin-sepiolite nanocomposites. Likewise Alcantara et al. (2012) have thoroughly studied the interaction between zein and sepiolite. These preliminary works on the interaction between silicates and protein stand as a starting point to address the development of rigid bionanocomposite foams, for bio-relevant applications. In parallel to the developments of protein-fibrous clay foams, the compatibilizations of other polymers such as polyvinyl alcohol (PVA) with sepiolite have allowed the development of highly structured bioactive foams (Figure 9.6) (Wicklein et al., 2013). Such materials, presenting tunable porosity (and thus tunable density) have proven to be extremely versatile since they combine an extremely versatile morphology to bio-compatibility as demonstrated by the authors by enzymatic activity assays.

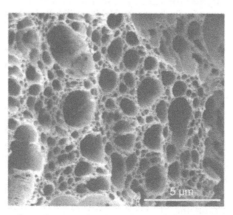

FIGURE 9.6   Cross-section of a polyvinyl alcohol-sepiolite hierarchically structured foam as imaged under FE-SEM.

As mentioned previously, bionanocomposites present one major advantage in foam processing: they often go through a hydrogel step which can be considered the ideal state for several foam processing techniques. Moreover, nanosized fibrous clays such as sepiolite and palygorskite act as skeletal reinforcements within the foam trabeculae which allow stabilizing foams that would otherwise not behave as self-supported structures.

Altogether the characteristics of nanosized fibrous clays such as their biocompatibility, their mechanical reinforcement behavior and, above all, their excellent affinity toward a great variety of biopolymers, allows us to foresee their growing application in the context of bionanocomposite foam materials.

9.6   CONCLUDING REMARKS

Natural sepiolite and palygorskite fibrous clays can be considered as paradigmatic nanomaterials for a wide field of applications from green plastics to biomedical uses. On the basis of their surface, chemical, mechanical and rheological properties they become attractive supports of drugs by themselves but their ability to assembly organic species of biological origin, such as biopolymers, opens new possibilities toward improved applications. In this sense, bionanocomposites based on sepiolite and palygorskite combined with polysaccharides, proteins and other biomolecules (e.g. phospholipids) give rise to biodegradable, biocompatible and biomimetic platforms able to introduce specific functionalities enabling their uses as bioplastics, adjuvant of vaccines, scaffolds and tissue engineering materials, DDS, removal of pollutants, etc. This is an emerging area of research still offering many opportunities for new applications of sepiolite and palygorskite fibrous clays, which were not long ago regarded just as "exotic clay minerals".

KEYWORDS

- **Biohybrid materials**
- **Bionanocomposites**
- **Bioplastic**
- **Freeze-drying**
- **Hybrid material**
- **Hydrogel**
- **Organoclay materials**
- **Photostabilization**

REFERENCES

1. Aguzzi, C.; Cerezo, P.; Viseras, C.; and Caramella, C.; *Appl. Clay. Sci.* **2007**, *36*, 22–36.
2. Alcantara, A. C. S.; Aranda, P.; Darder, M.; and Ruiz-Hitzky, E.; *J. Mater. Chem.* **2010**, *20*, 9495–9504.
3. Alcântara, A. C. S.; Darder, M.; Aranda, P.; and Ruiz-Hitzky, E., (in preparation).
4. Alcântara, A. C. S.; Darder, M.; Aranda, P.; and Ruiz-Hitzky, E.; *Appl. Clay. Sci.*, **2014**, *96*, 2–8. 5. Alcântara, A. C. S.; Darder, M.; Aranda, P.; and Ruiz Hitzky, E.; *Eur. J. Inorg. Chem.* **2012**, 5216–5224.
6. Alongi, J.; and Poskovic, M.; *E Polymers.* **2011**, *11*, 784–792.
7. Amorij, J.-P.; Hinrichs, W. L. J.; Frijlink, H. W.; Wilschut, J. C.; and Huckriede, A.; *Lancet. Infect. Dis.* **2010**, *10*, 699–711.
8. Anderson, T. H.; Min, Y. J.; Weirich, K. L.; Zeng, H. B.; Fygenson, D.; and Israelachvili, J. N.; *Langmuir.* **2009**, *25*, 6997–7005.
9. Ariga, K.; Hill, J. P.; Lee, M. V.; Vinu, A.; Charvet, R.; and Acharya, S.; *Sci. Technol. Adv. Mater.* **2008**, *9*, #014109.
10. Azami, M.; Orang, F.; and Moztarzadeh, F.; Nanocomposite bone tissue-engineering scaffolds prepared from gelatin and hydroxyapatite using layer solvent casting and freeze-drying technique. In: Proceedings of the International Conference on Biomedical and Pharmaceutical Engineering, Singapore, **2006**, 259–264.
11. Bazaka, K.; Crawford, R. J.; Nazarenko, E. L.; and Ivanova, E. P.; Bacterial extracellular polysaccharides. In: Bacterial Adhesion, Eds. D. Linke, and A. Goldman. Springer, Dordrecht, **2011**, 213–226.
12. Bilotti, E. et al. *Macromol. Mater. Eng.* **2010**, *295*, 37–47.
13. Boccaccini, A. R.; Stamboulis, A. G.; Rashid, A.; and Roether, J. A.; *J. Biomed. Mater. Res. Part B Appl. Biomater.* **2003**, *67B*, 618–626.
14. Bradley, W. F.; *Am. Min.* **1940**, *25*, 405–410.
15. Brandau, D. T.; Jones, L. S.; Wiethoff, C. M.; Rexroad, J.; and Middaugh, C. R.; *J. Pharm. Sci.* **2003**, *92*, 218–231.
16. Brauner, K.; and Pressinger, A.; *Min. Petrol.* **1956**, *6*, 120–140.
17. Carnero Ruiz, C.; Sugar-based surfactants. Fundamentals and applications. In: Surfactant Science series, Ed. Hubbard, A. T. CRC Press, Taylor & Francis Group, Boca Raton, FL, **2009**.
18. Carretero, M.; and Pozo, M.; Mineralogía aplicada: Salud y medio ambiente. Paraninfo, Madrid, 2007.
19. Carretero, M. I.; and Pozo, M.; *Appl. Clay. Sci.* **2009**, *46*, 73–80.
20. Clapp, T.; Siebert, P.; Chen, D.; Jones Braun, L.; *J. Pharm. Sci.* **2011**, *100*, 388–401.
21. Cohen, E.; and Joseph, T.; *Appl. Clay. Sci.* **2009**, *42*, 569–574.
22. Chaplin, M.; Hydrocolloid hydration, http://www.lsbu.ac.uk/water/hydro html. Accessed on December, 2013.
23. Chen, D.; and Kristensen, D.; *Exp. Rev. Vacc.* **2009**, *8*, 547–557.
24. Chivrac, F.; Pollet, E.; Schmutz, M.; and Averous, L.; *Carbohyd. Pol.* **2010**, *80*, 145–153.
25. Choy, J. H.; Choi, S. J.; Oh, J. M.; and Park, T.; *Appl. Clay. Sci.* **2007**, *36*, 122–132.
26. Darder, M.; Aranda, P.; and Ruiz-Hitzky, E.; *Adv. Mater.* **2007**, *19*, 1309–1319.
27. Darder, M.; Lopez-Blanco, M.; Aranda, P.; Aznar, A. J.; Bravo, J.; and Ruiz-Hitzky, E.; *Chem. Mater.* **2006**, 18, 1602–1610.
28. Denizeau, F.; Marion, M.; Chevalier, G.; and Cote, M. G.; *Cell. Biol. Toxicol.* **1985**, *1*, 23–32.
29. Donat, R.; Cilgi, G. K.; Aytas, S.; and Cetisli, H. J.; *Radioanal. Nucl. Chem.* **2009a**, *279*, 271–280.
30. Donat, R.; Esen, K.; Cetisli, H.; and Aytas, S. J.; *Radioanal. Nucl. Chem.* **2009b**, *279*, 253–261.

31. Duquesne, E.; Moins, S.; Alexandre, M.; and Dubois, P.; *Macromol. Chem. Phys.* **2007**, *208*, 2542–2550.
32. Esko, J. D.; Bacterial polysaccharides. In: Essentials of glycobiology, Varki, A.; Cummings, R. D.; Esko, J. D.; Freeze, H. H.; Stanley, P.; Bertozzi, C. R.; Hart, G. W.; and Etzler, M. E. Eds. Cold Spring Harbor Laboratory Press; Cold Spring Harbor (NY), **1999**, Ch. 21.
33. European bioplastics. http://en.european-bioplastics.org/bioplastics/. Accessed on December, 2013.
34. Fernandes, F. M.; Manjubala, I.; and Ruiz-Hitzky, E.; *Phys. Chem. Chem. Phys.* **2011**, *13*, 4901–4910.
35. Fernandes, F. M.; Ruiz, A. I.; Darder, M.; Aranda, P.; and Ruiz-Hitzky, E.; *J. Nanosci. Nano technol.* **2009**, *9*, 221–229.
36. Fratzl, P.; and Weinkamer, R. *Prog. Mater. Sci.* **2007**, 52, 1263–1334.
37. Fujimoto, Y.; Ray, S. S.; Okamoto, M.; Ogami, A.; Yamada, K.; and Ueda, K.; *Macromol. Rap. Commun.* **2003**, *24*, 457–461.
38. Fukushima, K.; Tabuani, D.; Abbate, C.; Arena, M.; and Ferreri, L.; *Poly. Degrad. Stab.* **2010**, *95*, 2049–2056.
39. Fukushima, K.; Tabuani, D.; and Camino, G.; *Mater. Sci. Eng. C Biomimetic Supramol. Syst.* **2009**, *29*, 1433–1441.
40. Fyrner, T.; et al. *Langmuir,* **2011**, *27*, 15034–15047.
41. Gatica, J. M.; and Vidal, H.; *J. Hazard. Mater*. **2010**, *181*, 9–18.
42. Gimenez, B.; Gomez-Guillen, M. C.; Lopez-Caballero, M. E.; Gomez-Estaca, J.; and Montero, P.; *Food. Hydrocolloids.* **2012**, 27, 475–486.
43. Giraud Guille, M. M.; Helary, C.; Vigier, S.; Nassif, N.; *Soft Matter.* **2010**, *6*, 4963–4967.
44. González-Alfaro, Y.; Aranda, P.; Fernandes, F. M.; Wicklein, B.; Darder, M.; Ruiz-Hitzky, E. *Adv. Mater.* **2011**, *23*, 5224–5228.
45. Hederos, M.; Konradsson, P.; and Liedberg, B.; *Langmuir.* **2005**, *21*, 2971–2980.
46. Heegaard, P.; et al. *Arch. Virol.* **2011**, *156*, 183–202.
47. Herrera, J. I.; et al. *Biomaterials* **1995**, *16*, 625–631.
48. Hrenovic, J.; Tibljas, D.; Ivankovic, T.; Kovacevic, D.; and Sekovanic, L.; *Appl. Clay. Sci.* **2010**, *50*, 582–587.
49. Ito, T.; Kobayashi, M.; Koide, N.; Sugafuji, H.; and Yamato, H.; Japan Patent, 20011278810, 2001.
50. Joshi, G. V.; Kevadiya, B. D.; Patel, H. A.; Bajaj, H. C.; and Jasra, R. V. *Int. J. Pharm.* **2009**, *374*, 53–57.
51. Kikuchi, M.; Ikoma, T.; Itoh, S.; Matsumoto, H. N.; Koyama, Y.; Takakuda, K.; Shinomiya, K.; Tanaka, J.; *Compos. Sci. Technol.* **2004**, 64, 819–825.
52. Kose, A. A.; Karabagli, Y.; Kurkcuoglu, M.; and Cetin, C.; *Wound. Compend. Clin. Res. Pract.* **2005**, *17*, 114–121.
53. Koutsopoulos, S.; Kaiser, L.; Eriksson, H. M.; Zhang, S.; *Chem. Soc. Rev.* **2012**, *41*, 1721–1728.
54. Latgé, J.-P.; *Mol. Microbiol.* **2007**, *66*, 279–290.
55. Lee, C. H.; Singla, A.; and Lee, Y.; *Int. J. Pharm.* **2001**, *221*, 1–22.
56. Lee, H.; Dellatore, S. M.; Miller, W. M.; and Messersmith, P. B.; *Science.* **2007**, *318*, 426–430.
57. Lee, H.; Scherer, N. F.; and Messersmith, P. B.; *Proc. Natl. Acad. Sci.* **2006**, *103*, 12999–13003.
58. Lehn, J. M.; *Science* **2002**, *295*, 2400–2403.
59. Li, A.; Liu, R. F.; and Wang, A. Q.; *J. Appl. Polym. Sci.* **2005**, *98*, 1351–1357.
60. Liao, S. S.; Cui, F. Z.; Zhang, W.; and Feng, Q. L.; *J. Biomed. Mater. Res. Part B Appl. Bio mater.* **2004**, *69B*, 158–165.

61. Linden, S. K.; Sutton, P.; Karlsson, N. G.; Korolik, V.; and McGuckin, M. A.; *Mucosal. Im munol.* **2008**, *1*, 183–197.
62. Liu, Q.; and Chen, D.; *Eur. Polym. J.* **2008**, *44*, 2046–2050.
63. Liu, X.; Smith, L. A.; Hu, J.; and Ma, P. X.; *Biomaterials.* **2009**, *30*, 2252–2258.
64. Liu, Y.; Wang, W. B.; and Wang, A. G.; *Desalination.* **2010**, *259*, 258–264.
65. Lizarbe, M. A.; Olmo, N.; Gavilanes, J. G.; *Biomaterials* **1987**, *8*, 35–37.
66. Lu, J. R.; Zhao, X. B.; and Yaseen, M.; *Curr. Opin. Coll. Interf. Sci.* **2007**, *12*, 60–67.
67. Lu, Q. J.; Ganesan, K.; Simionescu, D. T.; Vyavahare, N. R.; *Biomaterials.* **2004**, *25*, 5227–5237.
68. Luckham, P. F.; and Rossi, S.; *Adv. Coll. Interf. Sci.* **1999**, *82*, 43–92.
69. Maa, Y.-F.; Zhao, L.; Payne, L. G.; and Chen, D. J.; *Pharm. Sci.* **2003**, *92*, 319–332.
70. Martínez-Frías, P., Industrial Engineering Degree Dissertation. Autonomous University of Madrid, Madrid, 2008.
71. Mu, B.; Kang, Y.; and Wang, A.; *J. Mater. Chem. A.* **2013**, *1*, 4804–4811.
72. Mukherjee, D.; May, M.; and Khomami, B.; *J. Coll. Interf. Sci.* **2011**, *358*, 477–484.
73. Oh, J.-M.; Biswick, T. T.; Choy, J.-H.; *J. Mater. Chem.* **2009**, *19*, 2553–2563.
74. Ojijo, V.; Ray, S. S. *Prog. Poly. Sci.* **2013**, *38*, 1543–1589.
75. Olmo, N.; Lizarbe, M. A.; Gavilanes, J. G.; *Biomaterials* **1987**, *8*, 67–69.
76. Oscarson, D. W.; Van Scoyoc, G. E.; and Ahlrichs, J. L.; *Clay. Clay. Min.* **1986**, *34*, 74–80.
77. Pacwa-Plociniczak, M.; Plaza, G. A.; Piotrowska-Seget, Z.; and Cameotra, S. S. *Int. J. Mol. Sci.* **2011**, *12*, 633–654.
78. Pan, G.; Zhang, M. M.; Chen, H.; Zou, H.; and Yan, H.; *Environ. Pollut.* **2006**, *141*, 195–200.
79. Pang, C.; Liu, Y. H.; Cao, X. H.; Hua, R.; Wang, C. X.; and Li, C. Q. *J. Radioanaly. Nucl. Chem.* **2010**, *286*, 185–193.
80. Patil, V.; Venkatesh, M. D.; and Krishnappa, G. K.; and Srinivasa Gouda, R. N.; *Ind. J. Anim. Sci.* **2004**, *74*, 845–847.
81. Peek, L. J.; Middaugh, C. R.; and Berkland, C.; *Adv. Drug. Deliv. Rev.* **2008**, *60*, 915–928.
82. Perez-Castells, R.; Alvarez, A.; Gavilanes, J.; Lizarbe, M. A.; Martinez Del Pozo, A.; Olmo, N.; and Santaren, J;. Adsorption of collagen by sepiolite. In: Proceedings of the international clay conference denver, 1985, Schultz, L. G.; van Olphen, H.; Mumpton, F. A. Eds, The Clay Minerals Society, Bloomington, IN, USA, 1987, 359–362.
83. Petty, M. C.; Langmuir-blodgett films. An introduction. Cambridge University Press, Cambridge, **1996**.
84. Privé, G.G.; *Methods.* **2007**, *41*, 388–397.
85. Pum, D.; Toca-Herrera, J. L.; and Sleytr, U. B.; *Int. J. Mol. Sci.* **2013**, *14*, 2484–2501.
86. Rapuano, R.; and Carmona-Ribeiro, A. M.; *J. Coll. Interf. Sci.* **1997**, *193*, 104–111.
87. Roh, J. D.; et al. *Biomaterials.* **2008**, *29*, 1454–1463.
88. Ruiz-Hitzky, E.; *J. Mater. Chem.* **2001**, *11*, 86–91.
89. Ruiz-Hitzky, E.; Organic-inorganic materials: From intercalation chemistry to devices. In: Functional hybrid materials. Eds. Gómez-Romero, P.; Sanchez, C., Wiley-VCH, Weinheim, **2004**, 15–49.
90. Ruiz-Hitzky, E.; and Aranda, P.; *J. Sol Gel Sci. Technol.* **2014**, *70*, 307–316.
91. Ruiz-Hitzky, E.; Aranda, P.; Alvarez, A.; Santarén, J.; and Esteban-Cubillo, A.; Advanced materials and new applications of sepiolite and palygorskite. In: Developments in palygorskite-sepiolite research. A new outlook on these nanomaterials, Eds. Galán, E.; and Singer, A.; Elsevier B. V., Oxford, UK, **2011a**, 393–452.
92. Ruiz-Hitzky, E.; Aranda, P.; and Belver, C.; Nanoarchitectures based on clay materials In Manipulation of nanoscale materials: An introduction to nanoarchitectonics, Ed. K. Ariga; Royal Society of Chemistry, **2012a**, 87–111.

93. Ruiz-Hitzky, E.; Aranda, P.; and Darder, M.; Bionanocomposites. In: Kirk-othmer enciclopedia of chemical technology. Hoboken, NJ: John Wiley & Sons, **2008**, 1–28.
94. Ruiz-Hitzky, E.; Aranda, P.; Darder, M.; and Fernandes; F. M.; Fibrous clay mineral–polymer nanocomposites. In Handbook of clay science. Part a: Fundamentals, Eds. Bergaya, F.; Lagaly, G. Amsterdam & Oxford: Elsevier, **2013a**, 721–741.
95. Ruiz-Hitzky, E.; Aranda, P.; Darder, M.; and Ogawa, M.; Chemical Society Reviews **2011b**, *40*, 801–828.
96. Ruiz-Hitzky, E.; Aranda, P.; Darder, M.; and Rytwo, G.; *J. Mater. Chem.* **2010a**, *20*, 9306–9321.
97. Ruiz-Hitzky, E.; Aranda, P.; and Serratosa, J. M.; Clay-organic interactions:Organoclay complexes and polymer clay nanocomposites. In: Handbook of layered materials, Eds. Auerbach, S. M.; Carrado, K. A.; and Dutta, P. K., Marcel Dekker, **2004**, 91–154.
98. Ruiz-Hitzky, E.; Darder, M.; Aranda, P.; and Ariga, K.; *Adv. Mater.* **2010b**, *22*, 323–336.
99. Ruiz-Hitzky, E.; Darder, M.; Aranda, P.; Martin del Burgo, M. Á.; and del Real, G.; *Adv. Mater.* **2009**, *21*, 4167–4171.
100. Ruiz-Hitzky, E.; Darder, M.; Fernandes, F. M.; Wicklein, B.; Alcantara, A. C. S.; and Aranda, P.; *Prog. Poly. Sci.* **2013b**, *38*, 1392–1414.
     Ruiz-Hitzky, E.; Darder, M.; Wicklein, B.; Alcantara, A. C. S.; and Aranda, P.; *Adv. Polym. Sci.* (in press)
101. Ruiz-Hitzky, E.; et al. Advanced biohybrid materials based on nanoclays for biomedical applications. In Nanosystems in engineering and medicine, Eds. Choi, S. H.; Choy, J. H.; Lee, U.; and Varadan, V. K. 8548, Spie-Int Soc Optical Engineering, Bellingham, **2012b**, 85480D.
102. Ruiz-Hitzky, E.; and Fripiat, J. J.; *Clay. Clay Min.* **1976**, *24*, 25–30.
103. Rytwo, G.; Mendelovits, A.; Eliyahu, D.; Pitcovski, J.; Aizenshtein, E.; *Appl. Clay. Sci.* **2010**, *50*, 569–575.
104. Salgado, A. J.; Coutinho, O. P.; and Reis, R. L.; *Macromol. Biosci.* **2004**, *4*, 743–765.
105 Sandberg, T.; Karlsson Ott, M.; Carlsson, J.; Feiler, A.; and Caldwell, K. D.; *J. Biomed. Mater. Res. Part A.* **2009**, *91A*, 773–785.
106. Santarén, J.; and Alvarez, A.; *Ind. Min.* **1994**, April, 101–114.
107. Santarén, J.; Sanz, J.; and Ruiz-Hitzky, E.; *Clay. Min.* **1990**, *38*, 63–68.
108. Sedaghat, M. E.; Ghiaci, M.; Aghaei, H.; and Soleimanian-Zad, S.; *Appl. Clay. Sci.* **2009**, *46*, 131–135.
109. Sleytr, U. B.; and Messner, P.; Crystalline cell surface layers (s layers). In: Encyclopedia of microbiology, Schaechter, M., Ed., Vol. 1, Academic Press/Elsevier Science, San Diego, CA, USA, **2009**, 89–98.
110. Srinivasan, R.; *Adv. Mater. Sci. Eng.* **2011**, 872531.
111. Stubbs, G. W.; Smith, H. G.; and Litman, B. J.; *Biochimica et Biophysica Acta.* **1976**, *426*, 46–56.
112. Suárez, M.; and García-Romero, E.; Advances in the crystal chemistry of sepiolite and palygorskite. In: Developments in Palygorskite-Sepiolite Research. A New Outlook on these Nanomaterials. Eds. Galán, E.; and Singer, A., Elsevier B.V., Oxford, UK, **2011**, 33–65.
113. Szabo, T.; et al. *Clays. Clay. Min.* **2008**, *56*, 494–504.
114. Takahashi, T.; and Yamaguchi, M. J.; *Phenom. Mol. Recognit. Chem.* **1991**, *10*, 283–297.
115. Tsai, F. C.; et al.; *J. Appl. Polym. Sci.* **2012**, *126*, 1906–1916.
116. Vert, M.; et al. *Pure. Appl. Chem.* **2012**, *84*, 377–408.
117. Viçosa, A. L.; Gomes, A. C. O.; Soares, B. G.; and Paranhos, C. M.; *Exp. Poly. Lett.* **2009**, *3*, 518–524.
118. Vidyarthi, U.; Zhdan, P.; Gravanis, C.; and Lekakou, C.; Gelatine-hydroxyapatite nanocomposites for orthopaedic applications. In: World Congress on Engineering **2007**, Vol. 1 and 2, **2007**, 1251–1256.

119. Viseras, C.; Aguzzi, C.; Cerezo, P.; and Bedmar, M. C.; *Mater. Sci. Technol.* **2008**, *24*, 1020–1026.

120. Viseras, C.; Cerezo, P.; Sanchez, R.; Salcedo, I.; and Aguzzi, C. *Appl. Clay. Sci.* **2010**, *48*, 291–295.

121. Vu, B.; Chen, M.; Crawford, R. J.; and Ivanova, E. P.; *Molecules* **2009**, *14*, 2535–2554.

122. Wang, Q.; Wang, W.; Wu, J.; and Wang, A. J.; *Appl. Poly. Sci.* **2012**, *124*, 4424–4432.

123. Wang, Q.; Wu, J.; Wang, W.; and Wang, A. J.; *Biomat. Nanobiotechnol.* **2011a**, *2*, 250–257.

124. Wang, W. B.; Zheng, Y. A.; and Wang, A. Q.; *Polym. Adv. Technol.* **2008**, *19*, 1852–1859.

125. Wang, X. H.; Zheng, Y.; and Wang, A.Q. J.; *Hazardous. Mater.* **2009**, *168*, 970–977.

126. Wang, Y.; Silvent, J.; Robin, M.; Babonneau, F.; Meddahi-Pelle, A.; Nassif, N.; and Guille, M. M. G.; *Soft Matter.* **2011b**, *7*, 9659–9664.

127. Wicklein, B., PhD dissertation. Autonomous University of Madrid, Madrid, 2011.

128. Wicklein, B.; Aranda, P.; Ruiz-Hitzky, E.; and Darder, M.; *J. Mater. Chem. B.* **2013**, *1*, 2911–2920.

130. Wicklein, B.; Darder, M.; Aranda, P.; and Ruiz-Hitzky, E.; *Langmuir.* **2010**, *26*, 5217–5225.

131. Wicklein, B.; Darder, M.; Aranda, P.; and Ruiz-Hitzky, E.; *ACS. Appl. Mater. Interf.* **2011**, *3*, 4339–4348.

132. Wicklein, B.; et al. *Eur. J. Inorg. Chem.* **2012**, 5186–5191.

133. Wittayanukulluk, A.; Jiang, D.; Regnier, F. E.; and Hem, S. L.; *Vaccine.* **2004**, *22*, 1172–1176.

134. Xie, Y.; Wang, A.; and Liu, G.; *Poly. Compos.* **2010**, *31*, 89–96.

135. Xu, Z. P.; Zeng, Q. H.; Lu, G. Q.; Yu, A. B.; *Chem. Eng. Sci.* **2006**, *61*, 1027–1040.

136. Yang, H. L.; et al. *Energy. Build.* **2011**, *43*, 386–392.

137. Yoshida, N.; and Sato, M.; *Appl. Microbiol. Biotechnol.* **2009**, *83*, 791–798.

138. Yuan, Q.; Lu, W.; and Pan, Y. *Polym. Degrad. Stab.* **2010**, *95*, 1581–1587.

139. Zhang, R.; and Ma, P.X. J.; *Biomed. Mater. Res.* **1999**, *45*, 285–293.

140. Zhang, R. Y.; and Ma, P. X.; *Macromol. Biosci.* **2004**, *4*, 100–111.

141. Zhu, X. L.; Eibl, O.; Scheideler, L.; Geis-Gerstorfer, J.; *J. Biomed. Mater. Res. Part A.* **2006**, *79A*, 114–127.

142. Zou, X. H.; et al. *Chem. Eng. J.* **2011**, *167*, 112–121.

CHAPTER 10

BIOPOLYMER NANOCOMPOSITES: POLYLACTIC ACID/HALLOYSITE NANOTUBES COMPOSITES

RANGIKA T DE SILVA, POORIA PASBAKHSH, and KHENG LIM GOH

CONTENTS

10.1 INTRODUCTION

At present, petrochemical-based plastics occupy a leading place in many industries, but the emission of CO_2 and green-house gases during the processing of these plastics leads to global warming. Moreover, the nonbiodegradability of these plastics creates issues in terms of waste disposal. For example, in the US, 400,000 tons of hard garbage is generated as waste, and plastics take 30 percent of its volume (Kumar et al., 2012). The inability to dispose of plastics has become a large threat to the stability of the global ecosystem. Hence, environmental concerns over plastic materials have raised interest in the use of biodegradable alternatives in many industries including the packaging industry (Rhim et al., 2009, Fukushima et al., 2009b). Renewable resource-based biodegradable plastics can be derived in many ways and chemically synthesized biopolymers (poly (glycolic) acid, poly (lactic acid) (PLA), poly (vinyl alcohol)) are among them. PLA has gained enormous attention as a replacement for conventional petrochemical-based packaging plastics due to its unique properties (Rhim et al., 2006).

PLA is a biopolymer derived from agricultural resources such as corn and sugar beet. It is a biocompatible, biodegradable and compostable thermoplastic polymer, and is used in packaging, biological and textile applications, and as composites or coating materials (Najafi et al., 2012, Fukushima et al., 2009, Papageorgiou et al., 2010). PLA is slightly expensive compared to regular commodity packaging materials. According to Cargill Dow LLC (one of the pioneer developers of commercial PLA), the gross production of PLA is 140,000 tons per year and its cost is expected to decrease by increasing the supply to meet the growing demand (Auras et al., 2004). Even though the cost of PLA is slightly higher when compared to regular commodity packaging materials, it has been introduced as an advanced material for some applications where the unique properties of PLA attract attention over its cost factor. For instance, the biocompatibility of PLA helps to prevent mold growth on cheese, which expands shelf life, and so, PLA has been used in packaging films (Plackett et al., 2006). Very recently, Econ-Core Industries have developed PLA-based thin, laminated films for automotive interiors and they are conducting further research on this application (2010). Additionally, PLA is introduced as an advanced material for pharmaceutical applications (Kumar et al., 2012). Due to its high bio-compatibility, PLA films have been subjected to research for tissue engineering applications as well, but the limitations of the physical properties of PLA have slowed down the process of its use commercially (Nair et al., 2004). As we can observe in society today, commonly-used grocery bags require strength to carry loads. Generally, they are made out of commodity petroleum-based plastics, but earlier or later they have to be replaced with eco-friendly materials due to environmental concerns. Therefore, to maintain eco-stability, replacing petroleum-based plastics with bioplastics has become an urgent necessity.

However, PLA has poor mechanical, thermal properties (glass transition temperature is around 59°C) and barrier properties compared to regular commodity packaging film materials, which restricts its usage in a wider range of applications in the packaging industry (Rhim et al., 2009, Fukushima et al., 2009, Najafi et al., 2012, Bleach et al., 2002). Therefore, a great deal of attention has been paid to reinforce the properties of PLA by incorporating different types of fillers. Additionally, various fillers such as organo-modified montmorillonite (O-MMT) (Fukushima et al., 2009; Jeszka et al., 2010; Najafi et al., 2012; Papageorgiou et al., 2010; Rhim et al., 2009), microcrystalline cellulose (MCC) (Petersson and Oksman, 2006), silica (Papageorgiou et al., 2010), multiwalled carbon nanotubes (MWCNTs) (Papageorgiou et al., 2010), and biphasic calcium phosphate (BSP) (Bleach et al., 2002), and sepiolite (Fukushima et al., 2009) have been used to enhanced the thermo-mechanical properties of PLA nanocomposites. The mechanical properties for some of these PLA nanocomposite films are summarized in Table 10.1. However, even though there have been studies carried out to reinforce PLA using different fillers, natural mineral nanotubes have gained much attention to use as nanofillers to significantly enhance the thermo-mechanical characteristics of PLA by lowering the cost and design more sustainable materials in order to compete with petro-chemical based plastics for packaging applications. Among these novel fillers, halloysite nanotubes (HNTs) and sepiolite have been found to be the suitable candidates for this purpose. In this study, preparation and characterization of PLA/HNTs nanocomposites have been discussed and reviewed in order to give comparative and explanatory outputs of these nanocomposites versus to the other types of fillers used in this area.

TABLE 10. 1 Mechanical Properties of PLA Nanocomposites Reinforced with Selective Fillers.

Type of Nano Filler(s)	Preparation Method	Filler Loading	Tensile Stress (MPa)	Young's Modulus (GPa)	Elongation at Break (%)	Reference
		Neat PLA	28.5 ± 3.8	1.7 ± 0.2	186 ± 30	Peterson et al. (2006), Petersson and Oksman, (2006)
(i) MCC (ii) Bentonite	Solution castin	5% (w/w) bentonite	42.0 ± 4.3	2.6 ± 0.3	46 ± 20	
		5% (w/w) MCC	31.9 ± 2.8	1.5 ±0.2	157 ± 30	
O-MMTs (three types)		Neat PLA	50.6 ± 1.1	-	3.0 ± 0.1	Rhim et al. (2009)
(1) Cloisite Na⁺	Solution casting	5 pph Cloisite Na⁺	40.8 ± 1.1	-	2.5 ± 0.3	Rhim et al. (2009)
(2) Cloisite 30B		5 pph Cloisite 30B	40.9 ± 4.3	-	3.1 ± 0.3	
(3) Cloisite20A		5 pph Cloisite20A	45.3 ± 1.3	-	2.7 ± 0.3	

TABLE 10. 1   *(Continued)*

Type of Nano Filler(s)	Preparation Method	Filler Loading	Tensile Stress (MPa)	Young's Modulus (GPa)	Elongation at Break (%)	Reference
Bacterial cellulose nanowhiskers (BCNW)	Electrospinning	1 %(w/w)	59.04 ± 1.94	2.01 ± 0.07	5.33 ± 1.66	Martinez-Sanz et al. (2012), Martinez-Sanz et al. (2012)
		2 %(w/w)	61.36 ± 1.59	2.16 ± 0.08	3.33 ± 0.58	
		3 %(w/w)	60.89 ± 4.59	2.16 ± 0.13	4.44 ± 1.08	
BCNW	Melt blending	1%(w/w)	47.01 ± 3.58	1.95 ± 0.08	5.82 ± 2.47	Martinez-Sanz et al. (2012), Martinez-Sanz et al. (2012)
		2 %(w/w)	46.29 ± 3.09	2.05 ± 0.07	2.54 ± 0.22	

10.2   PLA/CLAY NANOCOMPOSITES

This section aims to highlight the types of clay fillers, apart from HNTs, that have been incorporated into PLA composites. As applications of PLA nanocomposites are mostly based on their biocompatibility and biodegradability, nano-clays such as MMTs, kaolinite and sepiolite are used extensively. Pure MMTs did not seem to be a satisfactory filler due to its relatively high OH content on the surfaces, which generates a polar repulsion with the hydrophobic polymer PLA (the filler-matrix interaction was poor) and led to a decrease in mechanical properties (Table 10.1) (Rhim et al., 2009). Therefore, it was deemed that modification of MMTs was necessary to achieve satisfactory results. Papageorgiou et al. added organically modified MMTs (O-MMTs) (modified with dimethyl, dihydrogenated tallow quaternary ammonium chloride salt) to reinforce the PLA using the solution casting method and reported improvements in mechanical and thermal properties due to (i) interactions between the functional groups of the O-MMT and end hydroxyl groups of PLA (ii) good dispersion and dispersion patterns (O-MMTs had dispersed in an intercalated manner) (Papageorgiou et al., 2010). Fukushima et al. also studied PLA/O-MMTs composites prepared by melt compounding and reported improvements in flexural modulus after introducing two types of O-MMTs (Cloisite 30B and Nanofil 804) into PLA and stated that such improvements were due to the aforementioned factors (Fukushima et al., 2009). However, Jeszka et al. (2010) adopted a different approach to the dispersion of O-MMTs (Cloisite 30B) whereby they synthesized PLA/O-MMTs by solution casting to investigate dielectric properties and found that the dispersion of O-MMTs in the PLA matrix was poor based on the interfacial polarization. This poor dispersion could have been due to lack of vigorous stirring (using either mechanical or magnetic stirring methods) except when ultra-sound sonication was used in the synthesis of composites. However, thermo-mechanical reinforcement of PLA through the addition of O-MMTs has been well reported in the literature (Chang et al., 2003, Sinha Ray et al., 2003). Apart from thermo-mechanical properties, other performance measures such as barrier properties and biodegradability have also been enhanced in PLA composites with the addition of O-MMTs. The degradation rate of PLA/O-MMTs composites prepared by melt blending was studied by Fukushima et al. (2009), who reported that the incorporation of O-MMTs increased the degradation rate due to the presence of hydroxyl groups from the modifier used. Sinha Ray et al. (2009) showed that the presence of O-MMTs in the PLA matrix decreases the oxygen permeability, since the dispersion of the fillers acts as a barrier to hinder the diffusion of gasses.

Kaolinite also studied reinforcing filler for PLA. Matusik et al. (2011) synthesized PLA/kaolinite using the solution casting method and observed enhancements in mechanical properties after adding 1wt % of kaolinite (modulus and tensile strength increased by 31% and 80%, respectively). However, the mechanical properties suggested the filler dispersion is inhomogeneous, since the statistical interquartile ranges were large. Sepiolite (SEP) is another promising nano-clay filler to reinforce PLA. As reported by Sabzi et al. (2013), at lower loadings of SEP, the strength of melt compounded PLA composites increase when compared to pure PLA. An interesting phenomenon was observed with the steady shear viscosity of the samples; pure PLA showed pseudo-Newtonian (constant) viscosity while the PLA/SEP composites showed shear thinning behavior, which was attributed to changes in the morphological structure (filler alignment). Fukushima et al. (2009) stated that SEP is a potential filler to reinforce thermo-mechanical properties of PLA, since SEP can easily disperse in the PLA matrix without bringing about any surface modifications. According to their dynamic mechanical-thermal analysis, storage modulus at temperatures before and after the glass transition temperature had increased by 25 percent and 97 percent, respectively. This was attributed to a better interfacial interaction between SEP and PLA. PLA/SEP composites can provide promising results for flammability performances too; one of the studies showed that the heat release capacity tends to decrease with addition of SEP compared to pure PLA samples, which demonstrates the suitability of the material in high temperature applications (Hapuarachchi and Peijs, 2010).

10.2.1 PERFORMANCE OF PLA BASED COMPOSITES REINFORCED BY HNTS

To date, only very little research has been conducted in the PLA/HNTs nanocomposite area. This section discusses the physical and chemical properties of PLA/HNTs, based on their recent studies. The reported PLA/HNTs composites were mainly synthesized by melt compounding techniques such as internal mixing and melt extrusion. In addition, characteristics of solvent casted PLA/HNTs films are reported here too (De Silva et al., 2013). Table 10.2 provides a detailed summary of the characteristics of PLA/HNTs nanocomposite films.

TABLE 10.2 Summary of Mechanical Properties of PLA/HNT Composites.

HNTs and Type of Modifiers (if any)	Preparation Method	Filler Loading	Tensile Stress (MPa)	Young's Modulus (GPa)	Elongation at Break (%)	Reference
(i) Quaternary ammonium salt treated HNTs (QM-HNTs)	Melt blending	Neat PLA	≈62	2.05	–	Murariu et al. (2012)
		6% QM-HNTs	73	2.80	–	
(ii) 3-methacryloxy-propyltrihoxy silane (m-HNTs)		6% m-HNTs	65	2.75	–	
		Neat PLA	54.48 ± 1.62	2.92 ± 0.014	3.6±0.11	
(i) Unmodified HNTs	Melt extrusion	2% HNTs	57.98 ± 1.24	3.148 ± 0.018	3.3±0.11	Prashantha et al. (2013)
(ii) Quaternary ammonium salt with benzoalkonium chloride (QMB-HNT)		4% HNTs	60.67 ± 1.44	3.263 ± 0.016	3.0±0.009	
		6% HNTs	62.30 ± 2.21	3.378 ± 0.020	2.6±0.12	
		2% QMB-HNTs	61.66 ± 1.52	3.288 ±0.015	3.6±0.08	
		4% QMB-HNTs	65.97 ± 1.72	3.482 ± 0.018	3.4±0.11	
		6% QMB-HNTs	70.32 ± 1.92	3.644 ± 0.021	3.2±0.11	

TABLE 10.2   *(Continued)*

HNTs and Type of Modifiers (if any)	Prepa- ration Method	Filler Loading	Tensile Stress (MPa)	Young's Modulus (GPa)	Elonga- tion at Break (%)	Reference
Unmodified HNTs	Injection molding	Neat LPA	55.2 ± 1.2	–	7.0±0.3	Liu et al. (2013)
		5 Phr HNTs	66.7 ± 1.4	–	14.3±0.5	
		10 Phr	68.6 ± 0.5	–	15.7 ± 0.8	
		20 Phr	69.5 ± 0.8	–	15.8 ± 0.8	
		30 Phr	74.1 ± 0.9	–	5.7 ± 0.4	
		40 Phr	75.1 ± 0.7	–	5.0 ± 0.2	
(i) Unmodified HNTs (ii) 3-methacryloxy-propyltrihoxy silane (m-HNTs)	Melt blending	Neat PLA	48.0 ± 2	1.5 ± 0.1	4.0 ± 0.5	Gorrasi et al. (2013)
		3% HNTs	37.0 ± 5	1.6 ± 0.16	2.8 ± 0.4	
		6% HNTs	40.0 ± 7	1.77 ± 0.13	2.8 ± 0.3	
		12% HNTs	47.0 ± 8	1.73 ± 0.11	3.0 ± 0.9	
		3% m-HNTs	47.0 ± 5	1.55 ± 0.1	4.0 ± 0.8	
		6% m-HNTs	40.0 ± 6	1.65 ± 0.09	2.9 ± 1.0	
		12 m-HNTs	40.0 ± 4	1.67 ± 0.11	3.0 ± 0.4	

10.2.1.1   MECHANICAL PROPERTIES

TENSILE PROPERTIES

Murariu et al. (2012) reported improvements in mechanical properties of PLA after adding quaternary ammonium salt-treated HNTs (QM-HNTs). In the same study, the effect of surface modified HNTs with 3-methacryloxypropyltrimethoxy silane (TM-SPM) (henceforth, referred to as m-HNT) on PLA properties has been evaluated. As reported, both tensile strength (σ) and elastic modulus (E) was improved by increasing the QM-HNTs content up to 6 percent, but at high QM-HNTs concentrations, both the aforementioned properties decreased, due to the inevitable aggregation of HNTs. A similar trend was observed in the mechanical properties of PLA/m-HNTs composites. With addition of 6 percent of m-HNTs, σ and E increased to 65 MPa and 2.75GPa, respectively. As reported, the elongation at break (E_b) did not show any appreciable change with addition of both types of surface treated HNTs.

Prashantha et al. (2013) prepared PLA composites with unmodified and modified HNTs using a master batch dilution process with a single-screw extruder. Quaternary ammonium salt with benzoalkonium chloride was selected to modify HNTs (QBM-HNTs). From the tensile properties, σ increased by 14.4 percent and 29.1 percent and E increased by 15.6 percent and 24.7 percent with addition of 6 percent of unmodified and modified HNTs, respectively. Prashantha et al. (2013) validated these tensile findings using micromechanical models; E was validated using the Halpin-Tsai model while σ was validated with a Pukanszky model. The results showed that the experimental results are compatible with the empirical theoretical models. In addition, the Pukanszky model was used to access the degree of interfacial interaction of unmodified and modified HNTs separately and proved that interfacial interaction is stronger in m-HNTs compared to unmodified HNTs.

Liu et al. (2013) synthesized PLA/HNTs composites by a two-rolled mill followed by injection molding and demonstrated that HNTs are suitable as a filler to reinforce PLA even without any sophisticated surface modification processes. Mechanical properties are gradually increased when introducing HNTs into the PLA matrix. For example, σ and E of PLA composites with HNTs of 30 phr increased by 34 percent and 116 percent, respectively. These improvements were attributed to the dispersion of HNTs and hydrogen bonding between PLA and HNTs. E_b of composites less than 20 phr HNTs was reported higher than the pure PLA as the small amount of HNTs could have acted as a plasticizer.

Gorrasi et al. (2013), on the other hand, synthesized PLA/HNTs films by employing the melt mixing procedure using an internal mixer (Brabender bench scale kneader—50EHT model). The thickness of the prepared samples was approximately 0.2 mm. In their study, the effect of modified HNTs (m-HNTs) (surface treated with 3-(trimethoxysilyl) propyl methacrylate (TMSPM)) was also evaluated. As opposed to the findings of all mechanical properties in other studies mentioned above (Liu et al., 2013, Murariu et al., 2012, Prashantha et al., 2013), in this study, tensile strength did not show any improvement with addition of HNTs as well as m-HNTs, whereas the elastic modulus increased with HNTs loading. E_b decreased with addition of HNTs, but E_b of PLA/m-HNTs was slightly higher than PLA/HNTs films. Additionally, in Russo's findings, σ fluctuates with addition of HNTs (σ was high for some and low for some compositions), while the modulus barely increased and stated that the dispersion of HNTs within the PLA matrix was poor (Russo et al., 2013).

Since there are contradicting statements in the literature regarding the interfacial adhesion and compatibility of PLA with HNTs, De Silva et al. (2013) have also investigated the feasibility of HNTs as a filler to reinforce PLA. A solution casting method was adopted to prepare PLA/HNTs nanocomposite films. Synthesised films are illustrated in Figure 10.1. As revealed by tensile results (Figure 10.2), mechanical properties showed improvements with addition of HNTs into PLA, thereby supporting previous studies (Liu et al., 2013, Murariu et al., 2012, Prashantha et al., 2013). For instance, σ of pure PLA films was 36.66 MPa and it increased to 52.75 MPa with addition of 5 percent of HNTs, which is a 43.8 percent increment. As illustrated in Figure 10.2a, it was noted that the optimum σ value of PLA/HNTs films was achieved with 5 percent of HNTs. The increase in σ was attributed to the possible interactions between the HNTs and PLA. Thereafter, σ values tended to slightly decrease at 7.5 percent and 10 percent of HNTs compared to the σ of 5 percent, but it should be noted that those values are still higher than σ of pure PLA films. This could have occurred due to the aggregation of HNTs at high concentrations. One-way ANOVA and Tukey's Post Hoc tests were conducted to examine the statistical significance of the results, which revealed statistically significant differences in σ between (i) 0 and 5 percent, (ii) 0 and 7.5 percent as well as between (iii) 5 and 10 percent of HNTs. These results suggest that 5 percent was the optimum filler concentration. E also displays a similar trend as σ (Figure 10.2b). At optimum concentration (5% of HNTs), E increased by 57.75 percent with respect to pure PLA films. The stiffening of PLA composite films with HNTs can be attributed to an increase in the structural rigidity of the PLA polymer chains arising from the interaction between adjacent HNTs and PLA. It appears that E_b tended to decrease when HNTs loading was increased (Figure 10.2c). According to ANOVA and Tukey's Post Hoc tests, there is a statistically significant difference in E_b between pure PLA and all other compositions, but there is no appreciable change in E_b of PLA/HNTs films among the other concentrations. Therefore, it was stated that E_b decreases with addition of low concentrations of HNTs and remains unchanged even at higher concentrations.

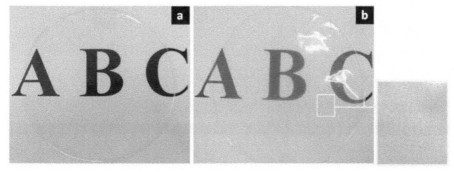

FIGURE 10.1 Synthesised PLA films with (a) 0 percent and (b) 5 percent HNTs, by the solvent casting method (De Silva et al., 2013).

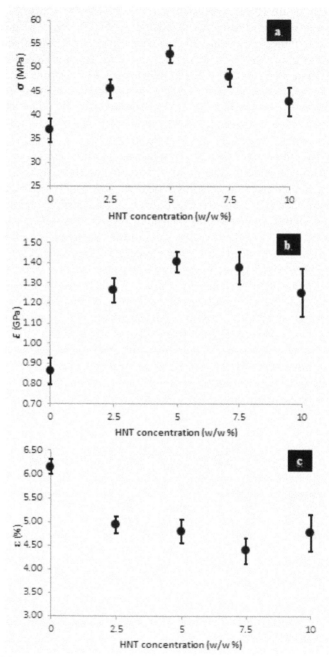

FIGURE 10.2   Mechanical properties of PLA/HNT films; (a) tensile strength (σ), (b) elastic modulus (E) and (c) elongation at break (ε) vs HNT concentration (De Silva et al., 2013).

IMPACT ENERGY, STORAGE MODULUS, AND FLEXURAL STRENGTH AND FLEXURAL MODULUS

As put forth by Murariu et al. (2012), the Izod impact strength of the composites remain constant (did not decrease) with addition of QM-HNTs. Also, impact strength significantly improved with addition of m-HNTs; 2.8 KJ/m² for pure PLA and 3.5 KJ/m² for PLA with 12 percent of m-HNTs. As reported, the improvement in the impact strength provokes bridging of HNTs to absorb energy during the crack propagation as well as the interfacial interaction between HNTs and PLA. Furthermore, the storage modulus of the composites improved with addition of HNTs and it demonstrated the suitability of this material for high temperature applications.

In agreement with the study by Murariu et al. (2012), Prashantha et al. (2013) found increments in impact strength by incorporating both modified and unmodified HNTs. QBM-HNTs added composites showed better improvements in impact strength compared to PLA/HNTs composites. Flexural strength and modulus also increased with addition of HNTs. But flexural properties did not improve significantly with addition of unmodified HNTs compared to QBM-HNTs, which was attributed to the de-bonding of HNTs from the PLA matrix. For all the mechanical properties, QBM-HNTs incorporated composites show better improvements than unmodified HNTs. This is due to the better dispersion and stronger interfacial interaction of QBM-HNTs compared to unmodified HNTs.

According to the DMA results of Liu et al. (2013), the storage modulus of the composites improved significantly compared to neat PLA samples. For instance, PLA composites with 40 phr HNTs showed 143 percent increment in storage modulus (at 37°C), which can be attributed to the stiffened interphase generated due to the reinforcing effect (restricting the mobility of polymer chains) of HNTs and formation of HNTs networks at high loadings within the matrix. As mentioned before, these aspects are affected by dispersion, concentration and degree of exfoliation of HNTs.

10.2.1.2 MORPHOLOGICAL PROPERTIES

According to the morphological study of De Silva et al. (2013), HNTs dispersion is fairly good at moderate loadings such as 5 percent of HNTs (Figures 10.3a and 10.4a). Furthermore, the number of micro-voids (created due to the pulled out nano tubes during the tensile testing) and pores (removal of matrix during tensile testing due to the applied force) of composites with 5 percent of HNTs are less compared to composites with 10 percent of HNTs (comparing Figure 10.4b, c). In addition, large clusters due to the aggregation of HNTs could be observed in the micrographs of composites with 10 percent of HNTs (Figure 10.4c; magnified area). This shows that the interfacial interaction between PLA and HNTs are not strong enough at high loadings of HNTs compared to low and moderate loadings. The cluster formation can be avoided and the dispersion of HNTs can be improved by modifying the surface of HNTs. This was well proven by Prashantha et al. (2013); as illustrated in Figure 10.5, the dispersion of HNTs improved after incorporating HNTs modified with quaternary ammonium salt with benzoalkonium chloride. Furthermore, Murariu et al. (2012) studied the effect of two types of modified HNTs (QM-HNTs and m-HNTs) on PLA and stated that both types of modified HNTs were well dispersed within the PLA matrix and dispersion of m-HNTs at high loadings (12% of HNTs) were better compared to QM-HNTs. Supporting evidence can be found elsewhere (Gorrasi et al., 2013). However, as reported by Russo et al. (2013), according to the morphological analysis, dispersion of HNTs seemed to be poor (even at low loading such as 3% of HNTs by weight) with traces of aggregates.

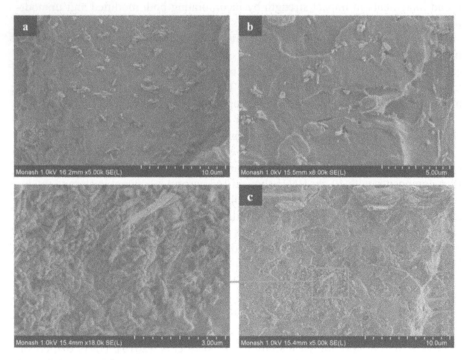

FIGURE 10.3   SEM images of PLA/HNT films with (a) and (b) 5 wt% HNTs and (c) 10 wt% of HNTs (De Silva et al., 2013). Marked arrows illustrate the micro-voids.

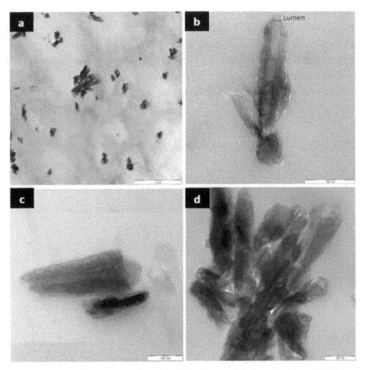

FIGURE 10.4 TEM micrographs showing (a) the spatial distribution of HNTs in the PLA matrix at 5 percent HNT concentration and (b, c, d) the structural morphology of HNTs at higher magnifications (De Silva et al., 2013).

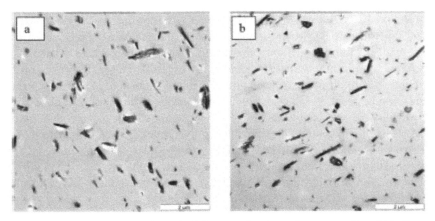

FIGURE 10.5 TEM images of PLA/HNT composites by melt compounding; (a) before and (b) after modifying HNTs (Prashantha et al., 2013).

Liu's morphological observations are very interesting (Liu et al., 2013). The SEM micrographs of Liu et al.'s fractured surface shows only the end face of the tubes (white color circular dots), which was attributed to the orientation of the fillers; i.e. fillers aligned in one direction. The orientation of the fillers was confirmed by the TEM micrographs (Figure 10.6; filler direction is marked by an arrow). When the filler loading increases, HNTs tend to aggregate as illustrated in Figure 10.6 (as circled).

FIGURE 10.6   TEM image of aligned HNTs in PLA matrix (Liu et al., 2013).

10.2.1.3   THERMAL PROPERTIES AND CRYSTALLINITY

De Silva et al. (2013) observed improvements in thermal stability with addition of HNTs (Table 10.3). For instance, as HNTs concentration increased, the temperature at 50 percent weight loss of the PLA/HNTs film increased appreciably from that of pure PLA film. Furthermore, the remaining weight percentage at maximum temperature significantly increased with addition of HNTs. This increase was attributed to the setup of diffused degraded products into the lumen of HNTs. Supporting evidence can be found in Liu et al. (2013). As mentioned by Murariu et al. (2012), TGA revealed that the thermal stability of PLA did not change much with addition of QM-HNTs. Temperatures corresponding to maximum thermal degradation (T_D) decreased with addition of QM-HNTs and as reasoned by the authors, this reduction is due to the hydration of residual water and degradation of the ammonium modifier. Surprisingly, as reported by Russo et al. (2013), the thermal stability decreased with HNTs loading due to the invoked catalytic phenomena.

TABLE 10.3   Summary of TGA results of De Silva et al.'s work (De Silva et al., 2013).

HNTs Concentration (w/w %)	Temperature at 50% Weight Loss (°C)	Remaining Weight (%) at Maximum Temperature
0	355	0.2
2.5	353	3.8
5	368	4.9
7.5	374	7.5
10	368	14.8
Pure HNTs	-	85

Murariu's differential scanning calorimetric (DSC) study revealed that addition of QM-HNTs does not change the glass transition (T_g) and melting point (T_m) temperatures of the composites (Murariu et al., 2012). In addition, the degree of crystallinity does not change, which showed that QM-HNTs do not act as an effective nucleating agent for PLA polymer. With addition of m-HNTs, thermal stability shows slight improvements, but there are no changes to T_g and T_m of the composites compared to pure PLA. With m-HNTs, a higher degree of crystallinity and nucleating ability were revealed. According to the DSC results of Prashantha et al. (2013), the T_g and T_m did not change for both types of composites, which supports Murariu et al. (2012) findings. But pure PLA has a single peak at the T_m point and the peak splits into two (exhibits a double melting point) with addition of HNTs (modified and unmodified). Such melting behavior could be due to the melting of the crystals which formed in the cold crystallization stage during heating, followed by recrystallization. In contradiction to Prashantha et al. (2013), double-peaked melting points were not observed in the DSC results of Russo's study (Russo et al., 2013). Furthermore, DSC results showed the nucleating effect of HNTs (Prashantha et al., 2013, Russo et al., 2013).

As revealed by the dynamic mechanical analysis (DMA) results of Prashantha et al. (2013), for PLA/HNTs and PLA/QBM-HNTs composites, the peak of the loss factor (tan δ) becomes broadened and shifted toward higher temperature. This shift can be attributed to the reduction in the segment motion of polymer matrix. It was noted that the DMA conducted by Liu et al. (2013) revealed an increase in T_g with addition of HNTs in contrast to other findings; T_g of PLA was 66.9°C and increased to 82.5°C at 40 phr loading. This phenomenon confirmed the interfacial interaction of PLA and HNTs and the reinforcing effect of the fillers. This could have occurred due to the enhanced interphase between the PLA and aligned HNTs.

As reported by Liu et al. (2013) the X-ray diffraction (XRD) results revealed that intercalation of PLA chains into the interlayers of HNTs had not occurred, since the interlayer distance ($2\theta = 12.26°$) of the HNTs was unchanged. The diffraction peak at $2\theta = 20°$ had disappeared for all the composite samples due to the PLA-HNTs interaction. In contradiction to Liu's argument, the diffraction angle at 20° was visible for all composites in Gorrasi et al. (2013) study, which could be due the poor filler-matrix interaction (which was also confirmed by mechanical and morphological properties). Generally there has been little work on the X-ray analysis of biopolymer-HNTs composites and this area can be of interest for future studies.

10.2.1.4   CHEMICAL PROPERTIES

Fourier transform infrared (FTIR) results reported by Liu et al. (2013) (Figure 10.7) showed that hydrogen bonding causes the filler-matrix interaction. The peaks appearing at 3,621 and 3,695 cm^1 wavenumbers (vibrational band of inner hydroxyl groups and hydroxyl groups at the octahedral surface of HNTs) may shifted to 3,625 and 3,698 cm^1 with addition of 40 phr of HNTs. It is stated that this particular shift could be due to hydrogen bonding interactions between the carbonyl groups of PLA (C=O) and hydroxyl groups of HNTs. But to the best of our knowledge, if there is hydrogen bonding interaction between carbonyl groups of PLA, the band at 1,756 cm^1 should show a band shift. But in the illustrated figure, there is no such noticeable band shift. Furthermore, the band at 536 cm^1 corresponding to the deformation of Al-O-Si of HNTs has shifted to lower frequencies, indicating an occurrence of hydrogen bonding between the filler and matrix. Another band shift was observed at 3,501 cm^1 which was assigned to terminal hydroxyl groups of PLA and this shift was attributed to the hydrogen bonding interaction with outer surface siloxane (Si-O-Si) groups of HNTs. But it was difficult to analyze peaks within 1,300–500 cm^1, since bands for siloxane groups overlapped with the bands of PLA within that range.

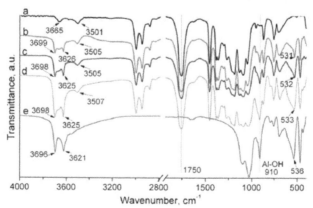

FIGURE 10.7   FTIR spectra of (a) PLA, PLA with HNT content of (b) 5, (c) 20 and (d) 40 phr, and (e) pure HNT (Liu et al., 2013).

De Silva et al. (2013) have also studied the FTIR spectra to identify the inter-action between PLA and HNTs. In line with previous studies, the vibrational band 3,648 cm ^{1}, corresponding to end chain hydroxyl groups of PLA, shifted to higher frequencies with addition of HNTs; it shifted to 3,659 and 3,652 cm ^{1} with 5 and 10 percent of HNTs, respectively (Figure 10.8, inset 1) despite the interaction of end hydroxyl groups of PLA with HNTs via hydrogen bonding. Furthermore, the binding energy of PLA with 5 percent of HNTs was higher than with 10 percent of HNTs, which was due to the amount of band shifts explained earlier. The intensity of the peak at 1,043 cm ^{1} of PLA increased with addition of HNTs. The initial absor-bance value of the peak 1,043 cm ^{1} of PLA/HNT0 was 0.11 and increased to 0.125 and 0.155 for PLA/HNT5 and PLA/HNT10, respectively (Figure 10.8, inset 2). In general, when considering the FTIR spectra of composites, the intensity of the peaks of the matrix will not change drastically. Even if it is changed, there would only be a slight decrease in matrix peak intensity, because the weight percentage of the matrix decreases when adding fillers. However, it would not increase unless there is an overlap with another peak (possibly from the filler). Hence, the shift in the 1,043 cm ^{1} band could be due to the overlapping of the vibrational band at 1,030 cm ^{1} of HNTs which corresponds to the outer surface Si-O groups, which has shifted (red shift). Therefore, this shift implies that the external surface Si-O groups could form hydrogen-bonds with the end hydroxyl groups of PLA.

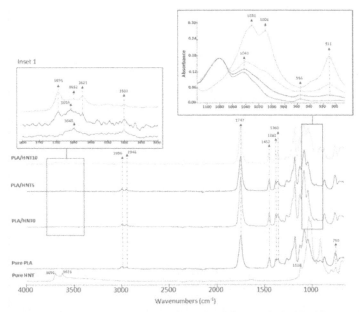

FIGURE 10.8   FTIR spectra of PLA/HNT films (De Silva et al., 2013). The notation PLA/HNT *x*has been used to describe the figure, where *x* denotes the HNT concentration of the films.

10.2.1.5 BARRIER PROPERTIES

Barrier properties in biopolymeric composites are defined by measurement of per-meability of gas or water vapor. As observed by Gorrasi et al. (2013), water sorption was accessed using the Flory-Huggins model and it decreased with HNTs loading due to the presence of hydrophilic HNTs. When the diffusion coefficient (D_o) is plot-ted against the filler loading, it is possible to note that D_o decreases drastically up to 3 percent of HNTs while the drop is less steep from 3 to 12 percent. For PLA/m-HNTs films, D_o decreases drastically up to 12 percent. This finding demonstrates that the silane treatment of HNTs remarkably improved HNTs dispersion in the PLA matrix, since diffusion depends directly upon the barrier effect created by the fillers based on filler dispersion. Water vapor permeability of the composite has decreased with HNTs loading and m-HNTs addition showed a very linear decreasing trend. As depicted, the barrier properties were improved due to the filler dispersion. As re-ported by Russo et al. (2013), the diffusion coefficient of PLA/HNTs films reduced with respect to pure PLA film and it was attributed to the tortuous path created by the dispersion of fillers which increased the effective diffusion path length. In ad-dition, water absorption was increased compared to pure PLA and, this increment was due to the low aspect ratio, poor dispersion, and the structure of HNTs (lumen which may entrap water).

10.2.1.6 SURFACE WETTABILITY

De Silva et al. (2013) assessed the surface wettability of PLA/HNTs films by em-ploying contact angle (θ) measurements. θ decreased with addition of 2.5 and 5 percent of HNTs. This reduction was attributed to an increased amount of polar groups within the composite due to the presence of OH groups of HNTs. As a result, the surface energy of the composite films increased and therefore, θ values tended to slightly decrease. But when filler loading increased (7.5 and 10 percent), the surface roughness of the films increased (since they were solvent-casted films). As surface roughness increases, the material's surface energy reduce (interfacial tension be-tween polymer and water reduces) and therefore, θ increased slightly compared to pure PLA films.

10.3 CONCLUSION AND PROSPECTS

The studies showed that HNTs are a suitable nano-filler to reinforce PLA, where it enhances the tensile, impact, thermal and barrier properties significantly. The hy-droxyl groups and siloxane groups of HNTs could possibly interact with the end hy-droxyl or carbonyl groups of PLA to provide a stronger interphase which improves the abovementioned physical properties. With these improved physical properties, the PLA/HNTs nanocomposite can possibly be used in packaging applications such as biodegradable heat-resistant food storage containers and cups, over-wrapping and lamination films, and blister packaging. However, the compatibility of PLA/HNTs nanocomposites with food item should be scientifically tested. Not only packaging applications, but also these nanocomposites may be used as in pharmaceutical applications and in interior panels for automobile applications. Moreover, heat retardant and flammability tests should be carried out to investigate the heat resistance of PLA/HNTs nanocomposites. Although both PLA and HNTs are derived from natural resources, there is no evidence for the biodegradability of PLA/HNTs nanocomposites in the literature. Furthermore, rheological properties can be examined to understand the dispersion of HNTs within the PLA. Finally, further investigations (both synthesis and characterization techniques) can be carried out with melt-blown PLA/HNTs films, which could possibly be used as grocery bags. With the knowledge gathered from these fundamental studies and the suggested future work on PLA/HNTs nanocomposites usage of PLA may expand in the near future.

KEYWORDS

- **Barrier properties**
- **Biodegradability**
- **Differential scanning calorimetric (DSC)**
- **Elastic modulus**
- **Multiwalled carbon nanotubes (MWCNTs)**
- **PLA/Clay nanocomposites**
- **Wettability of biocomposite membranes**

REFERENCES

1. Auras, R.; Harte, B.; and Selke, S.;*Macromol. Biosci.* **2004**, *4*(9), 835–864.
2. Bleach, N.C.; Nazhat, S. N.; Tannera, K.E.; Kellomaki, M.; and Tormala, P.;*Biomaterials.* **2002**, *23*,1579–1585.
3. Chang, J.-H.; An, Y. U.; and Sur, G. S.; *J. Pol. Sci. Part B: Pol. Phys.* **2003**, *41(1)*, 94–103.
4. De Silva, R. T.; Pasbakhsh, P.; Goh, K. L.; Chai, S.-P.; and Shen, J.; *J. Compos. Mater.* **2013**,doi: 10.1177/0021998313513046.
5. Fukushima, K.; Abbate, C.; Tabuani, D.; Gennari, M.; and Camino, G.; *Poly. Degrad. Stab.* **2009a**, *94(10)*, 1646–1655.
6. Fukushima, K.; Tabuani, D.; and Camino, G.; *Mater. Sci.Eng. C.* **2009b**, *29(4)*, 1433–1441.
7. Gorrasi, G.; Pantani, R.; Murariu, M.; and Dubois, P.; *Macromol. Mater. Eng.* **2013**, doi: 10.1002/mame.201200424.
8. Hapuarachchi, T. D.; and Peijs, T.; *Compos. Part A: Appl. Sci. Manuf.* **2010**, *41*(8), 954–963.
9. Jeszka, J. K.; Pietrzak, L.; Pluta, M.; and Boiteux, G.; *J. Non Crystall. Solids.* **2010**, *356(11–17)*, 818–821.

10. Liu, M.; Zhang, Y.; and Zhou, C.; *App. Clay Sci.* 2013, *75–76*, 52–59.
11. Martinez-Sanz, M.; Lopez-Rubio, A.; and Lagaron, J. M.; *Biomacromolecules.* 2012, *13*(11), 3887–3899.
12. Matusik, J.; Stodolak, E.; and Bahranowski, K.; *Appl. Clay. Sci.* **2011**, *51*(1–2), 102–109.
13. Murariu, M.; Dechief, A.-L.; Paint, Y.; Peeterbroeck, S.; Bonnaud, L.; and Dubois, P.; *J. Poly. Environ.* **2012**, *20* (4), 932–943.
14. Nair, L. S.; Bhattacharyya, S.; and Laurencin, C. T.; *Exp. Opin. Biol. Ther.* **2004**, *4(5),* 659–668.
15. Najafi, N.; Heuzey, M. C.; Carreau, P. J.; and Wood-Adams, P. M.; *Poly. Degrad. Stab.* **2012**, *97(4),* 554–565.
16. Papageorgiou, G. Z.; Achilias, D. S.; Nanaki, S.; Beslikas, T.; and Bikiaris, D.;*Thermochimica Acta.* **2010**, *511* (1–2), 129–139.
17. Petersson, L.; and Oksman, K.; *Compos. Sci.Technol.* 2006, *66*(13), 2187–2196.
18. Plackett, D. V. et al.; *Pack. Technol.Sci.* **2006**, *19*(1), 1–24.
19. Prashantha, K.; Lecouvet, B.; Sclavons, M.; Lacrampe, M. F.; and Krawczak, P. *J. Appl. Poly. Sci.* **2013**,*128*(3), 1895–1903.
20. Rhim, J.-W.; Hong, S.-I.; and Ha, C.-S.;*LWT   Food Sci. Technol .*2009, *42*(2), 612–617.
21. Rhim, J.-W.; Mohanty, A. K.; Singh, S. P.; and Ng, P. K. W.; *J. Appl. Poly. Sci.* **2006**, *101*(6), 3736–3742.
22. Russo, P.; Cammarano, S.; Bilotti, E.; Peijs, T.; Cerruti, P.; and Acierno, D.; *J. Appl. Poly. Sci.* **2013**, doi: 10.1002/APP.39798.
23. Sabzi, M.; Jiang, L.; Atai, M.; and Ghasemi, I; *J. Appl. Pol. Sci.* **2013**, *129*(4), 1734–1744.
24. Kumar, S.; and Gupta, S. K.; *Middle East J. Sci. Res.* **2012**, *12*(5), 699–706.
25. Sinha Ray, S.; Yamada, K.; Okamoto, M.;and Ueda, K.; *Polymer* **2003**, *44*(3),

CHAPTER 11

HALLOYSITE-POLY(LACTIC-CO-GLYCOLIC ACID) NANOCOMPOSITES FOR BIOMEDICAL APPLICATIONS

MINGXIAN LIU, PENG AO, QI PENG, BINGHONG LUO, and CHANGREN ZHOU

CONTENTS

11.1   INTRODUCTION

The empty lumen of halloysite nanotubes (HNTs) enables them to have a high loading and protection ability for small molecules such as, proteins, DNA, drugs, anticorrosive agents etc. Furthermore, HNTs are nontoxic and have little effect on the
adhesion, growth, and differentiation of cells (Vergaro et al., 2010). Also, they are
environmentally friendly materials. These characters make HNTs ideal candidate as
components of biomaterials especially as carriers for drug controlled release. The
drugs released from the HNTs can last 30~100 times longer than those from the
drugs themselves or other carriers. Coating of biocompatible polymers on the drug-
loaded HNTs surfaces can enhance the biocompatibility of the tubes and further
retard the rate of drug release (Lvov and Abdullayev, 2013). Various drugs can be
bonded on the inner or outer surface of HNTs via physical or chemical interactions.
Drug loaded HNTs-polymer nanocomposites can be presented as a formulation of
powders, suspensions, bulk materials, and fibrious scaffolds, which have many potential applications in tissue engineering, wound dressing, and drug/DNA delivery
systems.

Many synthesized and natural biodegradable polymers are compounded with
HNTs for the preparation of the nanocomposite for biomedical applications. Among
them, poly (lactic-co-glycolic acid) (PLGA) has received wide attentions due to
its many unique properties (Jain, 2000). PLGA is a biodegradable and biocompatible copolymer which has already been approved for therapeutic devices by Food
and Drug Administration (FDA). PLGA is synthesized by means of random ring-
opening copolymerization of glycolic acid and lactic acid (Figure 11.1). Depending
on the ratio of lactide to glycolide used for the polymerization, different forms of
PLGA can be obtained. They are usually identified in regard to the used monomers'
ratio (e.g. PLGA 75:25 identifies a copolymer whose composition is 75% lactic acid
and 25% glycolic acid). All PLGAs are amorphous rather than crystalline and show
a glass transition temperature in the range of 40–60°C. Unlike the homopolymers of
lactic acid (polylactide, PLA) and glycolic acid (polyglycolide, PGA) which show
poor solubilities, PLGA can be dissolved by a wide range of common solvents,
including chlorinated solvents, tetrahydrofuran, acetone or ethyl acetate. Therefore,
PLGA is easily solution-processed into different forms of materials such as nanofibers, microspheres and porous scaffolds. With many advantages of PLGA-based
nanomaterails, they can be useful for many different biomedical applications such
as wound dressing, drug delivery system, and tissue engineering scaffolds. However, due to the weak mechanical property of PLGA and the need for multifunc-
tionality for practical applications, it is essential to modify PLGA with enhanced
mechanical properties and desired functionality. Due to the individual features of
PLGA and HNTs, it is feasible to combine them to make an effective nanocomposite
for biomedical applications.

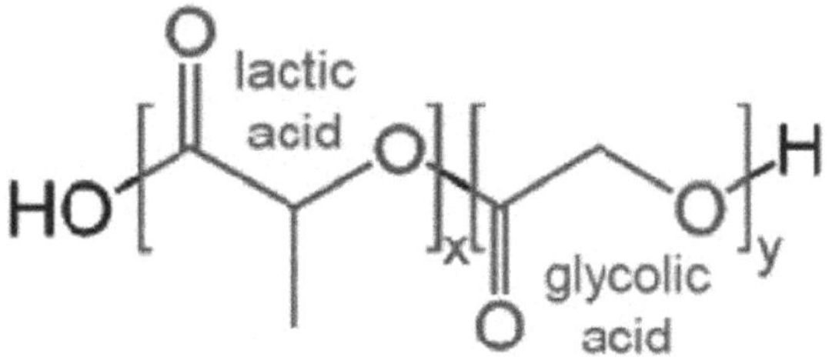

FIGURE 11.1 The chemical formula of PLGA. x and y indicate the number of times each unit repeats.

In this chapter, we review advances in the applicability of the HNTs-PLGA nanocomposites for biomedical applications with emphasis on the structure and properties of the composite systems. The characteristics of HNTs for biomedical applications are firstly summarized. Then functionalizaiton methods for HNTs are introduced for the formation of PLGA nanocomposites. The fabrication approaches, morphology, and performance for HNTs-PLGA are then described in detail. Lastly, the potential applications of the HNTs-PLGA nanocomposites are discussed.

11.2 CHARACTERISTICS OF HNTS FOR BIOMEDICAL APPLICATION

11.2.1 TUBULAR AND POROUS MICROSTRUCTURE

HNTs have empty lumen microstructures (Figure 11.2) and also are porous on their surfaces which ensure them a high aspect ratio and drug loading ability. The pores of HNTs arise not only from their hollow interiors and surface defects, but also from the voids created when they pack together with other tubes (Churchman et al., 1995). Many chemically and biologically active substances could be loaded into the pores of HNTs by vacuum or by simply immersing them in their solutions. Different types of drugs such as tetracycline, khellin and nicotinamide adenine dinucleotide have been loaded into the HNTs by soaking with the saturated drug solution under vacuum (Lvov and Abdullayev, 2013). The drug released from the HNTs could last 30~100 times longer than that from the drug themselves or other carriers. Coating of polymers on the drug-loaded HNTs surface further retards the drug release rate.

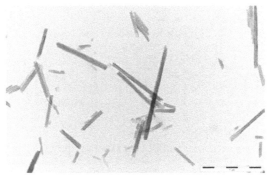

FIGURE 11.2   TEM image for purified HNTs.

11.2.2   GOOD DISPERSION IN AQUEOUS SOLUTIONS

HNTs have good water dispersion abilities compared with other nanoparticles. This has been attributed to their small dimensions, relatively weak tube-tube interactions and intrinsic hydrophilicity. Raw HNTs can be dispersed readily in water by mechanical stirring or ultrasonic treatment (Liu et al., 2012, 2007). The milk-white HNTs dispersion with 5 wt.% concentration is stable for several weeks without sedimentation (Lu, et al., 2013a). The zeta potential of HNTs aqueous dispersion shows that HNTs are negatively charged above pH 2.4 (Vergaro et al., 2010). High zeta potential is a condition of good colloidal stability of dispersed nanotubes. Therefore, they are readily mixed with water soluble polymers in aqueous solution for preparing nanocomposites (Liu et al., 2013a, Cavallaro et al., 2011). Also the water-soluble drugs can be mixed with the HNTs-polymer dispersions and subsequently processed into drug loaded nanocomposite systems (Lvov and Abdullayev, 2013).

11.2.3   BIOCOMPATIBILITY

Recent researches demonstrate that HNTs are biocompatible and could be potentially used as biomaterials (Vergaro et al., 2010, Liu, et al., 2013a, Qi, et al., 2012, Zhou, et al., 2010). For example, Vergaro et al. (2010) found human epithelial adenocarcinoma cell and breast cancer cell could adhere, grow and be taken up by HNTs up to concentrations of 75 µg/mL. HNTs-polymer nanocomposites, such as HNTs-poly(lactic-co-glycolic acid) nanocomposites fibers (Qi et al., 2012), HNTs-polyvinyl alcohol nanocomposite films (Zhou et al., 2010) and HNTs-chitosan 3D scaffolds (Liu et al., 2013a) showed good cytocompatibility. However, the in vivo biosafety of HNTs should be further assessed to expand their practical applications to clinical treatments.

11.2.4 CROSSING THE PLASMA MEMBRANE

One of the key advantages of tubular nanoparticle is the possibility of effective crossing of biological barriers, which would allow their use in the delivery of therapeutically active molecules into the cells. Functionalized carbon nanotubes (CNTs) could be taken up by different cells, and this has been attributed to their unique structures. The cylindrical shape and high aspect ratio of CNTs could allow their penetration through the plasma membrane, similar to a "nanosyringe" (Kostarelos et al., 2007). HNTs have a similar morphology to CNTs, so in theory they could penetrate the plasma membrane to enter the cells. In fact, the uptake of functionalized HNTs has been confirmed and it has also been employed as a siRNA carrier for drug delivery systems targeting tumor cells (Vergaro et al., 2010, Shi et al., 2011). The properties which enable crossing the plasma membrane have made HNTs potentially applicable as nanovectors for therapeutic agent delivery.

11.2.5 REINFORCING ABILITY

For many medical applications, the mechanical properties of the material are important. For instance, the required strength of the substitution scaffolds or tissues of the real biological systems is in 50~200 MPa (Karageorgiou and Kaplan, 2005), so, reinforcing polymers using different strategies is necessary. HNTs have an excellent reinforcing effect on polymers which has been partially attributed to their high aspect ratio and their nanoscale dimensions (Du et al., 2010). Moreover, dispersion of HNTs in the polymer matrix is readily realized due to the relatively weak tube-tube interactions. Also, interfacial interactions between HNTs and polymers could be tailored by chemical or physical approaches. Since, many polymers suffer from low mechanical properties for biomedical applications HNTs could be employed as promising fillers for them. The results are shown that almost, all the polymers could be mechanically reinforced by HNTs even at low loading (5 wt.%). More importantly, the strength, modulus, stiffness and impact resistance of polymers could simultaneously be increased by HNTs.

11.2.6 EASY, AVAILABLE AND CHEAP

HNTs are available economically in thousands of tons from natural deposits in countries such as China, United States of America, Brazil, France, Spain and New Zealand. Compared with carbon nanotubes (CNTs) and other synthesized nanotubes, the price of HNTs is relatively low for industrial scale applications. For example, the price of HNTs deposits in China is~US$250 /ton. The low price and unique structures make HNTs-polymer nanocomposites cost competitive for biomedical applications.

11.3   FUNCTIONALIZATION OF HNTS FOR PLGA NANOCOMPOSITE

11.3.1   PURIFICATION

Before being used for the preparation of polymer composites, raw halloysite should firstly be purified to remove the impurities in the mineral. This has always realized by dispersing them into water in the presence of surfactants. The surfactants are added to increase the stability of halloysite in water and help their dispersion. Sodium hexameta phosphate and Tween 80 are the most commonly used surfactants. The purified HNTs exhibit much more uniform tubular morphology compared with the raw halloysite.

11.3.2   SILANIZATION

To improve the dispersion of HNTs in the hydrophobic PLGA and increase the interactions with fat-soluble drugs, surface hydrophobic treatment of HNTs should be performed. The silanization of HNTs has been performed via condensation reactions between hydrolyzed silanes and the surface aluminol/silanol groups of HNTs. The silanization reactions could take place in either toluene or water/alcohol mixtures. For example, the grafting of 3-aminopropyltriethoxysilane (ASP) on HNTs was performed in chloroform/methanol mixture at 120°C with magnetic stirring (Haroosh, et al., 2013).The ASP modified HNTs showed improved dispersion in PLA relative to unmodified HNTs (Haroosh et al., 2013), since, ASP can react with the HNT surface that helping in turn to reduce the adhesion between the modified HNT.

11.3.3   GRAFTING POLYMERIZATION ON HNTS

To increase interfacial bonding between HNTs and the PLGA matrix, surface-grafting poly(L-lactide)(PLLA) or poly(D,L-lactide) on HNTs was performed by bulk ring-opening polymerization of D,L-lactideusing stannous octoate as catalyst and HNTs as a co-initiator or L-lactide oligomer under microwave irradiation (Luo et al., 2013b, Li et al., 2013). Figure 11.3 shows the reaction mechanism for the grafting reaction of PLLA on HNTs. The optimal reaction conditions for grafting polymerization and the structure and properties of the surface-grafted HNTs (g-HNTs) were characterized by various approaches. Results showed that the grafted PLLA chains on the surfaces of HNTs, as inter-tying molecules, played an important role in improving the adhesive strength between the nanotubes and the polymer matrix. The enhanced interaction among g-HNTs and the PLLA matrix resulted in a better tensile strength and modulus compared to the pristine PLLA and HNTs-PLLA.

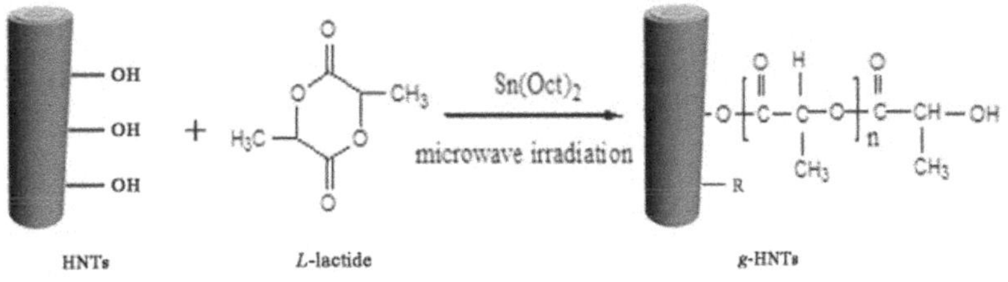

R=PLLA

FIGURE 11.3   The grafting polymerization reactions of PLLA on HNTs surface.

11.4   APPROACHES FOR THE FABRICATION OF HNTS–PLGA NANOCOMPOSITES

11.4.1   ELECTROSPINNING

Electrospinning is an effective and economical fabrication method for three-dimensional fibrous structures. The electrospun nanofibers create a fabric network that has a high porosity, small pore sizes, and very large surface-to-volume ratio. This has many biomedical applications such as drug delivery, artificial organs, wound dressing, and medical prostheses. Generally, dichloromethane (DCM) or tetrahydrofuran has been employed to dissolve the PLA or PLGA along with the additional solvent dimethyl formamide (DMF) to enhance the electric conductivity of the solution. The use of the mixed solvent was able to generate nanofibers with a uniform distribution of fiber diameters and smooth fiber morphology (Zhao et al., 2013). Generally, the concentration of PLAs in solution for electrospinning was 12 wt.%/v (Dong et al., 2013, 2011), while the concentration of PLGA's solution was 25 wt.-%/v (Qi et al., 2010, 2013).Then raw HNTs or functionalized HNTs powder were added into the corresponding solutions under magnetic stirring or ultrasonic treatment. If the HNTs-PLGA or PLA composites were designed to be used as drug delivery systems, the drugs should firstly be loaded into HNTs via application of a vacuum. Later, the drug-loaded HNTs were mixed with the PLGA or PLA solutions using a similar procedure. The electrospinning method for HNTs-PLGA or PLA nanocomposites is similar to most other systems, and could be controlled by the electrical potential, the distance between the tip of the needle and the collector, feeding speed, and the solution viscosity. The optimum condition to achieve well define fibers for the HNT-PLGA system is using the electrical potential of 20 kV, the distance between the tip of the needle and the collector of 15 cm, feeding speed of 1.0 mL h ^{1} by a syringe pump and the solution concentration of 25% (w/v) (Qi et al., 2010).

11.4.2   MELT MIXING

Melt processing is a simple and easy method for standard industrial facilities, so, it is a capable approach for the production of HNTs-PLGA or PLA composites. PLA or PLGA is thermoplastics so they can be mixed with HNTs in a melt. The melt mixing method always leads to a good dispersion state for HNTs in the matrix. When processing the nanocomposite, the polymer chains could interact with HNTs due to the strong shear force which leads to interfacial compatibility. For example, in our previous work, we prepared high performance HNTs-PLA nanocomposites via melt mixing using an open two-roll mill (Liu et al., 2013b). By the melt mixing method, HNTs had a preferred orientation in the PLA matrix which was of benefit for the enhancement of the mechanical properties. However, one of the disadvantages of the melt processing was an unexpected polymer degradation and oxidation under high temperatures and the strong shear force, which could lead to deterioration in polymer properties (Tsuji and Fukui, 2003).

11.4.3   EMULSION EVAPORATION

Drug loaded polymer microspheres could be prepared by the evaporation of an organic solvent from dispersed oil droplets containing both polymer and drug. Firstly, the drug for encapsulation was dissolved in water and then this aqueous phase was dispersed in a polymer organic solution (usually the solvent was dichloromethane, DCM). This was the first W/O emulsion. Dispersion of the first emulsion in a stabilized aqueous medium (usually using poly(vinyl alcohol) as stabilizer) formed the final O/W emulsion. Finally, polymer microspheres were formed as the DCM evaporates and the polymer hardens, trapping the encapsulated drug (Freiberg and Zhu, 2004). When preparing HNTs-PLA or PLGA nanocomposite microspheres, the HNTs could be mixed with the drug in the first step. And then, this aqueous dispersion mixture could be treated with the same procedures with the drug-loaded composites microspheres being the resulting product.

HNTs could also be used as particulate emulsifiers for the preparation of drug-carrying PLGA microspheres based on a Pickering emulsion (Wei et al., 2012). The stable oil-in-water emulsion was formed using HNTs as the emulsifier and dichloromethane (CH_2Cl_2) solution of PLGA as an oil phase. The HNTs-coated PLGA microparticles were prepared via the evaporation of CH_2Cl_2 from the emulsion, and then bare-PLGA microparticles were prepared by removal of the HNTs. The scanning electron microscope (SEM) with energy dispersive analysis system (EDS) results confirmed the adsorption of HNTs only at the surface of the PLGA microparticles. After washing with acid aqueous solution, there was a negligibly small amount of HNTs on the bare-PLGA microparticles. The release curves of ibuprofen (IBU) loaded in bare-PLGA microparticles were closely fitted by the Weibull equation and the release followed Fickian diffusion. The combined system of a Pickering emulsion and solvent volatilization has opened up a new route to fabricate a variety of polymer microparticles.

11.5 MORPHOLOGY OF HNTS–PLGA NANOCOMPOSITES

11.5.1 NANOFIBERS

HNTs-PLGA or PLA composites prepared via electrospinning showed a morphology based on nanofibrous mats. The morphological structures of electrospun HNTs-PLA nanocomposites demonstrated that at added HNTs contents of below 3 wt % the HNTs-PLA nanocomposites yielded relatively small fiber diameters (Dong, et al., 2011). It can be seen this decrease could be attributed to enhancements of both electrical conductivity and viscosity of prepared solutions due to the inclusion of tubular clay. A significant dimensional variation was observed for fibers when the HNTs loading was 3 and 5 wt %. The fiber diameter of 3 wt% HNTs included PLA nanocomposites was bigger than that of pure PLA but the clay dispersion was uniform. However, when 5 wt% HNTs were included into PLA, large HNTs agglomerates and a "slurry-like" unfavorable structure coexisted, along with nonwoven entangled fibers. Interestingly, the influence of HNTs on the morphology of the PLGA fiber was less significant compared with that of PLA (Zhao et al., 2013). Adding different loadings of HNTs into PLGA, the morphology of PLGA nanofibers did not significantly change when compared with PLGA nanofibers, except that the mean diameter increased with the HNTs loading. The incorporation of negatively charged HNTs into the PLGA solution might result in a decrease of the surface charge density of the spinning jet, leading to the formation of nanofibers with a larger diameter. The different influence of HNTs on the fiber morphology between PLA and PLGA might arise from the different solution properties and the electrospinning equipment.

In addition of drugs into the HNTs-PLGA nanocomposites also could change the morphology of the electrospun nanofibers. Qi et al. (2010, 2013) found that the diameter of the tetracycline hydrochloride (TCH)-HNTs-PLGA composite fibers became much smaller when compared to that of pure PLGA and HNTs-PLGA nanofibers (Figure 11.4). The author attributed the decreased fiber diameter to the increase of the surface charge density of the spinning jet by the addition of the cationic TCH.

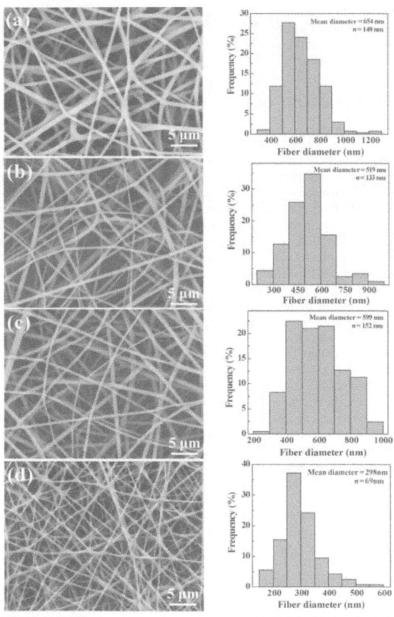

FIGURE 11.4 SEM micrographs of the electrospun (a) PLGA, (b) drug-loaded TCH/HNTs/PLGA nanofibers (1 wt. % TCH and 3 wt.% HNTs relative to PLGA), (c) drug-loaded TCH/HNTs/PLGA nanofibers (2 wt. % TCH and 3 wt.% HNTs relative to PLGA), and (d) TCH/PLGA nanofibers, and the corresponding fiber diameter distribution histograms (Qi et al., 2013).

11.5.2   BULK MATERIALS

HNTs-PLA or HNTs-PLGA nanocomposites can be processed into bulk materials by compression molding, injection molding or extrusion. For example, we prepared the uniformly dispersed HNTs-PLA nanocomposites by an injection molding method (Liu et al., 2013b). TEM images demonstrated the good dispersion and orientation of HNTs in PLA (Figure 11.5). The orientation of HNTs in the PLA was induced by the shearing force during processing. All the tubes were uniformly dispersed in the PLA matrix even with high HNTs content (40 phr). The hydrogen bonding interactions between HNTs and PLA were considered as the origin for the good dispersion of HNTs. The good dispersion and orientation of HNTs were of benefit for the enhancement of the mechanical properties and thermal properties of PLA. Although no study on HNTs-PLGA nanocomposites prepared by melt mixing has been found in the literature, we have successfully prepared them in our laboratory.

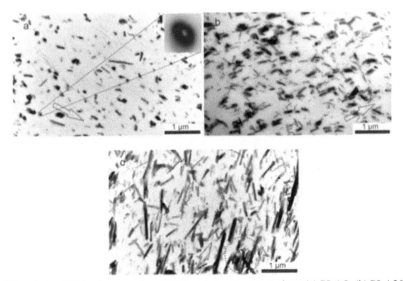

FIGURE 11.5   TEM micrographs of HNTs-PLA nanocomposites: (a) PLA5; (b) PLA20 and (c) PLA40. The arrows represent the orientation of the tubes. The circle in (c) represents the HNTs aggregates. The inset in (a) represents the end faces of a tube in the PLA matrix (Liu et al., 2013b).

11.5.3 MICROSPHERES

Polymer microspheres present a flexible platform for applications in diagnostics, drug release and bio-separations. HNTs were firstly mixed with PLA or PLGA in the DCM solution, and then the mixture was dipped into the aqueous solution. Finally, the HNTs-PLA or PLGA nanocomposite microspheres were formed when the solvent evaporated. The size of the microspheres could influence the release profiles of drugs. The rate of drug release was found to decrease with increasing sphere size. To investigate the effect of HNTs on the morphology of the PLA microsphere, a SEM experiment was conducted (Figure 11.6). It could be seen that the addition of HNTs had a slight effect on the morphology of the PLA microsphere. The diameters of the microsphere were comparable for all the samples even at relatively high HNTs loading. The pure PLA microsphere exhibited smooth surfaces, while the HNTs-PLA nanocomposite microspheres showed much rougher surfaces with HNTs embedded in the PLA matrix. From the morphology results, it could be speculated that the HNTs-PLA nanocomposite microspheres have good drug loading and release properties.

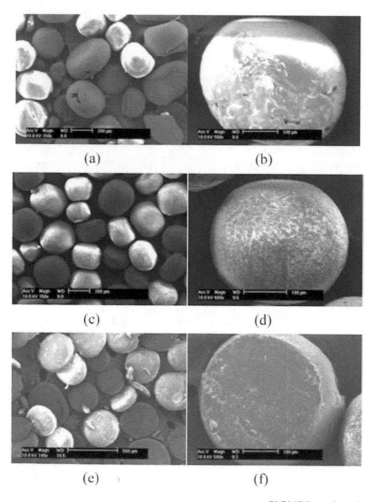

(a)   (b)

(c)   (d)

(e)   (f)

FIGURE 11.6   *(Continued)*

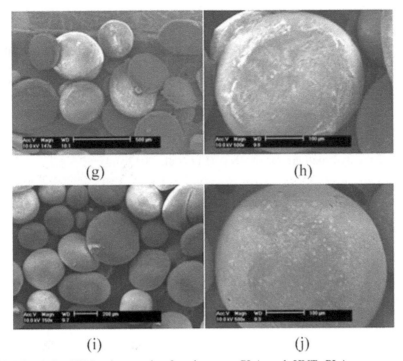

(g) (h)

(i) (j)

FIGURE 11.6 SEM micrographs for the pure PLA and HNTs-PLA nanocomposite microspheres. (a) and (b) pure PLA, (c) and (d) PLA+5 phr HNTs, (e) and (f) PLA+10 HNTs, (g) and (h) PLA+20 HNTs, (i) and (j) PLA+40 HNTs; (unpublished work).

11.6 PROPERTIES OF HNTS–PLGA NANOCOMPOSITES

11.6.1 MECHANICAL REINFORCEMENT

The incorporation of HNTs into polymers can effectively increase the mechanical properties, which is attributed to the efficient transfer of external load from the polymer matrix to the uniformed dispersed HNTs. HNTs also can significantly improve the mechanical properties of PLGA or PLA (Qi et al., 2010). Figure 11.7 compare the tensile properties of electrospun PLGA nanofibers and HNTs-PLGA nanofibers (Zhao et al., 2013). Compared with pure PLGA fibrous mats, the breaking strength, Young's modulus, and failure strain significantly increased with the addition of a small amount of HNTs. It also could be seen that the mechanical properties were not very sensitive to the HNTs loading for the HNTs in the experiment wide of 1~5 wt.%. The enhancement of the mechanical property of the PLGA nanofibrous mats by HNTs was ascribed to the alignment of HNTs in the PLGA fibers. In another study, HNTs could significantly improve both tensile, flexural and impact properties of PLA, and the increase was proportional with the loading of HNTs (Liu et al.,

2013b). For instance, the tensile strength, flexural strength and flexural modulus of HNTs-PLA nanocomposites with 30 phr HNTs were 74.1 MPa, 108.3 MPa and 6.56 GPa, which were 34, 25, and 116 percent higher than those of the neat PLA respectively. The toughness of PLA was improved by HNTs especially, when the loading of HNTs was less than 20 phr. HNTs could act as plasticizer of PLA or PLGA, which promotes the mobility of the polymer chains and distributes the impact energy during the fracture. The reinforcing effect of HNTs on PLGA and PLA could be attributed to (i) the uniformly distributed rigid nanotubes in PLA matrix and (ii) the hydrogen bonding interactions between the carbonyl groups (C=O) of PLA or PLGA and the hydroxyl groups of HNTs (Liu et al., 2013b).

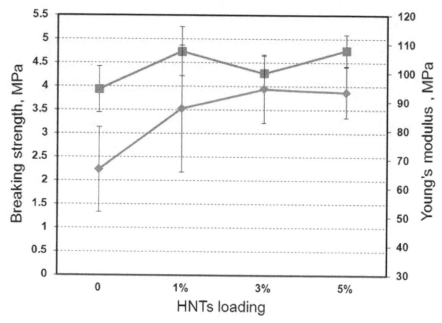

FIGURE 11.7 Comparison of tensile properties of electrospun PLGA nanofibers and HNTs-PLGA composite nanofibers. Data adapted from reference of (Zhao et al., 2013).

11.6.2 DRUG LOADING AND RELEASE

The drug release profiles suggested that the HNTs-PLGA nanocomposite nanofibers could be used to hinder the release of drugs from the HNTs to the solution. For example, TCH embedded in the PLGA nanofibers exhibited an obvious initial burst release. Within the first 24 h, 83.8 percent of the drug was released. The drug release reached to a plateau after 48 h. By contrast, the drug-loaded HNTs-PLGA fibrous mats displayed a sustained release profile within a month, with only 36 and 32 percent of TCH released after 28 day, respectively (Qi et al., 2010). The sustained release profile of drug from the HNTs-PLGA nanocomposite fibers was ascribed to the fact that the drug was firstly released from the HNTs carrier to the PLGA matrix, and then from the PLGA matrix to the release medium, thereby significantly reducing the diffusion rate of the drug.

11.6.3   CYTOCOMPATIBILITY

The HNTs-PLGA or PLA nanocomposites must be biocompatible for biomedical applications. Qi et al. studied the biocompatibility of the electrospun HNTs-PLGA composite nanofibers using the mouse fibroblast cells (Qi et al., 2012; Qi et al., 2010). The 3-(4,5-dimethylthiazol-2-yl)-2,5-diphenyl tetrazolium bromide (MTT) assay of cell viability and SEM observation of cell morphology results showed that HNT-PLGA nanofibers were able to promote cell attachment and proliferation as well as the pure PLGA. This suggested that the incorporation of HNTs within PLGA nanofibers did not compromise the biocompatibility of the PLGA nanofibers. The protein adsorption experiment showed that HNT-PLGA allowed more protein adsorption than those without HNTs, which could provide sufficient nutrition for cell growth and proliferation (Yah et al., 2012). By further incorporation of the drugs into the HNTs-PLGA nanocomposite, they also exhibited good cytocompatibility. It should be noted that the cytocompatibility should be further confirmed using other cell lines and also an in vivo experiment.

11.6.4   HEMOCOMPATIBILITY

For application in tissue engineering scaffolds or cardiovascular or blood-contacting medical devices, hemocompatibility of materials is necessary. Zhao et al. investigated the hemocompatibility of the HNTs-PLGA nanocomposite fiber mate by hemolytic and anticoagulant assays (Zhao et al., 2013). The results showed that HNTs-PLGA nanocomposites displayed similarly good anticoagulant properties and negligible hemolytic effects to human red blood cells as the pure PLGA. This was ascribed to the fact that the HNTs were well dispersed within the PLGA nanofibers, and the surface properties of HNTs-PLGA nanofibers did not show appreciable changes when compared with pure PLGA nanofibers. The good anticoagulant property of the HNTs-PLGA fibrous mats implied that they could potentially be used as blood-contacting materials.

11.6.5   ANTIBACTERIAL PROPERTIES

Antibacterial activity is related to materials that nearby kill bacteria or slow down their growth, without being generally toxic to surrounding tissue. The antimicrobial property of the drug loaded HNTs-PLGA nanocomposites was assessed through incubating S. aureus as a model bacterium in liquid and on solid media, respectively (Qi et al., 2013, 2011). The results showed that the TCH-loaded HNTs and TCH-loaded HNTs-PLGA composite nanofibers and nanofibrous could effectively inhibit bacterial growth. In contrast, TCH-free PLGA and HNTs/PLGA mats did not show an obvious bacterial inhibition effect. Therefore, the antibacterial activity of the materials was ascribed to the sustained release of the antibacterial drug TCH.

11.7 POTENTIAL APPLICATION OF HNTS–PLGA NANOCOMPOSITES

Due to many unique properties of the HNTs-PLGA and HNTs-PLA nanocomposites, they have many potential applications in biomedical areas. Tissue engineering is the use of a combination of cells, engineering and materials methods, and suitable biochemical and physio-chemical factors to improve or replace biological functions. For this purpose, a porous scaffold should be used as a shape on which cells can grow. Also, the scaffold should be cytocompatibile, bioactive and mechanically strong. HNTs can significantly reinforce PLGA or PLA. Meanwhile, HNTs have less influence on the biodegradation and biocompatibility of PLGA. HNTs-PLGA nanocomposites in the form of nanofiber mats or microspheres can potentially be used as tissue engineering scaffolds. Due to the good drug encapsulation and sustained release properties of the HNTs, the HNTs-PLGA nanocomposites also can be used as drug or gene carriers in the formulation of powders, bulk materials, and fibrous scaffold. Further, applications of HNTs-PLGA nanocomposite include wound dressing materials and absorbable antiadhesive membranes.

11.8 CONCLUSIONS

The natural tubular nanomaterial, HNTs, have unique combinations of tubular structure, large aspect ratio, natural availability, rich functionality, good biocompatibility and high mechanical strength. These characteristics of HNTs give those exceptional good mechanical, thermal, biological properties and also a low price for the HNTs-polymer nanocomposites. Hence, HNTs are promising nanoparticles for PLGA polymer composites for biomedical applications. The state of dispersion and the interfacial interactions determine the properties of the PLGA nanocomposites. Functionalization of HNTs benefits for the enhancement of interfacial bonding of the composite systems. HNTs can be mixed with PLGA or PLA using both solution mixing and melt mixing approaches nanofibers, bulk materials and films microspheres of the nanocomposites can be obtained. HNTs-PLGA or PLA nanocomposites exhibit substantially improved mechanical properties. The nanocomposites also show good biocompatibility and sustained drug release abilities; therefore, they have promising applications in tissue engineering, and as drug vehicles, wound dressings, and absorbable antiadhesive membranes. One of the obstacles for the biomedical application is the possible nonbiodegradation of HNTs *in vivo*. Therefore, further cyto-and tissue compatibility should be assessed to avoid any harm from the tubes to the body. In conclusion, although the preliminary research results for HNTs-PLGA nanocomposites seem promising, there is a long way to go before their practical application in clinical treatments.

ACKNOWLEDGEMENTS

This work was financially supported by the Guangdong natural science funds for distinguished young scholar (S2013050014606), the foundation for the author of Guangdong excellent doctoral dissertation (sybzzxm201220).

KEYWORDS

- **Biomedical applications**
- **Cytocompatibility**
- **Drug-loaded HNTs**
- **Electrospinning**
- **Hemocompatibility**
- **Polymerization**
- **Ultrasonic treatment**
- **Zeta potential**

REFERENCES

1. Cavallaro, G.; Lazzara, G.; and Milioto, S.; *Langmuir.* **2011**, *27*, 1158–1167.
2. Churchman, G. J.; Davy, T. J.; Aylmore, L. A. G.; Gilkes, R. J.; and Self, P. G.; *Clay. Min.* **1995**, *30*, 89–98.
3. Dong, Y.; Bickford, T.; Haroosh, H. J.; Lau, K. T.; and Takagi, H.; *Appl. Phys. A.* **2013**, *112*, 747–757.
4. Dong, Y.; Chaudhary, D.; Haroosh, H.; and Bickford, T.; *J. Mater. Sci.* **2011**, *46*, 6148–6153.
5. Du, M. L.; Guo, B. C.; and Jia, D. M.; *Poly. Int.* **2010**, *59*, 574–582.
6. Freiberg, S.; and Zhu, X.; *Int. J. Pharm.* **2004**, *282*, 1–18.
7. Haroosh, H. J.; Dong, Y.; Chaudhary, D. S.; Ingram, G. D.; and Yusa, S.; *Appl. Phys. A.* **2013**, *110*, 433–442.
8. Jain, R. A.; *Biomaterials.* **2000**, *21*, 2475–2490.
9. Karageorgiou, V.; and Kaplan, D.; *Biomaterials.* **2005**, *26*, 5474–5491.
10. Kostarelos, K. et al. *Nat. Nanotechnol.* **2007**, *2*, 108–113.
11. Li, J. J.; Luo, B. H.; Zhang, J. X.; Huo, R. Q.; and Zhou, C. R.; *Appl. Mech. Mater.* **2013**, *275–277*, 1742–1745.

12. Liu, M.; Guo, B.; Du, M.; and Jia, D.; *Appl. Phys. A*. **2007**, *88*, 391–395.
13. Liu, M. X.; Wu, C. C.; Jiao, Y. P.; Xiong, S.; and Zhou, C. R.; *J. Mater. Chem. B*. **2013a**, *1*, 2078–2089.
14. Liu, M. X.; Zhang, Y.; Wu, C. C.; Xiong, S.; and Zhou, C. R.; *Int. J. Biol. Macromol*. **2012**, *51*, 566–575.
15. Liu, M. X.; Zhang, Y.; and Zhou, C. R., *Appl. Clay Sci.* **2013b**, 75–76, 52–59.
16. Luo, B.-H. et al.; *J. Biomed. Nanotechnol*. **2013a**, *9*, 649–658.
17. Luo, Z.et al.; *Langmuir.* **2013b**, *29*, 12358–12366.
18. Lvov, Y.; and Abdullayev, E., *Prog. Polym. Sci.* **2013**, *38*, 1690–1719.
19. Qi, R. L.; Cao, X. Y.; Shen, M. W.; Guo, R.; Yu, J. Y.; and Shi, X. Y.; *J. Biomater. Sci. Polym. Ed.* **2012**, *23*, 299–313.
20. Qi, R. L.; Guo, R.; Shen, M. W.; Cao, X. Y.; Zhang, L. Q.; Xu, J. J.; Yu, J. Y.; and Shi, X. Y.; *J. Mater. Chem.* **2010**, *20*, 10622–10629.
21. Qi, R. L.; Guo, R.; Zheng, F. Y.; Liu, H.; Yu, J. Y.; and Shi, X. Y.; *Colloid Surf., B***2013**, *110*, 148–155.
22. Qi, R. L.; Zheng, F. Y.; Liu, H.; Yu, J. Y.; and Shi, X. Y.; Antibacterial Activity of Antibiotic-loaded ElectrospunHalloysite/poly(lactic-co-glycolic acid) Composite Nanofibers. In: 2011 International Forum on Biomedical Textile Materials, Proceedings, **2011**, 21–25.
23. Shi, Y. F.; Tian, Z.; Zhang, Y.; Shen, H. B.; and Jia, N. Q.; *Nanoscale. Res. Lett.* **2011**, 6, 1–7.
24. Tsuji, H.; and Fukui, I.; *Polymer* 2003, *44*, 2891–2896.
25. Vergaro, V. et al.; *Biomacromolecules* 2010, *11*, 820–826.
26. Wei, Z. J.; Wang, C. Y.; Liu, H.; Zou, S. W.; and Tong, Z., J.; *Appl. Polym. Sci.* **2012**, *125*, E358–E368.
27. Yah, W. O.; Xu, H.; Soejima, H.; Ma, W.; Lvov, Y.; and Takahara, A., *J. Am. Chem. Soc.* **2012**, *134* 12134–12137.
28. Zhao, Y. L.; Wang, S. G.; Guo, Q. S.; Shen, M. W.; and Shi, X. Y.; *J. Appl. Polym. Sci.* **2013**, *127*, 4825–4832.
29. Zhou, W. Y.; Guo, B. C.; Liu, M. X.; Liao, R. J.; Rabie, A. B. M.; and Jia, D. M.; *J. Biomed. Mater. Res., A***2010**, *93A*, 1574–1587.

CHAPTER 12

CURRENT RESEARCH ON CHITOSAN-HALLOYSITE COMPOSITES

KAVITHA GOVINDASAMY, POORIA PASBAKHSH, and KHENG LIM GOH

CONTENTS

12.1 INTRODUCTION

Chitosan, a derivative of chitin, is widely used in the biomedical field, waste-water treatment and food industries. In 1811, chitin was discovered in mushrooms by Professor Henrni Braconnt and in 1820s, chitin was successfully isolated from insects (Bhatnagar and Sillanpää, 2009). It has been reported that chitin is the second most abundant natural polysaccharide after cellulose (Sun et al., 2010).

Chitin consists of β (1–4) linked 2-acetomido-2-deoxy-β-D-glucose units (Barbara, 2004), otherwise known as β (1–4)–N-acetyl-D-glucosamine (Rinaudo, 2006). Basically, chitin appears to have an ordered crystalline micro fibril structure that is observed in any form of exoskeleton or arthropods, corals, cell walls of fungi, yeast, mushrooms, etc. (Kumar, 1999). Foremost commercially available sources of chitin are crab and shrimp shells, which are in large quantities as waste products in the seafood industry (Jayakumar et al., 2009). It has been reported that 150,000 tons of shrimps, 85,000 tons of crab and 25,000 tons of lobster are processed yearly in the United States (Mathur and Narang, 1990). The edible meat in shrimps, crab and lobsters comprise only 15–20 percent; the waste material can yield over 15,000 tons of chitin.

Besides that, chitin is found in the exoskeleton of marine zooplankton, which includes coral and jellyfish. Insects, such as butterflies and ladybugs, also have chitin in their wings and the cell walls of yeast, mushrooms and other fungi also contain this substance (Shahidi and Abuzaytoun, 2005). Because chitin is not readily dissolved in common solvents, it is often converted to its deacetylated derivative, chitosan (Pillai et al., 2009). Chitosan is often preferred to chitin owing to its solubility in acidic, neutral and alkaline solutions (Jayakumar et al., 2009). Chitosan contains 2-acetomido-2-deoxy-β-D-glucopyranose residues and is often identified by its degree of deacetylation (DD), a percentage measurement of free amine groups along the chitosan backbone (Figure 12.1) (Roberts, 1992). Deacetylation sets the amino groups in the polymer chain free and together with the hydroxyl groups; it makes the chitosan highly reactive. Thus, positively charged chitosan interacts with negatively charged molecules (Kumar, 1999; Di Martino et al., 2005).

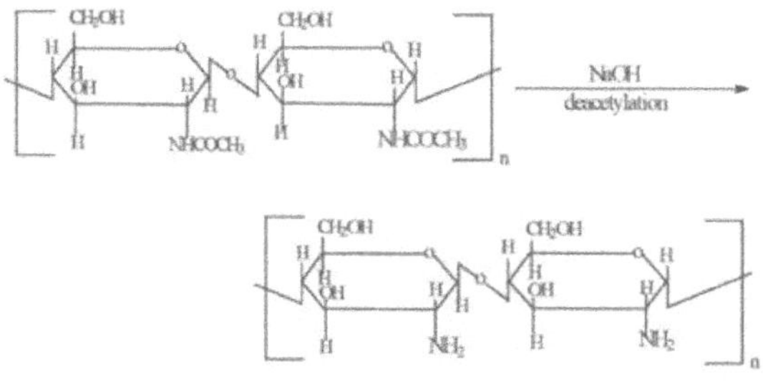

FIGURE 12.1 Schematic changes in the structural features when chitin is deacetylated to form chitosan (Majeti N.V, 2000).

Chitin and chitosan are biodegradable, biocompatible, and nontoxic polymers which display adsorption properties (Kumar, 1999; Shahidi and Abuzaytoun, 2005; Riva et al., 2011). The commercial usage and attention to research of chitin and chitosan has vastly increased, as they consist of 6.89 percent of nitrogen (Majeti N.V, 2000), which is often the reason these natural polysaccharides are chosen over synthetically substituted cellulose (Muzzarelli, 1973).

Chitosan is manufactured in many countries such as USA, India, Japan, Russia, Italy, etc. (Singla and Chawla, 2001). The demand for chitosan has increased tremendously and it is sold commercially in solution, fine powder, flaked, beads and fiber forms. Chitosan has been shown to have superior characteristics compared to chitin giving flexibility for many uses (Table 12.1) (Aranaz et al., 2009) and thus chitosan is considered for a wide range of applications which include waste water filtration (Yan et al., 2011), biosensors (Leedy et al., 2011), wound dressings (Jayakumar et al.), tissue engineering (Khor and Lim, 2003; Juliano et al., 2011; Riva et al., 2011), agriculture (Jayakumar et al., 2009; Bang et al., 2011; Di Pierro et al., 2011), etc. For each application, the properties of chitosan vary, mainly with changes in degree of deacetylation and molecular weight (Babu et al., 2013). Some of the commercial products which are available for biomedical and pharmaceutical purposes which include GNC Chitosan Plus, Aunew Chitosan, Genesis Nutrition Super Chitosan Fat Block 120 Caps, etc.

TABLE 12.1   Chemical and Biological Properties of Chitosan (Dutta et al., 2011).

Chemical properties	1. Linear polyamine
	2. Reactive amino groups
	3. Reactive hydroxyl groups
	4. Chelates many transitional metal ions
Biological properties	1. Biocompatible
	2. Biodegradable, to normal body constituents
	3. Safe, nontoxic
	4. Appreciable binding affinity to mammalian and microbial cells
	5. Hemostatic
	6. Fungistatic
	7. Immunoadjuvant

Many researchers (Wang et al., 2007; Liu et al., 2011; Mohamed et al., 2011) have paid attention to combining chitosan with other polymers or/and micro/nano-fillers in order to improve the process ability and properties of chitosan because chitosan has proven to have fairly low mechanical strength, low thermal stability and gas barrier properties, which restrict its use in a wide range of potential applications. (Tang et al. 2008; Hong et al., 2011; De Silva et al., 2012). In order to enhance the mechanical, thermal and gas barrier properties of chitosan, different fillers such as carbon nanotubes, hydroxyapatite, halloysite nanotubes, etc, have been used to prepare chitosan nanocomposites (Zhang et al., 2008; Ge et al., 2009; Peter et al., 2010; Khunawattanakul et al., 2011). Halloysite nanotubes (HNTs) have been studied extensively as the mineral has been found to be economical and has been mined in abundance from deposits in countries such as China, New Zealand, America and Turkey (Levis and Deasy, 2002; Hong and Mi, 2006; Ismail et al., 2009). HNTs have advantages whereby the outer surface is composed of siloxane and hydroxyl groups and as a result HNTs have a greater ability to disperse than other natural silicates (montmorillonite and kaolinite). Furthermore, hydrogen bonds can form between the amine and hydroxyl groups of chitosan and the Si-O bonds of HNTs (Shchukin et al., 2005; Sun et al., 2010). As well as being inexpensive and abundantly available, HNTs are also durable, and have been reported to have high mechanical strengths and to be biocompatible. Naturally occurring nanomaterials hold great promise for improving polymer characteristics and for creating a wide range of new capabilities in the formulation of both environment-friendly nanocomposite materials and cutting-edge implantable medical devices (Wagner et al., 2005). Unrefined halloysite clay has traditionally been used in fine ceramics and as a suspension agent in glaze preparations. As with kaolin, halloysite is also used in some Asian medicinal practices. The naturally formed nanotubes are not always found in pure deposits and can typically show a range in the observed widths and lengths of the rolled tubes, therefore making separation and refinement the key to their use in composites and as controlled delivery devices.

This chapter reviews some of the very recent research work on chitosan/HNTs composites. This review also focuses on the possible applications of chitosan/HNTs nanocomposites in the fields of waste-water treatment, the food industry, and in agriculture and biomaterials.

12.2   CHITOSAN/HALLOYSITE COMPOSITES

A few recent studies have introduced the preparation and characterization of chitosan/HNTs composite films/membranes and coatings. According to these studies incorporating HNTs into the chitosan has resulted in significant improvements in mechanical, thermal and barrier properties. Some of these chitosan/HNTs composites are used for applications such as tissue engineering (Liu et al., 2012, 2013), waste-water treatment (Zheng and Wang, 2009; Zhai et al., 2013), drug delivery (Liu et al., 2013) and wound healing (Liu et al., 2012). Most of these studies have shown that HNTs are nontoxic and are safe for use in the above mentioned applications. HNTs offer advantages of biocompatibility and lower cost compared to other fillers such as carbon nanotubes. The various methods used to fabricate chitosan-HNTs membranes/films include solution casting (De Silva et al., 2012; Liu et al., 2013), electrophoretic deposition (Deen et al., 2012) and freeze drying (Liu et al., 2012).

12.2.1 MECHANICAL PROPERTIES

Liu et al. (2012), have incorporated HNTs into solvent-casted chitosan films and conducted a series of mechanical analysis (Liu et al., 2012). It was found that under the influence of 7.5 percent HNTs loading, the tensile strength and elastic modulus improved by 134 percent (54.2 MPa) and 65 percent (1240 MPa), respectively, compared to pure chitosan. It was understood that the increase in mechanical properties of chitosan occurred when nanoparticles were used. However, as the HNTs content was increased further (above 7.5%), both the strength and modulus decreased. This reflected the poorer dispersion of the nanoparticles across the polymer matrix. HNTs aggregates were revealed in micrographs (as discussed in Section 12.2.4 of this chapter). These aggregates lead to a concentration of stress at a particular point which caused the decrease in strength and modulus. However, it also meant that the strength and modulus were higher in comparison to these properties of pure chitosan.

De Silva et al. (2013) developed a solution casting approach to fabricate membranes of chitosan reinforced by HNTs in order to study the effects of concentration of HNTs on the tensile properties of chitosan-HNTs membranes (De Silva et al., 2012). Different HNTs concentrations, namely 2, 5, 10 and 15 (w/w%), have been used in the fabrication of these membranes and it was found that the elastic modulus and strength of the membranes increased with HNTs concentration. The strength and elastic modulus of chitosan membranes were approximately 60 MPa and 0.43 GPa, respectively, which was increased to 82 MPa and 0.52 GPa, respectively at 5 (w/w%) HNTs loading. Whereas the elastic modulus of the membrane fluctuated with no appreciable change from 5 to 15 (w/w%), its strength decreased from 5 to 10 (w/w%) and fluctuated with no appreciable change thereafter. Scanning electron micrographs of the fractured morphology of all membranes revealed that HNTs contributed to the failure of a membrane by increasing the withdrawal and agglomeration of tubes upon addition of more HNTs. A particle recruitment model (Goh and Tan, 2012) predicts that (1) HNTs aggregates occurred in all concentrations and all membranes yielded a wide range of aggregate sizes (parameterized by diameter, see Figure 12.2a, d) and the largest aggregates are about 4 μm in diameter; (2) the mean aspect ratio of HNTs aggregates at 5 (w/w%) is smaller than at 15 (w/w%) (see Figure 12.2b, e); (3) this suggests that the aggregates in the former membrane are shorter than those in the latter membrane; (4) there was no appreciable variation in the elastic modulus of the HNTs aggregates in all the membranes (see Figure 12.2c, f). It was speculated that the higher strength of chitosan/HNTs membranes at 5 (w/w%) (Compared to that at 15 (w/w%)) is attributable to a more effective stress transfer mechanism between the chitosan and the shorter HNTs aggregates. In the membranes containing higher HNTs concentration, the long HNTs aggregates formed within these membranes possess little ability to provide reinforcement to the chitosan matrix but fragment easily under stress into smaller aggregates or into individual HNTs.

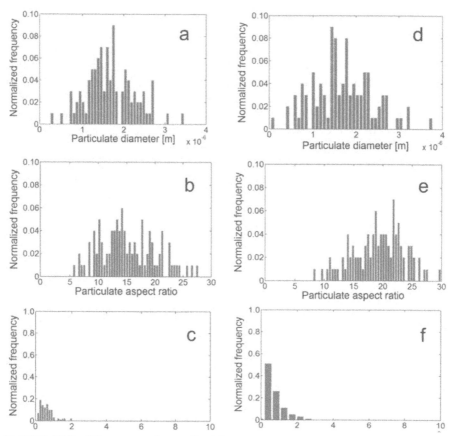

FIGURE 12.2   Histograms of normalized frequency versus HNTs particulate diameter (a, d), aspect ratio (b, e) and elastic modulus (c, f) derived from a particle recruitment model of the chitosan-HNTs composite membrane for HNTs concentration at 5 (w/w%) and 15 (w/w%) (left histograms, 5%; right histograms 15%). Here, "particulate" refers to either HNTs or HNTs aggregates.

In another study, Khoo et al. (2013) have prepared membranes using polyvinyl alcohol (PVA), chitosan and HNTs by cross-linking these polymer composites with glutaraldehyde (Khoo et al., 2013). The mechanical tests have shown contradictory results compared to other studies. Although low filler loadings at 0.25 (w/w%) and

0.5 (w/w%) of HNTs resulted in increasing tensile strength, it was also reported that adding more HNTs (1 (w/w%) and above) resulted in reductions in tensile strength. Having these polymer composites cross-linked, has caused stronger covalent bonding which led to an increase in crystallinity which has contributed to the increase in tensile strength. The reduction in tensile strength was reported to be due to agglomerations of the HNTs. These acts as stress concentrators of stress whereby stress applied to the films was unable to be transferred from the matrix to the filler, resulting in early failure of the films.

12.2.2 THERMAL PROPERTIES

De Silva et al. (2013), reported that the addition of HNTs led to a significant increase in thermal stability (shown in Table 12.2) (De Silva et al., 2012). When compared to different concentration of HNTs in chitosan membranes, the remaining weight drastically increased with higher incorporation of HNTs (shown in Table 12.2). The two main reasons that were discussed are (1) char residue of chitosan/HNTs acted as a barrier and hindered the escape of volatiles, and (2) the physical characteristics of HNTs mean, that the lumens entrapped products of chitosan which caused a delay in mass transport therefore, thermal stability increased significantly.

TABLE 12.2 Summary of TGA Results for Chitosan in Different HNTs Composition (De Silva et al., 2012).

HNTs Composition ((w/w %))	Temperature at 5% Weight loss (°C)	Temperature at 50% Weight Loss (°C)	Remaining Weight (%)	Temperature at Maximum Weight (loss rate/°C)
0	45	370	35	295
2	50	375	38	302
5	45	380	37	300
10	50	396	39	301
15	45	455	44	297

Another study by Zheng and Wang (2010), has prepared chitosan as the backbone to graft poly (acrylic acid) to form granular hydrogel composites with HNTs particles being embedded within the polymeric networks (Zheng and Wang, 2009). It was clearly found at 141 and 311°C, chitosan grafted with poly (acrylic acid) without HNTs had lost 10 and 20 percent weight respectively. However, when HNTs were added, the temperatures observed for weight loss at 10 percent and 20 percent were 178 and 313°C. Results have supported De Silva's work which found that addition of HNTs can improve the thermal stability of polymeric networks. It was reasoned that the incorporation of HNTs provides a barrier effect so that some small molecules generated during the thermal decomposition cannot permeate because of the obstacles due to HNTs they have to bypass (Zheng and Wang, 2009).

12.2.3 CHEMICAL PROPERTIES

Chitosan is a polycationic polymer which is a positively charged. The HNTs have negatively charged surfaces due to the isomorphous substitution of Al^{3+} for Si^{4+}(Joussein et al., 2005). Hence, when HNTs are introduced into chitosan, this results in an electrostatic attraction. It has also been reported that HNTs bond with chitosan through a hydrogen bonding interaction whereby the amine and the hydroxyl groups of chitosan interact with the Si-O bonds of HNTs (Darder et al., 2003). This was supported by FTIR analysis by Liu et al. (2013), whereby there was a slight shift in the absorbance band of NH_2 peaks from 1,545 to 1,548 cm ^{1} (Liu et al., 2013). Furthermore, two peaks at 3,620 cm ^{1} and 3,695 cm ^{1} appearing in the spectrum of the chitosan/HNTs nanocomposite with relatively higher HNTs concentrations describe the Al_2-OH stretching bands of HNTs. Moreover, Sun et al. (2010), also have attained similar results of Al_2-OH stretching bands at 3698–3618 cm ^{1} in membranes of chitosan/HNTs nanocomposites based on horseradish peroxidase (HRP) (Sun et al., 2010). After enzyme immobilization into the HNTs matrix, the amide I and II bands at 1,659 and 1,534 cm ^{1} respectively are nearly the same as those of native HRP membranes. Therefore the authors have suggested that HNTs help to retain the essential features of HRP upon being immobilized. Another study by Zhai et al. (2013), who have prepared chitosan/HNTs membranes by incorporating HRP cross linked with glutaraldehyde (GTA), found O-H stretching bands of chitosan/HNTs composites were broadened and displaced to lower frequencies (from 3,488 to 3,570 cm ^{1}) (Zhai et al., 2013). This shift was attributed to the formation of hydrogen bonds between the lattice hydroxyls and organic groups. Like Sun et al's study, Zhai et al's study also found that chemical interaction between chitosan/HNTs indicates that the immobilized enzyme retains the essential features of the native structure on the support. De Silva et al's study reported that the peak at 1,118 cm ^{1} which refers to the perpendicular Si-O stretching appears in membranes with 15 (w/w%) of HNTs and does not appear in chitosan membranes with lower concentration of HNTs (De Silva et al., 2012). It was suggested that this is due to the HNTs aggregates, which was confirmed in the SEM micrographs. In addition, the amide I band for chitosan at peak 1,653 cm ^{1} shifted to lower wavelengths of 1,647 and 1,650 cm ^{1} for membranes consisting of 5 (w/w%) and 15 (w/w%) of HNTs. Tang et al. (2009), confirmed that the shift indicates the occurrence of hydrogen-bonding between chitosan and fillers (Tang et al., 2009).

12.2.4 MORPHOLOGY

Transmission electron microscopy analysis (TEM) revealed that the morphology of HNTs appears to be hollow and open-ended tubular structures. As the research incorporates chitosan/HNTs nanocomposites based on horseradish peroxidase, the HNTs has enhanced enzyme immobilization thanks in large part to a high surface area. This suggests that HNTs can be used as a good host for the immobilization of biomolecules. Deen et al. (2012), on the other hand prepared chitosan/HNTs/hydroxyapatite films through electrophoretic deposition (EPD) (Deen et al., 2012). The SEM images of chitosan with 0.3 g L ^{1} and 0.6 g L ^{1} of HNTs in films showed continuous distribution of HNTs, crack-free without agglomeration of HNTs (Figure 12.3). The increasing concentration of HNTs resulted in an increasing HNTs content in the deposits. In addition, the author suggested that a large inner diameter of HNTs with a consistent deposition pattern would be beneficial factors when it is being applied for drugs, proteins and other functional materials.

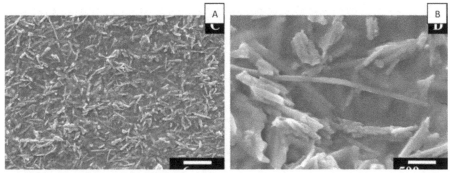

FIGURE 12.3 SEM micrographs of composite films prepared from 0.5 g.L ^{1} chitosan solutions containing (A and B) 0.6 g. L ^{1} HNTs (Deen et al., 2012).

Liu et al. (2012), on the other hand, investigated the state of dispersion of HNTs in a chitosan matrix (Liu et al., 2012). The scanning electron micrographs revealed that HNTs were uniformly dispersed throughout the matrix. It was demonstrated that the strength and modulus (referring to mechanical properties) increased with the presence of 7.5 percent HNTs and this was attributed to the interfacial interactions between HNTs and chitosan. Contradicting Deen et al.'s study, when concentrations of HNTs were higher than 7.5 percent, formation of aggregates of about 1μm × 1μm were observed which resulted in a decrease in tensile and modulus properties (Figure 12.4) (Liu et al., 2012). However, the concentration of HNTs incorporated in Deen et al.'s study was far lower than in Liu et al.'s. The micrographs explained the mechanical results that were mentioned in Section 12.2.1. In another similar study conducted by De Silva et al. (2012), distribution of HNTs with 2, 5, and 15 (w/w %) were evident as white dots in the scanning electron micrographs (Figure 12.5).

HNTs were embedded through the chitosan matrix suggesting good interaction between chitosan and HNTs. However, dispersions rates of 2 and 5 (w/w %) were better than 15 (w/w %) (Figure 12.3). Agglomerates of HNTs (circled in Figure 12.3), were observed which explain lower stress transformation that resulted in inferior tensile results obtained in membranes with 15 (w/w %) concentrations.

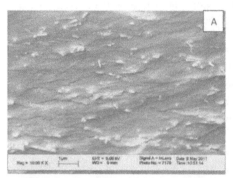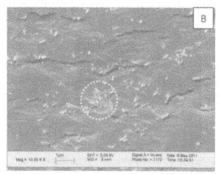

FIGURE 12.4   SEM photos of the fracture surfaces of the chitosan/HNTs nanocomposite with 7.5 percent (A), and 10 percent (B) HNTs (the *white circle* region represents the HNTs aggregates) (Liu et al., 2012).

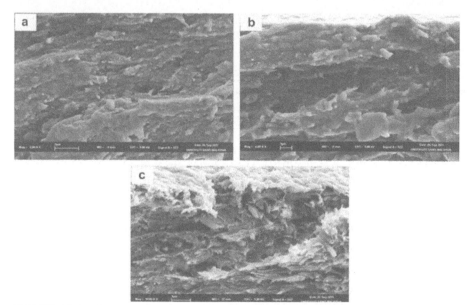

FIGURE 12.5   Scanning electron micrographs of typical fractured morphology of chitosan-based membranes reinforced by (a) 2 (w/w%), (b) 5 (w/w%) and (c) 15 (w/w%) of HNTs (De Silva et al., 2012).

12.2.5 BIOCOMPATIBILITY OF CHITOSAN/HNTS NANOCOMPOSITE MEMBRANES/FILMS

As far as biomaterial is concern for tissue engineering purposes or drug delivery applications, biocompatibility and nontoxicity of the material is a critical issue. Biocompatibility is studied by introducing a substrate which supports the cellular activity in order to induce or enhance tissue regeneration without causing any undesirable systemic response or toxicity. It was found by incorporating HNTs into other types of polymer membrane/films (Qi et al., 2010; Zhou et al., 2010; Abdullayev and Lvov, 2013; Liu et al., 2013) that HNTs are noncytotoxic to various cells. In an *in vitro* study conducted by Liu et al. (2013), mouse N1H3T3 cells were implanted on pure chitosan membranes and chitosan/HNTs nanocomposites (Liu et al., 2013). Both chitosan and chitosan/HNTs showed well expanded cells that are typically spindle shaped. Results also revealed the formation of tight intercellular junctions with adjacent cells. However, although both the films revealed similar morphological characteristics, cell surfaces in chitosan/HNTs nanocomposites appeared to be flat as compared to those of cells anchored on pure chitosan. This was due to the surface roughness and the presence of Si in chitosan/HNTs nanocomposites which encouraged cell attachment.

12.2.6 APPLICATIONS OF CHITOSAN/HALLOYSITE COMPOSITES

As mentioned above incorporating HNTs into the chitosan matrix significantly improved mechanical, thermal and barrier properties. Chemical interactions between chitosan and HNTs enhance the properties of these chitosan/HNTs composite material/films for their use in many applications.

Liu et al. (2013) have produced bionanocomposites using chitosan reinforced with HNTs for tissue engineering applications (Liu et al., 2013). In order to introduce these bionanocomposites into the genre of tissue engineering, an *in vitro* test proved the feasibility of chitosan/HNTs for this application. In this study, NIH3T3 mouse fibroblast cell line was tested on the bionanocomposites. The results, whereby the cells could adhere and proliferate successfully, were positively inclined. Furthermore, incorporating HNTs into this polymer shows a nontoxic response with regard to using the cell lines employed. This provides indications that chitosan/HNTs bionanocomposites have a strong potential for the tissue engineering field.

Another study used the electrophoretic deposition method to produce chitosan/HNTs membrane as a material for the fabrication of biomedical implants (Deen et al., 2012). For this purpose, this study investigated the corrosion protection of stainless steel by electrochemical testing the fabricated chitosan/HNTs films in a simulated body fluid system. The addition of HNTs into the polymer film has resulted in an increase in corrosion potential and a reduction in anodic current. Results have supported the use of a barrier protective layer through prevention of ion diffusion providing corrosion protection for the stainless steel substrate. These promising materials are thereby indicated as having potential for fabricating biomedical implants.

In a study by Sun et al. (2010) on the other hand, HNTs incorporated into a chitosan matrix were developed for immobilization of horseradish peroxidase (HRP) for the purpose of an amperometric biosensor (Sun et al., 2010). A pair of well-defined redox peaks of HRP was obtained which exhibited fast direct electron transfer. On the other hand, chitosan/HNTs films facilitate the direct electrochemistry of HRP to catalyze the reduction of H_2O_2 Hence, chitosan/HNTs improved enzyme loading with the retention of bioactivity and furthermore, promoted direct electron transfer, which is credited to the tubular structure and high surface area of HNTs.

These studies that have used chitosan/HNTs films/membranes for many purposes have proved its versatility for several fields of applications.

12.3   CONCLUSION

Both chitosan and HNTs are materials which are abundantly available, obtainable at low cost and biocompatible. Another important attractive feature shown by many studies is their biological safety for its usage as a biomaterial. However, more studies are needed to determine the various applications of chitosan-halloysite nanotubes composite materials. Also, there are only a small number of studies that have been published using both materials in combination and projects should be carried out using these nanocomposites for various applications. However, future studies are needed to determine the factors that are important in applying chitosan/HNTs composites for commercial purposes. Thus, good research methods involving basic sciences that enable better understanding of the pathways and mechanisms involving chitosan and halloysite nanotubes are of paramount interest. This knowledge together with a little more development may pave new ways for this nanocomposite to be used in various fields.

KEYWORD S

- **Chitosan**
- **Deacetylation**
- **Electrophoretic deposition**
- **Polycationic polymer**
- **In vitro analysis**
- **Hydroxyapatite**
- **Tissue engineering**

REFERENCES

1. Abdullayev, E.; and Lvov, Y.; *J. Mater. Chem. B*. **2013**, *1*(23), 2894–2903.
2. Aranaz, I.; et al. *Curr. Chem. Biol.* **2009**, *3*(2), 203–230.
3. Babu, R.; O'Connor, K.; and Seeram, R.; *Prog. Biomater.* **2013**, *2*(1), 1–16.
4. Bang, S. H.; Hwang, I. C.;. Yu, Y. M; Kwon, H. R.; Kim, D. H.; and Park, H. J. *J. Microencap sulation* **2011**, *28*(7), 595–604.
5. Barbara, K.; *Enzy. Microb. Technol.* **2004**, *35*(2–3), 126–139.
6. Bhatnagar, A.; and Sillanpää, M.; *Adv Coll Interf. Sci.* **2009**, *152*(1–2), 26–38.
7. Darder, M.; Colilla, M.; and Ruiz-Hitzky, E.; *Chem. Mater.* **2003**, *15*(20), 3774–3780.
8. De Silva, R. T.; Pasbakhsh, P.; Goh, K. L.; Chai, S. P.; and Ismail, H.; *Poly Test,* **2012**.
9. Deen, I.; Pang, X.; and Zhitomirsky, I.; *Coll. Surf A: Physicochem Eng. Aspect,* **2012**, *410*(0), 38–44.
10. Di Martino, A.; Sittinger, M.; and Risbud, M. V.; *Biomaterials.***2005**, *26*(30), 5983–5990.
11. Di Pierro, P.; Sorrentino, A.; Mariniello, L.; Giosafatto, C. V. L.; and Porta, R.; *LWT Food Science and Technology,* **2011**, *44*(10), 2324–2327.
12. Dutta, P. K.; Rinki, K.; and Dutta, J.; Springer Berlin Heidelberg, **2011**; 244, 45–79.
13. Ge, B.; Tan, Y.; Xie, Q.; Ma, M.; and Yao, S.; *Sens. Actuat. B: Chem.* **2009**, *137*(2), 547–554.
14. Goh, K. L.; and Tan, L. P.; Springer Berlin Heidelberg, **2012**, *17*, 85–106.
15. Hong, H.-L.; and Mi, J.-X.; *Min Magaz.* **2006**, *70*(3), 257–264.
16. Hong, S. I.; et al. *Journal of Applied Polymer Science* 2011, *119* (5), 2742-2749.
17. Ismail, H.; Pasbakhsh, P.; Ahmad Fauzi, M. N.; and Abu Bakar, A.; *Polym Plas. Technol. Eng.* **2009**, *48(3)*, 313–323.
18. Jayakumar, R.; Prabaharan, M.; Nair, S. V.; and Tamura, H.; *Biotechnol. Adv.* **2009**, *28(1)*, 142–150.
19. Jayakumar, R.; Prabaharan, M.; Sudheesh Kumar, P. T.; Nair, S. V.; and Tamura, H.; *Biotech nol. Adv 29*(3), 322–337.
20. Joussein, E.; Petit, S.; and Churchman, J.; *Clay. Min.* **2005**, *40*(4), 383–426.
21. Juliano, C.; Galleri, G.; Klemetsrud, T.; Karlsen, J.; and Giunchedi, P.; *Int. J. Pharm.* **2011**, *420*(2), 223–230.
22. Khoo, W. S.; Ismail, H.; and Ariffin, A.; *Int. J. Poly. Mater.* **2013**, *62*(7), 390–396.
23. Khor, E.; and Lim, L. Y.; *Biomaterials.* **2003**, *24*(13), 2339–2349.
24. Khunawattanakul, W.; Puttipipatkhachorn, S.; Rades, T.; and Pongjanyakul, T.; *Int. J. Pharm.* **2011**, *407*(1–2), 132–141.
25. Kumar, M. R.; *Bull. Mater. Sci.* **1999**, *22*(5), 905–915.
26. Leedy, M. R.; Martin, H. J.; Norowski, P. A.; Jennings, J. A.; Haggard, W. O.; and Bumgard-ner, J. D., *Springer Berlin Heidelberg*, 2011, *244*, 129–165.
27. Levis, S. R.; and Deasy, P. B.; *Int. J. Pharm.* **2002**, *243*(1–2), 125–134.
28. Liu, H.; Nakagawa, K.; Chaudhary, D.; Asakuma, Y.; and Tadé, M. O.; *Chem. Eng. Res. Des,* **2011**, *89(11),* 2356–2364.
29. Liu, M.; Wu, C.; Jiao, Y.; Xiong, S.; and Zhou, C. *J. Mater. Chem. B* 2013.
30. Liu, M.; Wu, C.; Jiao, Y.; Xiong, S.; and Zhou, C.; *J. Mater. Chem. B,* **2013**, *1*(15), 2078–2089.
31. Liu, M.; Zhang, Y.; Wu, C.; Xiong, S.; and Zhou, C.; *Int. J. Biol. Macromol.* **2012**, *51*(4), 566–575.
32. Majeti N. V.; and Ravi, K.; *React. Funct. Poly.* **2000**, *46*(1), 1–27.
33. Mathur, N. K.; and Narang, C. K.; *J. Chem. Educ.* **1990**, *67*(11), 938.
34. Mohamed, K. R.; El-Rashidy, Z. M.; and Salama, A. A.; *Ceram. Int.* **2011**, *37*(8), 3265–3271.
35. Muzzarelli, R. A. A.; *Pergamon. Press.* **1973**, 83.
36. Peter, M.; et al. *Carbohyd. Polym.* **2010**, *79*(2), 284–289.
37. Pillai, C. K. S.; Paul, W.; and Sharma, C. P.; *Prog. Poly. Sci.* **2009**, *34*(7), 641–678.

38. Qi, R.; et al. *J. Mater. Chem.* **2010,** *20*(47), 10622–10629.
39. Rinaudo, M.; *Prog. Poly. Sci.* **2006,** *31*(7), 603–632.
40. Riva, R.; Ragelle, H.; Rieux, A.; Duhem, N.; Jérôme, C.; and Préat, V.; *Springer Berlin Hei delberg,* **2011;** *244*, 19–44.
41. Roberts, G. A. F.; Chitin Chemistry. Great Britain, The Macmillan Press, Basingstoke, 1992.
42. Shahidi, F.; and Abuzaytoun, R.; *Academic Press,* **2005;** Vol 49, 93–135.
43. Shchukin, D. G.; Sukhorukov, G. B.; Price, R. R.; and Lvov, Y. M.; *Small* **2005,** *1*(5), 510–513.
44. Singla, A. K.; and Chawla, M.; *J. Pharm. Pharmacol.* **2001,** *53*(8), 1047–1067.
45. Sun, X.; Zhang, Y.; Shen, H.; and Jia, N.; *Electrochimica. Acta.* **2010,** *56*(2), 700–705.
46. Tang, C.; Chen, N.; Zhang, Q.; Wang, K.; Fu, Q.; and Zhang, X.; *Poly. Degrad. Stab.* **2009,** *94*(1), 124–131.
47. Tang, C.; Xiang, L.; Su, J.; Wang, K.; Yang, C.; Zhang, Q.; and Fu, Q.; *J. Phys. Chem. B.* **2008,** *112*(13), 3876–3881.
48. Wagner, A. L.; Cooper, S.; and Riedlinger, M.; *Indus. Biotechnol.* **2005,** *1*(2), 190–193.
49. Wang, X.; Du, Y.; Luo, J.; Lin, B.; and Kennedy, J. F.; *Carbohyd. Poly.* **2007,** *69*(1), 41–49.
50. Yan, H.; Dai, J.; Yang, Z.; Yang, H.; and Cheng, R.; *Chem. Eng. J.* **2011,** *174*(2–3), 586–594.
51. Zhai, R.; Zhang, B.; Wan, Y.; Li, C.; Wang, J.; and Liu, J.; *Chem. Eng. J.* **2013,** *214*(0), 304–309.
52. Zhang, Y.; Venugopal, J. R.; El-Turki, A.; Ramakrishna, S.; Su, B.; and Lim, C. T.; *Biomateri als.* **2008,** *29*(32), 4314–4322.
53. Zheng, Y.; and Wang, A.; *J. Macromol. Sci., Part A,* **2009,** *47*(1), 33–38.
54. Zhou, W. Y. et al. *J. Biomed. Mater. Res. Part A* **2010,** *93A*(4), 1574–1587.

PART VI
**MECHANICAL PROPERTIES OF HALLOYSITE
AND OTHER NANOTUBES**

CHAPTER 13

MEASUREMENT OF THE ELASTIC MODULUS OF HALLOYSITE NANOTUBES USING ATOMIC FORCE MICROSCOPY

B. LECOUVET, C. BAILLY, and B. NYSTEN

CONTENTS

13.1   INTRODUCTION

In recent years, nanoscale materials (i.e., thin films, nanowires, nanotubes, etc.) have gained significant technological interest thanks to their outstanding intrinsic properties compared to those of bulk materials. The knowledge of their nanomechanical properties is extremely important to identify potential application areas. However, mechanical measurements of freestanding nanoscale objects are quite challenging due to the difficulties encountered in the setup of standard tensile or bending tests. Recently, several methods based on the use of atomic force microscopy (AFM) have been developed to study the mechanical behavior of nanostructured materials (Salvetat et al., 1999a; Wong et al., 1997; Yu et al., 2000). As an example, AFM has been used in contact mode to carry out three-point bending tests on nanotubes and nanowires suspended on nanoporous membranes or on silicon substrates (Cuenot et al., 2000; Ni et al., 2006; Niu et al., 2007; Salvetat et al., 1999b; Wu et al., 2005). In this configuration, the one-dimensional nanomaterial is considered as a beam (clamped or simply supported) lying over a pore and the vertical deflection is measured as a function of the force applied midway along the suspended length.

Halloysite is a naturally occurring aluminosilicate chemically similar to kaolinite, with a predominantly hollow tubular structure (Singh, 1996). The unique surface chemical properties of halloysite nanotubes (HNTs), coupled with their abundant availability and low price, make them potential candidates for reinforcing polymeric materials (Lecouvet et al., 2011a, 2011b; Prashantha et al., 2013) as well as in a huge variety of new applications including adsorption of contaminants (Liu et al., 2007), controlled release of chemical agents (Lvov and Price, 2008), and nanotemplating (Li et al., 2008). However, very little is known about the chemical and physical properties of halloysite.

Recently, Lu et al. (2011) have measured the Young's modulus of HNTs using transmission electron microscopy (TEM) with a bending stage and reported an average value of 130 ± 24 GPa. The elastic modulus showed a tendency to decrease with increasing diameter, probably because of the higher density of structural defects present in larger nanotubes. Guimaraes et al. (2010) have used the self-consistent charged density functional tight binding method to model the stiffness of single-walled HNTs and found values in the range of 230–340 GPa. Finally, the rule of mixtures has been selected to predict a Young's modulus of around 72 GPa for single HNTs in chitosan composite membranes (De Silva et al., 2013).

The aim of this section is to report the measurements of the elastic modulus of HNTs using a different approach based on a nanoscale three-point bending test. The size dependence of the elastic modulus is also discussed.

13.2 EXPERIMENTAL

Matauri Bay halloysite from New Zealand was purchased from Sigma-Aldrich. The aluminosilicate powder was first dispersed in ethanol and the mixture was ultrasonicated for 5 min. The clay suspensions were then filtered through polycarbonate (PC) nanoporous membranes (it4ip) that served afterwards as supports for the AFM measurements. The halloysite dispersion on the PC membranes was characterized by scanning electron microscopy (LEO 982, Zeiss). While most of the single nanotubes were lying on the membrane surface, some of them occasionally crossed pores corresponding to a suspended beam configuration at the nanoscale.

AFM experiments were carried out on an Agilent 5,500 microscope (Agilent Technology) operating in air at room temperature and equipped with a 100 μm closed-loop scanner. The cantilevers were silicon AC-mode probes from nanosensors. For each cantilever, the spring constant (k_{cant}) was calibrated using the thermal noise method with all values found to be around 25 N/m. Samples were first imaged at low magnification in intermittent-contact mode to identify nanotubes under the desired configuration (Figure 13.1a). Once a suspended nanotube was located, another image was taken at higher magnification (Figure 13.1b) to position the AFM tip exactly midway along the suspended length L, but also to measure accurately its dimensions. When a tube was not perfectly located in the middle of a pore, its minimum (L_{min}) and maximum (L_{max}) lengths were measured on the AFM image and an average elastic modulus was calculated based on its minimum and maximum values. The outer diameter (D_{out}) was defined as the maximum of the tube height profile with respect to the supporting membrane. Since both diameters are required to determine the elastic modulus of HNTs and AFM provides no information about the inner one (D_{in}), a distribution analysis of the tube dimensions was performed by TEM (Leo 922, Zeiss) on approximately 100 randomly chosen nanotubes (Lecouvet et al., 2013). As expected for natural minerals, only a rough correlation was found between diameters, which can be expressed by the following linear function:

$$D_{in} = 0.3784D_{out} \tag{13.1}$$

Therefore, the inner diameter of HNTs selected for force curve measurements was always calculated using Eq. (13.1).

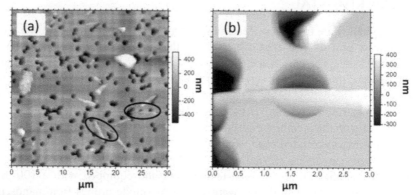

FIGURE 13.1   (a) Large scale AFM image of HNTs dispersed on a PC membrane; circles underline tubes lying over pores. (b) AFM image at higher magnification of a nanotube crossing a pore (Lecouvet et al., 2013) (Reproduced with the kind permission of IOPscience).

After positioning the tip midway of the suspended nanotube, a nanoscale three-point bending test was performed in contact mode following the procedure described in Landau and Lifshitz (1986). Figure 13.2 presents the force (F) versus nanotube deflection (δ) curves obtained on a silicon wafer, on a nanotube lying on the PC membrane and on a nanotube crossing a pore. The linear elastic behavior of HNTs in the range of applied forces is confirmed by the linearity and reversibility of approach and retraction curves. In addition, the force curve measured on a tube lying on the membrane indicates that there is no penetration of the AFM tip into the clay nanotube as revealed by the infinite slope of the curve in the contact zone (Figure 13.2). Finally, to ensure that the nanostructure does not roll during the measurements, images have been taken before and after each test (not shown here). Their superimposition confirms that the nanotube is in a stable configuration during the nanoscale three-point bending tests.

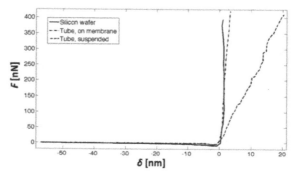

FIGURE 13.2 F vs. δ curves obtained on a silicon wafer (*solid line*), a nanotube lying on the PC membrane (*dashed and dotted line*), and a nanotube crossing a pore (*dashed line*).

13.3 RESULTS AND DISCUSSION

Based on the linear elastic beam theory within the limit of small deformations, the total deflection of a nanotube involves both bending (d_b) and shear (d_s) deformations and follows the superposition principle (Timoshenko and Gere, 1972):

$$\delta = \delta_b + \delta_s = \frac{FL^3}{\alpha EI} + \frac{f_s FL}{4GA} \tag{13.2}$$

where F is the applied force, I the cross-sectional moment of inertia, E is the elastic modulus, G is the shear modulus, A is the cross-sectional area, f_s is the shape factor (10/9 for a cylindrical beam), and α is a constant value depending on the clamping conditions of the beam at the pore edges. For a hollow cylinder, the cross-sectional moment of inertia is given by:

$$I = \frac{\pi\left(D_{out}^4 - D_{in}^4\right)}{64} \tag{13.3}$$

However, the shear contribution becomes negligible when the ratio between suspended length and D_{out} is larger than 10 (Lawrence et al., 2008). In this work, D_{out} of most of the nanotubes is below 120 nm and the pore diameter is around 1.2 μm. Therefore, the measured elastic modulus can be considered as a reduced modulus, taking into account only the contribution of tensile and compressive deformations to the nanotube deflection:

$$E_r = k_t \frac{L^3}{\alpha I} = \frac{\partial F}{\partial \delta} \frac{L^3}{\alpha I} \tag{13.4}$$

The nanotube deflection is also strongly influenced by its boundary conditions: simply supported or clamped beam (Timoshenko and Gere, 1972). The support conditions were defined by recording the stiffness profile of a tube along its suspended length. Figure 13.3a shows the experimental data in comparison with the expected behaviors for a simply supported or clamped ends beam (Cuenot et al., 2003; Landau and Lifshitz, 1986). Results are in better agreement with the stiffness profile expected for a simply supported beam as confirmed by the sharp decrease of k_t values when going away from the beam edges. The deviation from the model predictions at one end of the tube (underlined by circle) can be ascribed to the presence of impurities at the surface of the measured nanotube leading to artifacts (inset in Figure 13.3a). Based on this observation, the adhesion of the tubes on the membrane does not seem to be sufficient to prevent any liftoff during the bending test and HNTs can thus be regarded as simply supported beams ($\alpha = 48$).

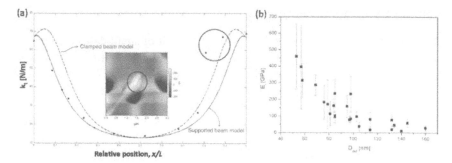

FIGURE 13.3 (a) Variation of the tube stiffness (k_t) along its suspended length: experimental points (*dots*) and expected behaviors for a simply supported (*solid line*) or clamped (*dashed line*) nanobeam. The inset shows impurities present at the nanotube surface. (b) Reduced elastic modulus (E_r) of HNTs as a function of D_{out} (error bars represent one standard deviation from the mean).

Figure 13.3b shows the reduced elastic modulus of HNTs as a function of the outer diameter for a set of 25 nanotubes with D_{out} varying between 50 and 160 nm. For each nanotube tested, between 3 and 5 force curves were measured and the standard deviations for each parameter were calculated. Unexpectedly, the elastic modulus increases from 10 to 460 GPa with decreasing diameter, with an average of 140 GPa. This last result is close to the elastic modulus recently reported for HNTs (130 GPa) using a TEM-based two-point bending method (Lu et al., 2011), as well as with values in the range of 230–340 GPa predicted for single-walled HNTs by self-consistent charge density functional tight binding calculations (Guimaraes et al., 2010).

The experimental error essentially arises from the uncertainty on the measurement of the stiffness and geometrical parameters of the nanotubes. The reproducibility of the method is given by the relative standard deviation of the tube stiffness, which is estimated to be around 16 percent. The relative error on each dimension of HNTs is evaluated by the following relation:

$$\frac{\Delta X}{X} = \frac{X_{max} - X_{min}}{X_{max} + X_{min}} \tag{13.5}$$

where X_{min} and X_{max} are respectively the minimum and maximum values of the geometrical parameter X measured on the AFM image. D_{out} is accurately measured with a small experimental error of 2%, whereas the uncertainty on L is of the order of 8 percent. There is also a large uncertainty on D_{in} (39%), which can only be estimated from D_{out} using Eq. (13.1). The maximum uncertainty on the reduced elastic modulus can thus be calculated to be smaller than 55 percent.

The unexpected steep jump of the elastic modulus for the small diameters (<50 nm) has already been reported in the literature for various nanostructures and may have two main origins. First, the concentration of structural defects in the multilayer structure is expected to increase with the number of constituting layers of the wall (Piperno et al., 2010). However, TEM did not reveal any structural defects for larger nanotubes. In addition, the elastic modulus is fundamentally related to the chemical bonding between the constituent atoms and is less sensitive to the defect concentration. As a consequence, this effect does not seem to be the key factor responsible for such behavior.

Second, surface properties of nanomaterials may substantially affect their mechanical properties (Gleiter, 2000). Indeed, the nanotube deflection results in its extension, and hence in an increase of its surface area. Cuenot et al. (2004) reported that the experimentally measured elastic modulus of polypyrrole nanotubes was an apparent modulus including the contribution of surface tension effects on the nanostructure stiffness, the surface tension being proportional to L^2/D_{out}^3 under bending load. Thus, for a constant suspended length, the contribution of surface tension effects to the intrinsic elastic modulus dramatically increases with decreasing diameter and the nanostructure behaves like a stiffer material. It should be emphasized that this calculation was proposed for a clamped configuration, whereas HNTs are simply supported here. Consequently, longitudinal displacements at the ends of the tubes are allowed and the surface extension is probably less pronounced than that expected for a clamped beam.

On the other hand, the mechanical model used to calculate the elastic modulus also has to be considered in detail to better understand the low values of the modulus obtained for larger tubes. This latter is based on the assumption that shear contribution to the total deflection can be ignored when $L/D_{out} \geq 10$ (Lawrence et al., 2008). However, shear deformations become significant when $L/D_{out} \leq 2\sqrt{E/G}$ (Timoshenko and Gere, 1972). In order to have a rough idea of the order of magnitude of the shear modulus of HNTs, additional bending tests have been performed on two larger nanotubes (200 and 220 nm) assuming that the total deflection comes only from shearing. A reduced shear modulus of 1.5 ± 0.26 GPa is obtained, which is approximately two orders of magnitude lower than the average value of E_r. This result clearly highlights the anisotropic behavior of halloysite with a nonnegligible contribution of shear deformations for large HNTs ($D_{out} \geq 120$ nm), increasing the total deflection of the tube, and hence lowering its apparent elastic modulus.

In order to describe shearing of halloysite during a three-point bending test, the aluminosilicate nanotube can be modeled as a sandwich material made of alternating layers of soft and hard matter in the radial direction (Vinson, 1999). Hard layers correspond to the covalently bonded alumina and silica layers, while the interlayer spacing, where weak H-bonds and van der Waals forces predominate, can be regarded as the soft matter. Hence, shear deformations result from the slip motions between adjacent layers due to the weak interactions in the radial direction. Based on this model, it is reasonable to accept that the contribution of shear deformations to the total deflection increases by increasing the number of aluminosilicate layers in the wall (i.e., the outer diameter).

13.4 CONCLUSIONS

The elastic modulus of HNTs has been measured using AFM nanoscale three-point bending tests. The stiffness of the tubes, and hence their elastic modulus have been deduced from force curve measurements using the bending equation assuming that shear deformations are negligible. The nanobeam has been identified as simply supported by recording the stiffness profile of a tube along its suspended length. An average elastic modulus of 140 GPa is obtained for a set of nanotubes with outer diameters ranging between 50 and 160 nm. Surprisingly, the elastic modulus strongly increases with decreasing diameter, with a steep jump below 50 nm. The size dependence of the elastic modulus has been attributed to: (i) surface tension effects dominating the mechanical properties of thinner tubes and giving rise to larger apparent stiffness and (ii) a nonnegligible contribution of shearing to the total deflection of larger tubes, lowering their apparent elastic modulus. This study furthers the understanding of the nanomechanical behavior of HNTs, which is essential to identify their potential reinforcing effects in polymeric materials.

ACKNOWLEDGMENTS

The authors gratefully acknowledge it4ip for providing PC membranes. We also acknowledge Ir. J. Horion and C. D'Haese for AFM measurements and Pr. T. Pardoen for enlightening discussions. We thank the FRS-FNRS for financial support.

KEWORDS

- **Aluminosilicates**
- **Elastic modulus**
- **Nanomaterials**
- **Nanoporous membranes**
- **Atomic force microscopy**
- **Shear modulus**

REFERENCES

1. Cuenot, S.; Demoustier-Champagne, S.; and Nysten, B.; *Phys. Rev. Lett.* **2000,** *85,* 1690–1693.

2. Cuenot, S.; Frétigny, C.; Demoustier-Champagne, S.; and Nysten, B.; *J. Appl. Phys.* **2003**, *93*, 5650–5655.
3. Cuenot, S.; Frétigny, C.; Demoustier-Champagne, S.; and Nysten, B.; *Phys. Rev. B*; **2004**, *69*, 165410.
4. Gleiter, H.; *Acta. Mater.* **2000**, *48*, 1–29.
5. Guimaraes, L.; Enyashin, A. N.; Seifert, G.; and Duarte, H. A.; *J. Phys. Chem. C,* **2010**, *114*, 11358–11363.
6. Landau, L. D.; and Lifshitz, E. M.; Theory of Elasticity. Oxford: Pergamon; **1986**.
7. Lawrence, J. G.; Berhan, L. M.; and Nadarajah, A.; *ACS Nano.* **2008**, *2*, 1230–1236.
8. Lecouvet, B.; Gutierrez, J. G.; Sclavons, M.; and Bailly, C.; *Polym. Degrad. Stab.* **2011a**, *96*, 226–235.
9. Lecouvet, B.; Sclavons, M.; Bourbigot, S.; Devaux, J.; and Bailly, C.; *Polymer.* **2011b**, *52*, 4284–4295.
10. Lecouvet, B.; Horion, J.; D'Haese, C.; Bailly, C.; and Nysten, B.; *Nanotechnology.* **2013**, *24*, 105704.
11. Li, C.; Liu, J.; Qu, X.; Guo, B.; and Yang, Z. J. *Appl. Polym. Sci.* **2008**, *110*, 3638–3646.
12. Liu, P.; and Zhang, L.; *Sep. Purif. Technol.* **2007**, *58*, 32–39.
13. Lu, D.; Chen, H.; Wu, J.; and Chan, C. M. *J. Nanosci. Nanotechnol.* 2011, *11*, 7789–7793.
14. Lvov, Y. M.; and Price, R. R.; Halloysite nanotubules, a novel substrate for the controlled delivery of bioactive molecules. In: Bio-Inorganic Hybrid Nanomaterials. Eds. Ruiz-Hitzky, E. et al. Weinheim, Germany: Wiley-VCH; 2008, 419–442.
15. Ni, H.; Li, X.; and Gao, H.; *Appl. Phys. Lett.* **2006**, *88*, 043108.
16. Niu, L.; Chen, X.; Allen, S.; and Tendler, S. J. B.; *Langmuir.* **2007**, *23*, 7443–7446.
17. Piperno, S.; et al. *Adv. Func. Mater.* 2010, *17*, 3332–3338.
18. Prashantha, K.; Lecouvet, B.; Sclavons, M.; Lacrampe, M. F.; and Krawczak, P.; *J. Appl. Polym. Sci.* 2013, *128*, 1895–1903.
19. Salvetat, J-P; et al. *Phys. Rev. Lett.* **1999a**, *82*, 944–947.
20. Salvetat, J-P; et al. *Adv. Mater.* **1999b**, *11*, 161–165.
21. Singh, B.; *Clays. Clay. Min.* **1996**, *44*, 191–196.
22. Timoshenko, S. P. and Gere, J. M.; Mechanics of Materials. New York: Van Nostrand; **1972**.
23. Vinson, J. R.; The behavior of sandwich structures of isotropic and composite materials. Lancaster: Technomic; **1999**.
24. Wu, B.; Heidelberg, A.; and Boland, J.; *J. Nat. Mater.* **2005**, *4*, 525–529.
25. Wong, E. W.; Sheehan, P. E.; and Lieber, C. M. *Science.* **1997**, 277, 1971–1975.
26. Yu, M-F; Lourie, O.; Dyer, M. J.; Moloni, K.; Kelly, T. F.; and Ruoff, R. S.; *Science.* **2000**, *287*, 637–640.

CHAPTER 14

MECHANICS OF HALLOYSITE NANOTUBES

KHENG LIM GOH, RANGIKA DE SILVA, and POORIA PASBAKHSH

CONTENTS

14.1   INTRODUCTION

Halloysite is a naturally occurring alumino silicate and its structural makeup is similar to kaolinite, that is, it features a laminar structure comprising repeating layers of one tetrahedral (silica) sheet and one octahedral (alumina) sheet. The two sheets are held together to form a single layer with a thickness of about 0.72 nm. Unlike kaolinite, halloysite is commonly seen in a tubular shape (Figure 14.1a, b), and for this reason, halloysite particles are commonly referred to as halloysite nanotubes (HNTs). Starting from a crystal lattice of a kaolinite sheet, it tends to an energetically stable structure by distorting (when hydrated) and deforming into a tube. The question on why it rolls instead of performing a tetrahedral rotation addresses the mechanics to correct for misfit of the octahedral (alumina) and tetrahedral (silica) sheets (Singh et al., 1996). Singh et al. (1996) presented an elegant model to explain that the rolling mechanism encounters lower resistance from the Si-Si repulsion (as compared to tetrahedral rotation) to correct the same amount of misfit.

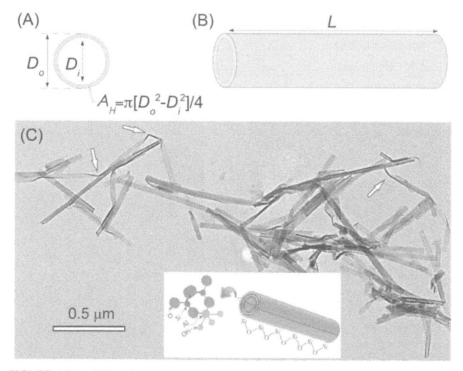

FIGURE 14.1   Halloysite nanotubes (HNTs); (A) Cross-sectional, (B) ongitudinal model, and (C) A transmission electron micrograph of HNTs (known as PATCH clay) (Pasbakhsh et al., 2013). Inset in C: schematic of HNT depicting the molecular makeup.

Electron micrographs reveal that the length (L) of HNTs ranges from 50 to 5,000 nm and the external diameter (D_o) ranges from 20 to 200 nm (corresponding to internal diameter, D_i, of 10–70 nm). Consequently, an empirical relationship between D_i and D_o, given by $D_i = 0.3784D_o$, has been proposed by Lecouvet et al. (2013) to account for the wide range of values of D_o and D_i. As for the slenderness of HNT, a simple analysis of the HNT aspect ratio, q ($= L/D_o$), to order of magnitude, reveals a range of values from 3 (lower bound [LB], obtained by setting $L = 50$ nm and $D_o = 20$ nm) to 250 (upper bound [UB], obtained using $L = 5,000$ nm and $D_o = 20$ nm). One important and immediate consequence of these estimates of q is that most reports on q values reveal values which are of the order of the LB or one order higher than LB (possibly <50 (Pasbakhsh et al., 2013)), suggesting that the lower the q, the more energetically stable is the structure. Correspondingly, Lecouvet et al. (see Chapter 13 of this book) pointed out that shear contribution to the total deflection of a HNT undergoing bending can be ignored when $L/D_o \geq 10$. Nevertheless, $q = 250$ (i.e.,UB) is purely illustrative and the existence of HNTs with high slenderness of order of the UB remains to be confirmed by experiment.

The overall tubular structure of HNT is stabilized by the binding force arising from chemical interactions such as covalent, hydrogen, and ionic bondings between the Al, Si, and OH (Singh et al., 1996). However, it is not clear how these interactions contribute to the mechanical properties. A few studies have sought to determine the mechanical properties of HNTs by computational modeling. Guimarães et al. (2010) used a "self-consistent charged density functional tight-binding" method to predict the Young's modulus, E_H, of halloysite. Experimentally, determining the E_H has been a challenge for researchers because of the length-scale of the instrumentation. Recently, Lu et al. (2011) measured the E_H using a bending stage, observed under a transmission electron microscope. The study revealed that the E_H ranges from 71 to 133 GPa, depending on the D_o and D_i. However, Lecouvet et al. (2013) showed that the E_H ranges from 10 GPa to 460 GPa and that E_H appears to decrease with increase in D_o. A previous study (De Silva et al., 2013) predicted that $E_H \approx 72.165$ GPa; this was derived by curve-fitting an equation, that is, $E = KE_H V_H + E_C V_C$ (the *rule of mixture* model for chitosan/HNT composites), to experimental data, where E and V represent the modulus and volume fraction while subscript H and C refer to halloysite and chitosan, respectively. Here, $V_H + V_C = 1$ and K, which describes the filler reinforcement efficiency for randomly oriented fillers, was estimated at 0.2. The predicted E_H in the De Silva et al. (2013) study was at the lower end of the range of magnitudes of the bending stiffness of HNT which were calculated using input data of force and displacement measured from experiments (Lu et al. 2011). Sheidaei et al. (2013) have adopted $E_H = 140$ GPa (an input parameter) for a multiscale computer simulation study of HNT reinforcing composites. From a structural property point of view, the structural stiffness of HNTs could depend on the mode of loading; findings from carbon nanotubes (CNTs) studies has demonstrated that the ratio of the bending to axially loaded stiffness of single-walled CNTs decreases with increasing wall thickness (DiBiasio et al., 2007). Nevertheless, to date, direct independent measurement of the material stiffness of halloysite has received little attention.

Clearly, HNTs are ideal for blending (due to the size effect) into low stiffness polymeric-based materials(e.g., epoxy (Deng et al., 2008), polypropylene (Sheidaei et al., 2013), chitosan (De Silva et al., 2013, Lu et al., 2012), and ethylene propylene-diene monomer (EPDM)(Ismail et al., 2008), to achieve composites with a bulk stiffness higher than those of the polymeric matrix materials. Chitosan composite films blended with HNTs resulted in higher tensile stress and stiffness (reflecting increases of 134 percent and 65 percent, respectively) compared to pure chitosan films, at an optimal concentration of 7.5 (w/w %) of HNTs (De Silva et al., 2013, Lu et al., 2012).Blending HNTs into a matrix material, like all nanoparticles, tend to lead to agglomeration (De Silva et al., 2013). One possible approach to uniform dispersion within the matrix materials is chemical modification of the HNT; Pasbakhsh et al. (2010) proposed that modification to the HNT using γ-methacryloxypropyltrimethoxysilane could lead to better dispersion in EPDM matrix and possibly even augmenting the mechanical properties of the composite (e.g., leading to higher strength). De Silva et al. (see Chapter 10 of this book) have provided a detailed analysis of the physical properties of polylactic acid-based (PLA) composite membranes reinforced by HNTs obtained by two methods, namely solution casting and melt blending. Unfortunately, the functional implications (e.g., stress uptake and resilience) of the structure and mechanical properties of HNT for reinforcing the composite materialare not well known.

In this chapter, simple order-of-magnitude models are provided to address the basic mechanics of HNTs with respect to the different modes of loading. Continuum mechanics (Gere and Timoshenko, 1996) was used to evaluate the mechanical parameters. In all cases, the values used for the input parameters were selected to consistency to ensure that the final estimate yields a minimum (or maximum) value. Of note, the minimum and maximum values are also designated as the LB and UB, respectively, of the parameter. Although this approach would appear somewhat oversimplified, it is nevertheless useful as (in most cases) the means to determine the values of the input parameters independently are yet to be developed.

14.2 MODES OF LOADING

14.2.1 AXIAL LOADING

In the present analysis, $z, \theta,$ and r represent the axial, circumferential, and radial directions, respectively, of the cylindrical coordinate system. Let $A_H (= \{[D_o/2]^2 \ [D_i/2]^2\})$ represent the cross-sectional area of a HNT (Figure 14.1a) and the numerical values of A_H were calculated; the minimum and maximum values of $A_H \approx 236$ nm^2 (using $D_o = 20$ nm and $D_i = 10$ nm) and 31,337 nm^2 (using $D_o = 200$ nm and $D_i = 10$ nm), respectively (upper and lower bounds of D_o and D_i are listed in Table 14.1. In other words, to order of magnitude, the LB and UB of A_H are 10^2 and 10^4 nm^2, respectively. This approach provides estimates for A_H that describes two possible extremes of HNT. At one extreme, the LB of A_H corresponds to a HNT with a small diameter and a thin wall; at the other extreme, the UB of A_H corresponds to a HNT with a large diameter and a thick wall. These extremes are intended to present HNTs that are structurally stable. The alternative approach evaluates the LB of A_H using the smallest values of D_i and D_o and the UB of A_H using the largest values of D_i and D_o. However, this leads to a HNT model with a large diameter but thin wall for the UB of A_H—from a structural view point, this could be relatively unstable as compared to a HNT with a large diameter and thick-walled HNT predicted using the first approach. Of course, this assumption remains to be confirmed. On the other hand, Lecouvet et al. (see Chapter 13 of this book) pointed out that as shear deformations result from the slip motions between adjacent layers due to the weak interactions in the radial direction, consequently the contribution of shear deformations to the total deflection increases with increase in the number of aluminosilicate layers in the wall (by increasing the outer diameter).

TABLE 14.1 Summary of the predicted limits of the mechanical and structural parameters of HNT. LB, lower bound; UB, upper bound.

Mechanical Parameter	Symbol	LB, UB
Young's modulus of halloysite	E_H	1 GPa, 1,000 GPa
Shear modulus of halloysite	G_H	0.4 GPa, 416.7 GPa
Effective axial stress in HNT, as applied to tensile-loaded HNT	σ_H	0.03 GPa, 4 GPa
Yield stress of halloysite	σ_y	10^3 MPa [a]
Effective residual axial strain in HNT, as illustrated for the case of tensile loading	ε_{res}	4×10^{-6}, 4×10^{-3}
Absolute shear stress (maximum) in HNT	τ^{abs}_{max}	1×10^{-1} GPa, 10 GPa
Elastic torque	T_e	15×10^{-5} pN m, 16 pN m
Elastic strain energy	U_e	9.0×10^{-24} J, 1.1×10^{-4} J
Yield torque	T_y	15×10^{-3} pN m, 16 pN m
Plastic torque	T_p	18×10^{-3} pN m, 20 pN m
Moment at yield point	M_y	31×10^{-4} pN.m, 33×10^{-1} pN m

TABLE 14.1 *(Continued)*

Mechanical Parameter	Symbol	LB, UB
Plastic moment	M_p	24×10^4 pN.m, 28×10^1 pN m
Applied force on composite	F_c	120 pN, 1 N
Axial stress within the wall of HNT, lumen-loaded case	σ_H	2.6×10^7 Pa, 2.0×10^8 Pa
Circumferential stress within the wall of HNT, lumen-loaded case	σ_θ	2.7×10^{10} Pa, 7.0×10^{10} Pa
Structural Parameter	Symbol	LB, UB
Inner diameter of HNT	D_i	10 nm, 70 nm[b]
Outer diameter of HNT	D_o	20 nm, 200 nm[b]
Length of HNT	L	50 nm, 5,000 nm[b]
Aspect ratio of HNT	q	3, 50[c]
Cross-section area of HNT	A_H	236 nm², 31,337 nm²
Polar moment of inertia of HNT	J_H	1.5×10^4 nm⁴, 1.6×10^8 nm⁴
Bending moment of inertia of HNT	I_H	7.4×10^3 nm⁴, 7.7×10^7 nm⁴
Change in volume of lumen, lumen-loaded case	ΔV	3.3×10^{16} nm³, 5.5×10^{20} nm³

[a]Only a single value was calculated

[b]These values are estimated based on data derived from experiments reported elsewhere

[c]Pasbakhsh et al. (2013)

FIGURE 14.2 Model of a HNT acted upon by axial loads, *F*. Here *F* acts on the two ends of the HNT, in the direction of the HNT axis, which defines the z-axis of the Cartesian coordinate system. At each end, *F* is applied across the shaded region. Throughout the discussion in this chapter, the Cartesian coordinate system is intended as a reference for the cylindrical coordinate system which is used to establish the stresses in the tube.

Consider tensile loads, F, acting at the ends of the HNT, in the direction along the HNT axis (Figure 14.2). The behavior of the HNT during tensile loading is described by the plot of stress versus strain for the HNT, which depicts a linearly elastic and perfectly plastic behavior (Figure 14.3). Let σ_H $(= F/A_H)$ represent the effective normal stress generated in a HNT. For the initial loading stage, we set $F = 1 \times 10^{6}$ N (which is equivalent to 0.1 mg) for illustrating situations corresponding to typical small loads (and small displacements as well). The HNT deforms elastically in response to F. The magnitude of σ_H is of the order of the ratio of F $(= 1 \times 10^{6}$ N) to A_H $(= 31,337$ nm², UB) giving 3×10^7Pa. Here, the designated magnitude of F follows from Lu et al. (2011). In the research done by Lu et al. (2011), HNT bending force as a function of HNT deflections (from 10 to 120 nm) was recorded; the bending force to yield the HNT was $\approx 1,000$ nN $(= 1 \times 10^{6}$ N) which corresponded to a HNT deflection of 40 nm. Here, our estimate of $F (\approx 1,000$ nN $(= 1 \times 10^{6}$ N)) is intended to address the maximum tensile force before HNT fails in tension. Thus, decreasing A_H increases σ_H dramatically. Recalling the LB and UB of A_H, it follows that decreasing A_H from 31,337 nm² to 236 nm² leads to an increase in σ_H from 3×10^7 Pa to 4×10^9 Pa (i.e., from 30MPa to 4×10^3MPa). Clearly, this represents an increase in σ_H of the order of 10^2. This simple exercise is intended to illustrate the sensitivity of the stress uptake in HNTs to A_H. In contrast, computer simulation of CNTs predicted that single-wall CNTs can sustain tensile stresses for as high as 4×10^4 MPa in the small strain regime before succumbing to yielding (Natsuki and Endo, 2004).

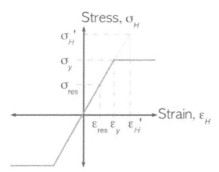

FIGURE 14.3   Linearly elastic and perfectly plastic analysis as illustrated by the graph of normal stress, σ_H, versus normal strain, ε_H, in the HNT.

As the applied load increases, the magnitude of the σ_H increases and eventually equals the yield stress, σ_y, of halloysite. Let δ represent the displacement in the F direction. At this stage, the corresponding effective normal strain ε_H $(=\delta / L)$ is equal to the yield strain(ε_y) of halloysite. Thereafter, as the applied load increases, the plastically deforming HNT continues to experience a stress of σ_y. When F is removed, the HNT responds elastically, generating a force,-F, in the opposite direction to F; correspondingly this generates a stress, $-\sigma_H'$. As a result the residual stress in the HNT is $\sigma_{res} = \sigma_H{}' - \sigma_y$. The corresponding residual strain is $\varepsilon_{res} = \sigma_{res} / E_H$. Thus, this leads to a permanent displacement (δ) of the HNT of the order of $\varepsilon_{res} L$.

   This paragraph presents an argument to estimate ε_{res}. We set $\sigma_y = 10^3$ MPa (to correspond to the UB of σ_H) and $E_H = 10^3$ GPa (UB). On evaluating the relationship $\sigma_y = E_H \varepsilon_y$, this leads to $\varepsilon_y \approx 0.001$. Let $\sigma_H{}'$ represent the elastic stress of the HNT (if the HNT had not yielded when the strain in the HNT equals $\varepsilon_H{}'$) in the axial direction. It is straightforward to see that $\sigma_H{}' = \sigma_y + \sigma_{res}$, where the magnitude of the σ_{res} is a fraction of σ_y (i.e., $0.001\sigma_y = 10^6$ Pa). Numerically, $\sigma_{res} = \sigma_H{}' - \sigma_y = 10^6$ Pa. Using $\sigma_{res} = 10^6$ Pa and $E_H = 1$ GPa (LB), thus ε_H is of the order of $\sigma_{res} / E_H \approx 4 \times 10^{-3}$. From a design point of view, a finite value of ε_{res} could upset the intended purpose of the HNT. Prediction of ε_{res} is important for understanding how HNT contributes to reinforcing the composite material when the material is subjected to subsequent loading regime.

14.2.2   TORSION

In this section, an argument is presented to estimate, to order of magnitude, the active elastic (T_e), yield (T_y), and plastic (T_p) torques generated within the HNT and the strain energy absorbed by the HNT under an increasing applied load. We will assume that T_e occurs at low torsional loads and T_p occurs at high torsional loads beyond the yield point. Let J_H represent the polar moment of inertia of a HNT, with respect to the HNT axis. Note that

$$J_H = [\pi / 32]\{D_o{}^4 - D_i{}^4\} \tag{14.1}$$

To order of magnitude, substituting $D_o = 20$ nm (LB) and $D_i = 10$ nm (LB) into Eq. (14.1) gives $J_H \approx 1.5 \times 10^4$ nm^4 (LB). Using $D_o = 200$ nm (UB) and $D_i = 10$ nm (LB), similar argument yields the UB of $J_H \approx 1.6 \times 10^8$ nm^4. Let v_H and G_H represent the shear modulus and Poisson's ratio, respectively, of halloysite; G_H is of the order of $E_H / \{2(1 + v_H)\}$ where v_H ranges from 0.2 (LB) to 0.3 (UB), assuming that halloysite behaves like a ceramic material. Setting $E_H = 1$ GPa (LB), we find $G_H \approx 0.4$ GPa (LB); similar argument gives the UB of $G_H \approx 416.7$ GPa. Let τ_o represent the shear stress on outer surface of the HNT (NB: at radial distance $\rho = D_o / 2$ from the axis). The LB and UB of τ_o are of order of $\tau^{abs}{}_{max}$ (i.e., 0.1 GPa and 10.0 GPa, respectively). (NB: $\tau^{abs}{}_{max}$ is obtained from analysis of stress in a small volume element in a solid; see Section 14.3.3.) These estimates of τ_o are conservative and are not unreasonable because single-wall CNTs can sustained shear stress of up to 20 GPa before buckling occurs (Chang et al., 2007).

FIGURE 14.4   Model of a HNT acted upon by torsional loads, T, about the HNT axis (*dashed line*). Here, the HNT axis defines the z-axis of the Cartesian coordinate system (see Figure 14.2).

During elastic torsional loading, as the HNT deforms elastically, the shear stress in the wall of the HNT varies linearly with distance, P, from the axis, such that

$$\tau = (2\rho / D_o)\tau_o \qquad (14.2)$$

Using the torsional formula, T_e is evaluated as follows. Starting from $T_e = \{2\pi / (D_o / 2)\}\tau_o \int_{D_i/2}^{D_o/2} \rho^3 d\rho$, the integration is evaluated (i.e., $T_e = (\pi / D_o)\tau_o \rho^4 \big|_{D_i/2}^{D_o/2}$), which eventually leads to

$$T_e = [\pi / 16]\tau_o (D_o^4 - D_i^4) / D_o \qquad (14.3)$$

Substituting $\tau_o = 0.1$ GPa (LB), $D_o = 20$ nm (LB), and $D_i = 10$ nm (LB) into Eq.(3) leads to $T_e \approx 1.5 \times 10^{16}$ N.m (LB). Applying the same argument by substituting $\tau_o = 10.0$ GPa (UB), $D_o = 200$ nm (UB), and $D_i = 10$ nm (LB) into Eq. (14.3) gives the UB of $T_e \approx 1.6 \times 10^{11}$ N.m. Thus the values of T_e range from 15×10^5 pN m to 16 pN.m. Here, the estimates for the range of magnitudes of T_e of HNT are not unreasonable based on the following justifications. Computer simulation of single-walled CNTs has predicted that CNTs can generate T_e of up to 600 eV (i.e., 9.7×10^{17} N.m) (Ertekin and Chrzan, 2005). Additionally, according to Eq. (14.3), it follows that multiwalled CNTs—given they are thicker than single-walled CNTs and possibly featuring larger D_o, all things being equal—could produce T_e that are larger than 9.7×10^{17} N.m. To some extent, the same conclusion applies to multiwalled HNTs. To continue the discussion, the carbon atoms within a CNT interact via covalent bonding but the heterogeneous atoms in HNT interact via ionic and covalent bonds. This suggests that the CNT is structurally stiffer as compared to HNT. Hence the LB of the T_e of HNT is one order of magnitude higher than that of the CNT.

The elastic strain energy of a torsionally deformed HNT is given by

$$U_e = \frac{T_e^2 L}{2G_H J_H} \qquad (14.4)$$

Substituting $L = 5,000$ nm (UB), $T_e = 1.6 \times 10^{11}$ N m (UB), $G_H = 0.38$ GPa (LB), and $J_H = 10^{32}$ m⁴ (LB) into Eq. (14.4) leads to $U_e \approx 1.1 \times 10^4$ J (UB). Similar argument leads to $U_e \approx 9.0 \times 10^{24}$ J (LB) by substituting $L = 50$ nm (LB),$T = 1.5 \times 10^{16}$ N m (LB), $G_H = 384$ GPa (UB), and $J_H = 10^{28}$ m^4 (UB) into Eq. (14.4). Thus the ratio of the maximum and minimum value of U_e is of the order of 10^{20}; the large variability in U_e, spanning several length-scales, suggest that HNT may be employed as torsional springs for a wide range of quantum mechanical systems.

Assuming a linearly elastic and perfectly plastic torsional loading, when yielding occurs,the reactive torque at yielding, T_y, is of order of $2\tau_y J_H / D_o$,where τ_y represents the yield stress in shear. Substituting the expression of J_H Eq. (14.1) leads to

$$T_y = \{\pi / 16\}\tau_y(D_o^{\,4} - D_i^{\,4}) / D_o \tag{14.5}$$

For simplicity, τ_y is of the order of the UB of τ_y. Substituting $\tau_y = 10^{10}$ Pa, $D_o = 200$ nm (UB), and $D_i = 10$ nm (LB) into Eq. (5) leads to$T_y \approx 1.6 \times 10^{11}$ N.m (UB). Similar argument gives the LB of $T_y \approx 1.5 \times 10^{14}$ N m by substituting $D_o = 20$ nm (LB), $D_i = 10$ nm(LB), and $\tau_y = 10^{10}$ Pa in Eq. (14.5). Thus, the values of T_y range from 15×10^3 pNm (LB) to 16.0 pN.m (UB). Although the UB of T_y is similar to that of T_e the LB of T_y is larger than that of T_e. One important and immediate implication is that when yielding occurs, the range of reactive torque that the HNT can generate is narrower than that during elastic loading. Given the crystalline nature, structural defects could occur in HNT. How these defects influence the structural properties (e.g., T_y), is not well understood. Drawing an analogy from defects study in single-walled CNTs under torsional loading (Huq et al., 2010), it is inferred that the influence of defects on the augmentation or diminution of the structural properties of HNTs depends on the defect orientation.

Beyond the yield point, further increase in the applied torque leads to a plastically deforming HNT. Integrating the expression for the plastic torque, $T_p = 2\pi \int_{D_i/2}^{D_o/2} \tau_y \rho^2 d\rho$, leads to

$$T_p = \{\pi / 12\}\tau_y(D_o^{\,3} - D_i^{\,3}) \tag{14.6}$$

Substituting $D_o = 20$ nm (LB), $D_i = 10$ nm (LB), and $\tau_y = 10^{10}$ Pa into Eq.(6)leads to$T_p \approx 1.8 \times 10^{16}$ N.m (LB). Similar argument leads to the UB of $T_p \approx 2.1 \times 10^{11}$ N m ($D_o = 200$ nm, $D_i = 10$ nm and $\tau_y = 10^{10}$ Pa). Thus, T_p ranges from 18×10^3 pN m to 21 pNm. It is important to emphasize that the LB and UBof T_p are only marginally larger than those of T_y but the range, on order of magnitude, is essentially the same. In other words, beyond the yield point, the HNT can respond by generating torques that are slightly higher than those of the T_y (i.e., a small advantage from design point of view).

The physical mechanisms addressing the elastic-plastic response of the HNT undergoing torsional deformation are not well established. From a design perspective, understanding the elastic-plastic response is of paramount importance because checks must be made to ensure that the HNT is not overloaded for the intended purpose. Of course it may be argued that since HNTs are very brittle structures, not all HNTs would be able to exhibit appreciable deformation in torsion before succumbing to torsional collapse. The authors note that PATCH HNTs (Figure 14.1c) could undergo appreciable torsional deformation before the tube collapses (Paskakhsh et al., 2013). Nevertheless, the predictions established in this section are intended to guide future experiments for investigating the torsional properties of HNTs.

14.2.3 BENDING

This section is intended to discuss the expressions for the elastic bending moment, M_e, the bending moment at yield point, M_y, and plastic bending moment, M_p, generated by the HNT under an increasing applied load. Figure 15.5 shows a HNT experiencing a bending moment, M, due to an applied load F. For the purpose of this discussion, the applied load is assumed to act on the HNT at one end of the HNT (i.e., at L distance from the other end and that M is of order of FL).

$M \langle$ $\rangle M$

FIGURE 14.5 Model of a HNT undergoing deflection. Here M represents the bending moment about the neutral axis (dashed line). The neutral axis lies along the x-axis of the Cartesian coordinate system (see Figure 14.2); the neutral axis is also perpendicular to the HNT axis, which defines the z-axis of the Cartesian coordinate system.

To begin the discussion, I_H, the moment of inertia of the cross-sectional area about the neutral axis of a HNT is first evaluated using

$$I_H = [\pi/64](D_o^4 - D_i^4) \tag{14.7}$$

(Lecouvet et al. Chapter in book and therein). Substituting the expression for I_H (from Eq. (14.7)) into the flexure equation $\sigma_H = yM/I_H$, where $y=D_o/2$, leads to

$$M_e = [\pi/32]\sigma_H\{D_o^4 - D_i^4\}/D_o \tag{14.8}$$

Thus, Eq. (14.8) predicts that M_e increases with increase in σ_H. Eventually, at sufficiently large F, $\sigma_H = \sigma_y$ ($= 4.2 \times 10^9$ Pa). Let M_y be the moment of inertia of the cross-sectional area about the neutral axis at the onset of yielding. Substituting $D_o = 20$ nm (LB), $D_i = 10$ nm (LB), and $\sigma_H = 4.2 \times 10^9$ Pa (UB) into Eq.(8), gives $M_y \approx 31 \times 10^4$ pN m (LB). On the other hand, substituting $D_o = 200$ nm (UB), $D_i = 10$ nm (LB), and $\sigma_H = 4.2 \times 10^9$ Pa into Eq. (14.8) leads to the UB of $M_y \approx 33 \times 10^1$ pN m.

To continue the discussion for plastic bending, note that the centroid of a HNT undergoing bending is found from $\bar{y} = \Sigma y A / \Sigma A$ giving $\bar{y} = \left\{ \frac{2D_o}{3\pi} \frac{\pi}{8} D_o^2 - \frac{2D_i}{3\pi} \frac{\pi}{8} D_i \right\} / \left[\frac{\pi}{8}(D_o^2 - D_i^2) \right]$. On further simplification, this leads to

$$\bar{y} = [2/(3\pi)]\{D_o^3 - D_i^3\}/[D_o^2 - D_i^2] \tag{14.9}$$

During plastic loading, assuming that the HNT behaves as a linearly elastic and perfectly plastic material (Section 14.2.1), it follows that the normal force generated across the crosssection on one side of the neutral axis, $[\pi/8]\{D_o^2 - D_i^2\}\sigma_y$, is equal but opposite in direction to the force on the other side. Thus the plastic moment, M_p, is of the order of $[\pi/8]\{D_o^2 - D_i^2\}\sigma_y$, multiplied by $\bar{y}$ (*Eq.*(9)), that is,

$$M_p = [1/12]\{D_o^3 - D_i^3\}\sigma_y \tag{14.10}$$

Substituting D_o = 20 nm (LB), D_i = 10 nm (LB), and σ_y = 4.2 × 10⁹ Pa into Eq. (14.10) gives $M_p \approx 24 \times 10^4$ pN.m (LB). Substituting D_o = 200 nm (UB), D_i =10 nm (LB), and σ_y = 4.2 × 10⁹Pa into Eq. (14.10) leads to the UB of $M_p \approx 28 \times 10^1$ pN.m.

The LB and UB of M_p are marginally smaller than those of M_y. This is to be expected since a buckled(plastically deforming) HNT collapses readily at lower reactive moments as bending progresses. Studies on HNTs subjected bending loads are often concerned with the bending stiffness of the HNT (Lecouvet et al. Chapter in book and therein). However, given the physical mechanisms regulating the elastic-plastic response of HNTs deflecting under a bending load are not well established from a design perspective, understanding the elastic-plastic response is of paramount importance because checks must be made to ensure that the HNT is not overloaded for the intended purpose.

14.3   HNT COMPOSITES

14.3.1   LOAD-SHARING CAPACITY

This section is intended to discuss the mechanics of HNTs reinforcing a composite material. For simplicity, consider an array of aligned HNTs embedded in a polymeric material to form a composite (Figure 14.6). An external load, P, acts on the composite.

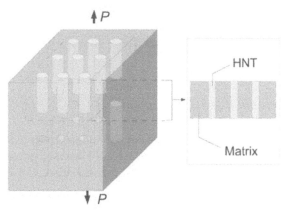

FIGURE 14.6 Model of HNT reinforcing composite. Left: three-dimensional view of HNTs reinforcing a composite material. Right (inset): front (plane) view of the model, showing only the top row of HNTs embedded in the matrix material. For simplicity, at the length-scale of the composite, it is assumed that the reinforcing effects of HNTs are equivalent to those arising from short (solid) fibers with an effective cross-sectional area given by A_H. Here, P represents the load acting on the top and bottom faces of the composite.

For simplicity, at the length-scale of the composite, it is assumed that the reinforcing effects of HNTs are equivalent to those arising from short (solid) fibers with an effective cross-sectional area given by A_H. Consider a section of the composite comprising n number of HNTs. Let P_H be the total force acting on the HNTs and P_M the force on the matrix material. The condition for the equilibrium of forces is $P_H + P_M = P$ (hereafter known as the equilibrium of forces condition). The corresponding condition for the compatibility of displacement requires that the longitudinal displacement of the HNTs equals the displacement of the matrix, that is, $P_H L/[nA_H E_H] = P_M L/[A_M E_M]$ or otherwise $P_M = (A_M/\{nA_H\})[E_M/E_H]P_H$ (here after known as the compatibility condition). Solving for the P_H and P_M simultaneously from the equilibrium of force and compatibility equations gives

$$P_H = nA_H \cdot [E_H / E']P \tag{14.11}$$

and

$$P_M = A_M \{E_M / E'\}P \tag{14.12}$$

where

$$E' = nA_H \cdot E_H + A_M E_M \tag{14.3}$$

The stresses $\sigma_M (= P_M/A_M)$ and $\sigma_H (= P_H/[nA_H])$ in the respective matrix and HNT are

$$\sigma_H = [E_H / E'] P \qquad (14.14)$$

and

$$\sigma_M = \{E_M / E'\} P \qquad (14.15)$$

Let $V (= \{nA_H + A_M\}L)$ represent the volume of the region shown in Figure 14.6 (inset). Additionally, let $V_H (= nA_H L/V)$ and $V_M (= A_M L/V)$ denote the respective volume fractions of HNTs and matrix such that $V_H + V_M = 1$. Equations (14.14) and (14.15) respectively give

$$\sigma_H = [E_H / E_{eff}] \sigma_{avg} \qquad (14.16)$$

and

$$\sigma_M = \{E_M / E_{eff}\} \sigma_{avg} \qquad (14.17)$$

where

$$E_{eff} = nV_H E_H + V_M E_M \qquad (14.18)$$

and $\sigma_{avg} (= P/\{nA_H + A_M\})$ is the average stress over the crosssection shown in the figure. To plot the graphs of σ_H (Figure 14.7a) and σ_M versus V_H (Figure 14.7b), the following estimates, to order of magnitude, are adopted: $A_{H'} = \pi(5\times10^{-7})^2$ m² (i.e., the effective area of the HNT), $A = 1.3 \times 10^9$ m², $E_H = 1 \times 10^9$ Pa, $E_M = 1 \times 10^7$ Pa, $\sigma_{avg} = 8 \times 10^2$ Pa, and $n\hat{I}$ [2, 150,000].Thus, the graphs show that as V_H increases from 1.0×10^5 to 0.8 (due to increases in n), σ_H decreases exponentially from 80.0 to 1.0 MPa; similarly, σ_M decreases exponentially from 0.8 to 0.01 MPa.

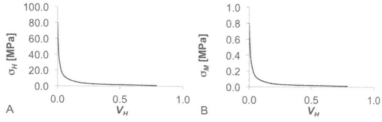

FIGURE 14.7 Graph of (A) HNT (σ_H) and (B) matrix (σ_M) normal stresses versus volume fraction of HNT, $V_{H'}$.

In reality, owing to the process of blending, the HNTs may end up randomly dispersed in the polymeric matrix material (see Chapter 8 of this book by Vahedi and Pasbakhsh). It may be assumed that the HNTs would attempt to realign in the direction of the applied load when the composite is loaded so that the arguments presented in the analysis can be applied to evaluate the composite elastic response to low external loads, although this remains to be confirmed.

14.3.2 ANALYSIS OF HNT PULL-OUT FROM COMPOSITES

This section is intended to discuss the mechanics of HNT pull-out from the matrix material when the applied load acting on a HNT reinforced composite material (Section 3.1) is large enough to initiate microscopic cracks around the HNT. For this purpose, it is believed that HNTs bridging the crack site of the matrix material would eventually be pulled out as the composite material deforms further. There are two key issues to note:(1) the force (P) that is required to pull out the HNT and (2) the elongation of the HNT, δ_E, just before the HNT begins to slip.

Figure 14.8 (inset) illustrates the HNT bridging a crack in the matrix. The length of the HNT on each side of the crack is respectively L' and L'' (i.e. $L = L' + L''$). L' may or may not be equal to L''; whether the portion with L' or L'' gets pulled out depends on the strength of adhesion to the matrix. For the purpose of illustration, suppose the portion with L' succumbed to pull-out. The frictional force along the embedded length L' is modeled as varying linearly from zero at point B to f_{max} (which parameterizes the force per unit length) at point A. To address the force (P) required to pull out the embedded HNT of length L', consider the following condition of the equilibrium of forces, $P - (1/2)f_{max}L' = 0$; the second term on the left of the equation represents the frictional force generated in response to P. On evaluating the equation of equilibrium, this leads to

$$P = [1/2]f_{max}L'. \tag{14.19}$$

To continue the discussion on the situation before slippage occurs, let $P_H(z)$ representthe effective force generated in the HNT at any point z along the HNT axis, measured from the origin at the crack site. The condition of equilibrium of forces requires that $P_H(z) + \{1/2\}[f_{max}z/L]z - f_{max}L/2 = 0$; on the left-hand side of the equation, the second term represents the average adhesive force generated whereas the third term represents the force required to pull the HNT out. Rearranging the terms leads to $P_H(z) = f_{max}L/2 - \{1/2\}[f_{max}z^2/L]$. Dividing both sides by $A_H E_H$, followed by integrating the expression of $P_H(z)$ with respect to z from the origin to L', that is,

$\int_0^{L'} P_H(z)/[A_H E_H]dz = \int_0^{L'} f_{max}L'/[2A_H E_H]dz - \int_0^{L'} f_{max}z^2/[2A_H E_H L']dz$ —noting that the

left-hand side of the equation is equal to δ_E and that only the terms on the right-

hand side of the equation need to be evaluated—leads to

$$\delta_E = f_{max} L'^2 / [3 A_H E_H]$$
(14.20)

Thus, δ_E is proportional to L' and f_{max}.

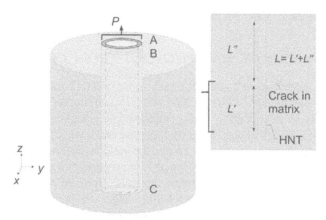

FIGURE 14.8 Model of a HNT partially embedded in the matrix. Left figure: A, B, and C are labels for the respective cross-sectional planes of the HNT. The distance of AB is infinitesimally small; BC = L', which represents the embedded length of the lower half of the HNT. To model HNT pull-out, a force, P, acts on A. Right figure (inset): the HNT within the crack region. The length of the HNT $L = L' + L''$, where L'' is the embedded length of the upper half of the HNT. The HNT axis defines the z-axis of the Cartesian coordinate system.

Clearly HNTs reinforcing a composite deform when the composite is subjected to low applied loads butthe extent to which a HNT deforms (i.e., δ_E) during pull-out is of immediate importance to composite design, because it could potentially be used to parameterize the critical point leading to the rupture of the composite. An argument is presented to estimate δ_E. First, an estimate for f_{max} is determined. Suppose P is equal to the force on the composite. To order of magnitude, $P = 1 \times 10^{-6}$ N, similar to that on the HNT as illustrated in Section 2.1, and $L' = 50$ nm (LB). Substituting the estimates of P and L' into Eq. (14.19) gives $f_{max} \approx 0.2$ N/m (LB); if one considers $L' = 5,000$ nm (UB) instead, this leads to $f_{max} \approx 19.6$ N/m (UB). Thus, the magnitude of f_{max} ranges from 0.2 N/m to 19.6 N/m. On the basis of the estimated values of f_{max}, it is revealed that $\delta_E = 1.3 \times 10^{-10}$ nm(LB) by substituting $E_H = 1,000$ GPa (UB), $A_H = 31,337$ nm² (UB), $f_{max} = 0.2$ N/m (LB),and $L = 50$ nm (LB) into Eq.(20). Applying asimilar argument to find the UB of δ_E gives 16 nm, by substituting $E_H = 1$ GPa (LB), $A_H = 235.6$ nm² (LB), $f_{max} = 19.6$ N/m (UB), and $L = 5,000$ nm (UB) in Eq.(14.20). Of note, the ratio of the maximum to the minimum δ_E is of the order of 10^{11}. It is argued that the large variability in δ_E indicates that the estimate of δ_E is sensitive to the input parameters, namely E_H, A_H, f_{max}, and L.

14.3.3 HNT AS PRESSURIZED NANOVESSELS

Recently several studies have investigated the use of HNTs as pressurized nanovessels encapsulating various substances such as drugs (Vergaro et al., 2012), enzymes (Lvov et al., 2002), and DNA (Rawtani et al., 2013) for medical applications. Of note, the cytocompatibility of HNTs has been well investigated; Vergaro et al. (2010) found that the addition of as much as 75×10^3 g/mL of HNT in cell culture did not kill the cells. Vergaro et al. (2010) also found that (human dermal fibroblasts) cells can adhere to HNT (very thin layer of halloysite of 300 nm) and the cells can maintain their cellular phenotype on HNTs. HNTs have also been proposed for use asself-healing, anticorrosive,and antimicrobial agents (Lvov et al., 2013 and Abdullayev et al., 2013) for smart composite materials. It is worth mentioning here that the Laplace's capillary pressure was found to be 180 atm for HNT of D_i= 15 nm (in water) which helps to suck entrapping agents into the HNT lumen (Lvov et al., 2013).

Here, the focus is concerned with the stresses generated in a HNT when the lumen of the HNT is filled with the matrix materials. When HNT is used as reinforcing particles for a weak polymeric matrix such as PLA, the matrix polymeric molecules may become polarized during the blending process, and diffusion of the polarized matrix molecules into the lumen of HNT, which has a net negative charge, may occur (Lvov et al., 2013). If HNT were thin-walled vessels, according to the criteria given by $D_i/(D_o-D_i) > 10$, the radial stress (σ_r) can be ignored, and the presence of the matrix material within the lumen generates a hydrostatic pressure (p_{lum}) on the inner walls of the HNT. The axial (σ_H) and circumferential (σ_θ) stresses within the HNT would be given by

$$\sigma_H = p_{lum}D_i \, / \, 2(D_o - D_i) \tag{14.21}$$

$$\sigma_\theta = p_{lum}D_i \, / \, (D_o - D_i) \tag{14.22}$$

by analogy to pressure vessels (Gere and Timoshenko 1996). Here, it is assumed that σ_H and σ_θ are constant throughout the wall of the HNT and these parameters contribute to the tension experienced by the wall of the HNT. Since $\sigma_H = \sigma_\theta / 2$, the respective LB and UB of σ_H can be easily determined from σ_θ. The magnitude of σ_H is half that of σ_θ; this would be expected since the bulk of the stress to resist the pressure would come from the component in the circumferential direction than in the axial direction. However, for $D_i = 10$ nm (LB) and $D_o = 200$ nm (UB), this yields $D_i/(D_o-D_i) \ll 10$. Thus, the thin-walled analysis would not be an appropriate approach for HNTs. Alternatively, applying the thick-walled analysis (Gere and Timoshenko, 1996) yields

$$\sigma_H = [p_{lum}D_i^2 - p_{ext}D_o^2] / \{D_o^2 - D_i^2\} \tag{14.23}$$

and

$$\sigma_\theta = [p_{lum}D_i^2 - p_{ext}D_o^2]/\{D_o^2 - D_i^2\} - D_i^2 D_o^2 \{p_{ext} - p_{lum}\}/\{D^2[D_o^2 - D_i^2]\} \quad (14.24)$$

Where p_{ext} is the pressure from the matrix surrounding the HNT, and radius $D/2$ represents a point in the HNT wall (maximum stress occurs when $D/2 = D_i/2$). A comparison of Eqs. (14.21) and (14.23) reveals that thick-walled analysis of HNT introduces an extra term (i.e., arising from the p_{ext}) to correct for the external pressure when determining the σ_H. On the other hand, a comparison of Eqs. (14.22) and (14.24) reveals that the analysis of σ_θ using the thick-walled approach is much more complicated than using thin-walled analysis. Of note, if $p_{ext} = p_{lum}$ we find $\sigma_\theta = \sigma_\theta$.

An argument to estimate σ_H and σ_θ using the thick-walled approach is presented as follows. Suppose $p_{lum} = 1$ GPa (Natsuki et al., 2006) and, for simplicity, $p_{ext} = 0.1 p_{lum}$. Substituting the values for $p_{ext}, p_{lum}, D_i = 10$ nm (LB), $D_o = 200$ nm (UB) (and, consider $D = (D_o - D_i)/2$) into Eq.(23) and (24) leads to $\sigma_H \approx 2.6 \times 10^7$ Pa (LB) and $\sigma_\theta \approx 7.0 \times 10^{10}$ Pa (UB), respectively; a similar argument leads to $\sigma_H \approx 2.0 \times 10^8$ Pa (UB) and $\sigma_\theta \approx 2.7 \times 10^{10}$ Pa (LB) by substituting similar values for p_{ext} ($= 0.1 p_{lum}$) and p_{lum} ($= 1$ GPa, Natsuki et al., 2006), $D_i = 10$ nm (LB) and $D_o = 20$ nm (LB) (and, consider $D = (D_o - D_i)/2$) into Eqs. (14.23) and (14.24), respectively. The overall analysis indicates that the $\sigma_\theta > \sigma_H$.

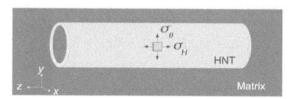

FIGURE 14.9 Model of a HNT reinforcing a matrix material. The lumen of the HNT is darkened to the same shade as the matrix to illustrate diffused matrix materials in the lumen. The HNT axis defines the z-axis of the Cartesian coordinate system.

For the square volume element shown on the HNT (Figure 14.9), the principal mimimum stress, σ_{min}, is zero in any element on the surface of the HNT subjected to a state of plane stress; the intermediate and principal (maximum) stress equal σ_H and σ_θ, respectively. An element oriented 45° with respect to this square element within this plane is described by a state of absolute maximum shear stress τ_{max}^{abs} ($=(=[\sigma_\theta - \sigma_{min}]/2 = \sigma_\theta/2)$ and an average normal stress, σ_{avg} ($=[\sigma_\theta + \sigma_{min}]/2 = \sigma_\theta/2$). It turns out that σ_{avg} is of the order of τ_{max}^{abs} (NB: recall the LB and UB of τ_o are 10^8 Pa and 10^{10} Pa, respectively, Section 14.2.2).

In practice, the HNT can experience an overall increase in size, arising from an expanded volume (ΔV) of the HNT lumen due to p_{lum}. According to small-strain analysis (Gere and Timoshenko 1996), ΔV is given by

$$\Delta V = \frac{4 p_{lum} \pi (D_i / 2)^4}{E_H (D_o - D_i)} [1 - v_H]$$

(14.25)

Substituting $v_H = 0.2$ (LB), $E_H = 1000$ GPa (UB), $D_o = 200$ nm (UB), $D_i = 10$ nm (LB), and $p_{\text{lum}} = 1$ GPa (Natsuki et al. 2006) into Eq. (14.25) gives $\Delta V \approx 3.3 \times 10^{16}$ nm³ (LB). On the other hand, substituting $v_H = 0.3$ (UB), $E_H = 1$ GPa (LB), $D_o = 20$ nm (LB), $D_i = 10$ nm (LB), and $p_{\text{lum}} = 1$ GPa (Natsuki et al., 2006) into *Eq.* (25) gives $\Delta V \approx 5.5 \times 10^{20}$ nm³ (UB). This simple calculation serves to illustrate the sensitivity of the ΔV to structural changes, namely the diameter of the HNT as well as the wall thickness of the HNT.

14.4 CONCLUSION

This chapter has provided basic arguments to determine order-of-magnitude estimates of the mechanical parameters of HNT undergoing different modes of loading, namely, axial tension and compression, torsion, and bending. The discussion has also covered the mechanics of HNT in the context of reinforcing a polymeric material. To summarize, the predicted limits (LBs and UBs) of the respective mechanical and structural parameters are shown in Table 14.1. As shown in the literature, the E_H has received a great deal of attention as compared to other mechanical parameters. However, from a materials design perspective, it is not sufficient to rely solely on E_H or a handful of HNT parameters for the intended application. In order for a design approach to be comprehensive, we suggest that additional mechanical (e.g., G_H, σ_H, σ_y ...) and structural (e.g., q, A_H, I_H ...) parameters of relevance to the design problem should also be taken into consideration. Finally, it is important to emphasize the illustrative nature of the LBs and UBs of the various parameters listed in Table 14.1 and that these values remain to be confirmed by experiments, and may even be refined based on the findings from experiments.

KEYWORDS

- **Axial stress**
- **Cartesian coordinate**
- **Elastic-plastic response**
- **Heterogeneous atoms**
- **Polymeric matrix material**
- **Volume fraction**
- **Young's modulus**

REFERENCES

1. Abdullayev, E. et al.; *ACS. Appl. Mater. Interf.* **2013**, *5*(10), 4464–4471.
2. Chang, T.; *Appl. Phys. Lett.* **2007**, *90*, 201910.
3. Deng, S.; Zhang, J.; Ye, L.; and Wu, J.; *Polymer.* **2008**, *49*, 5119–5127.
4. De Silva, R. T.; Pasbakhsh, P.; Goh, K. L.; and Chai, S.-P.; *Poly. Test.* **2013**, *32*, 265–271.
5. DiBiasio, C. M;and Cullinan, M. A.; *Appl. Phys. Lett.* **2007**, *90*, 203116.
6. Ertekin, E.;and Chrzan, D. C.; *Phys. Rev B.* **2005**, *72*, 045425.
7. Huq, A. M. A.; Bhuiyan, A. K.; Liao, K.; and Goh, K. L.; *Int. J. Mod. Phys. B.* **2010**,*24*, 1215–1226.
8. Ismail, H.; Paskabaksh, P.; Fauzi, A. M. N.; Abu, and Bakar, A. A.;*Poly. Plas. Technol. Eng.* **2009**, *48*, 313–323.
9. Gere J. M.; and Timoshenko, S.; Mechanics of Materials, 4th edition; Nelson Thornes: Cheltenham-UK, 1996.
10. Guimarães, L.; Enyashin, A. N.; Seifert, G.; and Duarte H. A.; *J. Phys. Chem.C.* **2010**, *114*, 11358–11363.
11. Lecouvet, B.; Horion, J.; D'Haese, C.; Bailly, C.; and Nysten, B.; *Nanotechnology.* **2013**, 24, 105704.
12. Liu, M.; Zhang, Y.; Wu, C.; Long, S.; and Zhou, C.; *Int. J. Biol. Macromol.* **2012**, *51*, 566–575.
13. Lu, D.; Chen, H.; Wu, J.; and Chan, C. M.; *J. Nanosci. Nanotechnol.* **2011**, *11*, 7789–7793.
14. Lvov, Y.; Price, R.; Gaber, B.; and Ichinose, I.; *Colloid. Surf. A: Physicochem. Eng. Aspect.* **2002**, *198–200(0)*, 375–382.
16. Natsuki, T.; and Endo, M.; *Carbon.* **2004**, *42*, 2147–2151.
17. Natsuki, T.; Hayashi, T.; and Endo, M.; *Appl. Phys. A.* **2006**, *83*, 13–17.
18. Pasbakhsh, P; Ismail, H.; Fauzi, A. M. N.; and Bakar, A. A.; *Appl. Clay. Sci.* **2010**, *48*, 405–413.
20. Rawtani, D.; and Agrawal, Y. K.; *Bio. Nano. Sci.* **2013**, *3*(1), 52–57.
21. Singh, B.; *Clay. Clay. Min.*; **1996**, *44*, 191–196.
22. S heidaei, A. et al.; *Compos. Sci. Technol.* **2013**, *80*, 47–54.
23. Vergaro, V.; Lvov, Y. M.; and Leporatti, S.; *Macromol. Biosci.* **2012**, *12*, 1265–1271.

PART VII
MODIFICATION OF NATURAL MINERAL NANOTUBES

CHAPTER 15

FUNCTIONALIZATION AND COMPATIBILIZATION OF HALLOYSITE NANOTUBES

VAHDAT VAHEDI and POORIA PASBAKHSH

CONTENTS

15.1   INTRODUCTION

Halloysites are dioctahedral 1:1 layered aluminosilicates with a chemical composition and a structure similar to that of kaolinite, dickite, or nacrite. They consist of adjacent sheets of gibbsite octahedra (Al (OH)$_3$) and tetrahedrally coordinated silicon dioxide. They also incorporate a monolayer of water molecules between their layers. Because of the mismatch between the sheets and the weak interactions between the adjacent layers, because of the interlayer water, the sheets are often rolled over to form a nanotubular structure with few layers (5–6) within the wall of the nanotubes (Figure 15.1). They are then known as halloysite nanotubes (HNTs). Unlike platy kaolinites, most of the Al-OH groups are mainly located between the rolled-up layers of HNTs and they are inaccessible in the ideal formula (Pavlidou and Papaspyrides, 2008; Deng et al., 2008; Joussein et al., 2005).

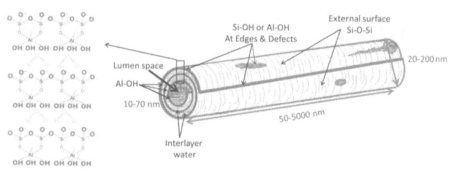

FIGURE 15.1   Schematic diagram of halloysite nanotubes structure.

As surface chemistry plays an important role in the performance of nanoparticles, better understanding of the surface chemistry of HNTs helps to improve their performance and to develop new applications. HNTs have a different chemistry on their outer and inner lumen surfaces. HNTs' internal lumen surface is composed of Al-OH octahedral sheet, whereas the external is composed of Si-O-Si groups. The reactive functional groups located on the inner sides of the tubes on HNTs surface are predominantly aluminols (Al-OH), compared to those aluminol (Al-OH) or silanol (Si-OH) groups located at surface defects or at the edges of the tubes. Because the aluminol and silanol groups are reactive and participating in various chemical reactions, siloxane (Si-O-Si) groups on HNTs external surface are supposed to be nonreactive and do not easily combine with other functional groups. Therefore the inner surface of HNTs is chemically active, whereas the outer surface is regarded as nonreactive. However, some hydroxyl groups (Al-OH or Si-OH) existing on external surfaces within the edges and defects (Figure 15.2) provide active sites for chemical reactions. Impurities, such as Fe$_2$O$_3$, quartz, alunite, and kaolinite, present in halloysite might affect its properties. X-ray fluorescence (XRF) and X-ray diffraction (XRD) analyses have been carried out extensively to trace the amount and type of the impurities present in various types of HNTs (Joussein et al., 2005; Pasbakhsh et al., 2013).

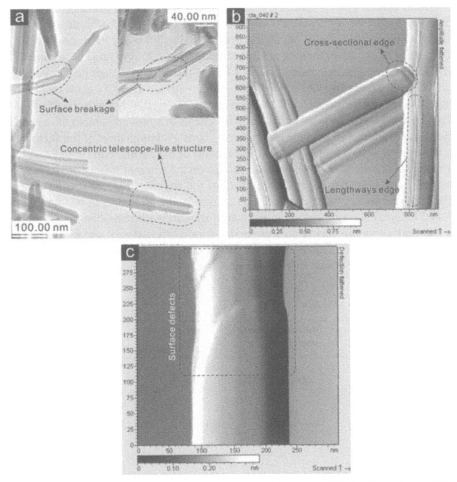

FIGURE 15.2 HNTs' surface morphology and defects: (a) TEM, (b) and (c) AFM micrographs of HNTs (Yuan et al., 2008).

The presence of polar OH groups on the lumen surface makes it highly hydrophilic, which may restrict their usage in the encapsulation of nonpolar chemicals for controlled release and nanocontainers applications (Yah et al., 2012). On the other hand, the relatively low content of functional groups on the outer surface of HNTs enables them to be easily separated from HNTs aggregates and well dispersed in a polymer-composite structure. Uniform dispersion of HNTs in the form of individual nanotubes in polymers has been reported by various researchers (Liu et al., 2007a; Ismail et al., 2009; Lin et al., 2011). However, the low activity of HNTs' external surface results in weaker interfacial interactions with polymer matrixes, in addition to lowering the efficiency of HNTs for reinforcement and for dispersion.

Modification of HNTs, both on inner and outer surfaces, can enhance their performance in various applications (Guo et al., 2008, 2009a, 2010b; Pasbakhsh et al., 2010; Lei et al., 2011; Liu et al., 2008b, 2007b, 2011,; Jia et al., 2009; Gârea et al., 2011; Rooj et al., 2010; Tang et al., 2011; Deng et al., 2009). Table 15.1 summarizes various modification methods of HNTs and the effects on their performances. The presence of reactive OH groups on the inner surface of HNTs allows modifying the inner surface through chemical bonding. The relatively low content of hydroxyl groups on the external surface of HNTs leads to a limited number of reactive sites for any number of possible chemical reactions. Although most of the chemical modifications have been based on the interaction of chemicals with surface silanol or aluminol groups, the amount of these groups on the external surface of HNTs is a matter of question (Yuan et al., 2008). Besides the chemical reactions with surface silanol or aluminol groups, physical interactions including hydrogen bonding with HNTs' external siloxane surface and electron transfer with transition metals, such as aluminum and iron, have also been used to modify the external surface of HNTs. In this chapter, various methods which have been used for functionalization and compatibilization of HNTs' surfaces are reviewed.

TABLE 15.1   Surface Modification of HNTs.

Ref.	Modifier	Interaction Mechanism	Matrix/ Guest Molecules	Remark
Du et al., (2006a)	γ MPS	Covalent bonds to surface hydroxyls	PP	30 °C higher $T_{5\% \, loss}$ at 10 phr modified HNTs compared to unmodified HNTs.
Ning et al., (2007)	QA	Ion exchange	PP	Not very much improved mechanical properties Improves dispersion
Liu et al., (2009)	CBS	Electron transfer	PP	Improves dispersion Higher tensile strength, flexural strength and flexural modulus, thermal stability Lower impact

TABLE 15.1 *(Continued)*

Ref.	Modifier	Interaction Mechanism	Matrix/ Guest Molecules	Remark
Liu et al., (2008b)	BBT	Electron transfer	PP	Formation of some ribbon-like or wire-like phase (fibrils) of BBT
				Much higher crystallinity
				Substantially increased tensile and flexural
Prashantha et al., (2011)	Unspecified QA	Electron transfer	PP	Slightly better dispersion
				Slightly increased mechanical performance
Guo et al., (2009b)	γ MPS	Covalent bonds to HNTs' surface hydroxyls	PA	Improves dispersion
		Covalent bond with the amine ends of the PA6 chains via Michael addition		More effective for improving the mechanical properties than unmodified HNT
Liu et al., (2011)	PMMA	Radical polymerization of MMA on γ MPS grafted HNT in water emulsion	PVC	Improves dispersion
				Better interfacial bonding
				Improves the strength, modulus, impact, thermal stability
Deng et al., (2009)	Potassium acetate	Hydrogen bonding	Epoxy	Reduces the size of the particle clusters
Deng et al., (2009)	AEAPS	Covalent bonds to HNTs' surface hydroxyls	Epoxy	Increases the tendency of particle agglomeration
Deng et al., (2009)	CTAC	Electron transfer	Epoxy	Increases the tendency of particle agglomeration
Tang et al., (2011)	Phenyl phosphonic acid (PPA)	Formation of P=O on HNT surface	Epoxy	Changes the morphology of most particles from nanotubes to nano-platelets
		Intercalation and unrolling HNT		Improves dispersion
				Significantly increased the fracture toughness

TABLE 15.1   *(Continued)*

Ref.	Modifier	Interaction Mechanism	Matrix/ Guest Molecules	Remark
Liu et al., (2007a)	Cyanate ester as a hardener	Covalent bonds to HNTs' surface hydroxyls Curing with epoxy resin	Epoxy	Uniform dispersion of individually separated HNTs
Liu et al., (2007b)	GPS	Covalent bonds to HNTs' surface hydroxyls Reaction with epoxy by glycidyl group	Epoxy	Improves dispersion
Rooj et al., (2010)	bis (triethoxysilyl propyl)-tetrasulphide	Covalent bonds to HNTs' surface hydroxyls Co-curing with NR	NR	Improves dispersion Better reinforcing effect
Pasbakhsh et al., (2009)	MA-gr-EPDM	Compatibilizer	EPDM	Improves dispersion Better reinforcing effect
Pasbakhsh et al., (2010)	γ MPS	Covalent bonds to HNTs' surface hydroxyls	EPDM	Improves dispersion Better reinforcing effect
Guo et al., (2009a)	SA	Dual mechanisms of grafting and complexation.	SBR	Improves dispersion Better reinforcing effect
Guo et al., (2008)	MAA	Dual mechanisms of grafting and complexation.	SBR	Improves dispersion Better reinforcing effect
Jia et al., (2009)	RH complex	Hydrogen bonding with HNTs Grafting onto SBR by the addition reaction	SBR	Improves dispersion Better reinforcing effect
Lei et al., (2011)	MimMP	Hydrogen bonding with HNTs Graft onto SBR via Thiol-ene reaction	SBR	Improves dispersion Better reinforcing effect
Lei et al., (2011)	BMimMS	Hydrogen bonding with HNTs Grafted onto SBR via Thiol-ene reaction	SBR	Improves dispersion Better reinforcing effect

TABLE 15.1 *(Continued)*

Ref.	Modifier	Interaction Mechanism	Matrix/ Guest Molecules	Remark
Zhao and Liu, (2008)	γ MPS	Covalent bonds to HNTs' surface hydroxyls	PS	Grafting of PS on HNTs via silane coupling agent
				Better interfacial bonding
Yuan et al., (2012)	APTES	Covalent bonds to HNTs' surface hydroxyls	Orange II	Improved encapsulation and release properties
Tan et al., (2013)	APTES	Covalent bonds to HNTs' surface hydroxyls	Ibuprofen	Improved encapsulation and release properties
Yuan et al., (2013)	Methanol	Covalent bonds to HNTs' surface hydroxyls	Cetyl-trimethyl-ammonium chloride	Improved interlayer surfactant loading of kaolinite for kaolinite-halloysite transformation

15.2   CHEMICAL FUNCTIONALIZATION

15.2.1   SILANE COUPLING AGENTS

Silane coupling agents are the most widely used modifiers for modification of HNTs (Yuan et al., 2008; Pasbakhsh et al., 2010; Liu et al., 2007b, 2011; Deng et al., 2009; Du et al., 2007). HNTs' surface could be functionalized by various chemical groups, such as amine, methacrylate, and epoxy, using different silane coupling agents such as amino silanes, 3-(methacryloxypropyl) trimethoxysilane (γ-MPS) (Du et al., 2007; Liu et al., 2011) and γ-glycidoxypropyltrimethoxy silane (GMS) (Liu et al. 2007b).

Amino silanes have been widely used to graft amine groups on HNTs' surface (Du et al., 2006b; Yuan et al., 2008; Joo et al., 2012; Barrientos-Ramírez et al., 2011; Yah et al., 2012; Zhu et al., 2012). Yuan et al. (2008) reported direct grafting of (3-Aminopropyl) triethoxysilane (APTES) on the lumen's hydroxyls, the edges, and defects of the outer surface. An evacuation pretreatment (cycles of vacuum application) was used to load chemicals into the HNTs' lumen. Joo et al. (2012) and Du et al. (2006b) reported grafting of APTES on the external surface of HNTs. Ramirez et al. (2011) grafted aminosilanes exhibiting two (3-(2-aminoethyl) aminopropyltrimethoxysilane, AEAPS) and three (3-[2-(2-aminoethyl) aminoethyl] aminopropyltrimethoxysilane, TAS) amino groups on HNTs' outer surface, and used these amine functionalized HNTs as a solid support for an atom transfer radical polymerization. Yah et al. (2012) reported the external and internal surfaces had been grafted by AE-APS and ODP beforehand, respectively. Zhu et al. (2012) grafted AEAPS on HNTs to immobilize gold nanoparticles on HNTs' surface.

In a general procedure of grafting of silane modifiers on HNTs' surface, HNTs were initially dispersed in a solvent using ultrasonic or mechanical mixing. Silane modifier was then added to HNTs suspension and the mixture was refluxed at specified temperature and time to perform the reaction. Subsequently, the reaction product was filtered using nano membrane or centrifuge. The modified HNTs were washed with the solvent to remove unreacted silane modifiers, and (vacuum) dried at high temperature. Table 15.2 provides more details about the modification methods used for grafting of silane coupling agents on HNTs' surface.

TABLE 15.2   Various Conditions Used for Silane Modification of HNTs.

Reference	Silane Modifier	Solvent	Reaction temperature / Time	Drying Temperature/Time
Du et al. (2007)	γ MPS	Ethanol	Room temperature / 15 min	70 °C
Liu et al. (2011)	γ MPS	Toluene	110°C /10 h	50°C / 24 h
Liu et al. (2007b)	GMS	Ethanol/water solution (50 vol. %))	Room temperature / 2 days	80°C / 6 h
Yuan et al. (2008)	APTES	Toluene	120 °C / 20 h	120 °C / Overnight
Joo et al. (2012)	APTES	Toluene	80 °C / 1 day	50 °C
Du et al. (2006b)	APTES	Ethanol/water solution (95 vol. %)	80 °C / 3 h	80 °C / 5 h
Barrientos-Ramírez et al., (2011)	AEAPS	Toluene	Room temperature / 2 days	70°C
Yah et al. (2012)	AEAPS	Toluene	75 °C / 20 h	100 °C
Zhu et al. (2012)	AEAPS	Ethanol/water solution (95 vol. %)	80° °C / 6 h	80 °C/ 12 hours
Barrientos-Ramírez et al. (2011)	TAS	Toluene	Room temperature/ 2 days	70 °C

Common silane coupling agents have a chemical structure of $(RO)_3SiR'X$, where RO is an hydrolyzable alkoxy group (such as methoxy and ethoxy), and X is an organofunctional group (such as amino, methacryloxy, and epoxy) (Table 15.3). This structure enables them to bond with organic and as well as inorganic materials and to couple them together (Plueddemann, 1982). The mechanism of reaction with inorganics usually involves hydrolysis of alkoxy groups with water molecules (from inorganic surface or added water) to form silanol groups (Figure 15.3). These silanols then react with hydroxyl groups on the inorganic surface to form an oxane bond by eliminating the water. The silanol groups of hydrolyzed-silane-coupling agent molecules can also react with each other to form an oligomerized or polymerized silane. These reactions result in a cross-linked network of silane molecules on the inorganic surface comprised of physical attachment, covalent bonding, two-dimensional self-assembly (horizontal polymerization), and silane multilayers (vertical polymerization) (see Chapter 18 of this book by Vahedi et al.).

TABLE 15.3 Chemical Structure of Silane Modifiers.

Sample	Composition	M.W. (g/mol)	Structure
APTES	$C_9H_{23}NO_3Si$	221.37	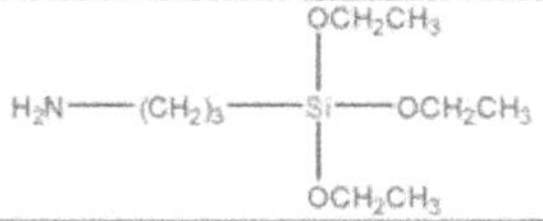 $H_2N-(CH_2)_3-Si(OCH_2CH_3)_3$
γ–MPS	$C_{10}H_{20}O_5Si$	248.35	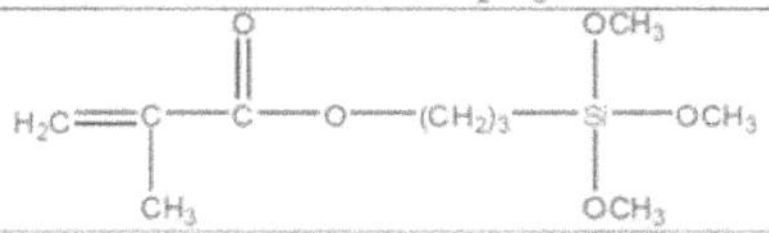 $H_2C=C(CH_3)-C(=O)-O-(CH_2)_3-Si(OCH_3)_3$
AEAPS	$C_8H_{22}N_2O_3Si$	222.36	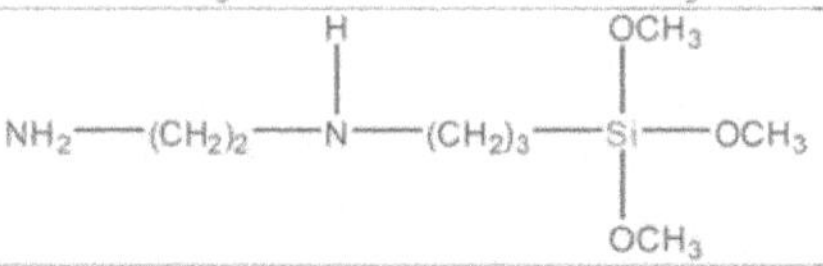 $NH_2-(CH_2)_2-NH-(CH_2)_3-Si(OCH_3)_3$
TAS	$C_{10}H_{27}N_3O_3Si$	265.43	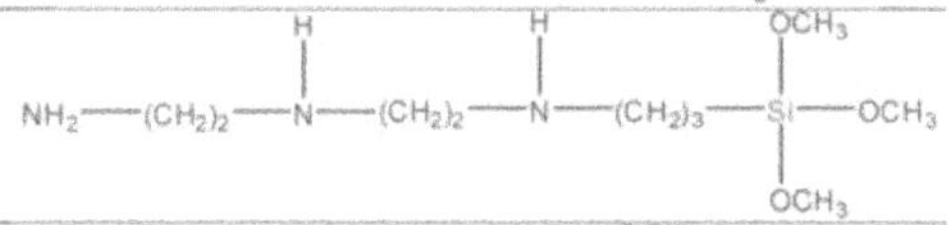 $NH_2-(CH_2)_2-NH-(CH_2)_2-NH-(CH_2)_3-Si(OCH_3)_3$
GMS	$C_9H_{20}O_5Si$	236.34	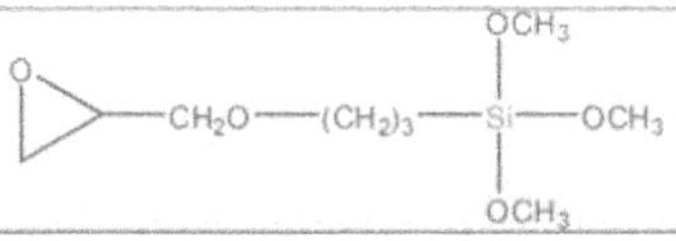 epoxide$-CH_2O-(CH_2)_3-Si(OCH_3)_3$

Increasing the reaction temperature promoted the reactivity and grafting of silane modifier (APTES) on HNTs (Yang et al., 2012). The type of hydrolyzable alkoxy group is also important in determining the reactivity of silane modifiers. Silane modifiers containing methoxy groups are more reactive compared to those containing ethoxy groups and their reaction can be conducted at lower temperatures (Plueddemann, 1982).

Pretreatment of HNTs before modification has considerable effect on their modi-fication with silane modifiers. Abundance of water molecules in the reaction medium causes extensive hydrolysis of silane-coupling agents, resulting in more olygomer-ization or even polymerization of silane modifiers (Yuan et al., 2008; Vrancken et al., 1992). Structural interlayer water of HNTs can promote the oligomerization or polymerization of silane modifiers. This may result in blocking of HNTs' lumen by silane modifiers, if the evacuation pretreatment is performed (Yuan et al., 2008). Drying the HNTs at elevated temperatures (400 °C), before modification, can re-move the interlayer water and reduce the resulting amount of oligomerized or po-lymerized silane, which results in the grafting of thinner layer of silane modifier on HNTs (Yuan et al., 2008).

15.2.2   OCTADEYLPHOSPHONIC ACID (ODP)

ODP is a linear alkyl phosphonic acid consisting of a nonpolar organic hydrophobic chain and inorganic hydrophilic groups in its chemical structure (Figure 15.4). Yah et al. (2012) selectively functionalized the lumen surface of HNTs by ODP in order to make the lumen surface to be hydrophobic. Approximately, 500 mg of HNTs was mixed with 2 mmol ODP in 500 mL of 4:1 Ethanol:H_2O solution. The pH of the so-lution was adjusted to four and evacuation pretreatment was applied to load chemi-cals into HNTs' lumen. The mixture was stirred for a week at a room temperature. Then the modified HNTs were washed and dried at 100°C overnight under vacuum.

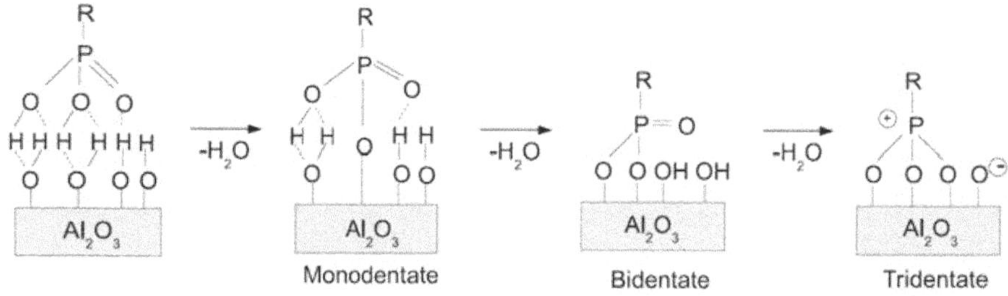

FIGURE 15.4   Possible reaction mechanism of phosphonic acid with Al-OH groups (Luschtinetz et al., 2008).

The grafting of ODP on HNTs lumen's surface took place through formation of Al-O-P bonds. The reaction mechanism involved stepwise acid–base condensation between phosphonic acid groups of ODP and Al-OH groups of HNTs to form mono-, bi-, and tri-dentate phosphorous bonds on the alumina surface (Fig. 4) (Luschtinetz et al., 2008). NMR and FTIR analysis of ODP modified HNTs showed that bonding was mostly bidentate and tridentate (Yah et al., 2012).

It has been argued that ODP was not grafted on the external surface of HNTs, because Si-O-P groups were not detected in XPS and FTIR results. This argument is based on the assumption that aluminol groups only existed on the inner surface of HNTs (Yah et al., 2012). However, it was reported that some Al-OH may be found at the defects of HNTs' external surface (Yuan et al., 2008). Consecutively, Yuan et al. (2008) grafted AEPTS on the external surface of ODP-grafted-HNTs and prepared bi-functionalized HNTs with hydrophobic inner and hydrophilic outer surface (Figure 15.5).

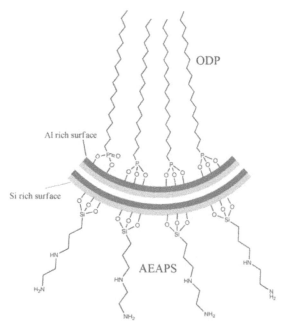

FIGURE 15.5 Schematic illustration of bi functionalized silica alumina oxide surface of HNTs by Octadeylphosphonic acid (ODP) and subsequent (3-(2-aminoethyl) aminopropyltrimethoxysilane (AEAPS) silylation (Yah et al., 2012).

15.2.3 CARBOXYLIC ACID

HNTs were functionalized by carboxylic acids through two-step reactions to prepare HNTs with pH-adjustable dispersion properties (Joo et al., 2012). To graft carboxylic acid moieties on HNTs' surface, Joo et al. (2012) initially prepared amine-functionalized HNTs by grafting APTES on HNTs' surface (Section 15.2.1). Amine-functionalized HNTs were reacted with succinic anhydride to create carboxylic moieties on HNTs' surface. For this purpose, the amine-functionalized HNTs were mixed with solution of succinic anhydride in dimethyl formamide (DMF) and stirred for 1 day. Then the modified HNTs were vacuum dried at 50 °C. The reaction mechanism involved a ring opening process which created an amide bond and formed carboxylate groups on HNTs surface (Figure 15. 6) (An et al., 2007).

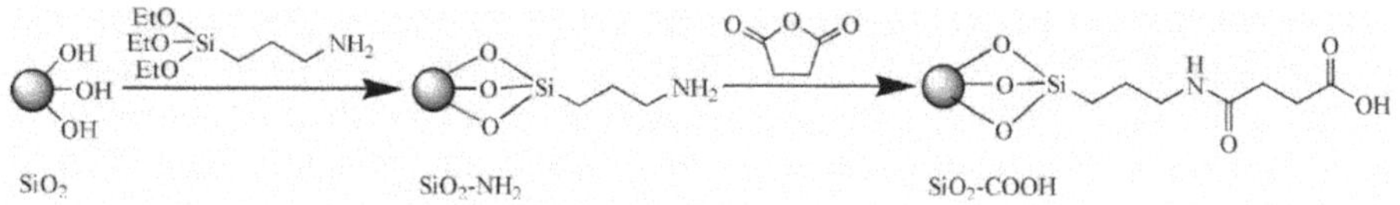

FIGURE 15.6 Possible reaction mechanism for functionalization of nanoparticles with carboxylic acid (Joo et al., 2012).

15.2.4 POLYPROPYLENE

Polypropylene (PP) was grafted on HNTs' surface to enhance the compatibility of HNTs with PP matrix (Du et al., 2006b). Du et al. (2006b) used amine (APTES) functionalized HNTs and maleic anhydride grafted PP (PP-g-MAH) to graft PP chains on HNTs' surface. The amine-functionalized HNTs were added to the solution of maleic anhydride grafted PP in xylene at 120°C and stirred for 3 h. Then resultant mixture was filtered, extracted by xylene, and vacuum dried at 100°C for 5 h.

The grafting mechanism is schematically illustrated in Figure 15.7. These amine groups of APTES-modified HNTs reacted with maleic anhydride groups of PP-g-MAH through an amidation reaction and bonded PP chains to HNTs surface. A cross-linked network of PP and HNTs was formed because the average functionality of amino and anhydride groups was greater than 2.

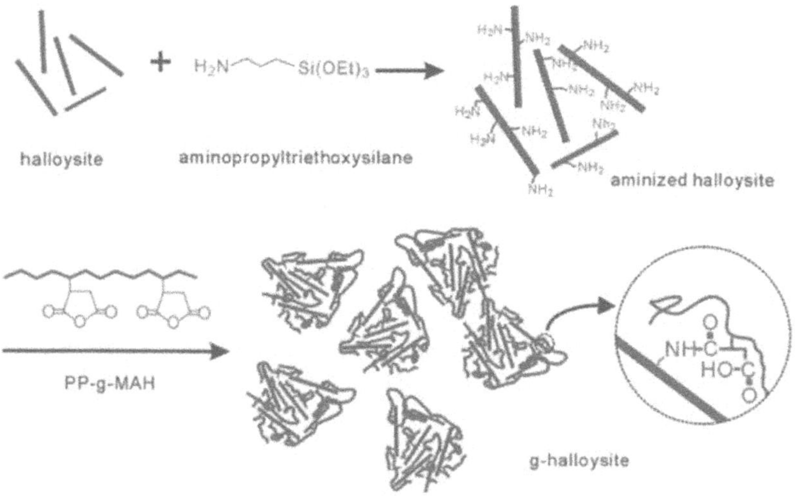

FIGURE 15.7 Possible mechanism for grafting polypropylene (PP) on HNTs (Du et al., 2006b).

15.3   SURFACE POLYMERIZATION

15.3.1   POLY METHYL METHACRYLATE (PMMA)

Liu et al. (2011) grafted PMMA on HNTs surface via in situ radical polymerization. HNTs were initially grafted by γ-MPS (Section 15.2.1). Then in-situ emulsion polymerization of MMA was carried out in the presence of γ-MPS functionalized HNTs. Polymerization reaction was conducted under nitrogen atmosphere at 80 °C for 3 h using sodium dodecyl sulfate (SDS) as an emulsifier and potassium persulfate (KPS) as an initiator in an aqueous medium. The polymerization product was extracted by chloroform for 24 h to remove ungrafted PMMA and vacuum dried at 50°C.

FTIR results showed that PMMA chains were covalently bonded to HNTs' surface through graft polymerization. TEM images (Figure 15.8) revealed that a layer of PMMA was formed around HNTs resulting in a rough and irregular surface in grafted nanotubes. Some parts of HNTs' lumen were filled by PMMA showing that graft polymerization took place on both outer and inner surfaces.

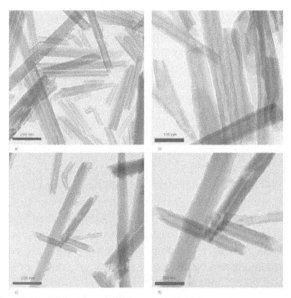

FIGURE 15.8   TEM photographs of HNTs and poly methyl methacrylate (PMMA)-grafted HNTs. (a) and (b) HNTs; (c) and (d) PMMA-grafted HNTs (Liu et al., 2011).

15.3.2   POLYSTYRENE (PS)

Polystyrene grafted HNTs was prepared through in situ polymerization of styrene in the presence of HNTs (Zhao and Liu, 2008; Zhao and Liu, 2007). Zhao et al. (2008)

used in situ bulk polymerization to prepare the PS/HNTs nanocomposite. The γ–MPS modified HNTs were mixed with styrene monomer and Azobisisobutyronitrile (AIBN) initiator and polymerization was conducted at 80°C for 20 h. The polymerization product was extracted by toluene to remove ungrafted PS and vacuum dried at 40°C. The FTIR results showed that PS was grafted to HNTs surface through copolymerization with C=C bonds of γ–MPS modified HNTs. Because the average functionality of γ–MPS modified HNTs is more than two, cross-linked network of PS/HNTs was formed which resulted in aggregation of HNTs. TEM micrographs showed that, besides the external surface of HNTs, their lumen surface was also grafted with PS.

Zhao et al. (2007) prepared core-shell HNTs/PS nanoparticles by soap-less micro-emulsion polymerization. Already modified HNTs were again modified with allyl alcohol (AA) by stirring a mixture of HNTs (0.58 g) and AA (1 ml) in water at 70°C for 5 h. AA modified HNTs were mixed with styrene (1 ml) in distilled water (50 ml) and an aqueous solution of KPS initiator was added drop wise. Then the mixture was stirred at 75°C for 7 h. The polymerization product was washed with water and dried. A seeded micro-emulsion polymerization mechanism was suggested for this grafting process. AA was absorbed on HNTs' surface probably by hydrogen bonding and formed an organic layer around HNTs. The AA modified HNTs entered the styrene phase in the emulsion polymerization medium because of the surface tension made by the nanoparticles of HNTs-encapsulated styrene. Copolymerization of styrene and AA in these nanoparticles resulted in a core-shell HNTs/PS nanocomposite. TEM observation showed that HNTs were fully dispersed in PS nanoparticles produced by this method and they were less agglomerated compared to unmodified HNTs.

15.3.3 POLYANILINE

Polyaniline is a polymer with high electrical conductivity which has attracted attention in the preparation of conducting nanoparticles. Zhang et al. (2008) coated HNTs with polyaniline via in-situ soap-less emulsion polymerization. Zhang et al. (2008) initially prepared anilinium chloride-adsorbed HNTs by mixing HNTs (3.00 g), aniline (1.50 mL), and different amounts of HCl 450 mL in water. The polymerization process was conducted by drop wise addition of 100 ml of initiator solution (aqueous solution of 6.80 g ammonium persulfate and 1.00 mL HCl) into the above mixture and stirring it for 12 h in an ice–water bath. The dark green powder product of polymerization was washed with water and vacuum dried at 40°C overnight. The reaction mechanism was proposed as follows: Anilinium chloride salts were formed from the reaction of aniline and HCl and the anilinium cations of the salt were absorbed to the surface of the negative HNTs in an acidic solution. This increased the affinity of HNTs towards aniline, such that the aniline monomer was absorbed on HNTs surface. Then copolymerization of aniline and anilinium chloride resulted in coating of polyaniline on HNTs' surface. TEM observations showed that polyaniline was only coated on the outer surface of HNTs.

15.4 PHYSICAL INTERACTIONS

15.4.1 IONIC LIQUIDS

Ionic liquids were used as interfacial modifiers in styrene butadiene rubber (SBR)/ HNTs nanocomposites (Guo et al., 2010b; Lei et al., 2011; Lei, 2010). Guo et al. (2010b) coated 1-butyl-3-methylimiazolium hexafluorophosphate (BMImPF6) on HNTs by stirring a suspension of HNTs and BMImPF6 in tetrahydrofuran–water solution at 50°C for 24 h. The modified HNTs was then filtered, washed, and dried at 80°C for 20 h. Lei et al. (2011) used 1-methylimidazolium mercaptopropionate (MimMP) and bis (1-methylimidazolium) mercaptosuccinate (BMimMS) as a compatibilizer of HNTs in their SBR compound formulation.

Ionic liquids (ILs) are organic salts with low melting point (usually <100°C) that consist of bulky asymmetric organic cations (e.g., imidazolium) and anions (e.g., PF6) (Table 15.4). ILs could be adsorbed on the surface of materials like single-walled carbon nanotubes (SWNTs), graphite, silica, mica, and kaolinite and modify their surface through physical interaction mechanisms such as H-bonding, π–π stacking, van der Waals forces, electrostatic forces. The XPS, NMR, and IR analysis showed that ILs could be bonded to HNTs' surface by different types of hydrogen bonding. BMImPF6 could be absorbed on HNTs by hydrogen boding between imidazolium cation and Al-O or Si-O groups on HNTs' surface. MimMP could interact with HNTs by Al-O/imidazolium cation, Si-OH/anion, and Al-OH/anion hydrogen bonding, whereas interaction between BMimMS and HNTs was identified as Si-O/ imidazolium cation, Si-OH/anion, and Al-OH/anion hydrogen bonding.

TABLE 15.4 Chemical Structure of ILs.

Sample	Composition	M.W. (g/mol)	Structure
BMimMS	$C_{12}H_{18}O_4N_4S$	314.4	
MimMP	$C_7H_{12}O_2N_2S$	188.2	
BMImPF6	$C_8H_{15}F_6N_2P$	284.18	

15.4.2   UNSATURATED ACIDS

Unsaturated acid could be adsorbed to the clay minerals by formation of hydrogen bonds between clay and the acid (Guo et al., 2008, 2009a, 2010a). Guo et al. used methacrylic acid (MAA) (Guo et al., 2008) and sorbic acid (SA) (Guo et al., 2009a) as compatibilizer of HNTs in their SBR compound formulation. MAA and SA could be bonded to HNTs' surface by hydrogen bonding between their carboxyl groups and Al-O or Si-O groups of HNTs. On the other hand, MAA and SA could be attached to SBR chains as an unsaturated rubber via free radical grafting. Moreover, MMA would react with zinc oxide during compounding of rubbers to form zinc methacrylate (ZDMA). The ZDMA could interact with HNTs through complexation between Zn^{2+} and the surface OH groups. This process may be described as follows

$$2\,(Al, Si\text{-}OH) + Zn^{2+} = (Al, Si\text{-}O)_2 Zn + 2H^+$$

15.4.3   COMPLEX OF RESORCINOL AND HEXAMETHYLENETETRAMINE

A complex of resorcinol and hexamethylenetetramine (RH) is a general adhesive in a rubber industry, which was recently used as reactive compatibilizer (Liu et al., 2006). Jia et al. (2009) used a complex of resorcinol and hexamethylenetetramine (RH) for modifying HNTs' surface in order to improve their interaction with an SBR matrix. To obtain RH modified HNTs, mixture of HNTs and RH was stirred vigorously at a room temperature for 30 min and then heated for 4 h at 60°C. XPS and FTIR data revealed that the interaction mechanism involved formation of hydrogen bonds between the oxygen atoms of Si–O bonds or hydroxyl groups on the surfaces of HNTs and the phenol hydroxyl groups of resorcinol in RH.

15.4.4   CONJUGATED ORGANIC COMPOUNDS

Conjugated organic compounds are electron-rich molecules, which can easily transfer their electrons to other compounds due to their chemical structure (Table 15.5). Liu et al. used 2,5-bis(2-benzoxazolyl) thiophene (BBT) (Liu et al., 2008b), 2,2'-(1,2-Ethenediyldi-4,1-phenylene) Bisbenzoxazole (EPB) (Liu et al., 2008a) and N-cyclohexyl-2-benzothiazole sulfenamide (CBS) (Liu et al., 2009) as a compatibilizer in their PP/HNTs nanocomposites. EPB, BBT and CBS could be adsorbed on HNTs via electron transfer mechanism. These chemicals contain N, O, or S atoms with lone electron pairs and conjugated benzene in their chemical structure (Table 15.5) which makes them capable of donating electrons. On the other hand, HNTs are supposed to be electron-deficient due to the presence of metal atoms such as aluminum and ferrous with capability to accept foreign electrons in their empty orbital.

TABLE 15.5 Chemical Structure of EPB, BBT and CBS.

Sample	Composition	M.W. (g/mol)	Structure
BBT	$C_{18}H_{10}N_2O_2S$	318.35	
EBP	$C_{28}H_{18}N_2O_2$	414.45	
CBS	$C_{13}H_{16}N_2S_2$	264.41	

15.4.5 QUATERNARY AMMONIUM

Quaternary ammoniums have been widely used to compatibilize layered silicates with polymer matrixes (Kozak and Domka, 2004; Brown et al., 2000). Ning et al. (2007) modified the surface of HNTs by methyl, tallow, bis-2-hydroxyethyl, quaternary ammonium (QA) and used it in a PP/HNTs nanocomposite. The researchers stirred HNTs in 5 percent QA at 80°C for 24 h. The modified HNTs were then filtered and dried at 60°C for 24h. Deng et al. (2009) used cetyl trimethyl ammonium chloride (CTAC) to modify HNTs surface for epoxy/HNTs nanocomposites. The researchers mixed HNTs (50 g) with solution of CTAC (60 g) and hydrochloric acid (25 ml) in distilled water at 60°C for 5 h. The modified HNTs were washed with distilled water and dried at room temperature for 4 days. The quaternary ammoniums can be absorbed on clay minerals by ion exchange mechanism (Kozak and Domka, 2004).

15.5 METAL DEPOSITION

15.5.1 SILVER

Liu et al. (2009) immobilized silver (Ag) nanoparticles onto the HNTs' surface via in situ reduction of $AgNO_3$ by the polyol process. Liu et al. (2009) initially prepared thiol-functionalized HNTs by ultrasonic mixing of HNTs (10 g) and mercaptoacetic acid (10 mL) in water (50 mL) for 24 h. Then the thiol-functionalized HNTs (0.5 g) were ultrasonically mixed with polyvinylpyrrolidone (PVP, 2.0 g), and ethylene glycol (EG, 50 g) in water (50 mL) for 24 h. After drop wise addition of 10 mL $AgNO_3$ solution (1.0 g $AgNO_3$) into the mixture for 30 min, the resulting mixture was stirred for 24 h at room temperature and the gray products (Ag-HNTs) were then centrifuged, washed with water, and dried.

Mercaptoacetic acid could be absorbed on the surface of HNTs through electro-static forces of carboxyl groups and functionalized HNTs with thiols. After addition of AgNO$_3$, silver atoms were immobilized on HNTs' surface by formation of S-Ag covalent bonds and formed silver nuclei on HNTs' surface. The excess silver ions, Ag+, reduced to silver metal, Ag, by reaction with PVP and EG. The reduced silver, including silver atoms or clusters in solution, deposited on immobilized silver nu-clei on the HNTs surface and silver nanoparticles were formed by growth of nuclei. XPS data confirmed the formation of S-Ag covalent bonds and TEM observations showed that silver nanoparticles with average diameters of 10 nm were uniformly dispersed on HNTs surface (Liu and Zhao, 2009).

15.5.2 GOLD

Cao et al. (2012) immobilized gold nanoparticles (AuNPs) on HNTs' surface for electrochemical sensing applications. Initially, Cao et al. (2012) prepared APTES grafted HNTs by following Yuan et al. (2008) (Section 15.2.1). Next, AuNPs solu-tion was prepared through citrate reduction of HAuCl$_4$ solution by refluxing the boiling solution of HAuCl$_4$ and sodium citrate for 15 min. Then APTES grafted HNTs and the AuNPs were mixed and sonicated for 30 min to prepare AuNPs/HNTs nanocomposite. TEM and FE-SEM observation showed that AuNPs, with an average diameter of 20 nm, were attached to HNTs surface. Immobilization of AuNPs on HNTs' surface was attributed to the electrostatic interactions between negatively charged AuNPs and positively charged NH$_3$ groups of APTES-grafted HNTs (Figure 15.9).

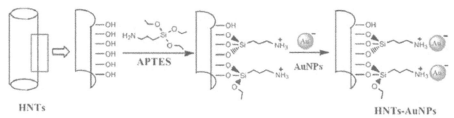

HNTs   HNTs-AuNPs

FIGURE 15.9 Possible mechanism for immobilization of gold nanoparticles (AuNPs) and on HNTs' surface (Cao et al., 2012).

Zhu et al. (2012) reported a green method for immobilization of gold nanoparticles (AuNPs) on HNTs surface for surface-enhanced Raman scattering substrates applica-tion. AEAPTS grafted HNTs (0.5 g) (section15.2.1) were mixed with HAuCl$_4$.3H$_2$O (0.06 g) in de-ionized water (45 mL) at 45°C for 1 h. Then solutions of tea poly-phenols (TP, 0.034 g) in de-ionized water (5 mL) were added to the mixture. After 15 min, prepared powder was washed with ethanol and vacuum dried at 80 °C. The following mechanism was suggested for this process. The lone electron pair of NH_2 groups in AEAPTES-grafted HNTs enabled them to form a chelating complex with Au^+ ions and immobilize Au^+ ions on HNTs' surface. After addition of TP, the phenolic hydroxyls of TP could form another chelating complex with immobilized Au^+ ions. Moreover, TP could reduce Au^+ ions to Au metal by oxidizing the phenolic hydroxyls to orthoquinone. AuNPs were grown by further diffusion of Au^+ ions to the Au-NH_2 complex region and reduction by TP complex. TEM observations showed that a large amount of AuNPs with diameters of 20–40 nm were attached to HNTs surface. HRTEM micrographs showed that a layer of TP with a thickness of about 3 nm was surrounding AuNPs. The amount of AuNPs immobilized on HNTs surface by this method was much higher than those immobilized by Cao's method.

15.5.3 TITANIUM DIOXIDE (TiO₂)

Papoulis et al. (2012) prepared TiO_2–HNTs photocatalytic nanocomposites by depositing anatase-TiO_2 on the HNTs surfaces using a titanium isopropoxide and hydrothermal process. Initially, Papoulis et al. (2012) mixed titanium tetraisopropoxide, $Ti(OC_3H_7)_4$, with hydrochloric acid, nanopure water, and ethanol to prepare 0.05 M TiO_2 sol dispersion. Then suspension of HNTs with water of approximately 1% wt was added to the TiO_2 sol dispersion and stirred until the TiO_2 content reached 70 %wt. The resultant HNTs–TiO_2 nanocomposite was washed with water and filtered by centrifuge. Then the TiO_2–HNTs nanocomposite was dispersed in a 1:1 water:ethanol solution and was hydrothermally treated in an autoclave at 180°C for 5 h. The final products were centrifuged and oven dried at 60°C for 3 h.

Wang et al. (2011) reported a one-step solvothermal method for preparation of TiO_2–HNTs nanocomposites. Wang et al. (2011) mixed HNTs (1 g) with a solution of butyl titanate (2 mL) in an isopropanol solution (40 mL) and adjusted the pH of the solution to four by adding 0.1 M hydrochloric acid. The mixture was then transferred to an autoclave and was kept under 160°C for 24 h in order to undergo solvothermal treatment. The resultant HNTs–TiO_2 nanocomposites were washed with water and ethanol and vacuum dried at 60°C for 12 h.

TEM and SEM observations showed TiO_2 nanoparticles with diameter of 3–10 nm were uniformly deposited on HNTs surface (Papoulis et al., 2012; Wang et al., 2011). TiO_2 nanoparticles covered the lumen of most HNTs and partially deposited within the lumen (Papoulis et al., 2012). ATR-FTIR indicated that hydrogen bonding occurred between TiO_2 and the outer surfaces of HNTs (Wang et al., 2011; Papoulis et al., 2012).

15.5.4 MAGNETIC HNTS

15.5.4.1 COBALT (CO)

Zhang et al. (2012) deposited cobalt (Co) nanoparticles on the surface of HNTs via electroless deposition to prepare magnetic Co-HNTs. Palladium (Pd) was initially immobilized on HNTs' surface by electroless deposition. HNTs (0.5 g) were mixed with sodium tetrachloride palladate (Na_2PdCl_4, 18 mg) and PVP (19 mg) in methanol (50 mL) and the mixture was stirred at room temperature for 5 h. Pd modified HNTs were separated by centrifuge and dried at 60 °C. Then electroless deposition of Co was conducted by mixing Pd-modified HNTs (50 mg) with the solution containing cobalt sulfate, sodium hypophosphite, ammonium chloride, and sodium citrate at 80°C. As soon as hydrogen bubbles formed in the mixture, the color of the mixture immediately turned to black. Finally, the black Co-coated HNTs were separated by centrifuge, washed with de-ionized water, and dried at 60°C.

Figure 15.10 illustrates the mechanism proposed for synthesis of Co-coated HNTs. The PVP–Pd (IV) complex was adsorbed on the surface of HNTs by electrostatic interaction and Pd (IV) ions were reduced to Pd metal by reaction with methanol

$$PdCl_4^{2-} + CH_3OH \xrightarrow{\text{HNTs}} Pd^0 + HCHO + 2H^+ + 4Cl^-.$$

The electroless deposition of Co took place by the following reaction around the Pd particles and Co nanoparticles formed on HNTs surface which was firmly held via the adhesion forces.

$$CoC_6H_5O_7^- + H_2PO_2^- + 3OH^- \xrightarrow[\text{Pd}]{} Co^0 + H_2PO_3^{2-} + C_6H_5O_7^{3-} + 2H_2O$$

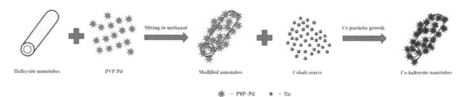

FIGURE 15.10 Possible mechanism for Co deposition on the surface of halloysite nanotubes (Zhang and Yang, 2012).

15.5.4.2 IRON OXIDE

Xie et al. (2011) deposited iron oxide nanoparticles on the HNTs' surface to prepare Fe_3O_4–HNTs magnetic sorbent. HNTs (0.5 g) were mixed with $FeCl_3, 6H_2O$ (1.165

g) and $FeSO_4,7H_2O$ (0.6 g) in 200 mL of water at 60°C under N_2 atmosphere. Then NH_3H_2O solution (20 mL, 8 molL^{-1}) was added drop wise and the pH was adjusted between 9 and 10. The mixture was stirred at 70°C for 4 h. The reaction product was washed with distilled water and dried at 100 °C for 3 h. TEM observations showed that clusters of iron oxide were formed by agglomeration of Fe_2O_3 nanoparticles with an average diameter of 10 nm and attached to HNTs surface. Some nanoparticles were found to penetrate inside HNTs lumen and deposit on the internal surface. The attachment of Fe_2O_3 nanoparticles on HNTs surface was attributed to a large surface area and hydroxyl content of HNTs which caused metal ions to absorb on HNTs' surface.

15.5.5   NICKEL

Barel et al. (1993) coated HNTs with microcrystalline nickel (Ni) film through electroless metallization. Barel et al. (1993) initially treated HNTs with tetrasodium ethylenediamine tetraacetate (EDTA) by stirring HNTs in 0.5 M EDTA aqueous solution for 24 h. EDTA treatment was found to be easier compared to conventional purification processes for clay minerals and importantly was less interfering with the subsequent metallization processes. HNTs were successively treated with acidic palladium chloride and stannous chloride solutions to activate the surface of the HNTs toward electroless plating. The light brown Pd-activated HNTs was immediately dispersed in 5 mL of water and mixed with 10 mL of electroless nickel boron plating bath (Niposit 468) at 65°C. Five minutes later, hydrogen bubbles were formed in the mixture, and the mixture color then turned to black. Approximately 15 min later, another 10 mL of Niposit 468 plating bath was added to the mixture, and then the plating process was continued at 65°C until hydrogen evolution dwindled and black-metalized particles started to precipitate (about 60–80 min). The nickel-coated HNTs were separated and washed with water and dry acetone.

The SEM results showed that a thicker and more uniform Ni film was deposited in the HNTs purified by EDTA procedure, whereas the HNTs purified by the conventional Calgon treatment (Churchman and Gilkes, 1989) were not metalized very well. Treatment of HNTs with palladium solution deposited small Pd particles on the HNTs' surface which acted as the catalyst for metal ion reduction and activated HNTs surface toward electroless plating. Approximately, 20-nm-thick Ni film completely covered the HNTs' external surface, but no evidence was observed for the deposition of Ni on lumen surface.

15.6   FINAL REMARKS

During the past 10 years, various kinds of modification methods have been performed on HNTs' surface, both inner and outer, to tune their surface chemistry with the requirements of different applications and enhance their performance. The external surface of HNTs is mainly covered by siloxane groups which are supposed to be chemically nonreactive, although the presence of hydroxyl groups in defects and edges made it possible to modify the external surface via chemical grafting. Moreover, physical interactions such as hydrogen bonding and electron transfer could be utilized to compatibilize HNTs with polymer matrixes. The internal surface of HNTs contains reactive aluminol groups which facilitate their chemical modification with polar or nonpolar moieties to adjust their encapsulation and release properties. Metal particles could be deposited on HNTs' surface to prepare new cermet materials with interesting electrical, magnetic, and photochemical characteristics. It is believed that all these mechanisms together could favor the current and potential applications of functionalized HNTs as nanocontainers, nanofillers, self- healing, catalyst support, etc.

KEYWORDS

- **Surface modification**
- **Silanization**
- **Surface polymerization**
- **Metal deposition**
- **Ionic liquids**

REFERENCES

1. An, Y.; Chen, M.; Xue, Q.; and Liu, W.; *J. Coll. Interf.* Sci. **2007**, *311* (2), 507–513.
2. Baral S.; Brandow S; and Gaber, B. P.; *Chem.Mater.* **1993**,*5*(9), 1227–1232.
3. Barrientos-Ramírez, S.; Oca-Ramírez, G. M. D.; and Ramos-Fernández, E. V.; Sepúlveda-Escribano A.; Pastor-Blas, M. M.; González-Montiel, A.; *Appl.Catalysis A: Gen.* **2011**.*406*(1–2), 22–33.
4. Brown, J. M.; Curliss, D; and Vaia, R. A.; *Chem. Mater.* **2000**,*12*(11), 3376–3384.
5. Cao, H.; Sun, X.; Zhang, Y.; and Jia, N.; *Anal. Biochem.* **2012**,430(2), 111–115.
6. Churchman, G.; and Gilkes, R.;*Clay Min.* *24*(4), 579–590.
7. Deng, S.; Zhang, J.; and Ye, L.; *Compos. Sci. Technol.* **2009**,*69*(14), 2497–2505.
8. Deng, S.; Zhang, J.; Ye, L.; and Wu, J;.*Polymer*, **2008**, *49*(23), 5119–5127.
9. Du, M.; Guo, B.; and Jia, D.; *Eur. Poly. J*, **2006a**, *42*(6), 1362–1369.
10. Du, M.; Guo, B.; Liu, M.; and Jia, D.; *Poly. J.* **2006b**, *38*(11), 1198–1204.
11. Du, M; Guo B; Liu M; and Jia, D.; *Poly. Poly. Compos.***2007**, *15*(4), 321–328.
12. Gârea, S. A.; Ghebaur, A.; Constantin, F.; and Iovu, H.; *Poly. Plast. Technol. Eng.***2011**,*50*(11), 1096–1102.
13. Guo, B.; Chen, F.; Lei, Y.; and Jia, D.; *J. Macromol. Sci. Part B*, **2010a**, *49*(1), 111–121.
14. Guo, B.; Chen, F.; Lei, Y.; Liu, X; Wan, J; and Jia, D.; *Appl. Surf. Sci.***2009a**,*255*(16), 7329–7336.
15. Guo, B; Lei, Y; Chen, F.; Liu, X.; Du, M.; and Jia, D.; *Appl. Surf. Sci.***2008**, *255*(5), 2715–2722.

16. Guo, B.; Liu, X.; Zhou, W. Y.; Lei, Y.; and Jia, D. J.; *Macromol. Sci. Part B.***2010b,** *49*(5), 1029–1043.
17. Guo, B.; Zou, Q.; Lei, Y.; and Jia, D.; *Polymer. J.* **2009b,** *41*(10), 835–842.
18. Ismail, H.; Pasbakhsh, P.; Ahmad Fauzi, M. N.; and Abu Bakar, A.; *PolyPlast. Technol. Eng.* **2009,**48(3), 313–323.
19. Jia, Z-x; Luo, Y-f.; Yang, S.-y.; Guo. B.-c.; Du, M-l; and Jia, D.-m. *Chin. J. Poly. Sci.* **2009,**27(6), 857–864.
20. Joo, Y.; Jeon, Y.; Lee, S. U.; Sim, J. H.; Ryu, J.; Lee, S.; Lee, H.; and Sohn, D. *J. Phys. Chem. C,* **2012,***116*(34), 18230–18235.
21. Joussein, E.; Petit, S.; Churchman, J.; Theng, B.; Righi, D.; and Delvaux, B. *Clay. Min.* **2005,** *40* (4), 383–426.
22. Kozak, M.; and Domka, L.; *J. Phys. Chem. Solids,* **2004,** *65*(2), 441–445.
23. Lei, Y.; Tang, Z.; Zhu, L.; Guo, B.; and Jia, D. *Polymer.* **2011,***52*(5), 1337–1344.
24. Lei, Y. D. *Exp. Poly. Lett.* **2010,**4(11), 692–703.
25. Lin, Y.; Ng, K. M.; Chan, C.-M.; Sun, G.; and Wu, J. ;*J. Coll. Interf. Sci.* **2011,** *358*(2), 423–429.
26. Liu, C.; Luo, Y.; Jia, Z.; Zhong, B.; Li, S.; Guo, B.; and Jia, D.; *Exp Polym Lett,* **2011,***5,* 591–603.
27. Liu, L.; Jia, D.; Luo, Y.; and Guo, B.J.;*Appl. Poly. Sci,* **2006,** *100*(3), 1905–1913.
28. Liu, M.; Guo, B.; Du, M.; Cai, X.; and Jia, D.; *Nanotechnology.* **2007a,**18(45):(art. No. 455703).
29. Liu, M.; Guo, B.; Du, M.; and Jia, D. *Poly. J.* **2008a,** *40*(11), 1087–1093.
30. Liu, M.; Guo, B.; Du, M.; Lei, Y.; and Jia, D.*J. Poly. Res.***2007b,***15*(3), 205–212.
31. Liu, M.; Guo, B.; Lei, Y.; Du, M.; and Jia, D.*A ppl. Surf. Sci.***2009,** *255*(9), 4961–4969.
32. Liu, M.; Guo, B.; Zou, Q.; Du, M.; and Jia, D.; *Nanotechnology.* **2008b,** *19*(20):(art. No. 205709).
33. Liu, P.; and Zhao, M.;*Appl. Surf. Sci.* **2009,***255*(7), 3989–3993.
34. Luschtinetz, R.; Oliveira, A. F.; Frenzel, J.; Joswig, J.-O.; Seifert, G.; and Duarte, H. A. *Surf. Sci.***2008,** 602(7), 1347–1359.
35. Ning, N.-y.; Yin, Q.-j.; Luo. F.; Zhang. Q.; Du. R.; and Fu. Q.; *Polymer.* **2007** *48*(25), 7374–7384.
36. Papoulis, D.;et al. *Appl. Catal. B: Environ.* **2012,** *132–133,* 416–422.
37. Pasbakhsh, P.; Churchman, G. J.; and Keeling, J. L.; *Appl. Clay. Sci.* **2013,** *74,* 47–57.
38. Pasbakhsh P.; Ismail, H.; Fauzi, M. N. A.; and Bakar, A. A.; *Poly. Test.* **2009,** 28(5), 548–559.
39. Pasbakhsh, P.; Ismail, H.; Fauzi, M. N. A.; and Bakar, A. A.; *Appl. Clay Sci.* **2010,** *48*(3), 405–413.
40. Pavlidou, S.; and Papaspyrides, C. D.; *Prog. Pol. Sci.* **2008,** *33*(12), 1119–1198.
41. Plueddemann, E. P.; Silane Coupling Agents. 2nd edition; New York: Plenum Press, **1982, 1991.**
42. Prashantha, K.; Lacrampe, M.; and Krawczak, P.; *Exp. Pol. Lett.* **2011,** *5*(4), 295–307.
43. Rooj, S.; Das, A.; Thakur, V.; Mahaling, R. N.; Bhowmick, A. K.; and Heinrich, G.; *Mater Des.* **2010,***31*(4), 2151–2156.
44. Tan, D.;et al. *Micro. Mesopor. Mater.* **2013,** *179*(0), 89–98.
45. Tang, Y.; Deng, S.; Ye, L.; Yang, C.; Yuan, Q.; Zhang, J.; and Zhao, C.; *Compos. Part A. Appl. Sci. Manuf.* **2011,** *42* (4), 345–354.
46. Vrancken, K. C.; Van Der Voort, P.; Gillis-D'Hamers, I.; Vansant, E. F.; and Grobet, P.; *J. Chem. Soc, Fara Trans.* **1992,** *88*(21), 3197–3200.
47. Wang, R.; Jiang, G.; Ding, Y.; Wang, Y.; Sun, X.; Wang, X.; and Chen, W.; *ACS Appl. Mater. Interf.* **2011,** *3*(10), 4154–4158.
48. Xie, Y.; Qian, D.; Wu, D.; and Ma, X.; *Chem. Eng. J.* **2011,** *168* (2), 959–963.

49. Yah, W.O.; Takahara, A.; and Lvov, Y. M.; *J. Am. Chem. Soc.* **2012**, *134*(3), 1853–1859.
50. Yang, S-q.; Yuan, P.; He, H-p.; Qin, Z.-h.; Zhou, Q.; Zhu, J.-x.; and Liu, D.; *Appl. Clay. Sci.* **2012**, *62–63* (0), 8–14.
51. Yuan, P.; Southon, P. D.; Liu, Z.; Green, M. E. R.; Hook, J. M.; Antill, S. J.; and Kepert, C. J.; *J. Phys. Chem. C.* **2008**, *112*(40), 15742–15751.
52. Yuan, P.; Southon, P. D.; Liu, Z.; and Kepert, C. J. *Nanotechnology.* **2012**, *23*(37), 375705.
53. Yuan, P.; Tan, D. Y.; Annabi-Bergaya, F.; Yan, W. C.; Liu, D; and Liu, Z. W.; *Appl. Clay Sci,* **2013.** (in press).
54. Zhang, L.; Wang, T.; and Liu, P., *Appl. Surf. Sci.* **2008,** 255(5), 2091–2097.
55. Zhang, Y.; and Yang, H.; *Appl. Clay. Sci.* **2012,** *56*(0), 97–102.
56. Zhao, M.; and Liu, P. J.; *Macromol. Sci. Part B, Phys.* **2007**, *46*(5), 891–897.
57. Zhao, M.; and Liu, P.; *J. Ther. Anal. Calorim.* **2008**, *94*(1), 103–107.
58. Zhu, H.; Du, M.; Zou, M.; Xu, C.; and Fu, Y. *Dal. Trans.* **2012,***41*(34), 10465–10471.

CHAPTER 16

MODIFICATION OF SEPIOLITE AND PALYGORSKITE NANOTUBES AND THEIR APPLICATIONS

GUSTAVE KENNE DEDZO and CHRISTIAN DETELLIER

CONTENTS

16.1   INTRODUCTION

Naturally occurring materials (clay minerals and biopolymers), although having interesting characteristics and properties, sometimes show significant limitations for their practical use. Among others, they can have compatibility issues with other materials or a poor performance in comparison to synthetic materials. The clay minerals are not an exception despite their increased implication in various fields of technology. They are increasingly used as a support material for the synthesis of specialized composites or as an adjuvant to substantially improve the mechanical properties of some materials. Their abundance, availability, and physicochemical properties account for their strong interest. In this field, undoubtedly, the smectites are the most popular clay minerals, because of the relatively easy modification processes required to obtain the appropriate changes in order to fit the intended use. Sepiolite and palygorskite, which belong to the family of fibrous clay minerals, are also widely used and sometimes are more suitable than smectites because of their unique structural properties. Like other clay minerals, they also have limitations that can be avoided by improving their properties and compatibility. Well chosen strategies can significantly modify the surface properties of these minerals and paving the way for various applications.

In this chapter, modifications of sepiolite and palygorskite are reviewed and discussed. The reactive sites and mechanisms involved during modifications are reviewed first, whereas the modification strategies with illustrative examples are reviewed later. Examples of applications of these materials are also provided.

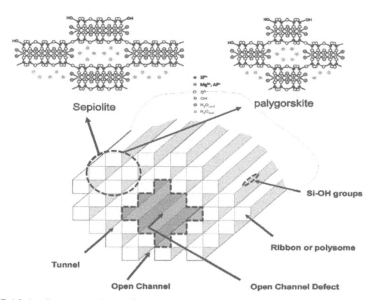

FIGURE 16.1   Structure of sepiolite and palygorskite.

16.2   RELATION BETWEEN STRUCTURE AND MODIFICATION SITES OF SEPIOLITE AND PALYGORSKITE

Structurally (Figure 16.1), sepiolite and palygorskite are composed of an octahedral sheet sandwiched by two sheets of continuous silica tetrahedral in the (a, b) plane. There is a periodic inversion at the silica tetrahedral sheets, which induces a reorientation of the apical oxygen. In the case of palygorskite, this inversion occurs after a series of two siloxane macrorings (every four silicon atoms) or three in the case of sepiolite (every six silicon atoms). This arrangement creates a structural discontinuity in the (a, b) plane, with the same periodicity as the inversion of the apical oxygen. This discontinuity induces the presence of tunnels with a rectangular cross section (3.7 × 10.6 Å for sepiolite and 3.7 × 6.4 Å for palygorskite) extending along the c-axis over micrometric distances. The structural blocks (consisting of an octahedral sheet between two tetrahedral sheets) defining these tunnels are also called polysomes or ribbons (Suárez and García-Romero, 2011). Another notable difference between sepiolite and palygorskite is that sepiolite has quasi-exclusive presence of magnesium in the octahedral sites, whereas palygorskite contains both magnesium and aluminum (Suárez and García-Romero, 2011; Guggenheim and Krekeler, 2011; Galan, 1996). The structure of these clay minerals has been largely discussed in recent reviews (Suárez and García-Romero, 2011; Krekeler and Guggenheim, 2011; Guggenheim, this volume).

To modify these minerals, material chemists take advantage of their chemical and structural properties: their ability to exchange cations, the presence of abundant silanol functions on the external surfaces, the space occupied by water in the tunnels, and the sites occupied by magnesium atoms at the border of the tunnels with a coordination sphere completed by water molecules.

16.2.1   THE CATION EXCHANGE CAPACITY (CEC)

Sepiolite and palygorskite have a low cation exchange capacity (20–30 meq/100 g) in comparison to montmorillonite (70–120 meq/100 g) (Weiss, 1958; Grim, 1968). The CEC is usually calculated based on isomorphic substitutions in the octahedral or tetrahedral sheets, resulting in charge deficits in the layer, which is compensated by cations present in the medium. Several studies have reported the use of sepiolite and palygorskite for the removal of heavy metals by adsorption, using their ability to accumulate cationic chemical species (Sanchez et al., 1999; Brigatti et al., 2000). Some authors clearly showed that heavy metals removal can occur through an isomorphic substitution between Mg (II) located along the tunnel (Figure 16.1) and metal cations in a solution (Kara et al., 2003; Brigatti et al., 2000). This property could also be exploited to modify these clay minerals by organic cations (Sabah et al., 2002; Lemić et al., 2005).

16.2.2   SURFACE SILANOLS

The formation of sepiolite fibers generally induces defects on the exposed atoms when breaking the repetition of the structure. The tetrahedral sheets are most affected through the apical oxygen atoms acting as linkers with a neighboring ribbon. The cleavage of this bond (Si-O-Si) creates charges that could be balanced by protonation or by reaction with hydroxide ions to form a silanol function (Galan, 1996). As a consequence, the outer surfaces of sepiolite or palygorskite needles are paved with silanol groups as shown in Figure 16.1. In an ideal structure, the distance between neighboring surface silanols along the c-axis is about 5 Å (Alvarez, 1984). Hermosín and Cornejo (1986) have determined the surface density of these silanols by reacting these clay minerals with diazomethane. They showed that the density of silanol groups per unit surface area is higher for palygorskite (4.5 Si-OH/100 Å^2) compared to sepiolite (2.2 Si-OH/100 Å^2) (Hermosin and Cornejo, 1986). This density difference results from a difference in the amount of defects in the respective structures (Serna and Vanscoyoc, 1979). These results, however, are not in full agreement with those of other studies (Serna et al., 1977; Ahlrichs et al., 1975). These highly reactive functions are widely used for the functionalization of silicate materials (Yah et al., 2012; Barrientos-Ramírez et al., 2011; Hoffmann et al., 2006; Sayari and Hamoudi, 2001), sepiolite, and palygorskite. They are also active in the formation of hydrogen bonds and are responsible for the hydrophilic nature of sepiolite and palygorskite (Alvarez, 1984).

16.2.3   SEPIOLITE AND PALYGORSKITE TUNNELS

In the natural state, the tunnels are filled by water molecules, which are of two types: (i) coordinated water molecules completing the coordination spheres of Mg(II) present at the border of the tunnels (two water molecules per magnesium) and (ii) zeolitic water molecules weakly retained in the tunnels through a network of hydrogen bonds (see Figure 16.1). Such an environment is ideal for guest compounds capable of forming hydrogen bonds or of interacting with magnesium cations through complexation, provided that their dimensions allow inclusion in the tunnel-confined space.

Structural defects such as the omission of single or multiple ribbons (open-channel defects) (Krekeler and Guggenheim, 2008) or the inclusion of other clay minerals (Guggenheim and Krekeler, 2011) constitute other modification sites, albeit much less abundant and distributed randomly.

16.3   SEPIOLITE AND PALYGORSKITE MODIFICATION STRATEGY

Considering the strength of the interactions involved, sepiolite and palygorskite could be modified either through physical interactions (adsorption of compounds on their external surfaces or in tunnels) or by grafting molecular units through co-valent bonds between the clay minerals surfaces and the modifier (Ruiz-Hitzky et al., 2011).

16.3.1 ADSORPTION ON EXTERNAL SURFACE

Several studies have reported the use of fibrous clay minerals for the removal of various pollutants by adsorption. Generally, heavy metals have been adsorbed (Brigatti et al., 2000; Kara et al., 2003; García-Sánchez et al. 1999); as well as pesticides (Rytwo et al., 2002) and dyes (Doğan et al., 2007; Rytwo et al., 2002; Uurlu, 2009; Santos and Boaventura, 2008). The proposed mechanisms related to these processes are cation exchange and weaker interactions between surface silanols and adsorbates, in the case of anionic and neutral compounds. Using these approaches, it is possible to permanently modify the surface of these minerals with surfactants to obtain materials with new properties for specific applications related to the presence of physisorbed compounds.

16.3.1.1 MODIFICATION BY SURFACTANTS

Cationic surfactants have been used to permanently modify the surfaces of sepiolite and palygorskite. Very few studies have reported the modification of these clay minerals with anionic surfactants. These are characterized by lower adsorption capacities compared to cationic surfactants (Sánchez-Martín et al, 2008; Özdemir et al., 2007, Tunç et al., 2012). A review of the work done on the adsorption of surfactants on sepiolite has been presented by Shuali et al. (2011). The modification methodology is involved in suspending sepiolite or palygorskite in a surfactant solution and in stirring, until the equilibrium has reached. The modified material is obtained after multiple washing, with centrifugation followed by drying (Özcan and Özcan, 2005; Sanchez-Martin et al., 2006). Sabah et al. (2002) has conducted a rigorous study about the adsorption of dodecyltrimethylammonium bromide, hexadecyltrimethylammonium bromide, and dodecylamine in an acidic medium on thermally activated sepiolite. The process is characterized by three main steps: (i) a cation exchange controlled adsorption process at low concentration ranges; (ii) a second step controlled by the interactions between adsorbed surfactant molecules; and (iii) a micelles formation at higher concentrations (Sabah et al., 2002). Other authors have also obtained similar results (Lemić et al., 2005).

These highly hydrophobic materials are successfully used for the adsorption of organic pollutants, mainly pesticides (Sanchez-Martin et al., 2006; Seki and Yurdakoç, 2005; Rodríguez-Cruz, 2008), dyes (Özcan and Özcan, 2005; Armagan et al., 2003; Özcan et al., 2006; Sarkar et al., 2011), and other chemical compounds widely used in chemical industry (Sarkar et al., 2012, 2010). The adsorption of a cationic surfactant on these minerals can induce anion exchange properties; especially when the amount of surfactant adsorbed is higher than the CEC of the clay mineral. Such materials interact strongly with anionic compounds (Armagan et al., 2003; Li et al., 2003). In general, these modifications increase the affinity between the pollutant and the clay mineral with better or equivalent efficiencies to those obtained with commercial activated carbon (Armagan et al., 2003).

On the other hand, the strong interaction between the surfactants and some active compounds allowed control of their release in suitable environments (Galán-Jiménez et al., 2013). Another important application of fibrous minerals modified by surfactants is their good compatibility with polymers. Indeed, the adsorption of surfactants could improve the dispersion of fibers as well as their compatibility with a given polymer matrix, if the surfactant is properly chosen (Ruiz-Hitzky et al., 2011; Verge et al., 2013).

Although not commonly, neutral surfactants have also been used for the modification of these clay minerals (Galán-Jiménez et al., 2013; Tunç et al., 2012). In this case, surface silanols are involved in establishing hydrogen bonds to stabilize the organic compounds on the surface of clay minerals. These composites have also applications in depollution and slow release (Galán-Jiménez et al., 2013).

16.3.1.2   OTHER COMPOUNDS

Interactions between fibrous clay minerals and enzymes are of interest. Enzymes are macro-biomolecules with functionalities that could stabilize them on the surfaces of sepiolite or palygorskite. Indeed, they may be cationic or anionic, and are able to establish multiple hydrogen bonds. When such a composite preserves the biocatalytic activity of the enzyme, it can be reused for several times, with several potential industrial applications. Such composites are common with other clay minerals (Fusi, et al., 1989; Mousty, 2004; Mbouguen et al., 2006). Sepiolite and palygorskite have also showed interesting behavior for the synthesis of this type of composite (Sedaghat et al., 2009; Garcia-Segura et al., 1987; Wicklein et al., 2011; Cabezas et al., 1991; De Fuentes et al., 2001; Cengiz et al., 2012; Luna et al., 2012).

Recently, Ruiz-Hitzky et al. (2009) have successfully used sepiolite as a support for the immobilization of influenza viral particles. The composite was prepared by immobilizing influenza viral particles on a sepiolite coated by a biopolymer (xanthan gum) which acted as a support favoring the adhesion of the viral particles, and also as a biointerface. Preliminary tests have shown that the virus/bionanocomposites are effective low-cost vaccine adjuvants that increase the immune response against influenza viruses with a similar efficiency to standard adjuvants (Ruiz-Hitzky et al., 2009).

The immobilization by physisorption provides low strengths of interaction between the organic compounds and the clay mineral surfaces, which limits the reuse of the composite over time, because of the release of the adsorbed compounds. Under certain conditions, this can be considered as an advantage (polymer composites, drug release ...), but in most cases it is a serious disadvantage (depollution materials, enzymes composites ...). Modifications made by grafting of the adsorbed units therefore become necessary.

16.3.2 MODIFICATION BY GRAFTING

As described above, the external surface of sepiolite or palygorskite is coated by highly reactive silanol groups suitable for the grafting of organic compounds. These surface functional groups are much more abundant on these minerals than on smectites (Hermosín and Cornejo, 1986; Ahlrichs et al., 1975; Sema et al., 1977). In several cases, coupling of organic molecules on sepiolite and palygorskite occurs via organosilanes with the formation of Si-O-Si bonds. Most often, the alkoxysilane (Doğan et al., 2008; Demirbaş et al., 2007; Alkan et al., 2005; Tartaglione et al., 2008; Marjanović et al., 2013) and chlorosilane (Aznar et al., 1992; Alkan et al., 2005) have been used.

Modification methods are usually simple and involve the reaction of organosilane with the clay mineral in a solvent. The solid is recovered by filtration and is washed several times with the synthesis solvent. A step involving the curing at 100–110 C is common in the synthesis process and a grafting process is completed by cross linking the neighboring molecules through Si-O-Si bonds to stabilize the grafted compounds. Aznar et al. (1992) have made an exhaustive study of the grafting mechanism of a series of organosilanes bearing a phenyl sulfonate group.

The formation of Si-O-C bonds by various organic reagents has also been reported (Frost and Mendelovici, 2006; Galan, 1996). One example is the reaction of sepiolite and palygorskite with a diazomethane to yield a methylated composite after a reaction with surface silanols (Hermosín and Cornejo, 1986). After performing several rounds of methylation, we can obtain a conversion rate of almost 100 percent. This was confirmed by the full disappearance of the IR vibration band of the silanol OH group. The low acidity of these minerals prevents the formation of polyethylene (encountered when performing the same reaction on other clay minerals). This reaction was used to quantify the amount of silanol groups on sepiolite and palygorskite (Hermosín and Cornejo, 1986). Fernandez Hernandez and Ruiz-Hitzky (1979) also reported the grafting of isocyanate (alkyl or phenyl isocyanates) on sepiolite through the formation of Si-O-CO-NH-R bonds. However, the silanol function was easily regenerated by breaking the Si-O-C bond in the presence of water.

Strategies to increase the density of silanols have been reported. They involve using HCl before or during the coupling reaction in order to increase the density of grafted compounds (Frost and Mendelovici, 2006; Fernandez Hernandez and Ruiz-Hitzky, 1979; Marjanović et al., 2013).

The functionalities that can be introduced into these minerals (especially through organosilanes) can provide a wide range of applications for the resulting nanohybrid materials:

(i)   These functionalized materials are used in sepiolite/polymer nanocomposites where they substantially increase the mechanical properties of the polymer; although present in small amounts (1–5%). The silane functional group has to be carefully chosen to ensure strong interactions with the polymer matrix. This will facilitate a good dispersion of the modified clay mineral. For example, this is the case for the glycidylsilane - sepiolite adduct. It provides improved performances of the nanocomposite prepared with epoxy polymers, with a substantial improvement in the mechanical properties (Franchini et al., 2009). Similar results, among others, were obtained with amine functionalized sepiolite dispersed in a poly (ε-caprolactone) matrix (Duquesne et al., 2007);

(ii) Functionalized material bearing groups suitable for complexation reactions (amine or thiol) have been used for the adsorption of heavy metals. They provided better performance compared to the unmodified clay minerals (Celis et al., 2000; Marjanović et al., 2013);

(iii) These functionalities were also used for preparing metal nanoparticles immobilized on sepiolite or palygorskite. Indeed, the presence of functional groups significantly increases the amount and the binding strength of nanoparticles precursors, and also their dispersion (Zhu et al., 2009; Letaief et al., 2011a, b).

Following the same idea, functionalizing the mineral surfaces could enhance the stability of an immobilized enzyme through the formation of covalent bonds (Huang et al., 2009).

16.3.3   TUNNELS MODIFICATION

The insertion of organic compounds into the tunnels of fibrous minerals has been the subject of several studies. The nanometric section and the length of this structure are very appealing because numerous potential applications in the field of nanomaterials could be considered. Unfortunately, this steric restriction also limits the number of compounds that can access this confined space and the results are sometimes controversial, because it could be difficult to demonstrate conclusively that the compounds are located in the tunnels and not on the surface, in external channels or in structural defects. However, it was shown that inert gas such as nitrogen or argon could be inserted into the tunnels of sepiolite or palygorskite (Jimenez-Lopez et al., 1978; Grillet et al., 1988). This is also the case of relatively small polar molecules such as ammonia, pyridine, acetone, and short-chain alcohols that can access the tunnels by replacing zeolitic water molecules or by filling the voids created after extraction of zeolitic water by heating or under the action of a dynamic vacuum (Inagaki et al., 1990; Weir et al., 2000; Kuang et al., 2004; Kuang and Detellier,

2004). Indeed, the removal of zeolitic water does not affect the structure of sepiolite tunnels, leaving a large vacant space for the compounds that can be inserted.

16.3.3.1 INSERTION OF SMALL ORGANIC MOLECULES
NH₃ INTERCALATION

A structural study of the composite obtained by insertion of ammonia in the tunnels of palygorskite was conducted by Vanscoyoc et al. (1979) using IR spectroscopy. These authors showed that NH_3 is not only able to replace the zeolitic water, but also the coordinated water to complete the coordination sphere of Mg (II) located in the border of tunnels. Other studies were performed with this compound with similar results (Liua et al., 2011; Serna and Vanscoyoc, 1979). In a review, Alvarez (1984) reported the use of this property in the case of sepiolite for the control of NH_3 in animal farms.

PYRIDINE INTERCALATION

The interaction of pyridine with the tunnel environment of sepiolite and palygorskite has been the subject of several studies (Shuali et al., 1991; Blanco et al., 1988; Kuang et al., 2003). Facey et al. (2005) showed by ^1H, ^{15}N and ^{13}C NMR spectroscopy that pyridine is capable of replacing the coordinated water molecules in the sepiolite tunnel at high temperature (400 °C), forming a stable nanomaterial. These results confirmed those obtained by Ruiz-Hitzky (2001) a few years earlier. The authors also showed that the pyridine molecules have a parallel orientation (to the (b, c) plane) in tunnels.

ACETONE INTERCALATION

Intercalation of acetone in sepiolite tunnels was confirmed (Kuang et al., 2006). A nanohybrid material with a structure similar to the parent sepiolite was obtained through the direct coordination of acetone molecules to the terminal Mg (II). The coordinated acetone molecule stabilizes the tunnel structure which collapsed at a temperature much higher than in the case of the sepiolite mineral (Kuang et al., 2006). These authors had already performed similar work with palygorskite with similar results (Kuang and Detellier, 2004).

    The co-intercalation into the sepiolite tunnels of 3-methyl cyclohex-2-en-1-one, the Douglas-Fir beetle antiaggregation pheromone, with methanol, ethanol, acetone, or benzene has also been described (Blank, 2011). Such systems could potentially be used to control the population of Douglas-Fir beetle in specific area by the slow release of pheromone intercalated in clay minerals tunnels.

ALCOHOLS INTERCALATION

The inclusion of methanol, ethanol, propanol, and butanol in sepiolite or palygorskite (Serna and Vanscoyoc, 1979; Kuang and Detellier, 2005) has also been reported. These alcohols can partially replace the coordinated water molecules. The amount of substitution was reduced when the chain length increased, so it is most probably because of steric hindrance.

16.3.3.2   INSERTION OF LARGER COMPOUNDS
INSERTION OF DYES, THE CASE OF MAYA BLUE

Indigo/sepiolite or indigo/palygorskite composites are probably the most studied nanocomposites in this area (Yacamán et al., 1996; Vandenabeele et al., 2005). They are prepared by thoroughly mixing the clay with small amount of indigo (less than 5%). Heating the mixture at 120–150 °C yields a very stable blue pigment, called Maya blue. The spectacular stability of this pigment is attributed to the clay mineral, which prevents the degradation of the dye. The location of the dye in the clay matrix is still under discussion (Sánchez del Río et al., 2011). Various experimental and computational studies (Fois et al., 2003; Chiari et al., 2003; Giustetto et al., 2005) have confirmed the insertion of the dye in the tunnels of palygorskite or sepiolite forming hydrogen bonds with the coordinated water molecules. Other studies demonstrate that indigo molecules adsorb on sepiolite and palygorskite by enabling the tunnel entrances to have strong interaction through hydrogen bonds. The indigo molecules being stacked at the entrance of the tunnels remain partially emptied (Hubbard et al., 2003).

OTHER COMPOUNDS

The high stability of Maya blue suggests that one could prepare other pigments based on a similar approach. Giustetto and Wahyudi (2011) prepared composites derived from palygorskite and some dyes (methyl red, alizarin, murexide, and Sudan red) and examined their stability in highly acidic, alkaline, or oxidizing conditions). Only the methyl red and alizarin form stable composites. However, only methyl red fits into the tunnels of palygorskite and the composite has a stability comparable to Maya blue. A significant increase in thermal stability and photostability of the composite obtained from methyl red was also observed by the same authors (Giustetto et al., 2012).

The inclusion of many other relatively bulky compounds could be achieved. 2,2-bipyridine was inserted in the tunnels of sepiolite (Sabah and Celik, 2002), methylene blue (Ruiz-Hitzky, 2001), and other dyes of various sizes (Rhodamine 6G, Pyronine Y, Styryl 698 et Styryl 722) (Martínez-Martínez et al., 2011). The insertion of these large compounds is limited by their tendency to form aggregates in solution. Consequently, it was recommended to operate with very dilute solutions and use important contact time to achieve satisfactory results (Martínez-Martínez et al., 2011).

16.4 CONCLUSION

Sepiolite and palygorskite are gifted with physicochemical characteristics suitable for the preparation of nanomaterials. Their prior modification is often necessary for their use in various applications. This involves modifications either by weak interactions (physisorption) or by the formation of chemical bonds (grafting). These modifications provide chemical environments allowing the use of these clay minerals for new applications, and also ensuring their compatibility with other materials such as polymers. The spectacular gain of stability of pigments combined with these clay minerals shows the important potential that exists when, by combining the properties of highly structured inorganic materials, such as these two clay minerals, with those of organic compounds, one can design new nanohybrid materials characterized by synergistic properties from their components.

ACKNOWLEDGMENTS

This work was financially supported by a Discovery Grant of the Natural Sciences and Engineering Research Council of Canada (NSERC). The Canada Foundation for Innovation and the Ontario Research Fund are gratefully acknowledged for infrastructure grants to the Centre for Catalysis Research and Innovation of the University of Ottawa.

KEYWORDS

- Sepiolite fibers
- Cationic surfactants
- Hydroxide ions
- Open-channel defects
- Modification methodology
- Organic compounds
- Physisorption

REFERENCES

1. Ahlrichs, J. L.; Serna, C.; and Serratosa, J. M.; *Clays. Clay. Min.* **1975**, *23*, 119–124.
2. Alkan, M.; Çelikçcapa, S.; Demirbaş, Ö.; and Dogan, M.; *Dyes. Pig.* **2005**, *65*, 251–259
3. Alkan, M.; Tekin, G.; and Namli, H.; *Micropor. Mesopor. Mater.* **2005**, *84*, 75–83.
4. Alvarez, A.; Sepiolite: properties and uses. In: Developments in Sedimentology, Palygor-skite—Sepiolite: Occurrences, Genesis and Uses. Eds. Singer, A.; Galan, E. **1984**, *37*, 253–287.
5. Aranda, P.; Kun, R.; Martín-Luengo, M. A.; Letaïef, S.; Dékány, I.; Ruiz-Hitzky, E.; *Chem. Mater.* **2008**, *20*, 84–91.
6. Armagan, B.; Ozdemir, O.; Turan, M.; and Çelik, M. S.; *J. Environ. Eng.* **2003**, *129*, 709–715.
7. Aznar, A. J.; Sanz, J.; and Ruiz-Hitzky, E.; *Colloid. Poly. Sci.* **1992**, *270*, 165–176.
8. Bakhtiary, S.; Shirvani, M.; and Shariatmadari, H.; *Micropor. Mesopor. Mat.* **2013**, *168*, 30–36.
9. Barrientos-Ramírez, S.; Montes de Oca-Ramírez, G.; Ramos-Fernández, E. V.; Sepúlveda-Escribano, A.; Pastor-Blas, M. M.; and González-Montiel, A.; *Appl. Catal. A Gen.* **2011**, *406*, 22–33.
9. Blanco, C.; Herrero, J.; Mendioroz, S.; and Pajare, J. A.; *Clays Clay Min.* **1988**, *36*, 364–368.
10. Blank, K.; Incorporation of organic molecules in the tunnels of the sepiolite clay mineral M.Sc. Thesis, University of Ottawa, 2011.
11. Brigatti, M. F.; Lugli, C.; and Poppi, L. *Appl. Clay. Sci.* **2000**, *16*, 45–57.
12. Cabezas, M. J.; Salvador, D.; and Sinisterra, J. V. J.; *Chem. Technol. Biotechn.* **1991**, *52*, 265–274.
13. Celis, R.; Carmen Hermosín, M.; and Cornejo, J.; *Environ. Sci. Technol.* **2000**, *34*, 4593–4599.
14. Cengiz, S.; Çavaş, L.; and Yurdakoç, K.; *Appl. Clay. Sci.* **2012**, *65–66*, 114–120.
15. Chiari, G.; Giustetto, R.; and Ricchiardi, G.; *Eur. J. Min.* **2003**, *15*, 21–33.
16. De Fuentes, I. E.; Viseras, C. A.; Ubiali, D.; Terreni, M.; and Alcántara, A. R. J. *Mol. Catal. B Enzym.* **2001**, *11*, 657–663.
17. Demirbaş, O.; Alkan, M.; Doğan, M.; Turhan, Y.; Namli, H.; and Turan, P. J. *Hazard. Mater.* **2007**, *149*, 650–656.
18. Doğan, M.; Özdemir, Y.; and Alkan, M.; *Dyes. Pigment.* **2007**, *75*, 701–713.
19. Doğan, M.; Turhan, Y.; Alkan, M.; Namli, H.; Turan, P.; and Demirbaş, O.; *Desalination.* **2008**, *230*, 248–268.
20. Duquesne, E.; Moins, S.; Alexandre, M.; and Dubois, P.; *Macromol. Chem. Phys.* **2007**, *208*, 2542–2550.

21. Facey, G. A.; Kuang, W.; and Detellier, C.; *J. Phys. Chem. B*. **2005**, *109*, 22359–22365.
22. Fernandez Hernandez, M. N.; and Ruiz-Hitzky, E;. *Clay. Clay Min.*. **1979**, *14*, 295–305.
23. Fois, E.; Gamba, A.; and Tilocca, A.; *Micropor. Mesopor. Mater*. **2003**, *57*, 263 272.
24. Franchini, E.; Galy, J.; and Gérard, J. F.; *J. Colloid. Interf. Sci*. **2009**, *329*, 38–47.
25. Frost, R. L.; and Mendelovici, E.; *J. Colloid. Interf. Sci*. **2006**, *294*, 47–52.
26. Fusi, P.; Ristori, G. G.; Calamai, L.; and Stotzky, G.; *Soil. Biol. Biochem*. **1989**, *21*, 911–920.
27. Galan, E.; *Clay Min*. **1996**, *31*, 443 453.
28. Galán-Jiménez, M. C.; Mishael, Y. -G.; Nir, S.; Morillo, E.; and Undabeytia, T.; *PLoS ONE*. **2013**, *8*(3), e59060. doi:10.1371/journal.pone.0059060.
29. Garcia-Segura, J. M.; Cid, C.; de llano, J. M.; and Gavilanes, J. G.; *Brit. Polym. J*. **1987**, *19*, 517–522.
30. Giustetto, R.; et al. *Phys. Chem. B*. **2005**, *109*, 19360–19368.
31. Giustetto, R.; Seenivasan, K.; Pellerej, D.; Ricchiardi,G.; and Bordiga, S.; *Micropor. Mesopor. Mat*. **2012**, *155*, 167–176.
32. Giustetto, R.; and Wahyudi, O.; *Micropor. Mesopor. Mat*. **2011**, *142*, 221–235.
33. Grillet, Y.; Cases, J. M.; Francois, M.; Rouquerol, J.; and Poirier, J. E.; *Clays. Clay. Min*. **1988**, *36*, 233–242.
34. Grim, R. E.; Clay Mineralogy; 2nd edition. McGraw-Hill: New York, **1968**.
35. Guggenheim, S.; and Krekeler, M. P. S.; The Structures and Microtextures of the Palygorskite–Sepiolite Group Minerals. In Developments in Clay Science, (Developments in Palygorskite-Sepiolite Research); Eds. Galàn, E.; and Singer, A..; **2011**, *3*, 3–32.
36. Guggenheim, S.; *This volume*. Phyllosilicates used as nanotube substrates in engineered materials: Structures, chemistries, and textures. In: Natural Mineral Nanotubes:Properties and Applications; Eds. Pasbakhsh, P.; Churchman, G. J. **2014**, xxx–yyy.
37. Hermosín, M. C.; and Cornejo, J. *Clays. Clay. Min*. **1986**, *34*, 591–596.
38. Hoffmann, F.; Cornelius, M.; Morell, J.; and Fröba, M.; *Angew. Chem. Int. Edit*. **2006**, *45*, 3216–3251.
39. Huang, J.; Liu, Y.; and Wang, X. J.; *Mol. Catal. B Enzyme*. **2009**, *57*, 10–15.
40. Hubbard, B.; Kuang, W.; Moser, A.; Facey, G. A.; and Detellier, C.; *Clays. Clay. Min*. **2003**, *51*, 318–326.
41. Inagaki, S.; Fukushima, Y.; Doi, H.; and Kamigaito, O.; *Clay. Min*. **1990**, *25*, 99–105.
42. Jimenez-Lopez, A.; et al. *Clay. Min*. **1978**, *13*, 375–385.
43. Kara, M.; Yuzer, H.; Sabah, E.; and Celik, M. S.; *Water Res*. **2003**, *37*, 224–232.
44. Katchalski-Katzir, E.; *Trends. Biotechnol*. **1993**, *11*, 471–478.
45. Krekeler, M. P. S.; and Guggenheim, S.; *Appl. Clay. Sci*. **2008**, *39*, 98–105.
46. Kuang, W.; and Detellier, C.; *Can. J. Chem*. **2004**, *82*, 1527 1535.
47. Kuang, W.; and Detellier, C.; Structuration of organo-minerals: nanohybrid materials resulting from the incorporation of alcohols in the tunnels of palygorskite. In: Proceedings of the 4th Conference on Access in Nanoporous Materials, Studies in Surface Science and Catalysis, **2005**, *156*, 451–456.
48. Kuang, W.; Facey, G. A.; and Detellier, C.; *Clays. Clay. Min*. **2004**, *52*, 635 642.
49. Kuang, W.; Facey, G. A.; and Detellier, C. J.; *Mater. Chem.*, **2006**, *16*, 179–185.
50. Kuang, W.; Facey, G. A.; Detellier, C.; Casal, B.; Serratosa, J. M.; and Ruiz-Hitzky, E. *Chem. Mater*. **2003**, *15*, 4956 4967.
51. Lemić, J.; Tomašević Čanović, M.; Djuričić, M.; and Stanić, T. *J. Colloid. Interf. Sci*. **2005**, *292*, 11–19.
52. Letaief, S.; Grant, S.; and Detellier, C.; *Appl. Clay Sci*. **2011**, *53*, 236–243.
53. Letaief, S.; Liu, Y.; and Detellier, C.; *Can. J. Chem*. **2011**, *89*, 280–288.
54. Li, Z.; Willms, C. A.; and Kniola, K.; *Clays. Clay. Min*. **2003**, *51*, 445–451.

55. Liua, H.; Lia, J.; Chena, D.; Changa, D.; Konga, D.; and Frost, R. L.; *Chem. Eng. J.* **2011**, *166*, 1017–1021.
56. Luna, D.; et al. *J. Int. J. Mol. Sci.* **2012**, *13*, 10091–10112.
57. Martínez-Martínez, V.; Corcóstegui, C.; Prieto, J. B.; Gartzia, L.; Salleresa S.; and Arbeloa, I. L. J. *Mater. Chem.* **2011**, *21*, 269–276.
58. Marjanović, V.; Lazarević, S.; Janković Častvan, I.; Jokić, B.; Janaćković, Dj.; and Petrović, R.; *Appl. Clay Sci.* **2013**, *80–81*, 202–210.
59. Mbouguen, J. K.; Ngameni, E.; and Walcarius, A.; *Anal. Chim. Acta.* **2006**, *578*, 145–55.
60. Mousty, C. *Appl. Clay. Sci.* **2004**, *27*, 159–177.
61. Özcan, A.; Öncü, E. M.; and Özcan, A. S.; *J. Hazard. Mater.* **2006**, *129*, 244–252.
62. Özcan, A.; and Özcan, A. S.; *J. Hazard. Mater.* **2005**, *125*, 252–259.
63. Özdemir, O.; Çinar, M.; Sabah, E.; Arslan, F.; and Çelik, M. S. J.; *Hazard. Mater.* **2007**, *147*, 625–632.
64. Rodríguez-Cruz, M. S.; Andrades, M. S.; and Sánchez-Martín, M. J. *J. Hazard. Mater.* **2008**, *160*, 200–207.
65. Ruiz-Hitzky, E.; *J. Mater. Chem.* **2001**, *11*, 86–91.
66. Ruiz-Hitzky, E.; Aranda, P.; Álvarez, A.; Santarén, J.; and Esteban-Cubillo, A.; Advanced Materials and New Applications of Sepiolite and Palygorskite. Developments in Clay Science, (Developments in Palygorskite-Sepiolite Research); Ed. Galàn, E.; and Singer, A. **2011**, *3*, 393–452.
67. Ruiz-Hitzky, E.; Darder, M.; Aranda, P.; Martin del Burgo, M. A.; and del Real, G. *Adv. Mater.* **2009**, *21*, 4167–4171.
68. Rytwo, G.; Tropp, D.; and Serban, C.; *Appl. Clay Sci.* **2002**, *20*, 273–282.
69. Sabah, E.; and Çelik, M. S.; *J. Colloid. Interf. Sci.* **2002**, *251*, 33–38.
70. Sabah, E.; Turan, M.; and Celik, M. S.; *Water Res.* **2002**, *36*, 3957–3964.
71. Sánchez del Río, M.; Doménech, A.; Doménech-Carbó, M. T.; Pascual, M. L. V. A.; Suárez, M.; and García-Romero, E.; The maya blue pigment. In: Developments in Clay Science, (Developments in Palygorskite-Sepiolite Research); Eds. Galàn, E.;and Singer, A.. **2011**, 3, 453–481.
72. Sanchez, A. G.; Ayuso, E. A.; and De Blas, O. J. *Clay Min.* **1999**, *34*, 469–477.
73. Sánchez-Martín, M. J.; Dorado, M. C.; del Hoyo, C.; and Rodríguez-Cruz, M. S.; *J. Hazard. Mater.* **2008**, *150*, 115–123.
74. Sanchez-Martin, M. J.; Rodriguez-Cruz, M. S.; Andrades, M. S.; and Sanchez-Camazano, M.; *Appl. Clay Sci.* **2006**, *31*, 216–228.
75. Santos, S. C. R.; and Boaventura, R. A. R. *Appl. Clay Sci.* **2008**, *42*, 137–145.
76. Sarkar, B.; Megharaj, M.; Xi, Y.; and Naidu, R.; *Chem. Eng. J.* **2012**, 185–186, 35–43.
77. Sarkar, B.; Xi, Y.; Megharaj, M.; Krishnamurti, G. S. R.; and Naidu R.; *J. Colloid. Interf. Sci.* **2010**, *350*, 295–304.
78. Sarkar, B.; Xi, Y.; Megharaj, M.; and Naidu R.; *J. Appl. Clay Sci.* **2011**, *51*, 370–374.
79. Sayari, A.; and Hamoudi, S.; *Chem. Mater.* **2001**, *13*, 3151–3168.
80. Sedaghat, M. E.; Ghiaci, M.; Aghaei, H.; and Soleimanian-Zad, S. *Appl. Clay Sci.* **2009**, *46*, 131–135.
81. Seki, Y.; and Yurdakoç, K. J. *Colloid. Interf. Sci.* **2005**, *287*, 1–5.
82. Serna, C. J.; and Vanscoyoc, G. E.; Infrared study of sepiolite and palygorskite surfaces. In: Proceedings of the International Clay Conference, Oxford; Eds. Mortland, M. M.; Farmer, V. C., Amsterdam: Elsevier 1979; 197–206.
83. Serna, C.; Vanscoyoc, G. E.; and Ahlrichs, J. L.; *Am. Min.* **1977**, *62*, 784–792.
84. Shuali, U.; Nir, S.; and Rytwo, G.; Adsorption of Surfactants, Dyes and Cationic Herbicides on Sepiolite and Palygorskite: Modifications, Applications and Modelling. In: Developments

in Clay Science, (Developments in Palygorskite-Sepiolite Research); Eds. Galàn, E.; Singer, A.; **2011**, *3*, 351–374.

85. Shuali, U.; Yariv, S.; Steinberg, M.; Muller-Vonmoos, M.; Kahr, G.; Rub, A.; *Clay Min.* **1991**, *26*, 497–506.

86. Suárez, M.; and García-Romero, E.; Advances in the crystal chemistry of sepiolite and palygorskite. In : Developments in Clay Science, (Developments in Palygorskite-Sepiolite Research); Eds. Galàn, E.; Singer, A. **2011**, *3*, 33 65.

87. Tartaglione, G.; Tabuani, D.; and Camino, G.; *Micropor. Mesopor. Mat.* **2008**, *107*, 161–168.

88. Tunç, S.; Duman, O.; and Kanci, B.; *Colloid. Surface. A.* **2012**, *398*, 37–47.

89. Uurlu, M;. *Micropor. Mesopor. Mat.* **2009**, *119*, 276–283.

90. Vandenabeele, P.; Bodé, S.; Alonso, A.; and Moens, L.; *Spectrochim. Acta A.* **2005**, *61*, 2349–2356.

91. Vanscoyoc, G. E.; Serna, C. J.; and Ahlrichs, J. L.; *Am. Min.* **1979**, *64*, 215–223.

92. Verge, P.; Fouquet, T.; Barrére, C.; Toniazzo, V.; Ruch, D.; and Bomfim, J. A.; *Compos. Sci. Technol.* **2013**, *79*, 126–132.

93. Weir, M. R.; Facey, G. A.; and Detellier, C.; *Stud. Surf. Sci. Catal.* **2000**, *129*, 551 558.

94. Weiss, A. Z. *Anorg. Allg. Chem.* **1958**, *297*, 257–286.

95. Wicklein, B.; Darder, M.; Aranda, P.; and Ruiz-Hitzky, E.; *Appl. Mater. Interf.* **2011**, *3*, 4339–4348.

96. Yacamán, M. J.; Rendon, L.; and Arenas, J.; Puche, M. C. S.; *Science*`**1996**, *273*, 223 225.

97. Yah, W. O.; Takahara, A.; and Lvov, Y. M. J.; *Am. Chem. Soc.* **2012**, *134*, 1853 1859.

98. Zhu, L.; Letaief, S.; Liu, Y.; Gervais, F.; and Detellier, C.; *Appl. Clay. Sci.* **2009**, *43*, 439–446.

CHAPTER 17

ORGANOPALYGORSKITES PREPARED FROM QUATERNARY AMMONIUM COMPOUNDS AND THEIR ENVIRONMENTAL USES

BINOY SARKAR and RAVI NAIDU

CONTENTS

17. 1   INTRODUCTION

Clay minerals are abundant in nature and have many industrial uses, such as in the ceramics, cement, paper, cosmetics, print, and drug industries. Clays are also extensively used in environmental remediation due to unique properties, such as high surface area, strong chemical stability, non-toxic nature, and the adsorptive and ion exchange properties (Churchman et al., 2006). Clays are generally produced by mining but can be increased in value by surface modification. Naturally occurring clay minerals are intrinsically hydrophilic in nature. As a result, clays have a good affinity for ionic contaminants, such as heavy metal cations, but do not significantly interact with hydrophobic organic contaminants. Clay minerals' surface modification with organic compounds, such as quaternary ammonium compounds (QACs) can produce modified clays with a high affinity for organic contaminants. Modified clay minerals thus prepared are known as organoclays (Boyd et al., 1988; Jordan and Williams, 1954; Sarkar et al., 2012c; Xi et al., 2005b). Depending on the type of organic compounds used for modification, organoclays can act as the adsorption sink for both hydrophobic organic contaminants and ionic metals and metalloids (Sarkar et al., 2012a, b, c).

For organoclay preparation, 2:1 type swelling clay minerals (e.g., smectite group) have been the primary choice for many researchers. These clay minerals have many interlayer hydrated cations which can be replaced by QAC cations in an easy exchange reaction. The chemical structure of such QACs is shown in Figure 17.1. In addition, the large surface area and fine particle size of thin smectite flakes make a good candidate for organic modification.

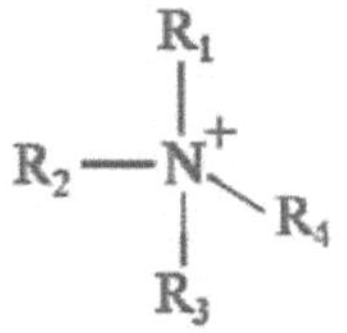

$$R_2 - \overset{\overset{\textstyle R_1}{|}}{\underset{\underset{\textstyle R_3}{|}}{N^+}} \diagdown R_4$$

FIGURE 17.1   Schematic chemical structure of a quaternary ammonium compound. The R groups may be the same or different alkyl or aryl groups of different C-chain lengths.

Palygorskite is also an important member of 2:1 type clay minerals. It is also known as attapulgite (Figure 17.2). It has a tubular particle shape, fine particle size, and internal channels, which create a very high surface area ($300 - 600$ m^2 g^{-1}) (Murray 2000). The tubular or fibrous shape is made by periodic reversal of the building blocks (SiO_4 tetrahedra) in a way that creates interlayer channels. The SiO_4 tetrahedra linkages at the edges of channels limit the expandability of palygorskite. Palygorskites have a moderately high structural charge due to considerable Al^{3+} substitution for Mg^{2+} and Fe^{2+} in the octahedral sheet (Sarkar et al., 2010b). The cation exchange capacity (CEC) value is lower than the smectite group minerals (~15 – 30 cmol (p$^+$) kg^{-1} compared to 100 cmol (p$^+$) kg^{-1} for smectites). The presence of silanol groups (Si-OH) on palygorskite external surfaces also enables interactions with cations through hydrogen bonding. By virtue of the charge and moderate CEC, palygorkite can also interact strongly with QACs, which can yield new classes of organopalygorskites.

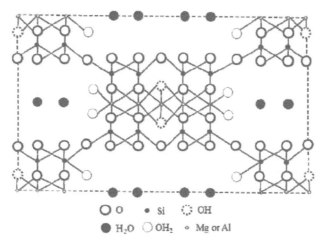

O O   • Si   ◌ OH
● H$_2$O   ◯ OH$_2$ ∘ Mg or Al

FIGURE 17.2   Schematic structure of palygorskite.

Palygorskite applications in environmental remediation have not been studied as extensively as that for smectitic clay minerals, but it can be an effective adsorbent for many pollutants (Chang et al., 2009a; Chen et al., 2007; Potgieter et al., 2006; Shirvani et al., 2006; Shuali et al., 2011; Zhang et al., 2009). Palygorskite-based adsorbents have an advantage over smectite or bentonite-based materials in environmental applications because of the greater permeability of palygorskites under natural flow conditions (Sarkar et al., 2012a, b, 2011b). In this chapter, we summarise our experience in modifying palygorskite with QACs, characterizing the modified materials, and successfully applying the modified clays to remediate diverse pollutants in contaminated waters and soils.

17.2   PREPARATION OF ORGANOPALYGORSKITES

Organopalygorskites can be prepared by the simple steps shown in Figure 17.3 (Sarkar et al., 2012a, b; Sarkar et al., 2011b; Xi et al., 2010). A mild acid pretreatment can expose greater numbers of silanol (Si-OH) groups on the surface of palygorskite and help in greater uptake of the added QAC. Understandably, less QAC is required to prepare a 100 percent CEC equivalent organopalygorskite than for an organobentonite with a similar level of QAC loading because palygorskites have a lower CEC. Usually, a pre-weighed amount of QAC (also commonly known as surfactant) is dissolved in deionised water heated at 70–80°C. Some of the QACs may have limited solubility in water. So, care should be taken so that the added QAC is properly dissolved in the solution before adding the clay mineral. The required mass of palygorskite is dispersed in the surfactant solution keeping the water/palygorskite mass ratio (v/w) at around 10–20. The mixture is stirred for 5–8 h at elevated temperature (70–80°C) to avoid spume (i.e. foam or scum) formation. Excess spume can inhibit uniform contact between the QAC molecules and the palygorskite surface. The palygorskite/QAC exchange reaction can be enhanced by sonication (Huang et al., 2007, 2008). Following adsorption of the QAC onto the palygorskite, the solid phase is separated from the liquid by centrifugation (at around 4,000 rpm for 10–15 min) and washed with deionized water for several times to make it free of halide anions (QACs are commercially available as halide salts). The washing step also ensures the removal of superficially attached QAC molecules from the organopalygorskite surface. The adsorbents are dried under mild conditions (60°C) and crushed into powder for storage. Several other methods using microwave radiation or plasma treatment may improve QAC adsorption to palygorskite, but would involve higher production costs.

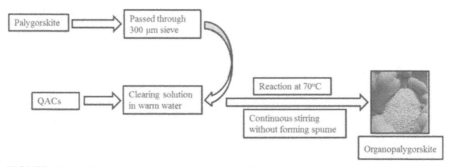

FIGURE 17.3   Steps to prepare organopalygorskite using QAC.

17.3   CHARACTERIZATION OF ORGANOPALYGORSKITES

Many techniques used in material science have been used to examine organopalygorskite preparation and structure. These include X-ray Diffraction (XRD), Fourier Transform Infrared (FTIR) spectroscopy, Thermogravimetric Analysis (TGA), Scanning Electron Microscopy (SEM), surface area analysis, and pore-size distribution. In addition, the surface charge behavior of this modified palygorskite materials that largely contribute to their affinity with environmental contaminants have also been studied.

17.3.1 X-RAY DIFFRACTION (XRD)

XRD is a useful tool for characterising organoclays. For organoclays prepared from swelling-type clay minerals, e.g. montmorillonite, the organic cation enters into clay mineral interlayers which can increase the d-spacing (001 plane). However, as palygorskite is a non-expandable type of clay mineral, normal XRD investigation reveals no apparent increase in the basal spacing (Figure 17.4). The interaction of QACs with palygorskite is largely confined to the clay mineral external surfaces and does not significantly change the structure (Chen and Zhao, 2009; Das et al., In press; Huang et al., 2007; Sarkar et al., 2011b; Xi et al., 2010). As palygorskite has a significant CEC (~15–30 cmol (p$^+$) kg^{-1}), it is possible that some QAC molecules exchange with the hydrated cations and enter the interlayer. However, to investigate this with greater precision, synchrotron-based powder diffraction techniques might be employed. Recently, there has been great interest in using this technique to characterize materials used in environmental remediation (Guggenheim and Krekeler, 2011; Post and Heaney, 2008; Sánchez del Río et al., 2009).

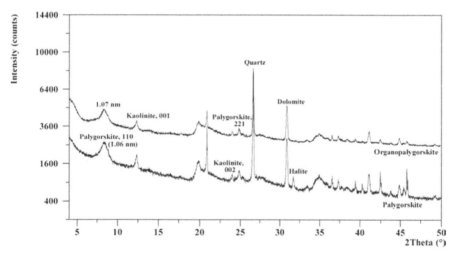

FIGURE 17.4 XRD patterns of palygorskite and organopalygorskite prepared with dimethyldioctadecylammonium bromide (DMDOA). CEC of the palygorskite was 17.0 cmol (p$^+$) kg^{-1}. DMDOA loading was equivalent to 200 percent of the CEC value.

17.3.2 FOURIER TRANSFORM INFRARED (FTIR) SPECTROSCOPY

Infrared spectroscopy is a very useful technique to examine the presence and orientation of organic QAC molecules on clay mineral surfaces (Sarkar et al., 2011a, 2012a, In press, 2010b, Xi et al., 2005a). The unmodified palygorskite usually contains a negligible quantity of organic material. Following modification, the QAC molecules attach to the palygorskite surfaces. As a result, the modified palygorskite provides signature IR bands because of the presence of organics (Figure 17.5). These bands are represented by symmetric (V_s) and asymmetric (V_{as}) stretching vibrations of the "–CH_2" bonds present in the alkyl chain of QAC. The positions of both the V_s (~ 2850) and V_{as} (~2920) bands shift towards lower frequency with the increase in QAC (DMDOA in this case) loading (Figure 17.5). At 200 percent CEC (CEC of palygorskite 17 cmol (p^+) kg^{-1}) equivalent loading the organopalygorskite exhibits band positions like pure DMDOA. This indicates formation of an ordered conformation of QAC molecules on the palygorskite surface and represents a solid-like surfactant environment (Sarkar et al. 2011a, 2012a; Xi et al. 2005a). As a result, a homogeneous chemical environment is produced on the organopalygorskite surface through which contaminant adsorption is enhanced (Sarkar et al., 2012a, 2011b).

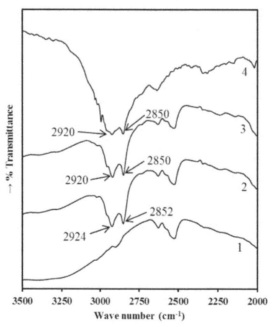

FIGURE 17.5 FTIR spectra of palygorskite and organopalygorskites in 3,500–2,000 cm^{-1} wave number regions. Spectra 1, 2, 3, and 4 represents unmodified palygorskite, organopalygorskite prepared with 100 percent CEC equivalent DMDOA, organopalygorskite prepared with 200 percent CEC equivalent DMDOA and pure DMDOA, respectively (CEC of palygorskite 17 cmol (p^+) kg^{-1}) (Modified from Sarkar et al., 2012a).

17.3.3   THERMOGRAVIMETRIC ANALYSIS (TGA)

Thermogravimetric analysis (TGA) has been successfully used to investigate the interaction of QAC molecules with clay minerals. In this technique, the changes in physical and chemical properties of a material are measured as a function of rising temperature and/or time. Thus, this technique provides information on the arrangement and conformation of the organic molecules within the clay mineral structure following organic modification.

The types of interaction between palygorskite and the organic QAC molecules could be distinguished by comparing the thermogravimetric (TG) and differential thermogravimetric (DTG) weight loss patterns of unmodified and organically modified palygorskites. As depicted in Figure 17.6, an unmodified palygorskite showed three major weight loss peaks during heat treatment. These peaks appear at around 64°C due to the removal of superficially adsorbed water, at 200–209°C because of the dehydration of tightly held zeolite water from the palygorskite structure and at 550–760°C because of the dehydroxylation and structural breakdown of palygorskite, respectively (Chang et al., 2009a; Guggenheim and van Groos, 2001; Sarkar et al., 2010b). A broad peak due to the dehydroxylation process indicated overlapping two-step changes, where the first step involves OH removal from octahedral Fe and Al at comparatively low temperatures followed by OH removal from Mg at higher temperature (Frost and Ding, 2003; Sarkar et al., 2010b). Unlike the unmodified palygorskite, the weight loss peaks because of the loss of superficially adsorbed water in organopalygorskites had much lower intensity and occurred at slightly higher temperature (78°C) (Figure 17.6b, c). This indicates that modifying the palygorskite with DMDOA gives rise to hydrophobic products. The peaks at temperatures between 315 and 507°C are due to the decomposition of adsorbed QAC molecules (Figure 17.6b, c). This involves two distinct phases which represents the relative affinity of QAC molecules for the palygorskite surface. The QAC molecules which bind to exchange sites, silanol groups, and aluminol groups exposed in external channels along palygorkite fiberrs decompose at comparatively higher temperature, whereas those bound to the neighbouring organic molecules decompose at relatively lower temperatures (Sarkar et al., 2010b, 2010c; Xi et al., 2004). The intensity of the peaks, which appear because of the decomposition of organic molecules, is higher in the organopalygorskite prepared with 200 percent CEC equivalent QAC (DMDOA in this case; CEC of palygorskite 17 cmol (p$^+$) kg^{-1}) than the product prepared with 100 percent CEC equivalent QAC. Finally, the peaks appear at around 750°C because of the dehydroxylation step in which the palygorskite structure collapses completely. The intensity of this peak in organopalygorskite is significantly greater than unmodified palygorskite (Sarkar et al., 2010b). This could be attributed to the overlapping of the dehydroxylation step with the decomposition of the remaining portion of QACs attached to the internal surface of palygorskite (Sarkar et al., 2010b). This also confirms a very strong interaction between QAC molecules and palygorskite surface in an organopalygorskite.

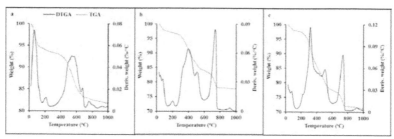

FIGURE 17.6   Thermogravimetric (TG) and differential thermogravimetric (DTG) weight loss patterns of (a) unmodified palygorskite, (b) organopalygorskite prepared with 100 percent CEC equivalent DMDOA, and (c) organopalygorskite prepared with 200 percent CEC equivalent DMDOA (CEC of palygorskite 17 cmol (p$^+$) kg^{-1}) (Modified from Sarkar et al., 2010b).

17.3.4   SCANNING ELECTRON MICROSCOPY (SEM)

Scanning electron microscopy has been a useful tool to investigate morphological changes in palygorskite due to the modification with QACs. Palygorskite fiberrs in organopalygorskites gradually become less entangled with increased QAC loading (Figure 17.7). Palygorskite fibers are usually flat, randomly oriented, and entangled in bundles. Because of attachment of the QAC molecules on the surface of the tubules, the individual fibers become more prevalent. As a result, the number of the cluster of fibers gradually diminishes. The SEM micrographs indicate fragmentation of the clustered fibers into individual tubules in the organopalygorskites which in turn might influence the interaction of contaminant ions or molecules on the surfaces (Sarkar et al., 2010b; Xi et al., 2010). However, greater water affinity of palygorskite might also contribute to morphology changes of the adsorbents.

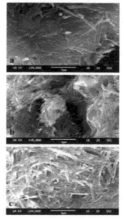

FIGURE 17.7   SEM images of (a) unmodified palygorskite, (b) organopalygorskite prepared with 100 percent CEC equivalent DMDOA, and (c) organopalygorskite prepared with 200 percent CEC equivalent DMDOA (CEC of palygorskite 17 cmol (p$^+$) kg^{-1}) (Modified from Sarkar et al., 2010a).

17.3.5   SURFACE CHARGE (ZETA POTENTIAL) BEHAVIOUR

Adsorbent material's charge behavior can affect environmental performance. Adsorbent charge behavior can be determined from zeta potential values measured of the material suspended in water. Like other clay minerals, palygorskite also contains intrinsic negative charge sites on surfaces indicated by a zeta potential of about –23 mV (Sarkar et al., 2012a, 2010b). Because palygorskite has a considerable number of surface variable charge sites, negative charge can vary with the pH of clay mineral suspensions (Naidu et al., 1994). In a pH range from 3 to 11, an Australian palygorskite had zeta potential values of –4 to –23 mV (Figure 17.8). However, because of palygorskite modification with QAC, these values changed. The degree of change solely depended on the QAC loading rates. For example, an organopalygorskite prepared with 100 percent CEC equivalent DMDOA (CEC of palygorskite 17 cmol (p^+) kg^{-1}) had zeta potential values from +3 to –16 mV at pH values from 3 to 11 (Figure 17.8). A DMDOA loading rate equivalent to 200 percent of the palygorskite CEC resulted in a zeta potential from +25 to +9 mV in the same pH range (Figure 17.8). The positive surface charge of organopalygorskites is created by DMDOA molecules adsorbed in excess of the palygorskite CEC (Bate and Burns, 2010; Sarkar et al., 2012a, 2012b). In this case, the QAC molecules totally cover internal and external palygorskite surfaces. The positive or less negative zeta potential values with increasing QAC concentrations (total organic carbon content) are attributed to the increased hydrophobic lateral interactions between QACs adding positive charge within the shear plane (Bate and Burns, 2010; Sarkar et al., 2011a, 2012a; Zadaka et al., 2010). Positive surface charge increases with increased QAC molecule carbon chain length used to prepare the organopalygorskite (Bate and Burns, 2010; Sarkar et al., 2012a).

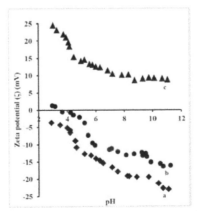

FIGURE 17.8 Surface charge behaviour of (a) unmodified palygorskite, (b) organopalygorskite prepared with 100 percent CEC equivalent DMDOA, and (c) organopalygorskite prepared with 200 percent CEC equivalent DMDOA (CEC of palygorskite 17 cmol (p^+) kg^{-1}) (Modified from Sarkar et al., 2012a).

17.3.6 SURFACE AREA AND PORE SIZE CHARACTERISTICS

The surface area and pore size characteristics of organoclay minerals are usually measured through N_2 adsorption and desorption experiments. An organopalygorskite can exhibit much lower BET (Brunauer–Emmett–Teller) surface area than its unmodified form. This is attributed to the pore/tunnel blocking effect by which the QAC molecules prevent easy penetration of N_2 into the organopalygorskite structure (Sarkar et al., 2011b; Wang et al., 2004). Organopalygorskite surface areas decrease as QAC loading is increased (Table 17.1). The larger the QAC molecule carbon chain size, the smaller the surface area (Table 17.1). It might also depend on the surfactant ion arrangement. Attachment of QAC molecules to clay mineral surfaces forms bi-dimensional porous networks of mesopores (Sarkar et al., 2011b; Wang et al., 2004). As a result, the palygorskite t-plot micropore area ($18.5\ m^2\,g^{-1}$) disappears after QAC modification (Sarkar et al. 2011b). The creation of larger pores is also reflected by the organopalygorskite average pore widths (Table 17.1). However, the organopalygorskite cumulative pore volumes gradually decrease because of pore-blocking effects (Table 1). It is interesting to note that contaminant adsorption by organopalygorskites does not depend on adsorbent surface area or pore geometry, but depends mainly on the homogeneous chemical environment of the modified palygorskite surface (Sarkar et al., 2011b).

TABLE 17.1 Surface area and pore characteristics of palygorskite and organopalygorskites (CEC of palygorskite 17 cmol (p^+) kg^{-1}) (Adapted from Sarkar et al., 2011b).

Clay Product	QAC	Loading Rate	Surface Area ($m^2\,g^{-1}$)	Pore Width ($\mathring{A}$)	Pore Volume ($cm^3\,g^{-1}$)
Palygorskite	-	-	97	122	0.26
Organopalygor-skite 1	Octadecyl trimethylam-monium bromide (ODTMA; C16)	100% CEC	44	159	0.27
Organopalygor-skite 2	Octadecyl trimethylam-monium bromide (ODTMA; C16)	200% CEC	26	180	0.21
Organopalygor-skite 3	Dioctadecyl dimethylam-monium bromide (DMDOA)	100% CEC	33	157	0.23
Organopalygor-skite 4	Dioctadecyl dimethylam-monium bromide (DMDOA)	200% CEC	24	167	0.18

17.4 ENVIRONMENTAL USE OF ORGANOPALYGORSKITES

Unlike QAC-modified smectite, montmorillonite, and bentonite, organopalygorskite's environmental application is less wide spread. However, this group of modified clay minerals have tremendous scope for application in cleaning up environmental contaminants (Table 17.2). The primary organopalygorskite contaminant adsorption mechanisms are hydrophobic interaction, electrostatic attraction, anion exchange, van der Waals interaction, and hydrogen bonding (Table 17.2). The contaminant adsorption capacity largely depends on QAC molecule type, QAC loading rates, and pollutant molecule chemical nature (Table 17.2). A greater QAC loading rate onto the palygorskite surface facilitates greater adsorption of hydrophobic organic contaminants (Sarkar et al., 2010b; Xi et al., 2010). It is worthwhile to note that most of the earlier studies were conducted in batch method. Therefore, the adsorption affinity of these adsorbents requires validation under real environmental conditions. Few researchers have conducted organopalygorskite contaminant adsorption studies under simulated environmental conditions. They studied pH, ionic strength, and temperature effects on adsorption and recommended that organopalygorskites could perform well under environmental pH, temperature, and salt concentrations (Sarkar et al., 2012a, 2010b; Xi et al., 2010). In some instances, a commercial organopalygorskite product was successfully applied to immobilize pollutants in contaminated soil. Research done by Cooperative Research Centre for Contamination Assessment and Remediation of the Environment (CRC CARE) in Australia demonstrated risk based remediation of perfluorooctanesulfonic acid (PFOS) and perfluorooctanoic acid (PFOA) in aqueous firefighting foam (AFFF) contaminated soils (Das et al., In press).

Organopalygorskite interaction with contaminants is mainly limited to the adsorbent surfaces. Because palygorskite is a non-swelling type of clay mineral, the contaminants cannot be intercalated in the interlayers. Hydrophobic interaction largely contributes to the adsorption of non-polar contaminants by organopalygorskites (Das et al., In press; Huang et al., 2008; Rodríguez-Cruz et al., 2008). In addition, adsorption through anion exchange, in which the halide anions associated with the QACs would be replaced by contaminant counter anions, might take place on organopalygorskite surfaces (Sarkar et al., 2012a, 2010b). The adsorption process might also be charge driven because organopalygorskite surfaces could generate positive charges which facilitate anionic contaminant binding (Li et al., 2003; Sarkar et al., 2012a, 2010b). However, non-specific type of adsorption might pose potential risk of releasing contaminants back into the environment from the organopalygorskite surfaces. Researchers found that organopalygorskites could release around 10 percent of adsorbed p-nitrophenol under simulated environmental conditions (Sarkar et al., 2012a). This release was less pronounced from an organopalygorskite prepared with a higher QAC loading (Sarkar et al., 2012a). Therefore, this aspect should be studied carefully before recommending such material for practical remediation applications.

TABLE 17.2 Removal of Environmental Contaminants by Organopalygorskites.

Contaminant	Modifying Amine Compound	Compound Loading	Adsorption Capacity (mg g⁻¹)	Adsorption System	Adsorption Mechanism	Reference
Orange II	Octadecyl trimethylammonium bromide	100% CEC	37	Batch adsorption	Electrostatic attraction and van der Waals interaction	Sarkar et al. (2011b)
		200% CEC	99			
	Dioctadecyl dimethylammonium bromide	100% CEC	38			
		200% CEC	96			
p-nitrophenol		100% CEC	26	Batch adsorption	Electrostatic attraction, anion exchange and hydrophobic interaction	Sarkar et al. (2010b)
	Dimethyl dioctadecylammonium bromide	200% CEC	42			
	Cetylpyridinium chloride	100% CEC	33			
		200% CEC	36			
p-nitrophenol	Dioctadecyl dimethylammonium bromide	100% CEC	117	Flow through reactor	Complex mechanism of physical sorption and chemisorption	Sarkar et al. (2012a)
		200% CEC	138			
	Octadecyl trimethylammonium bromide	100% CEC	12			
2,4-D[a]		200% CEC	42	Batch adsorption	Electrostatic attraction, hydrophobic interaction and van der Waals attraction	Xi et al. (2010)
	Dioctadecyl dimethylammonium bromide	100% CEC	9			
		200% CEC	25			
PFOS[b]	Oleylamine[c]	Not known	47	Batch adsorption	Hydrophobic interactions	Das et al. In (press)
PFOS	Dimethyl dioctadecylammonium bromide	200% CEC	90	Batch adsorption	Hydrophobic interactions	Our group's data (unpublished)
PFOA[d]	Dimethyl dioctadecylammonium bromide	200% CEC	100	Batch adsorption	Hydrophobic interaction and electrostatic attraction	Our group's data (unpublished)
p-nitrophenol	Hexadecyl trimethylammonium bromide (HDTMA) and sodium dodecyl sulphate (SDS)	200 mg g⁻¹ clay; HDTMA:SDS ratio 1:4	138	Batch adsorption	Both physical and chemical processes	Chang et al. (2009b)

TABLE 17.2 *(Continued)*

Contaminant	Modifying Amine Compound	Compound Loading	Adsorption Capacity (mg g⁻¹)	Adsorption System	Adsorption Mechanism	Reference
Penconazole	Dihexadecyl dimethylammonium	125% CEC	Not known $K_f^e = 1234$	Batch adsorption	Hydrophobic interaction	Rodriguez-Cruz et al. (2008)
Metalaxyl	Dihexadecyl dimetylammonium	125% CEC	Not known $K_f = 14$	Batch adsorption	Hydrophobic interaction	Rodriguez-Cruz et al. (2008)
Chromate	Hexadecyl trimethylammonium	520 mmol kg⁻¹	5	Batch adsorption	Anion exchange and electrostatic attraction	Li et al. (2003)
Congo red	Hexadecyl trimethylammonium bromide	100% CEC	189	Batch adsorption	Chemisorption	Chen and Zhao (2009)
Tannin	Octodecyl trimethylammonium chloride	2.44 mg g⁻¹ clay	208	Batch adsorption	Hydrophobic force and hydrogen bonding	Huang et al. (2008)
Reactive Red MF-3B	Octodecyl trimethyl ammonium chloride	2.44 mg g⁻¹ clay	91	Batch adsorption	Chemisorption	Huang et al. (2007)
Linuron	Octadecyltrimetylammonium bromide	125% CEC	Not known $K_f = 332$	Batch adsorption	Hydrophobic interaction	Sanchez-Martin et al. (2006)
Alachlor	Octadeyltrimetylammonium bromide	125% CEC	Not known $K_f = 292$	Batch adsorption	Hydrophobic interaction	Sanchez-Martin et al. (2006)
Atrazine	Octadecyltrimetylammonium bromide	125% CEC	Not known $K_f = 54$	Batch adsorption	Hydrophobic interaction	Sanchez-Martin et al. (2006)

[a] 2,4-D: 2,4-dichlorophenoxyacetic acid

[b] PFOS: Perfluorooctanesulfonic acid

[c] MatCARE™ – a commercial product developed by CRC CARE by modifying palygorskite with oleylamine

[d] PFOA: Perfluorooctanoic acid

[e] K_f: Freundlich sorption constant

17.4.1 COMPARISON OF REMEDIATION PERFORMANCE OF ORGANOPALYGORSKITES WITH OTHER MATERIALS

Currently, several groups of adsorbent materials are available in the market for re-mediating environmental contaminants. However, many of those materials have some inherent disadvantages (Table 17.3). In comparison to organopalygorskites, the adsorption capacity of chitosan-based adsorbents, calcined hydrotalcite, acti-vated carbon, and sludge might be greater, but these materials impart various issues for practical application. For example, the traditional method of extracting chitin from marine food products creates environmental problems as it generates large quantities of waste and the production involves acetylation. All of these materials involve much higher costs of production compared to organoclay minerals. The activated carbon-based adsorbents are not suitable for use under natural flow condi-tions over longer periods of time because of biological fouling and pore clogging. The organopalygorskites might also be compared to organo-smectites in Table 17.3.

TABLE 17.3 Comparison of organopalygorskites with some currently existing remediation materials.

Material	Disadvantage
Activated carbon	Mostly active at very low pH (2–3)
	Biological fouling is a problem
	Longevity is not good
	The higher the efficiency of the sorbent, the greater the cost of the ma-terial
	Sometime requires complexing agents to use as additives that include extra cost
	Non-selective sorbent
	Need high reactivation costs
	Reactivation process causes loss of carbon
Zero valent iron (ZVI)	Longevity is a problem
	Gradual loss in the permeability of the sorbent due to precipitation of chemical species when used in PRB
	Possibility of secondary contamination of Fe nanoparticles
Calcium polysul-phide	Very high application rate is needed (twice the stoichiometric amount)
	pH of the system needs to very high (9.5) for maximum efficiency
	62% of the material get lost within less than a year time
Hydrous TiO_2	pH needs to be very low (2) for getting maximum efficiency
	Adsorption capacity is not very high (about 5 mg/g)

TABLE 17.3   *(Continued)*

Material	Disadvantage
Ion exchange resin	Performance depends on the type of resin used, better the quality, higher the price
	Not environmental friendly because they are derived from petroleum based raw materials
	Performance depends highly on pH
	Poor contact with aquous solution
Chitosan based sorbent	Non porous materials
	Highly pH dependent
	Sometime may require chemical modification to achieve target level of remediation
Polymeric adsorbent	Very expensive ($40-65 per kg)
Anion exchanger (Dowex)	Expensive ($35.5 per kg)
Fe^{2+} based reductant	Not effective in alkaline pH condition

Keeping into account the disadvantages of some currently existing remediation materials, organopalygorskites may provide the following advantages over the other materials:

a)   Highly efficient in adsorption of numerous pollutants
b)   Application rate is not very high (5 g L^{-1})
c)   Efficient over a wide range of pH values
d)   Porous and permeable materials; can be used in filter
e)   Less expensive than anion exchangers and activated carbon

17.4.2   CONCERNS ABOUT USE OF ORGANOPALYGORSKITE AS A REMEDIATION MATERIAL

However, organopalygorskite adsorbents may impart problems such as; (a) toxicity to native micro-biota when applied to soils for contaminant fixation, (b) regeneration and disposal of spent adsorbents. QACs such as hexadecyltrimethyl ammonium bromide (HDTMA), octadecyltrimethyl ammonium bromide (ODTMA), and Arquad are known to be toxic to dehydrogenase activity in soils (reflects the oxidative activity or intensity of metabolism of the total microflora present in the soil) at concentration levels of 50, 100, and 750 mg kg^{-1} soil, respectively (Sarkar et al., 2010a).

These compounds can inhibit the nitrification activity (a soil function carried out by a specific group of micro-organisms called nitrifiers) even at lower concentrations (Sarkar et al., 2010a). Organoclays prepared with these QACs from bentonite show similar toxicity to soil micro-organisms and earthworms (Sarkar et al., 2013). The nature of QACs and soil properties influence the level of toxicity. For example, Arquad, either as a free compound or present in an organoclay, is less toxic to soil micro-organisms and earthworms as compared to HDTMA and ODTMA (Sarkar et al., 2013; Sarkar et al., 2010a). Therefore, evaluation of the toxicity of organo-palygorskites to microorganisms is required before use in stabilizing contaminants in soils. At the same time, emphasis is needed to dispose spent adsorbents properly. Bio-activation of organopalygorskites with micro-organisms, which are able to de-grade the contaminants, could be a green approach to achieve sustainable disposal (Sarkar et al., 2012c).

17.5   CONCLUSIONS

Organopalygorskites hold significant scope for application to remediate contami-nated waters and soils. The preparation of this material is easier and less expensive than many other commercially available materials. In this group of materials, the QAC molecules attach to the palygorskite by surface interaction and create a chemi-cal environment, which is congenial for contaminant immobilization either through hydrophobic bonding or electrostatic attraction. Additionally, anion exchange reac-tion can also take place to hold ionizable contaminants on the adsorbent surfaces. However, to prepare organopalygorskites, the type and loading of QACs should be selected carefully so that the final product does not pose (a) toxicity to the native bi-ota present in the environment and (b) potential risk of redispersal of adsorbed con-taminants back into the environment. Further research should also be concentrated on how these adsorbents could be regenerated in an environmental-friendly manner.

KEYWORDS

- Aluminol groups
- Aqueous firefighting foam (AFFF)
- Bentonite
- Differential thermogravimetric analysis
- Montmorillonite
- Nitrifiers
- Organopalygorskites
- Thermogravimetric analysis

REFERENCES

1. Bate, B.; and Burns, S. E. J.; *Colloid. Interf. Sci.* **2010**, *343*, 58–64.
2. Boyd, S. A.; Lee, J. F.; and Mortland, M. M.; *Nature.* **1988**, *333*, 345–347.
3. Chang, P. H.; Li, Z.; Yu, T. L.; Munkhbayer, S.; Kuo, T. H.; Hung, Y. C.; Jean, J. S.; and Lin, K. H. J.; *Hazard. Mater.* **2009a**, *165*, 148–155.
4. Chang, Y.; Lv, X.; Zha, F.; Wang, Y.; and Lei, Z. J.; *Hazard. Mater.* **2009b**, *168*, 826–831.
5. Chen, H.; and Zhao, J.; *Adsorption.* **2009**, *15*, 381–389.
6. Chen, H.; Zhao, Y.; and Wang, A. *J. Hazard. Mater.* **2007**, *149*, 346–354.
7. Churchman, G. J.; Gates, W. P.; Theng, B. K. G.; and Yuan, G.; Chapter 11.1 Clays and clay minerals for pollution control. In *Developments in Clay Science*; Eds. Theng, B. K. G., Bergaya, F., and Gerhard, L., Elsevier, 2006.
8. Das, P.; Victor, A. A. E.; Kambala, V.; Mallavarapu, M; Sarkar, B.; and Naidu, R.; *Water. Air. Soil. Pollut.*, doi: 10.1007/s11270-013-1714-y. (in press)
9. Frost, R. L.; and Ding, Z.; *Thermochim. Acta.* **2003**, *397*, 119–128.
10. Guggenheim, S.; and Krekeler, M. P. S.; Chapter 1—The structures and microtextures of the palygorskite–sepiolite group minerals. In: *Developments in Clay Science*; Eds. Emilio, G., Arieh, S., Elsevier, **2011**.
11. Guggenheim, S.; and van Groos, A. F. K.; *Clays. Clay. Min.* **2001**, *49*, 433–443.
12. Huang, J.; Liu, Y.; Jin, Q.; Wang, X.; and Yang, J. *J. Hazard. Mater.* **2007**, *143*, 541–548.
13. Huang, J.; Liu, Y.; and Wang, X. *J. Hazard. Mater.* **2008**, *160*, 382–387.
14. Jordan, J.; and Williams, F.; *Colloid Poly. Sci.* **1954**, *137*, 40–48.
15. Li, Z.; and Willms, C. A.; Kniola, K.; *Clays. Clay. Min.* **2003**, *51*, 445–451.
16. Murray, H. H.; *Appl. Clay Sci.* **2000**, *17*, 207–221.
17. Naidu, R.; Bolan, N. S.; Kookana, R. S.; and Tiller, K. G.; *Eur. J. Soil Sci.* **1994**, *45*, 419–429.
18. Post, J. E.; and Heaney, P. J.; *Am. Min.* **2008**, *93*, 667–675.
19. Potgieter, J. H.; Potgieter-Vermaak, S. S.; and Kalibantonga, P. D.; *Min. Eng.* **2006**, *19*, 463–470.
20. Rodríguez-Cruz, M. S.; Andrades, M. S.; and Sánchez-Martín, M. J.; *J. Hazard. Mater.* **2008**, *160*, 200–207.
21. Sanchez-Martin, M. J.; Rodriguez-Cruz, M. S.; Andrades, M. S.; and Sanchez-Camazano, M.; *Appl. Clay Sci.* **2006**, *31*, 216–228.
22. Sánchez del Río, M. et al.; *J. Mater. Sci.* **2009**, *44*, 5524–5536.
23. Sarkar, B.; Megharaj, M.; Shanmuganathan, D.; and Naidu, R.; *J. Hazard. Mater.*; **2013**, *261*, 793–800.
24. Sarkar, B.; Megharaj, M.; Xi, Y.; Krishnamurti, G. S. R.; and Naidu, R.; *J. Hazard. Mater.* **2010a**, *184*, 448–456.
25. Sarkar, B.; Megharaj, M.; Xi, Y.; and Naidu, R. *J.; Hazard. Mater.* **2011a**, *195*, 155–161.
26. Sarkar, B.; Megharaj, M.; Xi, Y.; and Naidu, R. *Chem. Eng. J.* 2012a, 185–186, 35-43.
27. Sarkar, B.; Naidu, R.; and Megharaj, M.; *Water. Air. Soil. Pollut.*, doi: 10.1007/s11270-013-1704-0. (in press)
28. Sarkar, B.; Naidu, R.; Rahman, M.; Megharaj, M.; and Xi, Y.; *J. Soils. Sedimen.* **2012b**, *12*, 704–712.
29. Sarkar, B.; Xi, Y.; Megharaj, M.; Krishnamurti, G.; Bowman, M.; Rose, H.; and Naidu, R.; *Crit. Rev. Environ. Sci. Technol.* **2012c**, *42*, 435–488.
30. Sarkar, B.; Xi, Y.; Megharaj, M.; Krishnamurti, G. S. R.; and Naidu, R. *J. Colloid. Interf. Sci.* **2010b**, *350*, 295–304.
31. Sarkar, B.; Xi, Y.; Megharaj, M.; Krishnamurti, G. S. R.; Rajarathnam, D.; and Naidu, R. *J. Hazard. Mater.* **2010c**, *183*, 87–97.
32. Sarkar, B.; Xi, Y.; Megharaj, M.; and Naidu, R.; *Appl. Clay Sci.* **2011b**, *51*, 370–374.

33. Shirvani, M.; Shariatmadari, H.; Kalbasi, M.; Nourbakhsh, F.; and Najafi, B.; *Colloids. Surf. A.* **2006**, *287*, 182–190.
34. Shuali, U.; Nir, S.; and Rytwo, G.; Chapter 15 - Adsorption of surfactants, dyes and cationic herbicides on sepiolite and palygorskite: Modifications, applications and modelling. In: Developments in Clay Science; Eds. Emilio, G., Arieh, S., Elsevier, **2011**.
35. Wang, C. C.; Juang, L. C.; Lee, C. K.; Hsu, T. C.; Lee, J. F.; and Chao, H. P.; *J. Colloid. Interf. Sci.* **2004**, *280*, 27–35.
36. Xi, Y.; Ding, Z.; He, H.; and Frost, R. L.; *J. Colloid. Interf. Sci.* **2004**, *277*, 116–120.
37. Xi, Y.; Ding, Z.; He, H.; and Frost, R. L.; *Spectrochim. Acta A.* **2005a**, *61*, 515–525.
38. Xi, Y.; Frost, R. L.; He, H.; Kloprogge, T.; and Bostrom, T. ;*Langmuir* **2005b**, *21*, 8675–8680.
39. Xi, Y.; Mallavarapu, M.; and Naidu, R.; *Appl. Clay Sci.* **2010**, *49*, 255–261.
40. Zadaka, D.; Radian, A.; and Mishael, Y. G.; *J. Colloid. Interf. Sci.* **2010**, *352*, 171–177.
41. Zhang, J.; Xie, S.; and Ho, Y. S.; *J. Hazard. Mater.* **2009**, *165*, 218–222.

CHAPTER 18

SURFACE MODIFICATION OF HALLOYSITE NANOTUBES: ROLE OF EXTERNAL HYDROXYL GROUPS

VAHDAT VAHEDI, POORIA PASBAKHSH, and SIANG-PIAO CHAI

CONTENTS

18.1 INTRODUCTION

Halloysite nanotubes (HNTs) are naturally occurring aluminosilicate clay minerals consisting of two adjacent sheets of gibbsite octahedra (Al $(OH)_3$) and tetrahedrally coordinated silicon dioxide (SiO_2), which are rolled up as a result of mismatch between these two sheets (Figure 18.1, Chapter 15). The rolled-up sheets occur within a few layers (5–6) in the wall of nanotubes resembling multi-walled carbon nanotubes (MWCNTs). These layers are separated by a monolayer of water molecules, which can be easily removed by heating (60–110 °C). Halloysite nanotubes are low cost and abundant. Compared with other nanofillers, used in polymer nanocomposites like CNTs and montmorillonite (MMT), the modification of HNTs and their dispersion is much easier to achieve because of their unique hollow tubular shape and crystalline structure. HNTs are non-toxic, environment-friendly, and biocompatible minerals. These unique properties have drawn attention to finding new potential applications for HNTs such as nanocontainers (Shchukin and Möhwald, 2007; Shchukin et al., 2008), sustained release (Levis and Deasy, 2003; Qi et al., 2010), immobilization (Machado et al., 2008; Zhai et al., 2010), cosmetics (Suh et al., 2011), nanoreactors (Shchukin et al., 2005; Tierrablanca et al., 2010), and reinforcement of polymeric matrixes (Du et al., 2006; Ning et al., 2007; Ye et al., 2007; Guo et al., 2008; Ismail et al., 2008; Deng et al., 2009; Guo et al., 2009; Jia et al., 2009; Handge et al., 2010; Tang et al., 2011; Soheilmoghaddam et al., 2013; Yin and Hakkarainen, 2013). Recent studies showed that HNTs can effectively improve mechanical and thermal properties of polymers such as polypropylene (PP) (Du et al., 2006; Ning et al., 2007), polyamide (PA) (Guo et al., 2009; Handge et al., 2010), ethylene propylene diene monomer (EPDM) (Ismail et al., 2008), styrene–butadiene rubber (SBR) (Guo et al., 2008; Jia et al., 2009), and epoxy (Ye et al., 2007; Deng et al., 2009; and Tang et al., 2011).

HNTs have a different chemistry on their outer and inner surfaces. The lumen (inner surfaces) of HNTs consists of AlOH octahedral sheets, and their external surface contains Si-O-Si groups. Unlike kaolinites, the reactive hydroxyl groups of HNTs, predominantly aluminols, are located on the inner side of HNTs, whereas some hydroxyl groups (AlOH or SiOH) exist on the external surface within the edges and defects (Joussein et al., 2005; Yuan et al., 2008). Siloxane groups at the external surface of HNTs are supposed to be non-reactive, which enable HNTs to be easily separated from aggregates as a result of lower-interfacial interactions. On the other hand, this lower reactivity of HNTs' outer surface may decrease their interaction with polymer matrixes, resulting in poorer mechanical properties compared with corresponding nanocomposites.

Surface treatment of HNTs could improve their performance in various applications such as polymer reinforcement (Deng et al., 2009; Pasbakhsh et al., 2010) and nanocontainers (Yah et al., 2012; Yuan et al., 2012). The treatment reactions that were used for surface modification of HNTs can be categorized into three different mechanisms: Chemical bonds with surface silanol or aluminol groups, electron transfer with transition metals such as aluminium and iron, and hydrogen bonding via an external siloxane surface of HNTs. Most modification mechanisms of HNT's outer surface have been based on the interaction of chemicals with surface aluminol or silanol groups (Du et al., 2007; Liu et al., 2007; Mu et al., 2007; Yuan et al., 2008; Deng et al., 2009; Pasbakhsh et al., 2010; Barrientos-Ramírez et al., 2011; and Yah et al., 2012;); although the amount of these groups on the external surface of HNTs is a matter of question (Yuan et al., 2008). The relatively low content of hydroxyl groups on the external surface of HNTs led to a limited presence of reactive sites for possible covalent bonding between HNTs and modifiers. The amount of these hydroxyl groups and their distribution on the external surface of HNTs can determine the quantity of grafted modifiers and the quality of the modification. There is little attention to this matter in the literature which is essential for improving the reinforcing efficiency of HNTs as nanofillers and for other applications as nanocontainers.

Silane modifiers are among those modifiers which have been widely used for surface treatment of HNTs (Du et al., 2007; Liu et al., 2007; Yuan et al., 2008; Deng et al., 2009; Pasbakhsh et al., 2010; Barrientos-Ramírez et al., 2011; and Yan et al., 2011). Yuan et al. (2008) reported direct grafting of APTES onto the lumen's hydroxyl groups. Ramirez et al. (2011) grafted aminosilanes exhibiting two (3-(2-aminoethyl)aminopropyltrimethoxysilane, DAS) and three (3-[2-(2-amino-ethyl) aminoethyl] aminopropyltrimethoxysilane, TAS) amino groups on the outer surface of HNTs. Yah et al. (2012) selectively functionalized the inner surface of halloysite by octadecyl phosphonic acid (ODP), through the formation of Al−O−P bonds. Consecutively, the researchers grafted the external surface of halloysite by silane-coupling agent (N-(2-Aminoethyl)-3-aminopropyltrimethoxysilane, AEAPS) through silanization with external silanol groups. It has been argued that ODP was not grafted on the external surface, because the aluminol groups were existed only on the inner surface of HNTs and not on the external surface (Yah et al., 2012).

Functionalization of fillers with epoxide groups was reported to improve the interaction between filler and polymer matrixes (Reddy et al., 2005; Reddy and Das, 2006; Zou et al., 2008; and Teng et al., 2012). Epoxide-grafted polymers were also used as compatibilizers for polymer blends (Holsti-Miettinen et al., 1995; Sailaja, 2006; and Ao et al., 2007). DGEBA (bisphenol A diglycidyl ether) was used by Liu et al. (2003) to functionalize nano silica particles through the ring-opening reaction between oxirane and silanol groups. This method avoids the drawbacks of organosilanes modification, such as high cost and the retention of residues of organosilanes in their products, because of incomplete dehydration and condensation reactions during the period of storage and usage.

In this study, the external surface of halloysite was grafted with DGEBA and APTES; the possible interaction mechanisms were investigated via the HNTs surface-functional groups; and the role of external-hydroxyl groups on the modification was studied. DGEBA was used as a modifier to functionalize the outer surface of HNTs by epoxide groups to increase the interaction of HNTs in an epoxy nanocomposite system not presented here. APTES was also used to modify the outer surface of HNTs, as a well-known silane-coupling agent, for comparison of properties of the grafted HNTs. NaOH treatment was used to increase the number of OH groups on the external surface of HNTs, because their external surface is supposed to be covered by siloxane and deficient for OH groups. The results of this work can be used for further investigation on reinforcement effects of surface-modified halloysite nanotubes used in polymer-based nanocomposite systems.

18.2   MATERIALS AND METHODS

Sodium hydroxide (NaOH) (R & M Chemicals, UK), bisphenol A diglycidyl ether (DGEBA), cobalt naphthenate (6%), and toluene (99.5 + %, A.C.S. reagent) were supplied by Sigma-Aldrich. γ-aminopropyltriethoxysilane (APTES) (E ACROSEAL 99%) was purchased from Acros Organics Co., Malaysia. Halloysite was supplied by Imery Tableware New Zealand and was used as received.

18.2.1   NAOH TREATMENT

NaOH treatment was conducted to increase the number of hydroxyl groups on the HNTs' external surface. For this purpose, 3 g of halloysite powder was dispersed in 80 mL of distilled water by magnetic stirring. The pH of the HNTs suspension was adjusted to 8 by adding about 6 mL of 0.1 M NaOH solution. To achieve a good dispersion of HNTs, ultrasonic mixing was performed for 30 min. Alkaline suspension of HNTs was kept under constant magnetic stirring at room temperature for 24 h. The resultant mixture was filtered by centrifugal separation and washed extensively with distilled water to remove the possible remnant of NaOH, until the pH of effluent reached 6.5. The separated solids were then dried overnight at 50°C and were used as NaOH-treated HNTs for APTES and DGEBA modification.

18.2.2   DRYING HNTS

In order to investigate the effect of over-drying on the HNTs interlayer water, HNTs powder was dried at 400°C for 12 h in a muffle furnace. These dried HNTs were used for modification of APTES and DGEBA.

18.2.3   APTES MODIFICATION

In order to modify the outer surface of halloysite by organosilane modifier (APTES), a general silanization procedure for silica based fillers was followed (Yuan et al., 2008). In a typical run, 0.6 g HNTs and 2 mL APTES were mixed in 25 mL toluene by magnetic stirring followed by 30 min ultrasonic to achieve a well-dispersed solution of HNTs. The suspension was then transferred to a three-necked flask, and refluxed at different temperatures (110°C, 80°C, and 32°C) for 24 h, whereas a calcium chloride drying tube was used to ensure a dry environment. The resultant mixture was filtered using nylon membrane with 0.45 μm pores and was extensively washed (6 times) with fresh toluene to remove the unreacted APTES from the filtrate, and then dried overnight at 50°C.

18.2.4   DGEBA MODIFICATION

In order to modify the outer surface of halloysite by DGEBA, 1 g HNTs and 1.5 g DGEBA were mixed in 50 mL toluene by magnetic stirring followed by 30 min ultrasonic mixing to achieve a well-dispersed solution of HNTs. The suspension was then transferred to a three-necked flask, and then 0.2 mL of catalyst i.e., cobalt naphthenate was added. This mixture was refluxed at different temperatures (110°C, 32°C) for 24 h, whereas a calcium chloride drying tube was used to ensure a dry environment. The resultant mixture was filtered, washed, and dried as described for APTES modification.

18.2.5   CHARACTERIZATION METHODS

To investigate the grafting of modifiers on HNTs surface, FTIR analysis was conducted before and after the modification process. A Nicolet iS10 FT-IR spectrometer was used to record the FTIR spectra of the samples in the range of 400–4000 cm^{-1}. The solid powders were used as a specimen by a Smart iTR diamond accessory without any preparation. Approximately, 32 scans were collected at a resolution of 4 cm^{-1} for each spectrum.

Approximately 10 mg of samples was used to perform thermogravimetric analysis (TGA) by the Universal V4.7A TA instruments at a heating rate of 10°C/min, under an atmosphere of nitrogen.

An HITACHI SU8010 field-emission scanning electron microscope (FE-SEM) equipped with an energy dispersive X-ray (EDX) was used to obtain electron micrographs of the samples. To prepare specimens for SEM analysis, 10 mg of halloysite was suspended in 3 mL of ethanol by vortex agitation for 2 min followed by 2 min sonication. A drop of suspension was dropped on carbon tape and air-dried for 10 min. Then the samples were coated with platinum using the QUORUM-Q150RS sputter coater instrument and transferred to FE-SEM for imaging.

To analyze the surface area and porosity of the samples, nitrogen adsorption–desorption isotherms were obtained with a Micromeritics ASAP 2020 surface area analyzer at a liquid-nitrogen temperature of −196°C. Before the nitrogen adsorption test, samples were outgassed to remove moisture and physisorbed chemicals from the HNTs. The samples were evacuated to 500 μm Hg with an evacuation rate of 5 mm Hg s^{-1} while heated to 90°C with a heating rate of 10°C min^{-1}. Then the samples were heated to 110°C with a heating rate 10°C min^{-1} and held at 110°C and 100 mm Hg for 6 h. The specific surface area was calculated by multiple-point Brunauer–Emmett–Teller (BET) over the relative pressure (P/P$_o$) range of the isotherms between 0.05 and 0.25. To calculate the total volume of pores and the pore size distribution, the Barrett–Joyner–Halenda (BJH) method by the Halsey equation and Fass correction were used.

18.3   RESULTS AND DISCUSSION

18.3.1   FTIR ANALYSIS

Figure 18.1(a), (b) shows the FTIR spectra of the HNTs before and after modification with APTES and DGEBA, respectively (summarized in Table 18.1). All characteristic peaks of HNTs were resolved in all of the modified and unmodified HNTs.

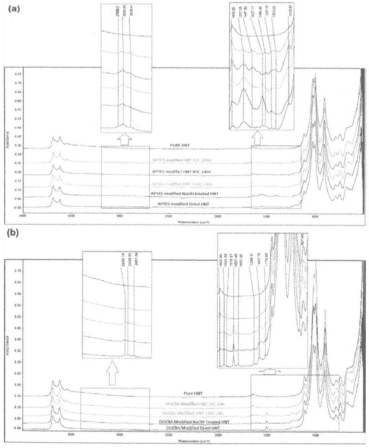

FIGURE 18.1   FTIR spectra of APTES (a) and DGEBA (b) modified HNTs.

TABLE 18.1 The assignments for major bonds of FTIR spectra (Frost and Vassallo, 1996; Bailey and McGuire, 2007; Yuan et al., 2008; Pasbakhsh et al., 2010; Barrientos-Ramírez et al., 2011).

Wave Numbers (cm^{-1})	Assignment
3692	O-H stretching of inner hydroxyl groups
3620	O-H stretching of inner-surface hydroxyl groups
3541	O-H stretching of water
2930	Symmetric stretching of CH_2
2869, 2930, 2969	C-H aliphatic stretching
1652	O-H deformation of water
1556	Deformation NH_2 vibration
1490	Deformation (scissoring) of CH_2
1411	Symmetric deformation of NH_2 in Si–O–H$\cdots NH_2$
1384	Deformation (wagging) of CH_2
1330	Si-C-H scissoring vibration
1195	C–N stretching
1114	Perpendicular Si-O-Si stretching
1025, 1008	In-plane Si-O stretching
940	O-H deformation of inner—surface hydroxyl groups
907	O-H deformation of inner hydroxyl groups
825	P-substituted phenyl ring
792	Symmetric stretching of Si-O
744	Perpendicular Si-O stretching
672	Perpendicular Si-O stretching

Compared with the unmodified HNTs, APTES-modified HNTs exhibited new peaks (Figure 18.1(a)). The presence of characteristic peaks of APTES in FTIR spectra after extensive washing of the APTES-modified HNTs with toluene can be a good indicator for successful grafting of APTES on the surface of HNTs.

To study the effect of temperature on grafting of APTES to the surfaces of HNTs the modification reaction was conducted at different temperatures. The reaction temperature used for silane modification of HNTs is varied extensively in different studies. Du et al. (2007) and Ramírez et al. (2011) conducted the grafting reaction of γ-MPS and DAS, respectively, on HNTs surfaces at room temperature, whereas Yuan et al. (2008) reported grafting of APTES on HNTs at 120°C. The FTIR results showed that the relative intensity of above-mentioned peaks was higher for the HNTs modified at higher temperatures. This indicated that the amount of grafted APTES increased with increase in a reaction temperature. This is in agreement with findings of Yang et al. (2012) utilizing higher temperatures (175–220°C) to promote the APTES grafting on kaolinite.

The relative intensity of APTES peaks in APTES-modified-dried HNTs has considerably decreased compared to APTES-modified HNTs. It is well-known that the presence of water molecules in a reaction medium of APTES can cause hydrolysis and olygomerization or even polymerization of APTES during the modification reaction (Vrancken et al., 1992; Yuan et al., 2008). Because of HNTs contain water in their structure, APTES could be oligomerized or polymerized with the APTES molecules bonded to OH groups on the surface of HNTs to form a cross-linked network around the HNTs surfaces consisting a number of possible interactions (Figure 18.2) including physical attachment, covalent bonding, two-dimensional self-assembly (horizontal polymerization), and multilayers (vertical polymerization). Drying the HNTs before APTES modification would remove the water molecules from the interlayer structure of HNTs resulting in grafting of a thinner layer of AP-TES on the HNTs surface. This variation in the thickness of grafted modifiers can affect the load-transfer mechanism and the reinforcing performance of the modified filler (Pavlidou and Papaspyrides, 2008). The length of grafted modifier chains on modified silicate fillers was reported to determine the morphology and mechanical properties of their PE (Wang et al., 2001; Osman et al., 2005), EPDM (Zheng et al., 2004), and SBR nanocomposites (Sadhu and Bhowmick, 2003).

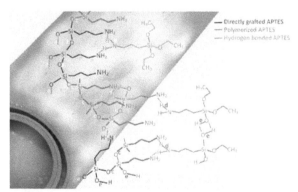

FIGURE 18.2 Various interactions in APTES network on grafted HNTs: (a) covalent bonding with OH groups on HNTs' surface (b) horizontal and (c) vertical polymerization with neighboring APTES (d) hydrogen bonded APTES; buried amino groups (e) Hydrolyzed APTES (f) protonated amino groups.

A noticeable increase in the relative intensity of APTES peaks was observed for HNTs which have been treated with NaOH. As no evacuation pretreatment was performed in this study, APTES molecules were most likely grafted on the OH groups of the external surfaces rather than the lumen of HNTs. Treatment of HNTs with alkaline solution increased the number of OH groups on the external surface of HNTs, resulting in more reactive groups on their surface and more grafted AP-TES. Transformation of siloxane to silanol groups in silicate fillers by reaction with water molecules has been reported by others (Dugas and Chevalier, 2003; Lygin, 2001; and Christy, 2011). Moreover, NaOH treatment may cause selective etching of the siloxane outer surface of HNTs and exposure of AlOH groups underneath the silicate sheets (Abdullayev et al., 2012; White et al., 2012). This change in surface groups in NaOH-treated HNTs could not be detected by FTIR spectra, because the total structure of HNTs remained unaffected. However, an increase in the grafted APTES indicated that more OH groups are available on the HNTs surface for the modification reaction.

DGEBA-modified HNTs (Figure 18.1(b)) exhibited new peaks in FTIR spectra compared to the unmodified HNTs. Major FTIR peaks of DGEBA could be observed in FTIR spectra of DGEBA-treated HNTs (at 110°C for 24 h). These observations indicated that DGEBA was successfully grafted on the HNTs. This modification process can be described by the ring opening reaction of DGEBA and OH groups of HNTs (Figure 18.3), as reported by Liu et al. (2003).

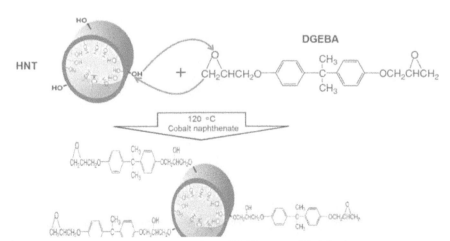

FIGURE 18.3   Possible reaction mechanism of DGEBA modified HNTs.

Compared with APTES modification, the effect of reaction temperature on the DGEBA-modified HNTs was more pronounced as the corresponding peaks of grafted DGEBA were insignificant in the HNTs modified at room temperature. Reaction of OH groups of HNTs with epoxide groups of DGEBA has a higher activation energy compared with their silanization reaction with APTES molecules.

No significant change in the relative intensity of grafted DGEBA peak (e.g., 1,507 cm ^{1}) was observed in dried HNTs, which indicated that the reaction between OH groups of HNTs and epoxide groups of DGEBA is not sensitive to water. So, drying of HNTs did not significantly affect the amount of grafted DGEBA on the surface of HNTs as it did in silane modifiers like APTES.

After NaOH treatment of HNTs, the relative intensity of grafted-DGEBA peak (e.g., 1,507 cm ^{1}) did not increase noticeably. It seems that NaOH treatment was not effective for increasing the amount of DGEBA grafted to HNTs surface, unlike what was observed in APTES treatment. This may be attributed to the lower reactivity of aluminol compared to silanol toward ring opening reaction with epoxide groups. No data is available on the kinetics of reaction between aluminol and silanol with epoxide groups. However, the difference between the reactivity of aluminol and silanol groups has been reported before (Strawn et al., 2004; Dias Filho and Do Carmo, 2006). According to Strawn et al. (2004) aluminol groups can be either proton acceptors or donors, and are even capable of forming covalent bonds with metal cations, whereas silanol groups act like weak acids and can easily dissociate the protons. The lower acidity of aluminol groups may result in a lower reactivity of aluminol groups with epoxide groups. Hence in NaOH-treated HNTs, although more OH groups were available on the surface, the increase in grafted DGEBA was not noticeable.

18.3.2 THERMOGRAVIMETRIC ANALYSIS

Thermogravimetric curves of halloysite samples before and after modification with APTES and DGEBA are presented in Figure 18.4(a), (b), respectively. For unmodified HNTs, two major weight losses occurred which have resulted from the evaporation of physically absorbed water (~100°C) and the dehydroxylation of structural OH groups of halloysite (425–600°C). However, for modified HNTs some new weight losses were observed, showing the effect of grafting of modifiers on HNTs.

Three more weight losses could be observed in APTES-modified HNTs (Figure 18.4(a)), which implied different stages of decomposition of grafted APTES (Table 18.2). The weight loss in the range of 100–220°C was attributed to physically absorbed hydrogen-bonded APTES molecules as the boiling temperature of APTES was around 217°C. Hydrogen-bonded APTES causes the amine groups to become confined within the APTES network (Figure 18.2(d)) and exposes ethoxy groups to the surface (Kim et al., 2011). Curing of APTES-grafted substrates (at 100–110°C) was suggested to stabilize grafted APTES, to activate the surface by exposing more amine groups to the surface, and to change the protonated amines to reactive neutral amine groups (Kim et al., 2008; Kim, 2010; and Kim et al., 2011).

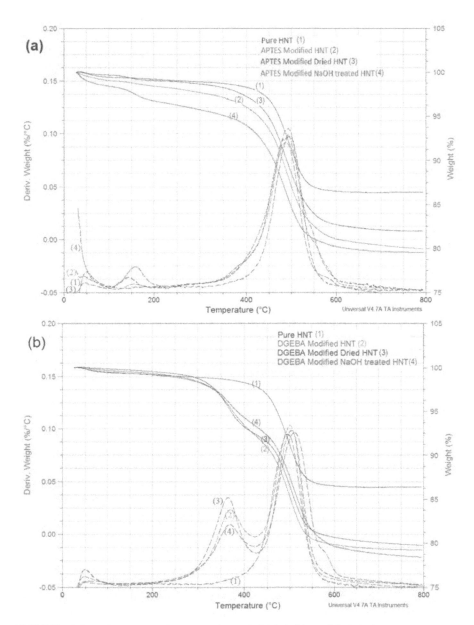

FIGURE 18.4 TGA curves of APTES (a) and DGEBA (b) modified HNTs.

TABLE 18.2   Weight loss (%) of APTES-grafted HNTs.

	Unmodified HNTs	APTES-Modified HNTs	APTES-Modi-fied-Dried HNTs	APTES-Modified-NaOH Treated HNTs
Room Tem.—100°C	0.56	0.93	0.12	1.58
100–220°C	0.23	0.65	0.33	1.7
220–325°C	0.39	0.99	0.56	1.29
325–425°C	0.98	2.39	2.14	2.65
100–425°C	1.6	4.04	3.03	5.64
425–600°C	11.33	13.96	11.70	12.57
Total	12.93	18	14.73	18.21

The weight loss between 220 and 325°C could be associated with the decomposition of oligomerized or polymerized APTES. The broad DTG peak in this region shows different stages of decomposition as a result of various kinds of bonding in cross-linked APTES network (defined in Figure 18.2). The weight loss from 325 to 425°C was attributed to the decomposition of grafted APTES to AlOH and SiOH on the halloysite surface. These APTES molecules are tightly packed under the polymerized APTES network and decomposed at higher temperatures that partially overlapped with dehydroxylation of structural OH groups.

NaOH-treated HNTs showed 10.9 percent increase in the amount of grafted APTES in the temperature range of 325–425°C, compared to pure HNTs, which confirmed that more APTES could be grafted on the surface of NaOH-treated HNTs because of the presence of more OH groups (Table 18.2). Moreover, the amount of oligomerized or polymerized APTES (220–325°C) and hydrogen-bonded APTES (100–220°C) in NaOH-treated HNTs increased by 30 and 161 percent, respectively. The total quantity of the introduced APTES, identified as a value of weight loss in the range 100–425°C of APTES-modified HNTs, increased by 39.85 percent after treatment of HNTs with NaOH.

For dried HNTs, the amount of grafted APTES in the temperature range of 325–425°C decreased by 10 percent. Although drying of halloysite at 400°C did not dehydroxylate the structural AlOH hydroxyl groups, it could reduce the hydroxyl group on the outer surface of halloysite (Del Rey-Bueno et al., 1989), resulting in lowering the amount of grafted APTES. In this case, 43 percent decrease in oligomerized or polymerized APTES, 15 percent decrease in hydrogen-bonded APTES, and 25 percent decrease in total quantity of the introduced APTES were noted.

In the case of DGEBA-modified HNTs (Figure 18.4(b)) only one more weight loss could be seen, compared to the unmodified HNTs. This weight loss, in the range of 250–425°C, was attributed to the decomposition of DGEBA molecules. For dried HNTs and NaOH-treated HNTs, no significant change in the amount of grafted DGEBA was observed which is in agreement with the FTIR results.

18.3.3   SCANNING ELECTRON MICROSCOPY

SEM images and EDX analysis of halloysite samples are presented in Figure 18.5. Hollow tubular shapes of halloysite could be observed in SEM micrographs. Defects on the surface and structure of halloysite are evident (shown by white circles in Figure 18.5(a)). These defects are sources of hydroxyl groups on the outer surface of HNTs that can be used for the modification of the external surface of HNTs.

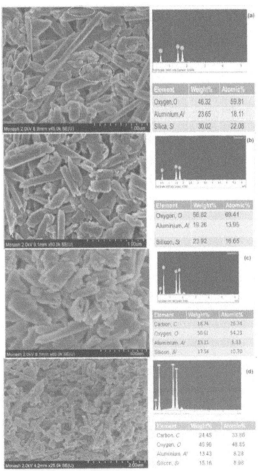

FIGURE 18.5   SEM micrographs and EDX data of halloysite samples: (a) untreated HNTs, (b) NaOH-treated HNTs, (c) APTES-modified NaOH-treated HNTs, and (d) DGEBA-modified NaOH-treated HNTs.

SEM micrographs of treated HNTs (Figure 18.5(b)–(d)) showed that the tubular shape of HNTs was preserved during modification. The external surface of modified HNTs seems to be rougher, especially in DGEBA-modified HNTs. More aggregation could be observed in modified HNTs, which could be attributed to the higher interaction between nanotubes because of the presence of polar groups.

EDX results showed that, for NaOH-treated HNTs (Figure 18.5(b)), the percentage of oxygen atoms on HNTs surface increased because of the presence of more OH groups. The presence of carbon atoms in APTES (Figure 18.5(c)) and DGEBA (Figure 18.5(d)) modified HNTs confirmed that APTES and DGEBA were successfully grafted on HNTs surface.

18.3.4   PORE SIZE AND SURFACE AREA OF HNTS

All modified and unmodified HNTs exhibited type IV nitrogen adsorption–desorption isotherms with an H3 hysteresis loop (Figure 18.6), which indicated mesoporous characteristics. This type of isotherm is identified by a concave curve at low P/P_0, an almost linear middle part, and a concave curve at high P/P_0 with hysteresis loops (Pierotti and Rouquerol, 1985). The hysteresis loops are generally caused by capillary condensation of nitrogen in porous structures in which the lower branch of an isotherm corresponds to the adsorption and the upper branch corresponds to the desorption of nitrogen.

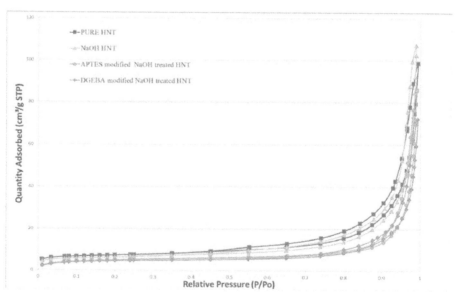

FIGURE 18.6   Nitrogen adsorption–desorption isotherms of modified and unmodified HNTs.

Analysis of pore size and surface area of HNTs are shown in Table 18.3. Compared with unmodified HNTs, considerable decreases in the specific surface area of APTES- and DGEBA-grafted HNTs could be observed. Attachment of modifiers to the surface of HNTs provided less space for nitrogen molecules to be adsorbed on the surface of HNTs, resulting in a reduction of the amount of adsorbed nitrogen and in lowering the measured surface area. This method fails to estimate the actual surface area provided by modified fillers for polymer reinforcement. Despite the lower BET surface area of HNTs after modification, adsorption of polymer chains to modified HNTs will be higher because of the better interaction between polymer and modified HNTs in a polymer-HNT nanocomposite system. On the other hand agglomeration of HNTs, as a result of higher interaction, among the nanotubes in modified HNTs could reduce the surface area of the HNTs. Partial blockage of tube openings would also reduce the surface area of HNTs after modification, although this blockage was less likely to happen as the reduction in t-plot external surface area (Table 18.3) was not significant. T-plot analysis data showed that the amount of external surface area (related to mesopores, macropores, and outside surface) decreased slightly after modification, but the micropores surface area (surface area of pores less than 2 nm) decreased quite noticeably. In unmodified and NaOH-treated HNTs, micropores surface area constitutes a large portion (41 and 36%, respectively) of the total surface area (S_{BET}), whereas in modified HNTs the contribution of micropores surface area to total surface area was quite small (less than 17%), because of the filling of surface micropores after the modification. This result supports the hypothesis that modification mostly would take place on the defects of HNTs external surface.

TABLE 18.3   Surface area and pore size data of pure and modified samples.

	HNTs	NaOH-Treat-ed HNTs	APTES-Modified HNTs	APTES-Modified-Dried HNTs	APTES-Modified-NaOH-Treated HNTs	DGEBA-Mod-ified NaOH-Treated HNTs
S_{BET} (m²/g)	24.57	21.45	13.78	12.84	13.03	15.72
S_{ads} of pores (1.7–300.0 nm diameter) (m²/g)	20.12	18.51	9.27	8.97	7.44	10.56
S_{des} of pores (1.7–300.0 nm diameter) m²/g	27.16	24.92	11.92	13.21	8.51	14.03
t-plot micropore area (m²/g)	10.14	7.63	1.05	1.77	0.82	2.66
t-plot external sur-face area (m²/g)	14.43	13.82	12.72	11.08	12.21	13.06

TABLE 18.3 *(Continued)*

	HNTs	NaOH-Treated HNTs	APTES-Modified HNTs	APTES-Modified-Dried HNTs	APTES-Modified-NaOH-Treated HNTs	DGEBA-Modified NaOH-Treated HNTs
V_{des} of pores (1.7–300.0 nm diameter) (cm³/g)	0.15	0.17	0.12	0.13	0.07	0.11
V_{des} of pores (in the range of lumen size) (cm³/g)	0.039	0.035	0.017	0.011	0.013	0.023
Pore space (%)	39.87	43.27	30.96	32.97	18.22	29.08
Lumen space (%)	10.22	9.06	4.49	2.81	3.6	6.006
Average pore size (nm)	22.58	26.72	39.95	38.40	32.85	31.90

The pore size distributions, calculated from the desorption branch of the isotherm by the BJH method, showed that unmodified and NaOH-treated HNTs had a multimodal distribution of pores comprising of four distinct population peaks (Figure 18.7). The peak centered at 3.9 nm was attributed to the defects of HNTs' surface and the peak at 10 nm could be primarily identified as the lumen of HNTs, respectively. Two peaks of larger sizes corresponded to the spaces between nanotubes in HNTs agglomerates.

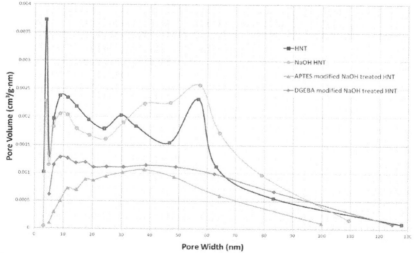

FIGURE 18.7   BJH pore size distribution curves (from desorption data).

In modified samples, the 3.9 nm peak had completely disappeared. Absence of this peak, corresponding to surface defects, indicated that modifier molecules were grafted onto the defects of the HNTs' surface and almost fully filled the surface micropores. The peak around 10 nm, assigned to the lumen of HNTs, was resolved in the modified HNTs indicating that the lumen space of HNTs was not blocked during the modification process. As reported by Yuan et al. (2008), the presence of air in the lumen of HNTs prevents the modifiers (APTES and DGEBA) from being penetrated into the lumen, leaving the lumen space unaffected. However, BJH pore distribution curves (Figure 18.7) showed a reduction in the population of pores in the range of lumen sizes and the population of larger pores of modified HNTs. Moreover, the pore size distribution of halloysites was broadened by modification. The pore distribution curves of modified samples were less segmented and their peaks were less distinct. The reduction in pores' volumes in modified HNTs may be attributed to formation of halloysite clusters with irregular and heterogeneous porosity and confinement of the lumen openings in halloysite clusters, as a result of the hindrance effect of grafted modifiers on the adsorption of nitrogen molecules, and the partial closure of nanotube openings by modifier molecules.

The average pore size of HNTs increased after modification with APTES and DGEBA (Table 18.3). This value is a weighted average of all pores, including surface micropores, lumen pores, and spaces between the nanotubes in HNTs' clusters. This increase in average pore size can be explained by the filling of surface micropores and broader pore size distribution curves which caused the average size to shift to higher values.

18.4 CONCLUSION

DGEBA were successfully grafted on HNTs to functionalize their surface with epoxide groups and the properties of DGEBA grafted HNTs were studied in comparison to those of silane-grafted HNTs. Treatment of HNTs with NaOH before modification could effectively enhance the grafting of modifiers on HNTs surface. Organosilane (APTES) functionalization of HNTs was a complex chemical reaction resulting in a cross-linked network of APTES molecules on HNTs surface. This network was formed by various kinds of attachment of APTES including direct grafting, vertical and horizontal polymerization, and hydrogen bonding. The morphology of the grafted-APTES network was highly sensitive to pretreatment (thermal and NaOH pretreatment) and the post-cure condition of HNTs which may affect its performance as a nanofiller.

On the other hand, modification of HNTs with DGEBA took place through a simple ring opening reaction between OH groups of HNTs and the epoxide ring of DGEBA. Compared with organosilane modification, this reaction is less sensitive to HNTs treatment condition. This method can be also used for functionalizing HNTs with double bonds or methacrylate groups using epoxide containing molecules such as allylglycidylether (AGE) and glycidylmethacrylate (GMA).

The results of this research showed that the chemistry of the external surface of modified HNTs is highly dependent on pretreatment, reaction, and post-cure conditions, which may affect their reinforcement efficiency in polymer matrixes. Our primary research on properties of epoxy/modified HNTs nanocomposites (not completed yet and not shown here) showed that the impact properties of epoxy/modified HNTs nanocomposites were highly affected by the surface chemistry of modified HNTs (impact energy of epoxy/modified HNTs was improved by 200% in some cases). Hence, tuning the surface chemistry of HNTs is important to exploit the full benefit of HNTs as nanofiller in polymeric matrixes.

KEYWORDS

- **Bisphenol A diglycidyl ether**
- **Halloysite nanotubes**
- **Surface modification**
- **Surface NaOH treatment**
- **γ-aminopropyltriethoxysilane**

ACKNOWLEDGEMENT

Financial support from Monash University Malaysia is gratefully acknowledged for the PhD scholarship scheme. This project was funded by a grant FRGS/2/2013/TK04/MUSM/03/1 from the Ministry of Higher Education, Malaysia and Advanced Engineering Platform (AEP) under the "Nanostructures and Advanced Materials for Sustainable Energy Production" sub-projects also from Monash University Malaysia.

REFERENCES

1. Abdullayev, E.; Joshi, A.; Wei, W.; Zhao, Y.; and Lvov, Y.; *ACS Nano.* **2012,** *6(8),* 7216–7226.
2. Ao, Y-H.; Tang, K.; Xu, N.; Yang, H-d.; Zhang, H-X.; *Poly. Bull.* **2007,** *59(2),* 279–288.
3. Bailey, J. R.; McGuire, M. M.; *Langmuir.* **2007,** *23(22),* 10995–10999.
4. Barrientos-Ramírez, S.; Oca-Ramírez, G. M. D.; Ramos-Fernández, E. V.; Sepúlveda-Escribano, A.; Pastor-Blas, M. M.; and González-Montiel, A.; *Appl. Cataly. A: Gen.* **2011,** *406(1–2),* 22–33.
5. Christy, A. A.; *Int. J. Chem. Environ. Eng.* **2011,** *2(1),* 27–32.
6. Del Rey-Bueno, F.; Romero-Carballo, J.; Villafranca-Sanchez, E.; Garcia-Rodriguez, A.; and Sebastian-Pardo, E. N.; *Mater. Chem. Phys.* **1989,** *21(1),* 67–84.
7. Deng, S.; Zhang, J.; and Ye, L.; *Compos. Sci. Technol.* **2009,** *69(14),* 2497–2505.
8. Dias Filho, N. L.; Do Carmo, D. R.; and Somasundaran, P.; Encyclopedia of Surface and Colloid Science. 2nd edition, New York: Ed Taylor & Francis; 2006, 209–228.

9. Du, M.; Guo, B.; Liu, M.; and Jia, D.; *Polym. J.* **2006**, *38(11)*, 1198–1204.
10. Du, M.; Guo, B.; Liu, M.; and Jia, D.; *Polym. Polym. Compos.* **2007**, *15(4)*, 321–328.
11. Dugas, V.; and Chevalier, Y.; *J. Colloid Interface Sci.* **2003**, *264(2)*, 354–361.
12. Frost, R. L.; and Vassallo, A. M.; *Clays Clay Minerals.* **1996**, *44(5)*, 635–651.
13. Guo, B.; Lei, Y.; Chen, F.; Liu, X.; Du, M.; and Jia, D.; *Appl. Surf. Sci.* **2008**, *255(5)*, 2715–2722.
14. Guo, B.; Zou, Q.; Lei, Y.; and Jia, D.; *Polym. J.* **2009**, *41(10)*, 835–842.
15. Handge, U. A.; Hedicke-Höchstötter, K.; and Altstädt, V.; *Polymer.* **2010**, *51(12)*, 2690–2699.
16. Holsti-Miettinen, R. M.; Heino, M. T.; and Seppälä, J. V.; *J. Appl. Polym. Sci.* **1995**, *57(5)*, 573–586.
17. Ismail, H.; Pasbakhsh, P.; Fauzi, M.; and Abubakar, A.; Polym. Test. **2008**, *27(7)*, 841–850.
18. Jia, Z-X.; Luo, Y-F.; Yang, S-Y.; Guo, B-C.; Du, M-L.; and Jia, D-M.; *Chin. J. Polym. Sci.* **2009**, *27(6)*, 857–864.
19. Joussein, E.; Petit, S.; Churchman, J.; Theng, B.; Righi, D.; and Delvaux, B.; *Clay Minerals.* **2005**, *40(4)*, 383–426.
20. Kim, J.; *PIKE Technol.* **2010**.
21. Kim, J.; Holinga, G. J.; and Somorjai, G. A.; *Langmuir.* **2011**, *27(9)*, 5171–5175.
22. Kim, J.; Seidler, P.; Fill, C.; and Wan, L. S.; *Surf. Sci.* **2008**, *602(21)*, 3323–3330.
23. Levis, S. R.; and Deasy, P. B.; *Int. J. Pharm.* **2003**, *253(1–2)*, 145–157.
24. Liu, M.; Guo, B.; Du, M.; Lei, Y.; and Jia, D.; *J. Polym. Res.* **2007**, *15(3)*, 205–212.
25. Liu, Y-L.; Hsu, C-Y.; Wang, M-L.; and Chen, H-S.; *Nanotechnology.* **2003**, *14(7)*, 813.
26. Lygin, V. I.; *Russ. J. Gen. Chem.* **2001**, *71(9)*, 1368–1372.
27. Machado, G. S.; de Freitas Castro, K. A. D.; Wypych, F.; and Nakagaki, S.; *J. Mole. Catal. A: Chem.* **2008**, *283(1–2)*, 99–107.
28. Mu, B.; Zhao, M.; and Liu, P.; *J. Nano. Res.* **2007**, *10(5)*, 831–838.
29. Ning, N-Y.; Yin, Q-J.; Luo, F.; Zhang, Q.; Du, R.; and Fu, Q.; *Polymer.* **2007**, *48(25)*, 7374–7384.
30. Osman, M. A.; Rupp, J. E. P.; and Suter, U. W.; *Polymer.* **2005**, *46(5)*, 1653–1660.
31. Pasbakhsh, P.; Ismail, H.; Fauzi, M. N. A.; and Bakar, A. A.; *Appl. Clay Sci.* **2010**, *48(3)*, 405–413.
32. Pavlidou, S.; and Papaspyrides, C. D.; *Prog. Polym. Sci.* **2008**, *33(12)*, 1119–1198.
33. Pierotti, R.; and Rouquerol, J.; *Pure Appl. Chem.* **1985**, *57(4)*, 603–619.
34. Qi, R.; et al. *J. Mater. Chem.* **2010**, *20(47)*, 10622–10629.
35. Reddy, C. S.; and Das, C. K.; *J. Appl. Polym. Sci.* **2006**, *102(3)*, 2117–2124.
36. Reddy, C. S.; Das, C. K.; and Narkis, M.; *Polym. Compos.* **2005**, *26(6)*, 806–812.
37. Sadhu, S.; and Bhowmick, A. K.; *Rubber Chem. Technol.* **2003**, *76(4)*, 860–875.
38. Sailaja, R.; *Compos. Sci. Technol.* **2006**, *66(13)*, 2039–2048.
39. Shchukin, D. G.; Lamaka, S. V.; Yasakau, K. A.; Zheludkevich, M. L.; Ferreira, M. G. S.; and Mohwald, H.; *J. Phys. Chem. C* **2008**, *112(4)*, 958–964.
40. Shchukin, D. G.; and Möhwald, H.; *Adv. Funct. Mater.* **2007**, *17(9)*, 1451–1458.
41. Shchukin, D. G.; Sukhorukov, G. B.; Price, R. R.; and Lvov, Y. M.; *Small.* **2005**, *1(5)*, 510–513.
42. Soheilmoghaddam, M.; Wahit, M. U.; Mahmoudian, S.; and Hanid, N. A.; *Mater. Chem. Phys.* **2013**, *141(2–3)*, 936–943.
43. Strawn, D. G.; Palmer, N. E.; Furnare, L. J.; Goodell, C.; Amonette, J. E.; and Kukkadapu, R. K.; *Clays Clay Minerals.* **2004**, *52(3)*, 321–333.
44. Suh, Y. J.; Kil, D. S.; Chung, K. S.; Abdullayev, E.; Lvov, Y. M.; and Mongayt, D.; *J. Nanosci. Nanotechnol.* **2011**, *11(1)*, 661–665.
45. Tang, Y.; Deng, S.; Ye, L.; Yang, C.; Yuan, Q.; Zhang, J.; and Zhao, C.; *Compos. Part A: Appl. Sci. Manuf.* **2011**, *42(4)*, 345–354.

46. Teng C-C; Ma C-CM; Yang, S-Y.; Chiou, K-C.; Lee, T-M.; and Chiang, C-L.; *J. Appl. Polym. Sci.* **2012**, *123(2)*, 888–896.
47. Tierrablanca, E.; Romero-García, J.; Roman, P.; and Cruz-Silva, R.; *Appl. Catal. A: Gen.* **2010**, *381(1–2)*, 267–273.
48. Vrancken, K. C.; Van Der Voort, P.; Gillis-D'Hamers, I.; Vansant, E. F.; and Grobet, P.; *J. Chem. Soc. Faraday Trans.* **1992**, *88(21)*, 3197–3200.
49. Wang, K. H.; Choi, M. H.; Koo, C. M.; Choi, Y. S.; and Chung, I. J.; *Polymer.* **2001**, *42(24)*, 9819–9826.
50. White, R. D.; Bavykin, D. V.; and Walsh, F. C.; *Nanotechnology.* **2012**, *23(6)*, 065705.
51. Yah, W. O.; Takahara, A.; and Lvov, Y. M.; *J. Am. Chem. Soc.* **2012**, *134(3)*, 1853–1859.
52. Yan, L.; Jiang, J.; Zhang, Y.; and Liu, J.; J. Nanopart. Res. **2011**, *13(12)*, 6555–6561.
53. Yang, S-Q.; Yuan, P.; He, H-P.; Qin, Z-H.; Zhou, Q.; Zhu, J-X.; and Liu, D.; *Appl. Clay Sci.* **2012**, *62–63(0)*, 8–14.
54. Ye, Y.; Chen, H.; Wu, J.; and Ye, L.; *Polymer.* **2007**, *48(21)*, 6426–6433.
55. Yin, B.; and Hakkarainen, M.; *Mater. Chem. Phys.* **2013**, *139(2–3)*, 734–740.
56. Yuan, P.; et al. *J. Phys. Chem. C* **2008**, *112(40)*, 15742–15751.
57. Yuan, P.; Southon, P. D.; Liu, Z.; and Kepert, C. J.; *Nanotechnology.* **2012**, *23(37)*, 375705.
58. Zhai, R.; Zhang, B.; Liu, L.; Xie, Y.; Zhang, H.; and Liu, J.; *Cataly. Commun.* **2010**, *12(4)*, 259–263.
59. Zheng, H.; Zhang, Y.; Peng, Z.; and Zhang, Y.; *J. Appl. Polym. Sci.* **2004**, *92(1)*, 638–646.
60. Zou, W.; Du, Z-J.; Liu, Y-X.; Yang, X.; Li, H-Q.; and Zhang, C.; *Compos. Sci. Technol.* **2008**, *68(15–16)*, 3259–3264.

PART VIII
NATURAL MINERAL NANOTUBES AS NANOREACTORS USED IN INDUSTRIAL AND AGRICULTURAL APPLICATIONS

CHAPTER 19

HALLOYSITE TUBULE NANOREACTORS FOR INDUSTRIAL AND AGRICULTURAL APPLICATIONS

ELSHAD ABDULLAYEV and YURI LVOV

CONTENTS

19.1   INTRODUCTION

Halloysite is a two-layered aluminosilicate which has a predominantly hollow tubu-lar structure in the submicron range and is chemically similar to kaolinite (Joussein et al., 2005). It is mined commercially from natural deposits in USA, New Zealand, China, Turkey, and Brazil (Bates et al., 1950). Dragon Mine in Utah has halloysite nanotubes (HNTs) with high purity exceeding 99 percent (Abdullayev et al., 2013). These minerals are formed from kaolinite over millions of years because of hydro-thermal processes (Tari et al., 1999). Layers are rolled into tubes because of the strain caused by lattice mismatches between adjacent silicon dioxide and aluminum oxide sheets (e.g., Joussein et al., 2005, Abdullayev et al., 2013).

Halloysite was discovered in the eighteenth century by Berthier (1826) and named in honor of the Belgian geologist Omalius d'Halloy who analyzed the min-eral. In early years it was very difficult to distinguish halloysite from other minerals, particularly from kaolinite. However, X-ray analysis has shown that it has a unique crystalline structure (Grim, 1968). Halloysite has been extensively used as a raw material in the ceramics industry, especially for the manufacture of porcelain and bone china (e.g., Wilson, 2004; Joussein et al., 2005). Having nanotubular struc-ture, HNTs can potentially be applied in several fields of nanotechnology. These multilayer tubes are commonly used for plastic composites, in controlled release applications (e.g., Price et al., 2001; Abdullayev et al., 2009; Abdullayev and Lvov, 2010; Abdullayev et al., 2011; Abdullayev and Lvov, 2011; Suh et al., 2011; Wei et al., 2012; Lvov and Abdullayev, 2013), and may be coated with metal by electro-less plating to make conductive fillers (Baral et al., 1993). Because of their porous structure and high catalytic activity, halloysite particles can be used in remediation of acid mine drainage, petroleum conversion in refining industries, as well as in the separation of liquids and gaseous mixtures. This paper summarizes the structure, physicochemical properties, and major application areas of HNTs.

19.2   HALLOYSITE STRUCTURE

Halloysite exists in nature as a white mineral that can easily be processed to obtain fine powder (Figure 19.1(a)). In some deposits, the mineral is colored from yellow-ish to brown or greenish because of trace amounts of metal ions such as Fe^{+3}, Cr^{+3}, and Ti^{+4}, that substitute Al^{+3} or Si^{+4} in the crystal lattice (Joussein et al., 2005). Scan-ning electron microscope (SEM) images of the halloysite samples clearly indicate that they have a tubular-shaped nature (Figure 19.1(b)).

Naturally, halloysite occurs as a hydrated mineral that consists of rolled sheets with the ideal formula of $Al_2Si_2O_5 (OH)_4 \cdot nH_2O$. Chemically, it is similar to kaolinite except for the presence of water molecules between the adjacent layers. Halloysite with $n = 2$ has a layer periodicity of 10 Å and is called halloysite—10 Å. Upon moderate heating of halloysite—10 Å, it loses interlayer water and is converted to halloysite—7 Å though complete removal of surface water takes place at about 300 °C (e.g., Bates et al., 1950; Tari et al., 1999; Joussein et al., 2005; and Abdullayev et al., 2013). Dehydrated halloysite has a formula of $Al_2Si_2O_5(OH)_4$ with interlayer spacing of about 7.2 Å (Joussein et al., 2005).

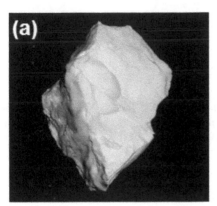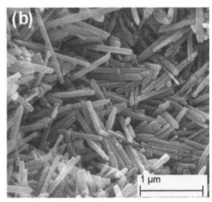

FIGURE 19.1   (a) Raw halloysite mineral from Dragon Mine (Applied Minerals, Inc.) and (b) SEM image from the rock, showing over 99 percent nanotube content. Reproduced with permission from (Abdullayev and Lvov, 2013). Copyright 2013 Royal Society of Chemistry.

Halloysite sheets are made of aluminum and silicon oxide layers (Figure 19.2(a)). Each deposit has HNTs with certain diameter and length. In general, the outside diameter of the tubes lies in the range from 50 to 200 nm, whereas the diameter of inner pores ranges from 10 to 40 nm (Joussein et al., 2005). The lengths of the nanotubes are about 0.5–1.5 μm (e.g., Bates et al., 1950; Abdullayev and Lvov, 2011; Lvov and Abdullayev, 2013). An important aspect of the HNT structure is the different surface chemistry at the inner and outer faces of the tubes (Baral et al., 1993; Veerabadran et al., 2007; Abdullayev and Lvov, 2011; Lvov and Abdullayev, 2013) (Figure 19.2(a)). Silicon dioxide is relevant to the outer surface, whereas the aluminum oxide is relevant to the inner face of the tubules. These layers have different dielectric and ionization properties, which is evident from electrical zeta-potentials in water (Figure 19.2(b)). Aluminum oxide has a positive charge up to pH value of 8.5, whereas silicon dioxide is negative above a pH of 1.5. The overall zeta potential of HNT is close to that of the silica. On the other hand, different charges at the inner and outer faces of the halloysite nanotubes allow for the selective internal loading of negatively charged molecules (e.g., Price et al., 2001; Abdullayev et al., 2009; Abdullayev and Lvov, 2010; Abdullayev et al., 2011; Abdullayev and Lvov, 2011; Suh et al., 2011; Wei et al., 2012; Lvov and Abdullayev, 2013).

(a)

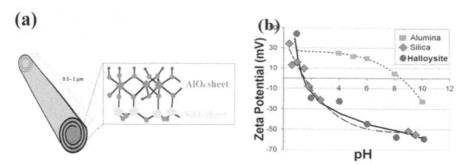

FIGURE 19.2 (a) Schematic representation of the HNT tubular structure and wall chemistry, (b) variation of the silica and alumina surface potentials by pH of the solution. Reproduced with permission from (Abdullayev and Lvov, 2013). Copyright 2013 Royal Society of Chemistry.

19.3 NANOTUBES FOR LOADING AND CONTROLLED RELEASE

HNT has been loaded with a large variety of chemical and biological active agents including biocides, corrosion inhibitors, metals, and salts. Figure 19.3 shows TEM images of HNTs loaded with metallic silver nanoparticles and nanorods. Silver was loaded into HNT using aqueous silver acetate solution. Nanotube lumen is partially filled with silver nanorods and nanoparticles. The loaded nanoparticles have a crystalline structure as indicated by high resolution TEM and XRD analysis (Abdullayev et al., 2011).

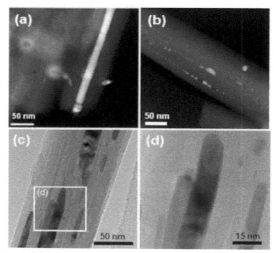

FIGURE 19.3 TEM images of the HNT loaded with silver. Reproduced with permission from (Abdullayev et al., 2011). Copyright 2011 American Chemical Society.

Recently, it has been explicitly shown that controlled release of corrosion inhibitors plays a critical role in delaying the rusting of steel in salty water solution (Joshi et al., 2010). Self-healing epoxy composites have been demonstrated by encapsulating diglycidyl ether of bisphenol A within HNT lumen. Self-healing effect has been achieved by leaking the loaded bisphenol monomer at the cracks and by cross linking with an amine-curing agent (Abdullayev et al., 2010, Abdullayev and Lvov, 2011).

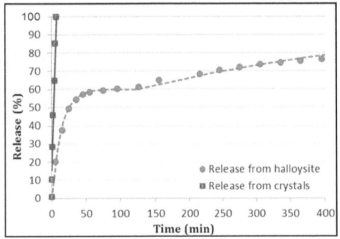

FIGURE 19.4 Brilliant Green dissolution curve in water (squares and solid line) and sustained release from HNT (circles and dashed line). Reproduced with permission from (Abdullayev and Lvov, 2013). Copyright 2013 Royal Society of Chemistry.

Release kinetics of the common antiseptic agent, Brilliant Green, from halloysite pores has been studied. Free Brilliant Green crystals completely dissolve in water within 5 min, but the HNT lumen extends its release over 8 h (Figure 19.4). According to the predicted release model, near complete (95%) release of Brilliant Green has been achieved within 30 h, which is 200 times slower than the release of free antiseptic. Most proteins diffuse from nanotube pores in a rather slow fashion with complete release lasting 50–500 h. This is because of their stronger interaction with lumen walls and lower diffusion coefficients of large macromolecules (e.g., Abdullayev and Lvov, 2013).

19.3.1 SYNTHESIS OF RELEASE STOPPERS

Synthesis of artificial caps at HNT tube endings allows better encapsulation and control over release rates. There are several ways for the synthesis of release stoppers like coating of nanotubes with polymers, layer by layer self assembly with polyelectrolytes, complexation of drug-loaded halloysite with chitosan, and so on. One of the most effective and simplest approaches involves the interaction of leaking agent-A and complex forming agent-B (from external solution). This method has been effectively implemented for the encapsulation of triazole and imidazole derivatives widely used in medicine and industry. These substances form thin films at tube ends while interacting with transition metal ions such as Cu (II) and Fe (II) (Figure 19.5). The film clots the tube endings and loaded agents are entrapped into the tube lumen (Abdullayev et al., 2009; Abdullayev and Lvov, 2011).

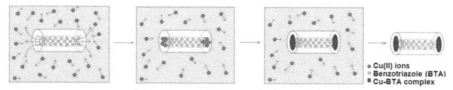

FIGURE 19.5   Illustration of stopper formation at halloysite tube endings by interaction of leaking benzotriazole with Cu (II) ions. Reproduced with permission from (Abdullayev et al., 2009). Copyright 2009 American Chemical Society.

Controllable benzotriazole release from halloysite pores was achieved by mixing loaded nanotubes and the aqueous solution containing Cu (II) ions. Original halloysite releases benzotriazole from the tube openings, and the leakage of Cu (II) ions is retarded because of the formation of Cu-BTA complexes. Release from capped tubes occurs through much smaller pores existing in the film. Tightness of encapsulation depends on the concentration of the Cu (II) ions in the solution, that is, higher concentration leads to tighter encapsulation (Abdullayev and Lvov, 2011). Decomposition Cu-BTA film with ammonia solution resumed the original release rate of the chemical agent (Abdullayev et al., 2009).

19.3.2   HNT LUMEN ENLARGEMENT

Loading efficiency of 5–10 wt % is easily achieved with raw HNT. Enlargement of the lumen diameter results in more space to be available for an internal loading of active agents. This has been achieved by selective etching of aluminum oxide layers from halloysite inner lumen by sulfuric acid yielding aluminum sulfate (Figure 19.6) (Abdullayev et al., 2012). Etching takes place in the inner pores because the external surface, that is, silicon dioxide, is inert toward sulfuric acid.

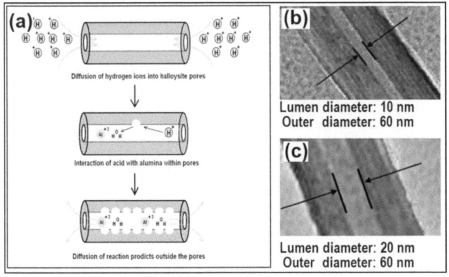

FIGURE 19.6 (a) Schematic of HNT lumen etching. TEM images of (b) original and (c) sulfuric acid treated halloysite (1 M acid, at 50 °C for 30 h). Reproduced with permission from (Abdullayev et al., 2012). Copyright 2012 American Chemical Society.

Enlarged lumen provides bigger space for loaded agents and hence improves loading efficiency. As much as 30 wt % of Epon 828 resin has been encapsulated within the lumen of the acid etched HNT, a threefold increase compared to untreated clay (Abdullayev and Lvov, 2013, Konnova et al., 2013). Similar observations have also been reported for benzotriazole (Abdullayev et al., 2012).

19.4 INDUSTRIAL APPLICATIONS

Having high surface area, elongated tubular shape with lumen which can serve as nanocontainer, and being environmental friendly, halloysite mineral can be extensively used in industry. Unlike other nano-tubular materials (such as boron nitride silica or carbon nanotubes) HNT is a readily available cheap material, which makes it attractive for many technological applications. It belongs to the family of the clay minerals and can substitute kaolinite, montmorillonite, and bentonite as nanofiller in composites. In paper production, modification of wood fibers by halloysite nanotubes was proven to increase the brightness and porosity of the paper sheets (Lu et al., 2007, Lvov, 2008). In the ceramic industry, halloysite was mainly used in the tableware market as an addition to bone china and porcelain products (Wilson, 2004). Halloysite also has several current potential applications in the field of nanotechnology. Several companies in the United States, New Zealand, and Turkey (such as Applied Minerals and Imerys) are supplying halloysite materials for industry. Halloysite nanotubes can be applied in the following major areas: (1) Polymer composites, (2) Protective coatings, (3) Concrete and cement, and (4) Heterogeneous catalysts.

19.4.1 POLYMER COMPOSITES

Polymer nanocomposites have increased strength (Ray and Okamoto, 2003), thermal resistance, flame retardancy (Du et al., 2006; Du et al., 2007; and Ismail et al., 2008), and decreased gas and liquid permeability (Ray and Okamoto, 2003). Two types of commonly used fillers are platy clays such as montmorillonite, hectorite and saponite (Ray and Okamoto, 2003), and carbon-based fillers like graphene, carbon fibers, and nanotubes (Velasco-Santos, 2005; Coleman et al., 2006). HNTs have advantages over carbon-based fillers, that is, they are much cheaper than carbon nanotubes and are non toxic. In spite of not being as strong as carbon nanotubes, HNTs still provide significant improvement in polymer tensile strength (5–10 wt % addition often doubles the composite strength) (Du et al., 2006; Du et al., 2007; and Ismail et al., 2008). Halloysite's advantage over platy clay minerals includes ease of blending with polymers. Clay layers are strongly stacked to each other by surface hydroxyl groups via ionic and hydrogen bonds, requiring surface modification and a costly exfoliation process to achieve nanodispersion within the polymer matrix (Ray and Okamoto, 2003). On the contrary, HNT has only few hydroxyl groups located at the edges of the tubes, and its tubule shape prevents tight particle stacking. It readily disperses in most of the polymers having moderate and high polarity without requiring any surface treatments (e.g., polyamides, polyethylene tereftalates, epoxy based polymers, and polysaccharides) (Du et al., 2010).

HNTs improved tensile strength and fracture behavior of epoxy nanocomposites (Figure 19.7). By adding 10 percent of HNTs, the critical stress intensity factor can be increased from 0.9 to 1.4 MPa'm$^{1/2}$ (Ye et al., 2007; Deng et al., 2009). Reasonably good dispersion of the HNTs within ethylene propylene diene monomer composite has been achieved even at the halloysite loading levels of 50 wt %. Tensile strength of this composite increased from 1.3 MPa (neat polymer) to 12.9 MPa (composite), while elongation at break increased by three times, which is generally unusual for polymer–clay composites (Ismail et al., 2008) . Similar results were also obtained on nanocomposites of HNTs with styrene-butadiene rubber (SBR) (Guo et al., 2009).

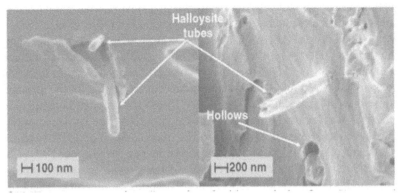

FIGURE 19.7   SEM images of HNT-epoxy composites. Reproduced with permission from (Deng et al., 2009). Copyright 2009 Elsevier.

HNTs have formed an effective network for the preservation of shape and structural integrity in phase change materials. Introducing 45 wt % HNTs into paraffin wax, a commonly used phase change material for thermal energy storage, preserved composite shape upon complete melting in a 95 °C oven (Figure 19.8(b)). On the contrary, the original wax was completely spread within a petridish, and its shape was totally lost above 72 °C (i.e., melting point, Figure 19.8(a)). This is related to the entangled network formed by halloysite nanotubes within paraffin matrix. Molten wax attached to the halloysite network and did not flow similar to a sponge filled with water. This new composite can be considered as a promising heat storage material because of its good thermal stability, high thermal/electrical conductivity, and the ability to preserve its shape during phase transitions.

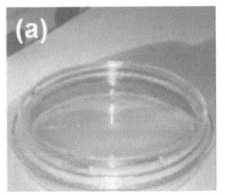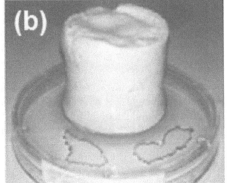

FIGURE 19.8   (a) Original paraffin wax and (b) 50 percent HNT-wax composite stored at 95 °C. Reproduced with permission from (Lvov and Abdullayev, 2013). Copyright 2013 Elsevier.

HNTs have been effectively used as heterogeneous nucleating agents for several polymers like polyethylene, isotactic polypropylene (Ning et al., 2007; Du et al., 2010), polybutylene terephthalate (Oburoglu et al., 2012), and polyamide 6 (Prashantha et al., 2011). Unlike other clay fillers, halloysite does not require organic surface modifiers like quarternary amine salts or organosilanes lowering the surface energy. The surface energy is believed to be responsible for the formation of crystalline spherulites as shown in Figure 19.9. Non-isothermal peak temperature of crystallization was increased from 110.7 to 119.6 °C, whereas half-time of crystallization was reduced from 0.525 to 0.458 min. upon addition of 30 wt % halloysite within isotactic polypropylene at 20 °C/min cooling rate (determined by DSC using Avrami-Jeziorni method). Activation energy of the crystallization increased from 206.2 to 267.3 kJ/moles, which is associated with the confinement effect of halloysite restricting polymer chain movements (Du et al., 2010). Similar effects have been reported for polyamide 6 and polybutylene terephthalate composites (Prashantha et al., 2011; Oburoglu et al., 2012). On the contrary, organo-modified montmorillonite did not show any crystal nucleation because of the reduction of surface energy by chemical treatment (Oburoglu et al., 2012). Nucleation capability in combination with fine dispersion, higher nanotube tensile strength, and interaction with polymer chains are believed to be the main reason behind tensile and impact strength improvement, whereas the enhanced nucleation also significantly reduces the cycle time for extruded polymer parts.

FIGURE 19.9 The final crystal morphology of pure polypropylene and polypropylene—10 percent HNT composites crystallized at 128°C. Reproduced with permission from (Ning et al., 2007). Copyright 2007 Elsevier.

HNT–polymer composites show increased flame retardancy and thermal stability. Maximum weight loss temperature of polypropylene increased from 351°C to 425°C, upon addition of 10 pecent halloysite. In cone calorimetry, 10 wt % HNT–PP composite delayed the ignition and generated less smoke (Du et al., 2007). Similar results were also obtained for HNT–polyamide (Lecouvet et al., 2011) and HNT–epoxy (Liu et al., 2008) nanocomposites. Improved flame retardancy is associated with enhanced charring, dehydration, and progressive formation of an inorganic rich layer at the composite surface.

19.4.2 PROTECTIVE COATINGS

Inorganic nanoparticles are widely used in current paint products. Some of these nanoparticles (e.g., rutile and clay) are added to improve some properties of the paint, whereas other nanoparticles (e.g., spherical silica) are incorporated just to reduce production costs (Lambourne and Strivens, 1999; Laurence, 2006). Halloysite nanotubes are promising for the paint industry because of being readily dispersed in a large variety of the liquid paint formulations and their ability to significantly improving their mechanical properties after curing.

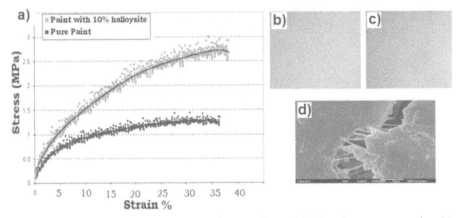

FIGURE 19.10 Mechanical properties of pure paint and HNT–paint nanocomposite. (a) Stress-strain curves and images of the (b) pure alkyd paint and (c) paint with 10 wt % HNTs after rapid deformation test. Reproduced with permission from (Abdullayev et al., 2011). Copyright 2011 American Chemical Society. (d) SEM image of a microcrack on halloysite epoxy composite. Reproduced with permission from (Ye et al., 2007). Copyright 2007 Elsevier.

Stress–strain curve of the HNT-based composite coating is shown in Figure 19.10(a) along with the original unmodified paint. The 10 wt % HNT–paint showed significant higher tensile strength and much better performance toward rapid deformation (Figure 19.10(b), (c)). Rapid deformation was caused by dropping a 0.2 kg metal bar onto painted ASTM A366 steel plates of 1 m height. Metal coated with original paint had a lot of cracks on it, whereas the same paint containing 10 wt % HNTs was intact (Abdullayev Abbasov et al., 2009). This is associated with an effective dissipation of the impact energy by clay nanotube de-bonding (pull-out) and bridging effects (Figure 19.10(d)) (Ye et al., 2007).

HNT lumens loaded with corrosion inhibitors provide sustained release for self-healing coatings (Abdullayev et al., 2009; Abdullayev and Lvov, 2010). Corrosion resistance, studied on painted copper strips (Figure 19.11), revealed the increased anticorrosion performance of the coatings containing inhibitor-filled HNTs. Strips were exposed to high-corrosive liquid containing 30 g/L of NaCl for 6 months. As it can be seen from Figure 19.11(a), a significant amount of green corrosion products was accumulated underneath the paint on conventional strips, whereas those coated with HNT–paint composite show no evidence of corrosion (Figure 19.11(b)) (Abdullayev, Abbasov et al., 2013). The reason for the increased corrosion resistance is the slow release of the inhibitor entrapped into the inner core of the tubes. Corrosion inhibitor slowly releases to the corrosive media once the paint is damaged and protects the metal.

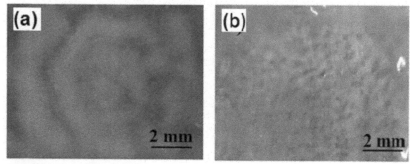

FIGURE 19.11   Copper strips coated with (a) pure paint and (b) paint–HNT nanocomposite. Reproduced with permission from (Abdullayev et al., 2013). Copyright 2013 American Chemical Society.

HNTs also significantly improved the anticorrosive performance of sol-gel coatings. Localized corrosion current densities of sol-gel coated and artificially scratched aluminum strips (2024 alloy) are shown in Figure 19.12. Activity of the anodic corrosion current at the sol-gel film is very high and is rapidly increased for several hours, indicating a fast pitting corrosion. On the contrary, the anodic current is completely suppressed on modified sol-gel coating in presence of HNTs, which clearly indicates an efficient corrosion protection (Shchukin and Möhwald, 2007; Lvov et al., 2008; and Abdullayev et al., 2009).

(A)   (B)

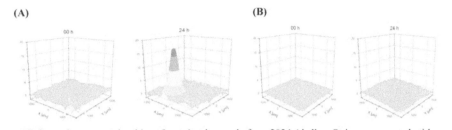

FIGURE 19.12   Corrosion current densities of metal strips made from 2024 Al alloy. Strips were coated with usual sol-gel coating (a) and with sol-gel containing HNTs (b) Reproduced with permission from (Lvov et al., 2008). Copyright 2008 American Chemical Society.

19.4.3 CONCRETE AND CEMENT

Concrete has many advantages as a construction material because of being mold-able, easily engineered, and rather cheap. The incorporation of clay minerals and chemical additives within cement results in high performance concrete with good workability and enhanced strength. HNT has favorable chemical composition and physical properties for enhancing the performance of cementitious composites. It is active as pozzolan, that is, it reacts with cement during hydration to form cementi-tious compounds with improved mechanical properties (John et al., 2011). Disso-ciated alumina from HNT pores is consumed to form cementitious minerals like calcium aluminosilicate. Silica from the halloysite outer surface reacts with calcium to form hydrated-calcium silicates (Kamruzzaman, 2006). The nanosize of particles (i.e., increased specific surface area) provides higher reactivity toward hydration process.

A study on concrete formulation based on type I Portland cement with 5 wt % silica fume and up to 3 percent untreated HNT revealed an increase in compression strength up to 24 percent and a decrease in flow up to 65 percent (Farzadnia, 2013). In another study, addition of 15 percent thermally treated halloysite within porous lightweight concrete prepared from Portland cement and expanded vermiculite re-sulted in an over 100 percent increase in compression strength (Kwon, 2008). Incor-poration of nanoclay promoted the formation of calcium silicate monohydrate, with higher consumption of calcium hydroxide and denser microstructures. HNTs filled up the voids and led to denser mortar. Enhanced consumption of calcium hydroxide, to form $CaSiO_3 \cdot H_2O$, is associated with the presence of SiO_2 on an outer surface of halloysite. Clay swelling with entrapped water leads to its expansion and results in filling the capillary pores. The entrapped water is gradually released and caused fur-ther hydration by increasing both the strength and density. Elongated tubular mor-phology of halloysite provides additional improvement in strength because of the bridging of cracks formed in the cement matrix (Figure 19.13).

 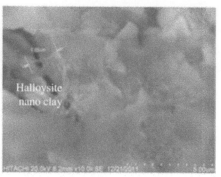

FIGURE 19.13 Scanning Electron Micrograph of the concrete prepared by type I Portland cement; control (left) and halloysite filled (right) samples. Bridging of gaps within concrete by HNT is clearly visible. Reproduced with permission from (Farzadnia et al., 2013). Copyright 2013 Elsevier.

Rabehi et al. (2012) studied the pozzolanic activity of Algerian halloysite treated at various temperatures in lime solution using X-ray diffraction and concluded that the clay calcinated at 750°C has remarkably high reactivity. Over 85 percent of the calcium hydroxide has reacted with HNT to form calcium silicate and aluminate that are responsible for the formation of high strength concrete. Similar data has also been reported by Kwon et al. (2008) on HNT samples from Korea. Incorporation of 10 wt % calcinated halloysite improved the resistance toward sulfuric acid etching by 32 percent, whereas the chloride ion diffusion was reduced by a factor of three. This is associated with reduced micro porosity and enhanced consumption of calcium hydroxide during pozzolanic reaction. In our recent research and development work with geopolymers, we demonstrated "smart" composites doped with halloysite. Adding these clay nanotubes into fly ash geopolymers at 5–6 wt % provided sustained release of active agents resulting in prolonged antibacterial protection without losing strength. Synergy in the improvement of mechanical properties, the ability to adjust setting time, and simplicity in geopolymers modification are attractive features of the work.

19.4.4 HETEROGENEOUS CATALYSTS

HNTs have been effectively used as a catalyst for synthesizing biofuels, converting hydrocarbons, and supporting various chemical processes. Blending of HNTs with commercial Y-zeolites provided a low cost alternative for the production of mesoporous cracking catalysts with low environmental footprint. Having larger pores (10–20 nm in diameter) with highly reactive Al^{3+} sites, halloysite provides cracking of heavier hydrocarbons and faster desorption of reaction products from active sites. Hydroxyl groups present on the inner octahedral alumina surface of halloysite create highly reactive sites with Hammett acidity below −8.2 (by acidity compares with 90 wt % sulfuric acid solution) (Henmi and Wada, 1974). A hydrocarbon conversion catalyst has been developed using acid-treated and MgO-doped halloysite clay. This catalyst showed higher performance toward the cracking of heavy hydrocarbons as compared with other natural clays (like kaolin and montmorillonite) (Robson, 1978). Over 1 million tons of pure halloysite was mined during 1947–1976 years only from Dragon Mine in Tintic district of Utah to meet the growing demand of the petrochemical industry (Wilson, 2004).

Pyrolysis of the vacuum gasoil and its mixture containing 10 percent cottonseed oil have been studied in a fluid catalytic cracking microreactor to determine the halloysite effect on activity and selectivity of the commercial Y-zeolite. Addition of HNTs increased the conversion rate by 7–10 percent, which contributed to the additional increase of the gasoline by 3–5 percent and light petroleum gases by 4–6 percent. Coke yield has been reduced by 7–15 percent. This is associated with the faster desorption of reaction products from nanotube mesopores. Linear diffusion path from the pores of larger diameter allows rapid removal of lighter hydrocarbons from the surface of the acidic sites, preventing their deep dehydroxylation and hence reducing coke formation (Abbasov et al., 2013).

Isomerization within HNT pores occured more efficiently, causing more branched hydrocarbons in produced gasoline. Preferential formation of alkyl benzenes (vs. benzene) upon halloysite addition is also indicative of enhanced chain branching, as branched olefins are precursors of the benzene derivatives. This is associated with the better adsorption efficiency of fatty acid chains at halloysite aluminol surfaces compared to zeolite. Indeed, Raman spectroscopy has shown that inner pores of halloysite favor carboxylic acids adsorption (Frost and Kristof, 1997).

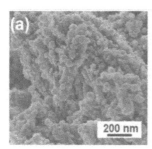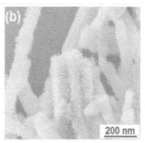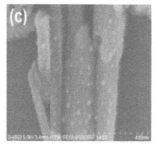

FIGURE 19.14 (a) Iron, (b) cobalt and (c) silver nanoparticles synthesized on HNTs. Reproduced with permission from (Abdullayev and Lvov, 2013). Copyright 2013 Royal Society of Chemistry.

HNT has also been used as template for metallic cobalt and iron nanoparticles for Fischer–Tropsch synthesis of artificial fuels from syngas (Figure 19.14). Cobalt-supported catalysts were prepared by a double solvent impregnation method using water–hexane mixture followed by calcination at 350 °C. Fischer–Tropsch synthesis was conducted on prepared catalyst at CO/H_2 ratio of 2, gas space velocity of 4 mL·hr $^{-1}$·g $^{-1}$, 1 MPa pressure and 230 °C. Carbon monoxide conversion exceeded 12 percent while C_{5+} selectivity was over 78 percent. HNTs have been found to be efficient in preventing migration and agglomeration of cobalt nanoparticles during reaction, which is a critical factor in increasing the produced gasoline yield (Chen et al., 2012).

FeNi nanoparticles have been synthesized on HNTs by immersing halloysite within aqueous solution of iron and nickel nitrates followed by calcination at 500 °C and reduction with KBH_4 and hydrogen gas. Nanoparticles had 2–4 nm diameter and were uniformly distributed on halloysite external surface. External electron-rich Si-O and hydroxide groups chelating iron and nickel oxides were thought to be responsible for effective stabilization of nanoparticles on halloysite surface and for preventing their aggregation. Catalyst treated by hydrogen gas was extremely efficient in decomposing phosphine gas into yellow phosphorus, resulting in 99 percent conversion at atmospheric pressure, 420 °C and PH_3 space velocity of 2,520 mL/(hr*g) (Tang et al., 2013).

A catalyst has been prepared for reduction of aromatic nitro compounds to corresponding amines using silver-supported halloysite nanotubes. Silver was synthesized from $AgNO_3$ with aqueous solution of polyvinyl pyrrolidone and ethylene glycol and deposited on halloysite surface using mercaptoacetic acid as chelating agent. Effective immobilization of silver using thermal decomposition of silver acetate was also demonstrated (Figure 19.14(c)). Approximately, 10 nm size silver nanoparticles were uniformly coated on HNT surface and found to be an efficient catalyst in reduction of nitrophenol to the corresponding aminophenol (Liu and Zhao, 2009).

19.5   FOOD AND AGRICULTURAL APPLICATIONS

Very few studies have been performed on the use of HNTs in agricultural and food applications. Potential exists in using HNT as nanocontainer for the encapsulation and subsequent-controlled release of herbicides, pesticides, and antibiotics to provide long-lasting effect. Controlled release of several herbicides including organic trichloro derivatives, isothiazoline, and chlorothanolil; and pesticides malathion, spectricide, and rotenone have been patented in United States (Price and Gaber, 1997).

Another prospective application of halloysite was found in the preparation of custom biohybrid composites and papers for food packaging. Papers have been prepared by modification of lignocellulose wood fibers with halloysite clay and polyelectrolytes using the layer by layer nanoassembly method (Figure 19.15(a), (b)). Porosity was found to increase by a factor of two with the deposition of two layers of halloysite on wood fibers. An increase in porosity was observed for SiO_2 and TiO_2 modified papers, but to a lesser extent (Figure 19.15(c)) (Lu et al., 2007). The increase of the paper porosity enhances the escape of ethylene and other gaseous products causing food rotting.

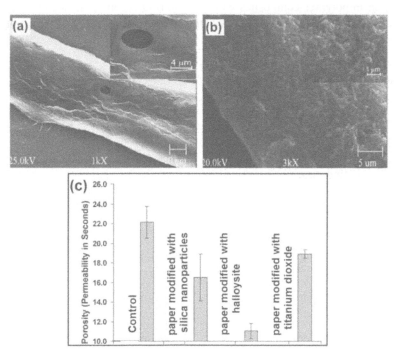

FIGURE 19.15 Scanning electron micrographs from the surface of the lignocellulose wood fibers (a) before and (b) after coating with four layers of HNTs. (c) Results of the porosity test (time needed to pass 100 cm^3 of air through handsheets) on paper prepared from halloysite, SiO_2 and TiO_2 modified wood fibers. Reproduced with permission from (Lu et al., 2007). Copyright 2007 Elsevier.

The addition of 4 wt % HNTs increased the tensile strength of the HNT–gelatin composites by 50 percent, whereas the elastic modulus increased by threefold. This indicates strong chemical interactions between gelatin and halloysites, resulting in restricted polymer chain movements. Nanocomposites became 30 percent more permeable to water vapor (Voon et al., 2012). Enhanced water permeability is desired in food packaging to reduce moisture induced rotting.

19.6 SUMMARY AND OUTLOOK

Halloysite nanotubes offer great promise for commercial applications, including nanoscale additives in polymers, nanocontainers for chemical storage and controlled release (anticorrosion and antimolding protection, flame-retardancy dentistry, and cosmetics), catalysts for hydrocarbon cracking, and many others. Unlike other types of polymer additives, naturally, halloysite is an abundant filler and provides better dispersion in polymer composites without requiring organomodification. There is no doubt that further investigations of halloysite properties will benefit various industries, particularly, polymer, cement, and fuel industries.

KEYWORDS

- **Cementitious composites**
- **Complexation**
- **Flame retardancy**
- **Protective coating**
- **Nanorods**
- **Organosilanes**
- **HNT lumen enlargement**
- **Sol-gel coating**

REFERENCES

1. Abbasov, V.; Mammadova, T.; Andrushenko, N.; Hasankhanova, N.; Lvov Y.; and Abdullayev, E.; *Fuel.* **2013,** *117A,* 552–555.
2. Abdullayev, E.; Sakakibara, K.; Okamoto, K.; Wei, W.; Ariga, K.; and Lvov, Y.; *ACS Appl. Mater. Interf.* **2011,** *3,* 4040–4046.
3. Abdullayev, E.; Abbasov, V.; and Lvov, Y.; *Proc. Petrochem. Oil Ref.* **2009,** *10,* 260–273.
4. Abdullayev, E.; Abbasov, V.; Tursunbayeva, A.; Portnov, V.; Ibrahimov, H.; Mukhtarova, G.; and Lvov, Y.; *ACS Appl. Mater. Int.* **2013,** *5,* 4464 4471.
5. Abdullayev, E.; Joshi, A.; Wei, W.; Zhao, Y.; and Lvov, Y.; *ACS Nano.* **2012,** *6,* 7216–7226.
6. Abdullayev, E.; and Lvov, Y.; *J. Mater. Chem.* **2010,** *20,* 6681–6687.
7. Abdullayev, E.; and Lvov, Y.; *J. Mater. Chem. B* **2013,** *1,* 2894–2903.
8. Abdullayev, E.; and Lvov, Y.; *J. Nanosci. Nanotech.* **2011,** *11,* 10007–10026.
9. Abdullayev, E.; and Lvov, Y.; *Polym. Mater. Sci. Eng.* **2011,** *104,* 246–247.
10. Abdullayev, E.; and Lvov, Y.; Proc. International Materials Science and Technology 2010 Conference. Houston, TX; October 17–21, **2010.**
11. Abdullayev, E.; Price, R.; Shchukin, D.; and Lvov, Y.; *ACS Appl. Mater. Interf.* **2009,** *1,* 1437–1443.
12. Baral, S.; Brandow, S.; and Gaber, B. P.; *Chem. Mater.* **1993,** *5,* 1227–1232.
13. Bates, T.; Hildebrand, F.; and Swineford, A.; *Am. Miner.* **1950,** *35,* 463–484.
14. Berthier, P.; *Ann. Chim. Phys.* **1826,** *32,* 332–335.
15. Chen, S.; Li, J.; Zhang, Y.; Zhang, D.; and Zhu, J.; *J. Nat. Gas Chem.* **2012,** *21,* 426–430.
16. Coleman, J.; Khan, U.; and Gun'ko Y.; *Adv. Mater.* **2006,** *18,* 689–706.
17. Deng, S.; Zhang, J.; and Ye, L.; *Comp. Sci. Tech.* **2009,** *69,* 2497–2505.
18. Du, M.; Gao, B.; and Jia, D.; *Eur. Polym. J.* **2006,** *42,* 1362–1369.
19. Du, M.; Guo, B.; and Jia D.; *Polymer Intern.* **2010,** *59,* 574–595.

20. Du, M.; Guo, B.; Liu M.; and Jia, D.; *Polym. Polym. Comp.* **2007**, *15*, 321–328.
21. Du, M.; Guo, B.; Liu, M.; and Jia. D.; *Polym. Polym. Comp.* **2007**, *15*, 321–328.
22. Du, M.; Guo, B.; Wan, J.; Zou, Q.; and Jia, D.; *J. Polym. Res.* **2010**, *17*, 109–118.
23. Farzadnia, N.; Ali, A. A.; Demirboga, R.; and Anwar, M. P.; *Cement. Concr. Res.* **2013**, *48*, 97–104.
24. Frost, R. L.; and Kristof, J.; *Clays Clay Mineral.* **1997**, *45*, 551–563.
25. Grim, R. E.; Clay Mineralogy. 2nd edition, New York: McGraw-Hill Book Company; **1968**.
26. Guo, B.; Chen, F.; Lei, Y.; Liu, X.; Wan, J.; and Jia, D.; *Appl. Surf. Sci.* **2009**, *255*, 7329–7336.
27. Henmi, T.; and Wada, K.; *Clay Mineral.* **1974**, *10*, 231–245.
28. Ismail, H.; Pasbakhsh, P.; Fauzi, A. M. N.; and Abu Bakar, A.; *Polym. Test.* **2008**, *27*, 841–850.
29. John, U. E.; Jefferson, I.; Boardman, D. I.; Ghataora, G. S.; and Hills, C. D.; *Eng. Geol.* **2011**, *123*, 315–323.
30. Joshi, A.; Abdullayev, E.; Vasiliev, A.; Volkova, O.; and Lvov, Y.; *Langmuir.* **2013**, *29*, 7439–7445.
31. Joussein, E.; Petit, S.; Churchman, J.; Theng, B.; Righi, D.; and Delvaux, B.; *Clay Miner.* **2005**, *40*, 383–426.
32. Kamruzzaman, A. H.; Chew, S. H.; and Lee, F. H.; *Ground Improv.* **2006**, *10*, 113–123.
33. Konnova, S.; et al. *Chem. Comm.* **2013**, *49*, 4208–4212.
34. Kwon, K. O.; Lee, M. G.; Kong, K. R.; and Kang, H. C.; *Resour. Proc.* **2008**, *55*, 115–119. Lambourne, R.; and Strivens, T. A.; Paint and Surface Coatings—Theory and Practice. 2nd edition, Woodhead Publishing; **1999**.
35. Laurence, M. C. K. W.; Fluorinated Coatings and Finishes Handbook. The Definitive User's Guide and Databook. William Andrew Publication/Plastics Design Library; **2006**.
36. Lecouvet, B.; Gutierrez, J.; Sclavons M.; and Bailly, C.; *Polym. Degr. Stab.* **2011**, *96*, 226–235.
37. Liu, M.; Guo, B.; Du, M.; Lei, Y.; and Jia, D.; *J. Polym. Res.* **2008**, *15*, 205–212.
38. Liu, P.; and Zhao, M.; *Appl. Surf. Sci.* **2009**, *255*, 3989–3993.
39. Lu, Z.; Eadula, S.; Zheng, Z.; Xu, K.; Grozdits, G.; and Lvov, Y.; *Coll. Surf. A Phys. Eng. Asp.* **2007**, *292*, 56–62.
40. Lvov, Y.; Bio-inorganic Hybrid Nanomaterials; Strategies, Syntheses, Characterization and Application. Hitzky, E. R.; Ariga K.; Lvov, Y.; eds. Weinheim: Wiley VCH Verlag GmbH & Co; **2008**.
41. Lvov, Y. M.; Shchukin, D. G.; Mohwald, H.; and Price, R. R.; *ACS Nano.* **2008**, *2*, 814–820. Lvov, Y.; and Abdullayev, E.; *Prog. Polym. Sci.* **2013**, *38*, 1690–1719.
42. Ning, N.; Yin, Q.; Luo, F.; Zhang, Q.; Du, R.; and Fu, Q.; *Polymer.* **2007**, *48*, 7374–7384.
43. Oburoglu, N.; Ercan, N.; Durmus, A.; and Kasgoz, A.; *J. Appl. Polym. Sci.* **2012**, *123*, 77–91.
44. Prashantha, K.; Schmitt, H.; Lacrampe, M. F.; and Krawczak, P.; *Comp. Sci. Tech.* **2011**, *71*, 1859–1866.
45. Price, R. R.; and Gaber, B. P.; Controlled Release of Active Agents Using Inorganic Tubules, US Patent 5 651 976 (1997), to United States of America as Represented by the Secretary of Navy.
46. Price, R.; Gaber, B.; and Lvov, Y.; *J. Microencap.* **2001**, *18*, 713–722.
47. Rabehi, B.; Boumchedda, K.; and Ghernouti, Y.; *Int. J. Phys. Sci.* **2012**, *7*, 5179–5192.
48. Ray, S.; Okamoto, M.; *Prog. Polym. Sci.* **2003**, *28*, 1539–1641.
49. Robson, H. E.; Synthetic Halloysite as Hydrocarbon Conversion Catalysts, US Patent 4 098 676 (1978), to Exxon Research & Engineering Co.
50. Shchukin, D.; and Möhwald, H.; *Adv. Funct. Mater.* **2007**, *17*, 1451–1458.
51. Suh, Y.; Kil, D.; Chung, K.; Abdullayev, E.; Lvov, Y.; and Mongayt, D.; *J. Nanosci. Nanotech.* **2011**, *11*, 661–665.
52. Tang, X.; Li, L.; Shen, B.; and Wang, C.; *Chemosphere.* **2013**, *91*, 1368–1373.

53. Tari, G.; Bobos, I.; Gomes, C. S. F.; and Ferreira, J. M. F.; *J. Coll. Interf. Sci.* **1999**, *210*, 360–336.
54. Veerabadran, N.; Price, R.; and Lvov, Y.; *Nano J.* **2007**, *2*, 215–222.
55. Velasco-Santos, C.; Martinez-Hernandez, A.; and Castano, V.; *Comp. Interf.* **2005**, *11*, 567–586.
56. Voon, H.; Bhat, R.; Easa, A.; Liong, M.; and Karim, A.; *Food Bioproc. Tech.* **2012**, *5*, 1766–1774.
57. Wei, W.; Abdullayev, E.; Hollister, A.; Lvov, Y.; and Mills, D.; *Macromol. Mater. Eng.* **2012**, *301*, 645–653.
58. Wilson, I. R.; *Clay Miner.* **2004**, *39*, 1–15.
59. Wilson, I.; *Indust. Miner.* **2004,** 54–61.
60. Ye, Y.; Chen, H.; Wu, J.; and Ye, L.; *Polymer.* **2007**, *48*, 6426–6433.

CHAPTER 20

SOME FURTHER INDUSTRIAL, ENVIRONMENTAL AND BIOMEDICAL APPLICATIONS OF HALLOYSITE NANOTUBES

B. LECOUVET

CONTENTS

20.1 POLYMER NANOCOMPOSITES

20.1.1 COMPARISON BETWEEN CLAYS

Halloysite is relatively new nanofiller in the field of polymer nanocomposites. In order to investigate the potential reinforcement of this tubular aluminosilicate in polymeric materials, it is of interest to compare the mechanical performance of halloysite nanocomposites with that of more conventional ones. In his PhD study, Van Es (2001) modelled and compared the mechanical reinforcement of unidirectionally aligned fibres and platelets (Figure 20.1), according to Halpin–Tsai's and Mori–Tanaka's equations. This graph shows that both models are equivalent for a fitted value of the shape factor ζ ($\zeta = 2/3$ l/t for platelets and $\zeta = (0.5 \text{ l/t})^{1.8}$ for rods). At a given aspect ratio, the reinforcing action of fibres (triangles) is higher than that of platelets (dots). Moreover, the upper limit (i.e., rule of mixture) is already reached at an aspect ratio of 100 for fibers and 1,000 for sheets.

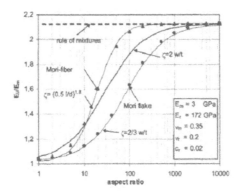

FIGURE 20.1 Comparison between the reinforcement effect of fibers and platelets as a function of the aspect ratio for unidirectional composites. Lines represent Halpin–Tsai's estimates. Black dots and triangles correspond to Mori–Tanaka's estimates for platelet and fiber reinforcement, respectively (Van Es, 2001).

An interesting study was performed by Hedicke-Höchstötter et al. (2009), who compared the mechanical and thermo-mechanical properties between PA6 nanocomposites containing pristine HNTs or organo-modified MMT with octadecyl-benzyl-dimethylammonium-chloride. Figure 20.2(a) shows the temperature dependence of the storage modulus (G') for neat polymer and nanocomposites with 2 wt % nanoclays (normalized to the silicate content). For both systems, the introduction of clay in PA6 increased the storage modulus at all temperatures. Moreover, this tendency was more pronounced for halloysite nanocomposites.

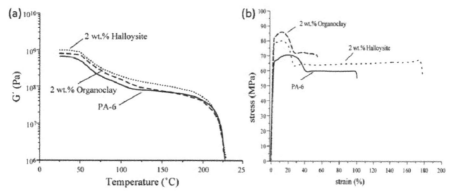

FIGURE 20.2 (a) Storage modulus G' as a function of temperature and (b) stress–strain curves for neat PA6 and nanocomposites containing 2 wt % clay (Hedicke-Höchstötter et al., 2009).

The authors ascribed this result to the higher intrinsic stiffness of the aluminosilicate nanotubes compared to the thinner and more flexible clay platelets. At higher clay content (5 wt %), there was almost no difference between both systems because the higher intrinsic stiffness of HNTs was counterbalanced by the limited interfacial interaction between pristine nanotubes and PA6 compared to organo-modified layered silicates.

The comparison of tensile properties of PA6/clay nanocomposites is presented in Figure 20.2(b). In both cases, the tensile strength was considerably enhanced by adding 2 wt % nanofillers. However, the use of organo-modified MMT resulted in a more significant increase of tensile strength because of the good interactions between layered silicates and the polymer matrix. The major difference between both systems concerned the elongation at break. Though the PA6 became brittle by adding the organoclay, the elongation at break was dramatically enhanced with the introduction of only 2 wt % HNTs into the matrix. The improved ductility of halloysite-filled nanocomposites was attributed to crack-bridging effects (see Figure 20.3) resulting from the elongated tubular shape of HNTs and the formation of hydrogen bonding with PA chains. As a conclusion, the authors mentioned that halloysite provides better mechanical reinforcement compared with organo-modified MMT, even without any chemical modification.

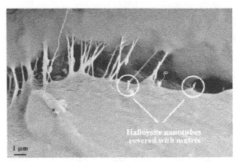

FIGURE 20.3   Scanning electron micrograph of a fracture surface of PA6/HNTs 2 wt % nanocomposite after a tensile test (Hedicke-Höchstötter et al., 2009).

Prashantha et al. (2011) examined mechanical performances of PP nanocomposites based on surface-treated HNTs in terms of tensile, flexural, and impact properties, and observed some differences with other PP nanocomposites reported in the literature. All investigated systems were prepared by melt compounding with the same filler fraction. Similar results were reported for moduli and strengths. Once again, the key feature of halloysite was the conservation of high elongation at break for nanocomposites, whereas this property dramatically decreased for other nanoparticles (e.g., CNTs and MMT). They also compared mechanical properties of PP nanocomposites at their optimum loadings (2 wt % CNTs, 4 wt % MMT and 6 wt % organo-modified HNTs (m-HNTs)). Beside the ductile behavior of PP/m-HNTs nanocomposites, a huge increase in impact strength was also observed for this system in contrast to PP/CNTs and PP/MMT nanocomposites. This difference was related to the ability of m-HNTs to disperse hydrophobic polymer in a uniform manner. Though CNTs tend to form aggregates in a non-polar matrix, the exfoliation of layered silicates is difficult to achieve because of the strong inter-layer ionic bonds; HNTs are more easily dispersed in PP because of the weak secondary interactions among the tubes (Du et al., 2008). Therefore, a homogeneous dispersion of halloysite in PP reduces the amount of aggregates that act as crack initiation sites and leads to a substantial improvement of impact strength (increased by 35% increases when adding 10 phr HNTs to PP).

20.1.2   RHEOLOGICAL PROPERTIES

Several factors influence the rheological properties of polymer nanocomposites: matrix molecular weight, filler loading, aspect ratio and shape of the nanoparticle, surface characteristics, dispersion, and interfacial interaction between polymer and filler (Liu et al., 2003; Du et al., 2004; and Abbasi et al., 2009). Therefore, rheological characterization is not only used as an indirect method to describe the filler dispersion state in polymer nanocomposites, but also to gain an insight into the processability characteristics of these materials (Shen et al., 2005; Cassagnau, 2009).

The rheological behavior of polymer nanocomposites has been widely reported in the literature (Médéric et al., 2005; Wu et al., 2006; and Ma et al., 2007). The evolution of the dynamic modulus and viscosity are usually used to characterize the viscoelastic properties of composite materials (Macosko, 1994). At high frequencies, the qualitative behavior of the viscoelastic properties is almost independent of the filler fraction. The equilibrium structure of the nanocomposite is distorted and the rheological response is mainly controlled by the polymer network formed by entanglements (Hsieh et al., 2004). On the other hand, at low frequencies, the viscoelastic response is dictated by the superposition of two networks: (i) the entangled polymer network and (ii) a network formed by filler–polymer–filler interactions (Pötschke et al., 2004). At low filler content, the rheological properties are dominated by the polymer chain entanglements. By increasing filler fraction, the relaxation process starts to be controlled by the polymer–particle interactions. Finally, above critical filler loading, so called the rheological percolation threshold, a polymer–filler–polymer structure is formed by entanglements between polymer chains and fillers when the distance between two adjacent nanoparticles is smaller than twice the radius of gyration of a polymer chain. The relaxation of polymer chains is hindered by the presence of nanofillers and a second rubber plateau is usually observed at low frequencies associated to a transition from a liquid-like to a solid-like behavior. Balberg (1987) demonstrated that the critical volume fraction for percolation of randomly distributed cylinders is proportional to the inverse of the aspect ratio.

Handge et al. (2010) performed rheological experiments on PA6/HNTs nanocomposites and studied the evolution of the linear viscoelastic properties as a function of the clay loading (Figure 20.4).

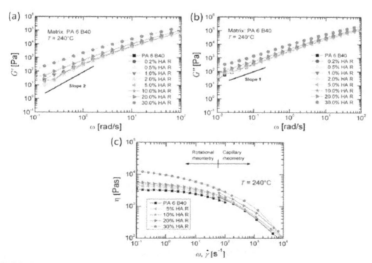

FIGURE 20.4   (a) Storage modulus (G'), (b) loss modulus (G'') and (c) complex viscosity (η) at 240°C as a function of frequency for neat PA6 and PA6/HNTs nanocomposites (Handge et al., 2010).

At 240°C, PA6 exhibited the usual behavior of homopolymer melt characterized by a terminal regime in the low frequency region ($G' \propto \omega^2$ and $G'' \propto \omega$) and a typical Newtonian viscosity plateau. Below 1 wt % HNTs, the effect of halloysite was negligible and the viscoelastic response of the composites was similar to that of the neat polymer. By gradually increasing the filler content, we can observe a monotonic increase of the complex viscosity and moduli at low frequencies. This effect was more marked for the storage modulus than for the loss modulus. Surprisingly, there was no tendency to the formation of a percolated network, even at 30 wt % HNTs. The reasons for the missing plateau might be the relatively low aspect ratio of HNTs as well as the higher filler agglomeration at larger clay fraction. As a consequence, highly loaded nanocomposites keep very good processability characteristics, quite similar to that of the neat polymer. Similar results were reported for PA12-HNTs (Lecouvet et al., 2011a) and poly (vinylidene fluoride)-HNTs nanocomposites (Tang et al., 2013).

Lee et al. (2013) studied the influence of halloysite on the melt rheological behavior of poly (ε-caprolactone) (PCL). Above a critical filler content of 7 wt %, the remarkable decrease of the slope of G' and G" curves at low frequencies combined to a more pronounced shear thinning behavior highlighted the formation of a three-dimensional network in the polymer melt. This solid-like behavior was ascribed to the uniform nanometer-scale dispersion of HNTs within PCL as well as the strong interphase attraction between polymer and clay nanofillers.

The viscoelastic properties of PP–HNTs nanocomposites prepared using a novel "one step" water-assisted extrusion process is described by Lecouvet et al. (2011b). In this study, different mixing strategies were tested and the best clay dispersion was achieved when PP–g–MA (i.e., compatibilizer, Plb) and water injection (W) were combined together. Interestingly, the morphological observations were in perfect agreement with the rheological measurements reporting the highest increase of the moduli and viscosity at low frequencies for PP-Plb-HNTs-W nanocomposites. For the later system, the authors investigated the influence of halloysite concentration on the rheological properties of PP–Plb matrix in the melt state. Figure 20.5(a) and 20.5(b) show, respectively, the storage modulus (G') and the complex viscosity (η^*) at 200°C for PP–Plb–HNTs–W nanocomposites. On one hand, G' and η^* decreased at high frequencies with increasing clay fraction, indicating shorter relaxation times of polymer chains for highly loaded composites. This result was confirmed by the lower molecular weights measured by size exclusion chromatography for nanocomposites, because of thermo-mechanical degradation during melt processing. On the other hand, a plateau appeared at low frequencies for G' starting from 16 wt % HNTs (Figure 20.5(a)) combined with a continuous rise of η^* (Figure 20.5(b). The percolated network structure formation of HNTs confirmed that they were uniformly dispersed in PP and was well-preserved at higher loadings, when water injection and compatibilizer were combined simultaneously.

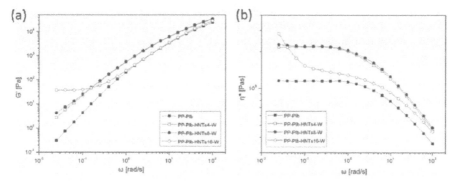

FIGURE 20.5 (a) Storage modulus (G') and (b) complex viscosity (η^*) at 200°C as a function of frequency for neat PP–Plb and its nanocomposites prepared using water-assisted extrusion (Polypropylene (PP), PP–g–MA (Plb), HNTs (H), Water injection (W)) (Lecouvet et al., 2011b).

20.2 CONTROLLED RELEASE

There is an increasing amount of research to develop inexpensive and easily processable multifunctional nanocontainers with potential applications in various areas such as biomedical, pharmaceutical, active corrosion-protection, nanoreactors, and fuel cells. Naturally occurring HNTs are promising cheap nanocontainers for the production of novel delivery systems for drugs and other chemical agents. The positively charged inner surface of HNTs is commonly used to selectively fill the lumen as well as the void spaces in the multilayered shell with negative macromolecules such as enzymes, biocides, pharmaceuticals, and other chemical species (Levis and Deasy, 2002; Vergaro et al., 2010a). Price et al. (2001) first demonstrated the potential of halloysite for the encapsulation of biologically active agents. Moreover, by means of an appropriate pre-treatment of the tubular aluminosilicate, both hydrophobic and hydrophilic agents can be embedded within its core lumen (Shchukin et al., 2008).

The first step in the elaboration of these new delivery systems is the nanotube loading. The procedure is well defined in the literature (Price et al., 2001; Veerabadran et al., 2007; and Abdullayev et al., 2009) and is schematically represented in Figure 20.6 for the entrapment of benzotriazole (i.e., an anticorrosive agent) into the lumen structure of HNTs. Halloysite is first mixed as a dry powder with a saturated solution of the specific active agent in water, acetone, or other alcohol solutions. The mixture is then soaked and transferred to a vacuum jar. Vacuum removes air bubbles from the lumen and this step is confirmed by a slight fizzing of the suspension. After the fizzing stops, vacuum is broken and the solution is penetrated into the core lumen. This process is usually repeated for several times to ensure the saturation of the inner cavity of the tubes. At the end, multiple centrifugations are used to separate HNTs from solution, and nanocontainers are washed with water for the removal of active material adsorbed at the external surface of the cylinders. This simple loading technique can be used to fill HNTs with a wide variety of chemical species.

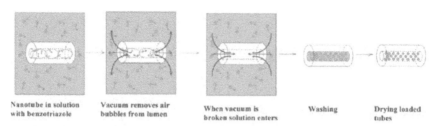

Nanotube in solution   Vacuum removes air   When vacuum is   Washing   Drying loaded
with benzotriazole   bubbles from lumen   broken solution enters   tubes

FIGURE 20.6   Loading procedure of HNTs with benzotriazole (Abdullayev et al., 2009).

As an example, Veerabadran et al. (2009) found that approximately 10 percent of the total volume of tubular halloysite can be filled with dexamethasone in deionized water slurry. The efficiency of the filling procedure can be calculated from absorption and luminescence of the encapsulated material and depends on a number of parameters including solvent, number of loading cycles, and the type of molecules to be loaded (Fix et al., 2009). The optimal solvent exhibits the lowest viscosity (i.e., high diffusion rate of solution into halloysite pores) combined with the highest active agent solubility. In their research work, Abdullayev et al. (2009) observed a substantial enhancement of the loading efficiency of HNTs when replacing water with acetone for benzotriazole entrapment, with filling the tube lumen volume for 90 percent instead of 40 percent. Increasing the number of cycles during the loading process enables to maximize the storage efficiency. As an example, Shchukin and Möhwald (2007a) filled HNTs with benzotriazole in an aqueous solution and observed that a maximum loading level of 80 percent was reached after the fourth loading step.

The entrapment step is followed by the solution retention in the tubes and its subsequent release. Figures 20.7(a) and 20.7(b) present the release profiles for two different active agents from HNTs in water; oxytetracycline (i.e., a water soluble antibiotic) (Price et al., 2001) and benzotriazole (Abdullayev et al., 2009), respectively. For comparison, the dissolution profile of benzotriazole powder in water was also added in Figure 20.7(b) and the vertical line confirmed a very fast dissolution process. In both cases, a two-step delivery process was observed with an initial burst release by desorption of the inhibiting species, located in thin nanotubes, onto the external surface at both extremities of the tubes as well as in natural gap defects at the end of the rolled aluminosilicate sheets (Abdullayev et al., 2009). Indeed, even if several centrifugation steps are performed to remove the most of exogenous material, it is not possible to eliminate all active agents which are located outside the tubes. This first step was followed by a slow and continuous diffusion process of the encased active solution.

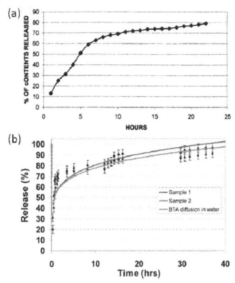

FIGURE 20.7 Release profiles of: (a) oxytetracycline (Price et al., 2001) and (b) benzotriazole (Abdullayev et al., 2009) from HNTs in water.

The release rate is also a function of different parameters such as the molecular weight and the charge of active agents. For example, Shchukin et al. (2005) reported that proteins had prolonged release time compared with the lower molecular weight drugs. In another study, the same research group demonstrated that the release rate of positively charged proteins was faster than for negatively charged ones, because of the positively charged internal surface of the tubes (Lvov and Price, 2007).

Other examples for fast releasing of chemical substances from HNTs in different physiological environments are described in the literature (Lvov et al., 2002; Veerabadran et al., 2007; Vergaro et al., 2010b; and Guo et al., 2012). Based on these observations and in view of developing innovative applications like controlled drug delivery to cancer cells or self-healing anticorrosion coatings, several approaches have been suggested to slow down and control the release rate of the active agents. One possibility is the blending of the solution with a more viscous polymer compatible with the active ingredient (Price et al., 2001). Another strategy, suggested by Abdullayev et al. (2009), is based on the formation of metal complex stoppers at tube extremities (see Abdullayev and Lvov, this volume).

In attempts to control and adjust the release rate of chemical species from HNTs, the layer-by-layer (LbL) self-assembly method has also been investigated to build organized polyelectrolyte nanoshells on the external surface of clay nanotubes, including both ends (Shchukin and Möhwald, 2007a). In this technique, HNTs are first loaded with the active agent and then coated via an alternate adsorption of polycations and polyanions (Figure 20.8) (Shchukin et al., 2008). This process is controlled by monitoring the zeta-potential of halloysite surface for each sequential layer adsorbed (Shchukin et al., 2006).

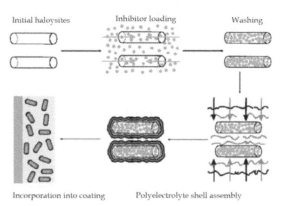

FIGURE 20.8   Fabrication of halloysite nanocontainers coated by a polyelectrolyte shell for anticorrosion purposes (Shchukin et al., 2008).

The two major advantages of this technique are: (1) the tunability of the shell thickness, which is a function of the number of polyelectrolyte layers grown on the aluminosilicate surface and (2) the permeability control, because polyelectrolyte layers are very sensitive to changes in chemical (e.g., pH, solvent, and local electrochemical potential of the surface) and physical (e.g., temperature and light) conditions of the surrounding environment (Shchukin et al., 2008; Andreeva and Shchukin, 2011).

Shchukin and Möhwald (2007a) demonstrated that pH modification in a corrosion media can trigger the release of corrosion inhibitor entrapped in HNTs. When the protective shell is made of two weak polyelectrolytes, a change of pH to acidic or alkaline states can significantly decrease the shell stability, and hence increase the release rate of encapsulated inhibitor into the surrounding media. For applications requiring a progressive release of the inhibitor, it is more interesting to combine weak and strong polyelectrolytes or to use two strong polyelectrolytes (Shchukin and Möhwald, 2007b). The release takes place only in corrosion areas where pH is modified (pH 10) and the process is self-regulated as the pH returns progressively to its initial value during the repair process. At the end, the electrolyte barrier closes and the release stops. HNTs coated with a pH-sensitive polyelectrolyte shell have been used as nanoadditives to develop novel self-healing (i.e., anticorrosive) coatings on the external surface of aluminium alloys (Andreeva et al., 2008; Shchukin et al., 2008; and Fix et al., 2009).

Veerabadran et al. (2009) used the LbL technique to assemble three polyelectrolyte bilayers on the nanotubular template and reported a strong influence of the polyelectrolyte molecular weight on the shell permeability properties. The increase in molecular weight markedly slowed down the release rate of dexamethasone (i.e., a synthetic corticosteroid) from the inner core of HNTs because of the higher stability and more dense packing of the protective shell. However, contradictory results were found for benzotriazole-loaded HNTs with no effect of the polyelectrolyte multilayer shell on the release rate of the corrosion inhibitor (Abdullayev et al., 2009). The authors believe that the small size of the benzotriazole molecules and the low density of the polylelectrolyte barrier are the main reasons to explain the low efficiency of the protective organic wall.

More recently, LbL shell assembly was applied to encapsulate resveratrol-loaded HNTs for controlled release and sustained killing of breast cancer cells (Vergaro et al., 2012). Two different polyelectrolyte couples were coated on halloysite surface. Both of them were biocompatible, whereas only one was biodegradable after internalization by cells. Biocompatibility was obviously one of the prerequisites for safe usage of halloysite in the delivery of biologically active substances in medical applications. For the non-biodegradable nanocarriers, polyelectrolytes could not be degraded by proteases and there was no drug release in the cells. On the contrary, biodegradable-coated HNTs exhibited high cytotoxicity after digestion of the polyelectrolyte shell by the intracellular proteases, with a further decrease in the release rate of resveratrol compared to the uncoated reference system.

There are also other interesting papers reporting on the use of HNTs and nanocontainers for controlled release of macromolecules (Byrne and Deasy, 2005; Veerabadran et al., 2007; Jafari et al., 2010; and Qi et al., 2010). From this selective literature review, it may be concluded that HNTs can be successfully used as cheap and easily processed nanocontainers for encapsulation and subsequent controlled release of various substances such as corrosion inhibitors, biocides, proteins, drugs, and other active agents for which sustainable release into a surrounding media improving their efficiency. Currently, the LbL deposition technology seems to be the most efficient technique to control permeability properties of loaded HNTs.

20.3   ADSORPTION OF CONTAMINANTS

Cationic dyes are extensively used in many industrial fields such as textile, leather, paper, cosmetics, pulp mills, printing, food, plastics, and dye manufacturing (Forgacs et al., 2004). However, the disposal of dye-containing industrial effluents is more and more criticized because many organic dyes represent serious environmental and human health issues because of their toxicity to microorganisms, non-biodegradability, and undesirable aesthetic aspects (Rai et al., 2005). Similarly, industrial discharges may also contain heavy metals that are harmful to both human beings and ecosystems (Volesky, 2001). Numerous treatment methods have been developed to eliminate chemical contaminants from wastewaters (Tunay et al., 1996; Rai et al., 2005). Among them, adsorption became the most popular technique because of its high efficiency and cost effectiveness (Forgacs et al., 2004). In recent years, clay nanoparticles have been tested as nanosorbents to remove organic dyes and heavy metal cations (Lopez Arbeloa et al., 1997; Ozdemir et al., 2006; Rytwo et al.; Vol-zone, 2007). Because of its unique surface chemical property, large surface area, high cation exchange capacity, and cheap and abundant resources, halloysite starts to attract a great deal of attention for decontamination of industrial effluents. This section briefly describes the ability of HNTs to remove dyes, metal ions, and other chemical contaminants from aqueous solutions.

Zhao and Liu (2008) first studied the adsorption behavior of methylene blue (MB) from aqueous solutions onto HNTs using a batch procedure. The removal rate and the adsorption capacity increased with increasing initial dye concentration. The authors believe that the driving force originates from the higher concentration gradient of dye molecules between bulk and clay surface. The dye uptake was very fast for the first 15 min and then equilibrium was established after 30 min. A maximum adsorption capacity of 84 mg/g of MB was achieved, which is similar to that reported for needle-like sepiolite but considerably higher than those for other clay nanoparticles (Ozdemir et al., 2006). By single factor experiments, the effects of pH and temperature were also investigated. As presented in Figure 20.9(a), the dye adsorption on HNTs was favored by shifting pH into the alkaline region. The adsorption equilibrium also quickly reached the higher pH value. It is well known that adsorption of cationic dyes on clay surface was significantly influenced by the surface charge of the adsorbent. The boundary layer of HNTs, mainly based on silica, is positive below pH 2.7 (i.e., zero point charge of HNTs) and becomes more negative when the pH is increased with ionization of the hydroxyl groups (Si/Al-OH → Si/Al-O⁻) (Luo et al., 2010a). Therefore, stronger electrostatic attraction between negatively charged adsorption sites and dye cations accelerates the MB uptake. This observation was confirmed for other cationic dyes (Luo et al., 2010a; Kiani et al., 2011), whereas pH did not seem to be a controlling parameter in adsorption of meth-yl violet (MV) onto HNTs (Liu et al., 2011). The adsorption rate of MB by HNTs markedly increased at lower temperatures (Figure 20.9(b)) and the equilibrium was also attained quicker. However, there is no comment about these last results and the opposite trend was usually observed for adsorption of other dyes onto HNTs (Kiani et al., 2011; Liu et al, 2011). Luo et al. (2010a) reported that the removal of neutral red (NR) by halloysite was endothermic. The increased adsorption capacity of malachite green (MG) onto HNTs at higher temperatures was associated with the lower viscosity of the solution, increasing the mobility and thus the diffusion rate of dye molecules at the external surface and in the internal pores of the tubes (Kiani et al., 2011).

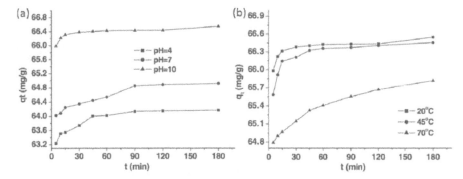

FIGURE 20.9 Effect of contact time and: (a) pH or (b) temperature on the adsorption capacity of MB onto HNTs from aqueous solutions (Zhao and Liu, 2008).

The adsorbent concentration also also an important role in the adsorption process (Luo et al., 2010a; Kiani et al., 2011; Liu et al, 2011). For example, the removal efficiency of NR onto HNTs was increased from 54 to almost 100 percent by increasing the adsorbent dose by a factor of three (Luo et al., 2010a). This is a direct consequence of the increase in the adsorbent specific surface area, and hence the greater number of adsorption sites available for cationic dyes. However, above a critical adsorbent loading, the removal efficiency was constant whereas the adsorption capacity was continuously decreased with increasing concentration. Therefore, an optimum amount of adsorbent has to be chosen to maximize the removal efficiency without sacrificing excess adsorption sites.

The overall adsorption process of dye molecules onto HNTs is spontaneous and follows the pseudo-second-order kinetic model (Zhao and Liu, 2008; Luo et al., 2010a; Kiani et al., 2011; and Liu et al, 2011) with low activation energies suggesting that physisorption is a rate-controlling step (Kiani et al., 2011).

Interestingly, HNTs that usually exhibit excellent dispersion properties in aqueous solutions tend to aggregate and form millimeter scale particles after MB or MG adsorption (Zhao and Liu, 2008; Kiani et al., 2011). This phenomenon can be explained by a change of the halloysite surface property from being hydrophilic to hydrophobic after dye adsorption. This leads to HNTs precipitation and enables reduction of processing costs by avoiding a centrifugation step commonly used to separate nanosorbents from aqueous solution.

Unfortunately, all cationic dyes do not lead to halloysite precipitation once adsorbed onto the clay surface. Recently, magnetic HNTs based on iron oxide (HNTs-Fe_3O_4) were synthesized by a chemical precipitation method to facilitate the separation process from aqueous solutions (Xie et al., 2011). Compared with natural HNTs, magnetic nanosorbents exhibited slight lower adsorption capacities for cationic dyes such as MB and NR, because of the lower amount of adsorption sites available on HNTs after adsorption of metal ions (Figure 20.10(a)).

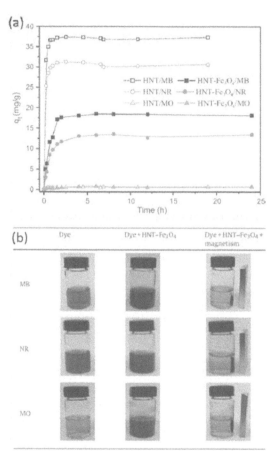

FIGURE 20.10 (a) Effect of contact time on adsorption of several dyes onto pristine and magnetic HNTs from aqueous solutions; (b) adsorption and magnetic separation of HNTs-Fe$_3$O$_4$ from the aqueous solution by a magnet (Xie et al., 2011).

However, the negatively charged surface of natural and magnetic HNTs restricted the removal of anionic methyl orange (MO) (Figures 20.10(a) and (b)) (Xie et al., 2011). At the end of the adsorption process, HNTs were separated from the solution by simply applying an external magnetic field (Figure 20.10(b). Similar results were reported for MV and magnetic HNTs successfully regenerated by high temperature calcinations (Duan et al., 2012).

In order to enhance the adsorption capacity of halloysite, activation and surface modification of the nanotubes were investigated. Luo et al. (2011b) prepared chemically activated HNTs by simple and low-cost treatment of pristine halloysite with HCl followed by NaCl to obtain an effective nano-adsorbent for MB. The zero point charge of HNTs dropped from 2.7 to 2.2 after the chemical activation. As for natural halloysite, the adsorption capacity of MB onto activated HNTs increased with increasing dye concentration, pH, and temperature. The maximum adsorption capacities of MB were around 91 and 104 mg/g at 298 and 318 K, respectively. The adsorption process was perfectly described by the pseudo-second-order kinetic model. The authors believe that the adsorption mechanism for activated HNTs primarily occurs by ion exchange between MB cations and protons of hydroxyl groups or sodium ions previously attached at the clay surface after saturation of sodium chloride. Furthermore, the enhanced adsorption properties at higher pH values indicate that electrostatic attraction between dye cations and nanotubes also contributes to the overall adsorption process. However, the authors did not compare the results with those obtained using pristine halloysite. Similar results were reported on the removal of MO by organo-modified HNTs with relatively high adsorption capacity of 92 mg/g (Liu R. et al., 2012).

More sophisticated designs were suggested to develop halloysite-based sorbents for organic dyes with high adsorption capacity and removal efficiency (Lu et al., 2006). As an example, Liu L. et al. (2012) prepared alginate-HNTs beads for MB removal. The hybrid beads exhibited an average diameter of about 1.5 mm and higher adsorption ability than neat alginate beads, under the same conditions, with a maximum value of 222 mg/g of MB at 298 K. Moreover, hybrid beads were easily recovered by elution and by keeping removal efficiency above 90 percent after 10 regeneration cycles.

There are also few relevant studies on the adsorption behavior of heavy metal elements onto HNTs (Choo and Sung, 1999; Kilislioglu and Bilgin, 2002; Jingmin et al., 2010; Jinhua et al., 2010; Luo et al., 2011b; Dong et al., 2012; and Li et al., 2012). Kilislioglu and Bilgin (2002) investigated the adsorption mechanism and kinetics of uranium onto HNTs and reported a spontaneous and endothermic adsorption reaction favored at high temperature.

In another work, Choo and Sung (1999) compared the adsorption ability of metal ions, contained in wastewaters from a gold-bearing metal mine, onto four different inorganic minerals including illite, zeolite, goethite, and halloysite. The adsorption capacity was mainly a function of the type of mineral sorbent and the contact time. Halloysite was identified as the most effective adsorbent of Fe and As among minerals with high adsorption capacity.

Chemical modification of halloysite was also performed to enhance its metal ions adsorption capacity (Jingmin et al., 2010; Jinhua et al., 2010; Luo et al., 2011b). Jinhua et al. (2010) studied the use of quaternary ammonium cation-grafted HNTs for Cr (VI) removal from aqueous solutions at different adsorbent doses, pHs, and ionic strengths. The adsorption process was very fast and the equilibrium was reached within 5 min. Moreover, the adsorption capacity can be decreased dramatically by increasing pH or ionic strength, with a maximum value at pH 3. The authors attributed the lower adsorption capacity at higher pH to: (i) the various forms of chromate anions present in the solution depending on pH and (ii) a change of the surface charge of HNTs, which is positive at low pH and becomes negative when pH increases. Therefore, the high adsorption capacity of modified HNTs at low pH results from an electrostatic interaction between chromate anions and positively-charged surface as well as from the interaction of metal ions with quaternary ammonium cations. The effect of ionic strength on Cr (VI) uptake was ascribed to be a competitive adsorption process between inorganic electrolyte (NO_3) and chromate anions. The adsorbent regeneration was achieved by elution at both high pH and ionic strength. In another study, Luo et al. (2011b) used silylation of HNTs to produce highly effective nano-adsorbents for Cr (VI) removal. Results corroborate well with those reported here above (Jinhua et al., 2010) and the main adsorption mechanism is believed to be electrostatic interaction between chromate anions and protonated amino groups grafted onto the aluminosilicate surface.

Halloysite also exhibits considerable potential for the removal of contaminants from wastewaters. Nayak and Singh (2007) used HNTs to eliminate phenol from aqueous solutions, and determined single phase and highly crystalline NaA zeolites, synthesized from pristine halloysite, displayed fast adsorption rate and high adsorption capacity for ammonium ions removal (Luo et al., 2010b).

Interestingly, halloysite was also used in cosmetics as a hypoallergenic skin cleanser to adsorb dead skin cells and aesthetically unpleasing oils (Sundararajan et al., 2011).

In conclusion, halloysite can be used as a low-cost and effective nano-adsorbent for the removal of residual dyes, heavy metal ions, and other water pollutants present in industrial effluents. Moreover, natural and modified HNTs can be easily separated, regenerated and reused, and considerably reducing the processing costs compared with traditional adsorption methods.

20.4   NANOTEMPLATING AND NANOREACTION VESSELES

Recently, halloysite has gained a great deal of interest in the synthesis of polymeric and metallic nanostructures (e.g., nanoparticles, nanowires, and nanotubes) using the tubular aluminosilicate as a nanotemplate. As all the template-based fabrication techniques require specific interactions between growing material and substrate, HNTs' surface has to be modified first.

Halloysite has been used as a substrate for the polymerization and growth of conducting polymer nanoparticles (Luca and Thomson, 2000; Zhang and Liu, 2008; Yang et al., 2010; and Liu et al., 2012). For instance, Luca and Thomson (2000) studied the intercalation and polymerization of aniline on the internal and external surfaces of Cu^{2+}-exchanged HNTs. Polyaniline (PANI) molecular wires were successfully synthesized inside the inner core of the clay nanotubes. More recently, Zhang and Liu (2008) prepared PANI nanotubes onto halloysite templates by insitu soapless emulsion polymerization. HNTs were first dispersed in an aqueous acidic solution of aniline. The mixture was ultrasonically irradiated with adsorption of anilinium chloride onto the external surface of the tubes. Then ammonium persulfate

(APS) was added as an oxidant for polymerization of aniline and the last step was the template dissolution by ultrasonic irradiation in an acidic solution of HCl/HF. The so-obtained PANI nanotubes exhibited uniform nanostructure and high electrical conductivity with complete removal of the inorganic template. In the same way, polypyrrole (PPy)/HNTs nanocomposites were prepared by insitu oxidative polymerization of the pyrrole monomer and PPy nanotubes derived after etching of the halloysite template by an acid mixture (Liu et al., 2012).

Yang et al. (2010) synthesized PPy/HNTs nanocomposites for electrochemical energy storage. In this study, HNTs were first silylated with APTES followed by insitu chemical oxidative polymerization of the monomer on the amine groups grafted onto the nanotube surface. Few papers also reported the use of insitu soapless emulsion polymerization to prepare other core-shell halloysite/polymer nanoparticles including polystyrene (PS) (Zhao and Liu, 2007) and poly(styrene-butyl acrylate-acrylic acid) (P-SBA) (Zhang et al., 2010).

Li et al. (2008) suggested another synthesis route to elaborate halloysite-based composite nanotubes by growing polymer brushes from HNTs via atom transfer radial polymerization (ATRP). The aluminosilicate support was first immobilized with the ATRP initiator by a two-step process including silylation of both internal and external surfaces of HNTs followed by their chemical reaction with a bromide compound. Then PS chains were grafted by ATRP of styrene onto the modified walls of HNTs and cross-linkers were added to avoid collapsing of polymer nanostructures after dissolution of the inorganic template. The authors mentioned that this technique can be used for the fabrication of non-woven porous fabric by casting the dispersion of polymer-coated HNTs with further thermal cross-linking. The same procedure was applied for the synthesis of cross-linked polyacrylonitrile (PAN)/HNTs composites.

The thickness of both inner and outer shells can be tuned by varying the polymerization time and the monomer concentration. Therefore, at longer reaction times (4.5 h) and/or higher styrene concentration, the inner core can be completely filled hence resulting in the formation of PS nanowires. The composition of the polymer layer can also be tailored to elaborate functional materials (Li et al., 2009). As an example, PS layers were converted into sulfonated PS (sPS) by sulfonation without sacrificing the coaxial tubular morphology. Sulfonic acid groups located on the polymer walls were able to interact with a broad range of materials (e.g., metallic, inorganic, and polymeric) and initiated their growth on the activated surface. Besides, sPS can also self-catalyze into carbon at high temperature. As an example, the authors reported the formation of carbon layers containing nickel nanoparticles at both surfaces of HNTs after the immersion of sPS/HNTs composites in a nickel nitrate aqueous solution with subsequent reduction at high temperature under H_2 atmosphere.

The ATRP technique was also advocated by Mu et al. (2008) to graft hyperbranched macromolecules with various components on the halloysite substrate via surface-initiated self-condensing vinyl (co)polymerization. The grafting of hyperbranched polymer and copolymer shells on the aluminosilicate walls was confirmed by a combination of FTIR, NMR, and thermogravimetric analyses. Similarly, di-block copolymer brushes were grown from the halloysite surface via a so-called reverse atom transfer radical polymerization (RATRP) process instead of the usual ATRP because of the high instability of the $CuCl_2$ catalyst when exposed to air and moisture (Wang L-P et al., 2008). XPS, SPM, and thermogravimetric experiments were employed to validate the organic grafting on HNTs.

Halloysite can also serve as a template for metallic nanoparticles to develop cermet composites and porous carbon materials. Fu et al. (2004, 2005a, and 2005b) fabricated a novel metallic/halloysite composite with enhanced magnetic properties through electroless deposition of nickel (Ni) nanoparticles on HNTs. The synthesis was carried out in two steps as follows: (1) HNTs were first activated by palladium (Pd) for the initiation of the electroless plating via insitu reduction of Pd ions by methanol on the aluminosilicate surface (Fu et al., 2005c). Activated HNTs were separated from solution by centrifugation and washed in water; (2) the so-obtained nanotubes were added to a plating bath of nickel sulphate and Ni nanoparticles si-multaneously deposited on the inner and outer surfaces of the inorganic template (Fu and Zhang, 2005a). After plating, the halloysite powder turned black. TEM re-vealed a homogeneous distribution of Ni nanoparticles on the external surface of the tubes (Figure 20.11(a)) as well as the formation of discontinuous metallic nanowires inside the lumen cavity (Figure 20.11(b)). This later result was attributed to the slow and incomplete removal of hydrogen from the inner cavity during the plating process.

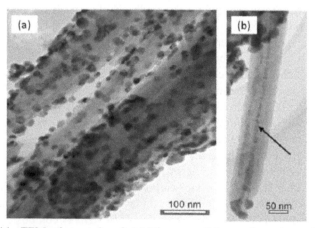

FIGURE 20.11 TEM micrographs of: (a) Ni nanoparticles on the outer surface of HNTs and (b) discontinuous Ni nanowires in the inorganic nanotemplate (Fu and Zhang, 2005a).

Post-annealing of the hybrid material had a great influence on the crystal structure of Ni nanoparticles, which switched from amorphous to face-centered cubic after complete crystallization at 673 K. Therefore, magnetic properties were dramatically enhanced after post-annealing, because of the combination of magneto-crystalline anisotropy and single magnetic domain effect (Fu and Zhang, 2005b).

There are also many other investigations of HNTs acting as nanotemplates and nanoreactors to elaborate metallic nanoparticles, nanowires, nanotubes, nanocoatings, etc. (Burridge et al., 2010; Adbullayev et al., 2011; Zhang et al., 2012; and Zhu et al., 2012). For example, silver/halloysite composites synthesized from aluminosilicate substrate by LbL deposition technique exhibited effective anti-microbacterial activity against Staphylococcus aureus bacteria (Burridge et al., 2010). In another study, silver nanorods were selectively synthesized inside the inner core of HNTs by thermal decomposition at 300°C of silver acetate previously loaded from an aqueous solution by pulling and breaking vacuum (Figure 20.12) (Adbullayev et al., 2011).

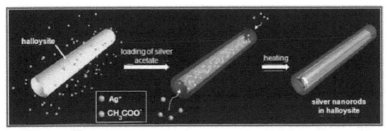

FIGURE 20.12 Illustrative synthesis of silver nanorods within the halloysite inner core (Adbullayev et al., 2011).

The incorporation of silver inside the tubes was attested by energy dispersive X-ray spectroscopy-STEM. This encapsulated configuration shielded the metal core from surrounding environmental effects such as oxidation or light degradation. Silver-loaded HNTs were used as additives to prepare antibacterial polymer paints (Adbullayev et al., 2011). The aluminosilicate shell enhanced the coated antimicrobial activity, reinforced its tensile strength, and avoided polymer degradation and change of paint colour usually observed for coatings based on unshelled silver nanoparticles.

Wang A. et al. (2008) proposed a novel cost-effective and straightforward processing method based on the template approach for the elaboration of mesoporous carbons with controlled pore structures, high specific surface areas, and large pore volumes. These materials were synthesized by polymerization and subsequent carbonization of a carbon precursor (furfuryl alcohol) embedded into halloysite matrix by impregnation at room temperature. After carbonization, HNTs were separated from the black powder by leaching with hydrogen fluoride. Morphological studies revealed that so-obtained carbons exhibited a predominant sheet-like structure with some tubular-like and rod-like nanoparticles reflecting the morphology and pore structure of the aluminosilicate template. These results indicated that: (i) the carbon precursor deposition mostly occurred on the external surface of HNTs during the impregnation step and (ii) the regular morphology of carbon tubes was partially damaged during the demineralization reaction. The efficiency of the mesoporous carbon structure, in terms of specific surface area and total pore volume, increased by working at higher impregnation pressure than the atmosphere pressure and with prolonged impregnation time. The mesoporous structure can also be tailored by adjusting the carbon precursor concentration (Huang et al., 2010). Following this concept, Liu et al. (2013) investigated the role of polypyrrole as a carbon precursor and found that carbon was mostly present in an amorphous state within the mesoporous sheets. In addition, the specific surface area of the mesoporous structure can be tuned by varying the ratio of halloysite content over the pyrrole monomer concentration.

20.5 SUPPORT FOR CATALYST IMMOBILIZATION

Mostly, the application of catalysts is hampered by their short lifetime because of low thermal stabilities and difficulties during recovery and recycling. Their immobilization on inorganic materials (e.g., zeolites and hydrotalcites) can effectively solve this problem and enhance their catalytic activity. However, these supports present some drawbacks related to diffusion limitations and low specific surface area (Corma et al., 2002). Because of its unique surface chemistry, tubular geometry, and large surface area, halloysite has proved to be an effective support for the immobilization of catalyst molecules such as metal nanoparticles (Liu and Zhao, 2009; Cai et al., 2011; and Wang et al., 2011), metallocomplexes (Machado et al., 2008), and enzymes (Zhai et al., 2010).

HNTs were used as the support to immobilize silver (Ag) nanocatalysts through insitu reduction of $AgNO_3$ by the polyol process (Liu and Zhao, 2009) (See also Abdullayev and Lvov, this volume). Ag/HNTs nanoparticles were then employed to catalyze the reduction of aromatic nitro compounds and the catalysts were easily recovered from the reacting mixture. In another work, ruthenium (Ru) nanoparticles were immobilized on HNTs by the wet impregnation technique and their catalytic activity in ammonia decomposition reaction was tested (Wang et al., 2011). Morphological analysis revealed a preferential location of metal nanoparticles on the outer surface of the tubes with higher aggregation by increasing the catalyst loading. During the catalytic reaction, Ru particles grew by sintering at high temperature. According to the authors, this undesired effect can be solved by confining catalysts inside the inner core of the tubes. Cai et al. (2011) studied the catalytic performance of halloysite-supported gold (Au) catalysts, elaborated through a deposition–precipitation procedure, for the selective oxidation of cyclohexene using molecular oxygen in a solvent-free system. The catalytic activity was found to be directly related to the metal content with a maximum cyclohexene conversion of 30 percent for 0.8 wt %-supported Au particles. This result was assigned to an apparent nano-size effect of Au catalysts during cyclohexene oxidation. Indeed, the size of Au particles increased with increasing metal concentration resulting in reduced catalytic activity. Therefore, an optimum metal content concentration had to be found.

Metalloporphyrins are commonly used as oxidants for hydrocarbons. However, one of the major problems in hydrocarbon catalytic oxidations is the poor thermal stability of the metallocomplexes to the reaction conditions. To overcome the catalytic activity limitation of these chemical species, Machado et al. (2008) investigated two different strategies to immobilize neutral, anionic, and cationic metalloporphyrins on HNTs under pressure and under magnetic/stirring reflux conditions. The presence of metalloporphyrins on the aluminosilicates was confirmed by solid state UV–vis and infrared spectroscopy. The pressurized system enabled more effective immobilization than the stirring/reflux technique because of the better contact between support and catalyst. In contrast with neutral metalloporphyrins, anionic and cationic species displayed high immobilization rates because of the good affinity of the halloysite surface for charged compounds. The cationic species preferentially immobilized on the negatively charged outer surface (SiO), whereas the anionic metalloporphyrins migrated to the positive immobilization sites $(Al(OH)_2^+)$ located at the edges and inside the lumen of the tubes.

Finally, Zhai et al. (2010) demonstrated the ability of halloysite for enzyme immobilization via the physical adsorption process. Two industrial enzymes were tested and exhibited higher thermal resistance once immobilized on an inorganic support. The catalytic performance of supported enzymes was also well preserved after recovering and recycling, with more than 55 percent initial activity retained after seven cycles.

KEYWORDS

- **Adsorbent concentration**
- **Anticorrosive agent**
- **Homopolymer melt**
- **Rheological properties**
- **Nanosorbents**
- **Polyelectrolytes**
- **Polymer-coated HNTs**
- **Rheological properties**

REFERENCES

1. Abbasi, S.; Carreau, P. J.; Derdouri, A.; and Moan, M.; *Rheol. Acta.* **2009**, *48*, 943–959.
2. Abdullayev, E.; Price, R.; Shchukin, D.; and Lvov, Y.; M. *Appl. Mater. Interf.* **2009**, *1*, 1437–1444.
3. Adbullayev, E.; Sakakibara, K.; Okamoto, K.; Wei, W.; Ariga, K.; and Lvov, Y.; *Appl. Mater. Interf.* **2011**, *3*, 4040–4046.
4. Andreeva, D. V.; and Shchukin, D. G.; *Mater. Today.* **2008**, *11*, 24–30.
5. Balberg, I.; *Phil. Mag. B* **1987**, *56*, 991–1003.
6. Burridge, K.; Johnston, J.; and Borrmann, T.; *J. Mater. Chem.* **2010**, *21*, 734–742.
7. Byrne, R. S.; and Deasy, P. B.; *J. Microencapsul.* **2005**, *22*, 423–437.
8. Cai, Z-Y.; Zhu, M-Q.; Dai, H.; Liu, Y.; Mao, J-X.; Chen, X-Z.; and He, C-H.; *Adv. Chem. Eng. Sci.* **2011**, *1*, 15–19.
9. Cassagnau, P.; *Polymer.* **2009**, *49*, 2183–2196.
10. Choo, C-O.; and Sung, I-H.; *J koSES.* **1999**, *4*, 57–68.
11. Corma, A.; Fornes, V.; and Rey, F.; *Adv. Mater.* **2002**, *14*, 71–74.
12. Dong, Y.; Liu, Z.; and Chen, L.; *J. Radioanal. Nucl. Chem.* **2012**, *292*, 435–443.
13. Du, M.; Guo, B.; Cai, X.; Jia, Z.; Liu, M.; and Jia, D.; *e Polymer.* **2008**, *130*, 1–14.
14. Du, F.; Scogna, R. C.; Zhou, W.; Brand, S.; Fischer, J. E.; and Winey, K. I.; *Macromolecule.* **2004**, *37*, 9048–9055.
15. Duan, J.; Liu, R.; Chen, T.; Zhang, B.; and Liu, J.; *Desalination.* **2012**, *293*, 46–52.
16. Fix, D.; Andreeva, D. V.; Lvov, Y. M.; Shchukin, D.; and Möhwald, H.; *Adv. Funct. Mater.* **2009**, *19*, 1720–1727.
17. Forgacs, E.; Cserháti, T.; and Oros, G.; *Environ. Int.* **2004**, *30*, 953–971.
18. Fu, Y.; Zhang, L.; and Zheng, J.; *Trans. Nonferrous Met. Soc. China.* **2004**, *14*, 152–156.
19. Fu, Y.; and Zhang, L.; *J. Solid State Chem.* **2005a**, *178*, 3595–3600.
20. Fu, Y.; and Zhang, L.; *J. Nanosci. Nanotechnol.* **2005b**, *5*, 1113–1119.
21. Fu, Y.; Zhang, L.; and Zheng, J.; *J. Nanosci. Nanotechnol.* **2005c**, *14*, 558–564.
22. Guo, M.; Wang, A.; Muhammad, F.; Qi, W.; Ren, H.; Guo, Y.; and Zhu, G.; *Chin. J. Chem.* **2012**, *30*, 2115–2120.
23. Handge, U. A.; Hedicke-Höchstötter, K.; amd Altstädt, V.; *Polymer.* **2010**, *51*, 2690–2699.
24. Hedicke-Höchstötter, K.; Lim, G. T.; and Altstädt, V.; *Compos. Sci. Technol.* **2009**, *69*, 330–334.
25. Hsieh, A. J.; Moy, P.; Beyer, F. L.; Madison, P.; and Napadensky, E.; *Polym. Eng. Sci.* **2004**, *44*, 825–837.
26. Huang, Z-H.; Wang, A.; Kang, F.; and Chuan, X.; *Mater. Lett.* **2010**, *64*, 2044–2046.
    Jafari, A. H.; Hosseini, S. M. A.; and Jamalizadeh, E.; *Electrochim. Acta.* **2010**, *55*, 9004–9009.
27. Jinhua, W.; Xiang, Z.; Bing, Z.; Yafei, Z.; Rui, Z.; Jindum, L.; and Rongfeng, C.; *Desalination.* **2010**, *259*, 22–28.
28. Jingmin, D.; Wang, J.; Zhang, B.; Zhao, Y.; and Liu J.; *Fresenius Environ. Bull.* **2010**, *19*, 2783–2787.
29. Kiani, G.; Dostali, M.; Rostami, A.; and Khataee, A. R.; *Appl. Clay Sci.* **2011**, *54*, 34–39.
    Kilislioglu, A.; and Bilgin, B.; *Radiochim. Acta.* **2002**, *90*, 155–160.
30. Lecouvet, B.; Gutierrez, J. G.; Sclavons, M.; and Bailly, C.; *Polym. Degrad. Stab.* **2011a**, *96*, 226–235.
31. Lecouvet, B.; Sclavons, M.; Bourbigot, S.; Devaux, J.; and Bailly, C.; *Polymer.* **2011b**, *52*, 4284–4295.
32. Lee, K-S.; and Chang, Y-W.; *J. Appl. Polym. Sci.* **2013**, *128*, 2807–2816.
33. Levis, S. R.; and Deasy, P. B.; *Int. J. Pharm.* **2002**, *243*, 125–134.

34. Li, C.; Liu, J.; Qu, X.; Guo, B.; and Yang, Z.; *J. Appl. Polym. Sci.* **2008**, *110*, 3638–3646.
35. Li, C.; Liu, J.; Qu, X.; and Yang, Z.; *J. Appl. Polym. Sci.* **2009**, *112*, 2647–2655.
36. Li, J.; Wen, F.; Pan, L.; Liu, Z.; and Dong, Y.; *J. Radioanal. Nucl. Chem.* **2012**, 1–8.
37. Liu, C.; Zhang, J.; He, J.; and Hu, G.; *Polymer.* **2003**, *44*, 7529–7532.
38. Liu, L.; Wan, Y.; Xie, Y.; Zhai, R.; Zhang, B.; and Liu, J.; *Chem. Eng. J.* **2012**, *187*, 210–216.
39. Liu, P.; and Zhao, M.; *Appl. Surf. Sci.* **2009**, *255*, 3989–3993.
40. Liu, R.; Zhang, B.; Mei, D.; Zhang, H.; and Liu, J.; *Desalination.* **2011**, *268*, 111–116.
41. Liu, R.; Fu, K.; Zhang, B.; Mei, D.; Zhang, H.; and Liu, J.; *J. Disp. Sci. Technol.* **2012**, *33*, 711–718.
42. Liu, Y.; Nan, H.; Cai, Q.; and Li, H.; *J. Appl. Polym. Sci.* **2012**, *125*, 638–643.
43. Liu, Y.; Cai, Q.; Li, H.; and Zhang, J.; *J. Appl. Polym. Sci.* **2013**, *128*, 517–522.
44. Lopez Arbeloa, F.; Tapia Estevez, M. J.; Lopez Arbeloa, T.; and Lopez Arbeloa, I.; *Clay Miner.* **1997**, *32*, 97–106.
45. Lu, X.; Chuan, X.; Wang, A.; and Kang, F.; *Acta Geol. Sin. Engl.* **2006**, *80*, 278–284.
46. Luca, V.; and Thomson, S.; *J. Mater. Chem.* **2000**, *10*, 2121–2126.
47. Luo, P.; Zhao, Y.; Zhang, B.; Liu, J.; Yang, Y.; and Liu, J.; *Water Res.* **2010a**, *44*, 1489–1497.
48. Luo, P.; Zhang, J-S.; Zhang, B.; Wang, J-H.; Zhao, Y-F.; and Liu, J-D.; *J. Hazard. Mater.* **2010b**, *178*, 658–664.
49. Luo, P.; Zhang, B.; Zhao, Y.; Wang, J.; Zhang, H.; and Liu, J.; *Korean J. Chem. Eng.* **2011a**, *28*, 800–807.
50. Luo, P.; Zhang, J-S.; Zhang, B.; Wang, J-H.; Zhao, Y-F.; and Liu, J-D.; *Ind. Eng. Chem. Res.* **2011b**, *50*, 10246–10252.
51. Lvov, Y; and Price, R.; Halloysite nanotubules, a novel substrate for the controlled delivery of bioactive molecules. In: Ruiz-Hitzky, E.; Ariga, K.; Lvov, Y.; eds. Bio-Inorganic Hybrid Nanomaterials. London, Berlin: Wiley; **2007**, Chapter 12, 419–441 p.
52. Lvov, Y.; Price, R. R.; Gaber, B.; and Ichinose, I.; *Colloids Surf. Eng.* **2002**, *198*, 375–382.
53. Ma, H.; Tong, L.; Xu, Z.; and Fang Z.; *Polym. Degrad. Stab.* **2007**, *92*, 1439–1445.
54. Machado, G. S.; Castro, K. A. D. F.; Wypych, F.; and Nakagaki, S.; *J. Mol. Catal. A Chem.* **2008**, *283*, 99–107.
55. Macosko, C. W.; Rheology: Principles, Measurements and Applications. New York: Wiley-VCH; **1994**.
56. Médéric, P.; Razafinimaro, T.; Aubry, T.; Moan, M.; and Klopffer, M. H.; *Macromole. Symp.* **2005**, *221*, 75–84.
57. Mu, B.; Zhao, M.; and Liu P.; *J. Nanopart. Res.* **2008**, *10*, 831–838.
58. Nayak, P. S.; and Singh, B. K.; *Desalination.* **2007**, *207*, 71–79.
59. Ozdemir, Y.; Dogan, M.; and Alkan, M.; *Micropor. Mesopor. Mater.* **2006**, *96*, 419–427.
60. Pötschke, P.; Abdel-Goad, M.; Alig, I.; Dudkin, S.; and Lellinger, D. *Polymer.* **2004**, *45*, 8863–8870.
61. Price, R. R.; Gaber, B. P.; and Lvov, Y.; *J. Microencapsulation.* **2001**, *18*, 713–722.
62. Prashantha, K.; Lacrampe, M. F.; and Krawczak, P.; *Express Polym. Lett.* **2011**, *5*, 295–307.
63. Qi, R-L.; Yu, J-Y.; and Shi, X-Y.; *J. Mater. Chem.* **2010**, *20*, 10622–10629.
64. Rai, H. S.; Bhattacharyya, M. S.; Singh, J.; Bansal, T. K.; Vats, P.; and Banerjee, U. C.; *Crit. Rev. Environ. Sci. Technol.* **2005**, *35*, 219–238.
65. Rytwo, G.; Nir, S.; Crespin, M.; and Margulies, L.; *J. Colloid Interface Sci.* **2000**, *222*, 12–19.
66. Shchukin, D.; Price, R.; Sukhorukov, G.; and Lvov, Y.; *Small.* **2005**, *1*, 510–513.
67. Shchukin, D. G.; Zheludkevich, M. L.; Yasakau, K. A.; Lamaka, S. V.; Ferreira, M. G. S.; and Möhwald, H.; *Adv. Mater.* **2006**, *18*, 1672–1678.
68. Shchukin, D. G.; and Möhwald, H.; *Adv. Funct. Mater.* **2007a**, *17*, 1451–1458.
69. Shchukin D. G.; and Möhwald, H.; *Small.* **2007b**, *3*, 926–943.

70. Shchukin, D. G.; Lamaka, S. V.; Yasakau, K. A.; Zheludkevich, M. L.; Ferreira, M. G. S.; and Möhwald, H.; *J. Phys. Chem. C* **2008**, *112*, 958–964.
71. Shen, L.; Lin, Y.; Du, Q.; Zhong, W.; and Yang, Y.; *Polymer.* **2005**, *46*, 5758–5766.
72. Sundararajan, G.; Mahajan, Y. R.; Joshi, S. V.; and Nisha, C. K.; *Nanotech. Insights.* **2011**, *2*, 19.
73. Tang, X-G.; Hou, M.; Zou, J.; and Truss, R.; *J. Appl. Polym. Sci.* **2013**, *128*, 869–878.
74. Tunay, O.; Kabdasli, I.; Eremektar, G.; and Orhon, D.; *Water Sci. Technol.* **1996**, *34*, 9–16.
75. Van Es, M.; PhD Thesis, Polymer-clay nanocomposites—the importance of particles dimensions. TU Delft, 2001.Veerabadran, N.; Price, R.; and Lvov, Y. M.; *Nano.* **2007**, *2*, 115–120.
76. Veerabadran, N. G.; Mongayt, D.; Torchilin, V.; Price, R. R.; and Lvov, Y. M.; *Macromol. Rapid Commun.* **2009**, *30*, 99–103.
77. Vergaro, V.; Abdullayev, E.; Lvov, Y. M.; Zeitoun, A.; Cingolani, R.; Rinaldi, R.; and Leporatti, S.; *Biomacromolecules.* **2010a**, *11*, 820–826.
78. Vergaro, V.; Abdullayev, E.; Lvov, Y. M.; Zeitoun, A.; Cingolani, R.; Rinaldi, R.; and Leporatti, S.; *Nanotech.* **2010b**, *3*, 395–396.
79. Vergaro, V.; Lvov, Y. M.; and Leporatti, S.; *Macromol. Biosci.* **2012**, *12*, 1265–1271. Volesky, B.; *Hydrometallurgy.* **2001**, *35*, 203–216.
80. Volzone, C.; Removal of metals by natural and modified clays. In: Wypych, F.; Satyanarayana, K. G.; eds. Clay Surfaces. Academic Press; **2007**.
81. Wang, L.; Chen, J.; Ge, L.; Zhu, Z.; and Rudolph, V.; *Energy Fuels.* **2011**, *25*, 3408–3416.
82. Wang, L-P.; Wang, Y-P.; Pei, X-W.; and Peng, B.; *React. Funct. Polym.* **2008**, *68*, 649–655.
83. Wang, A.; Kang, F.; Huang, Z.; Guo, Z.; and Chuan, X.; *Micropo. Mesopor. Mater.* **2008**, *108*, 318–324.
84. Wu, D.; Wu, L.; Wu, L.; and Zhang, M.; *Polym. Degrad. Stab.* **2006**, *91*, 3149–3155.
85. Xie, Y.; Qian, D.; Wu, D.; and Ma, X.; *Chem. Eng. J.* **2011**, *168*, 959–963.
86. Zhai, R.; Zhang, B.; Liu, L.; Xie, Y.; Zhang, H.; and Liu, J.; *Catal. Commun.* **2010**, *12*, 259–263.
87. Yang, C.; Liu, P.; and Zhao, Y.; *Electrochim. Acta.* **2010**, *55*, 6857–6864.
88. Zhang, Y.; Jiang, J.; Liang, Q.; and Zhang B.; *J. Appl. Polym. Sci.* **2010**, *117*, 3054–3059.
89. Zhang, J.; Zhang, Y.; Chen, Y.; Du, L.; Zhang, B.; Zhang, H.; Liu, J.; and Wang, K.; *Ind. Eng. Chem. Res.* **2012**, *51*, 3081–3090.
90. Zhao, M.; and Liu, P.; *J. Macromol. Sci. B* **2007**, *46*, 891–897.
91. Zhao, M.; and Liu, P.; *Micropor. Mesopor. Mater.* **2008**, *112*, 419–424.
92. Zhang, L.; and Liu, P.; *Nanoscale Res. Lett.* **2008**, *3*, 299–302.
93. Zhu, H.; Du, M.; Zou, M.; Xu, C.; and Fu, Y.; *Dalton Trans.* **2012**, *41*, 10465–10471.

CHAPTER 21

NANOTUBULAR MINERALS AS TEMPLATES AND NANOREACTORS

GUSTAVE KENNE DEDZO and CHRISTIAN DETELLIER

CONTENTS

21.1   INTRODUCTION

The concept of chemical reaction involves direct contact between chemical species to promote the formation of new chemical bonds and new chemical compounds. Therefore, this implies a surface contact as large as possible to obtain optimum yields. When the reactor size is reduced at the nanolevel, molecules are in contact in a restricted space. This substantially increases the probability of contact between them. In some cases these nanoreactors provide specific environments in which chemical reactions could take place selectively. These environments can also promote chemical reactions otherwise unfavoured (Shchukin and Sviridov, 2006). Such devices minimize the energy consumption if one considers the increase of yields and the significant reduction of external inputs to facilitate chemical reactions. Such environments created by microemulsions (using block copolymers and other amphiphilic chemical compounds) are used classically (Jang and Ha, 2002), but they are limited by their lower stability. Mesoporous silica and zeolites are also used. These are characterized by pores of well-defined sizes whose internal surfaces may in some cases be functionalized (especially mesoporous silicas), or modified by metal nanoparticles in order to perform specific reactions (Salavati-Niasari, 2009; Fang et al., 2012). Unlike microemulsions or micelles, materials allow reactions in extreme environments due to their better physicochemical stabilities.

Fibrous clay minerals, sepiolite and palygorskite particularly, have also been used in this field. Indeed, their structure and physicochemical properties make them good candidates for the building of systems where it is possible to control chemical reactions at the nanolevel. These two clay minerals have similar structures fully described in the literature (Krekeler and Guggenheim, 2011; Suárez and García-Romero 2011). Sepiolite and palygorskite are mainly characterized by tunnels with a rectangular section in the (a, b) plane and extending along the c axis. This particular morphology is induced by the periodical inversion of the apical oxygen of the tetrahedral sheet linked to the cation of the octahedral sheet. The section of these channels is larger for sepiolite (3.7 × 10.6 Å) than for palygorskite (3.7 × 6.4 Å). Another difference between two minerals is, the nature of the cations in the octahedral sheet: mainly magnesium for sepiolite, while a significant substitution of magnesium by aluminum occurs in palygorskite.

The research performed on the use of sepiolite and palygorskite as templates or nanoreactors is surveyed in this short chapter. This involves the use of the pristine minerals or of the minerals modified by various strategies, resulting in chemical reactions or in the preparation of nanomaterials. Their use as catalysts or as supports for catalysts is reviewed, as well as their use as reactors for the synthesis of polymers and nano-sized carbonaceous materials.

21.2   SEPIOLITE AND PALYGORSKITE IN CATALYSIS

Fibrous minerals have been extensively used as catalysts. However, in the native state, they show generally poor performance and require heat pre-treatments that increase their acidic character (Shuali et al., 1991). When modified by metal cations, metal nanoparticles or metal oxides, the resulting materials displayed properties allowing them to be used in heterogeneous catalysis, photocatalysis or electrocatalysis (He et al., 2011; Yang and Zhang, 2012).

21.2.1 CATALYSTS OBTAINED BY IMPREGNATION OR CATION EXCHANGE

Unlike other clay minerals, fibrous clay minerals display low acidic properties. It has been demonstrated that during the adsorption of ammonia on palygorskite, only a negligible amount of ammonium ion was observed (Vanscoyoc et al., 1979). Similarly, this mineral has almost no significant basic properties. Corma and Martín-Aranda (1991) have shown that it is possible to introduce strong basic sites (with $pK_b \leq 13.3$) in the structure of sepiolite by replacing the terminal magnesium cations located on the tunnels borders by alkali metal cations (Li^+, Na^+, K^+, and Cs^+). The modification procedure was achieved by simple cation exchange of magnesium by sodium cations, followed by the exchange of sodium with other cations. No calcination was applied before use. These materials showed noticeable basicity and were able to catalyze the condensation of benzaldehydes with ethyl cyanoacetate, ethyl acetoacetate, and ethyl malonate at moderate temperatures. Similarly, exchanging sepiolite magnesium cations by $[Pd (NH_3)_4]^{2+}$ ions gave rise to an important catalytic activity for the Suzuki coupling reaction in water (Shimizu et al., 2004). Sn-palygorskite was prepared by a simple procedure and was highly active as a catalyst for the Baeyer–Villiger oxidation of ketones in the presence of hydrogen peroxide as oxidant (Lei et al., 2005). Palygorskite exchanged with lanthanum cations that provided acidic catalysts (with Lewis and Bronsted acid sites) effective for dehydration of alcohols (Melo et al., 2002). Several researchers have used this impregnation method to modify sepiolite and palygorskite by different other metal cations to obtain the desired catalytic properties (Melo et al., 2000; Yang and Zhang, 2012).

21.2.2 SEPIOLITE AND PALYGORSKITE NANOREACTORS WITH METAL OXIDE

Metal oxides were also used to induce catalytic properties in sepiolite and palygorskite (Martín and Melo, 2006). In this case, after the impregnation step, the composite obtained was calcined at a high temperature (more than 400°C) to result in the formation of metal oxides on the minerals. However, the calcination step may be accompanied by the destruction of the minerals resulting from the irreversible dehydration of tunnels and their collapse. Salvador et al. (2002) have succeeded by preparing a nickel oxide supported sepiolite with the conservation of sepiolite structure, by this method, despite the high calcination temperature (400°C). Compared to natural sepiolite, this catalyst showed excellent activity for the oxidation of lindane

(Salvador et al., 2002). In another approach, sepiolite was pre-treated with a solution of HCl to exchange the magnesium cations by protons. This material, also called acid sepiolite, was used to prepare a copper oxide/sepiolite catalyst doped by rare earth (La, Ce, Pr, and Sm) metal oxides. It has been shown, that doping the catalyst by rare earth metals could significantly improve the catalytic reduction of NO in the presence of propylene (Ye et al., 2012).

Another approach is to obtain the metal oxide by a sol-gel process. In this case, the materials were prepared using TiO_2 and were mainly used as photocatalysts (Zhao et al., 2006; Zhao et al., 2007; and Nieto-Suárez et al., 2009). Aranda et al. (2008) have deposited TiO_2 nanoparticles on the surface of sepiolite by a sol-gel process, using cetyltrimethylammonium as template. After calcination (for removal of the template), the TiO_2 nanoparticles were uniformly distributed on the outer surfaces of sepiolite. Adding tetramethoxysilane into the reaction medium, a TiO_2/SiO_2-sepiolite composite were obtained. These composites developed excellent catalytic activity for the photo-oxidation of phenol, with a beneficial effect due to the presence of sepiolite (Aranda et al., 2008). This procedure has been used more recently by Bouna et al. with palygorskite (2011). TiO_2 nanoparticles were of comparable size, stable up to 900°C and exhibited excellent photocatalytic activity towards the degradation of Orange G. Still using the sol-gel method, it was possible to prepare multi metal oxides composites. This was the case for a material obtained by He et al. (2011). It was made of VO_x-WO_y/TiO_2 supported on palygorskite. It showed significant catalytic activity in the degradation of ortho-dichlorobenzene (He et al., 2011).

21.2.3 METALLIC NANOPARTICLES SUPPORTED ON SEPIOLITE AND PALYGORSKITE

Zerovalent metals are used as catalysts for various organic reactions. When they are in the form of nanoparticles, they are more effective, which is mainly due to their increased active surface area. However, significant losses are recorded in the recovery of these nanoparticles at the end of the reaction. This is the reason why it is necessary to immobilize them on supports. Sepiolite and palygorskite have been used for this purpose. The thermal and chemical stabilities of these minerals ensure an efficient use of the catalysts in drastic environments. In general, the metals were deposited directly on the clay mineral surfaces. Usually, the procedure included an adsorption or impregnation by a concentrated solution of metal ions (Cu, Ni, Pt, and Pd) followed by a reduction step. Quite often, the reduction was obtained at high temperature in a stream of H_2 used as reducing agent (Aramendía et al., 2000; Pecharromán et al., 2006; Rodríguez et al., 2010; and Liu et al., 2012) or in solution with a strong reducing agent like sodium borohydride (Frost et al., 2010). By immobilizing zerovalent iron nanoparticles on palygorskite, an important activity for the bleaching of methylene blue solutions was obtained. The performance of this composite was much superior to that of commercial zerovalent iron alone or of unmodified palygorskite (Frost et al., 2010). In the cases of direct deposition, the nanoparticles are not very abundant and the dispersion is not always well achieved. This procedure however yielded significant results.

Another approach is to use a functionalized mineral having an excellent affinity for the metal ions used as the nanoparticles precursor. Zhu et al. (2009) have grafted an aminosilane onto the surface of sepiolite to improve its affinity for the gold nanoparticles precursor Au $(en)_2 Cl_3$ (en = ethylenediamine). The reduction of the adsorbed metal ion complex was achieved with sodium borohydride. A second approach was to synthesize the material by reacting the functionalized sepiolite with a gold hydrosol previously prepared by reacting the precursor (Au $(en)_2 Cl_3$) with tetrakis (hydroxymethyl) phosphonium chloride. Both synthesis pathways, which are summarized in Figure 21.1, have yielded gold nanoparticles of average diameter <5 nm and well-dispersed on sepiolite fibres. This good dispersion resulted from the presence of N-Au interactions (Zhu et al., 2009). The material obtained by the hydrosol route was successfully used for the oxidation of several alcohols (Letaief et al., 2011).

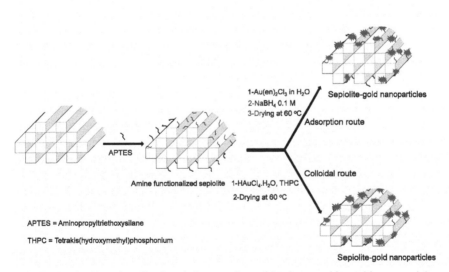

FIGURE 21.1 Functionalized sepiolite or palygorskite decorated by gold nanoparticles.

21.3 SEPIOLITE AND PALYGORSKITE SUPPORT FOR BIOCATALYSTS

Enzymes are specialized macromolecules used by living organisms for the synthesis of specific compounds by conversion of other compounds. This is known as biocatalysts because the presence of enzymes enables or significantly accelerates chemical reactions. Nowadays, there are many industrial applications involving enzymes, where they are used to produce various chemical compounds. Several reviews report a large number of studies on the immobilization of biocatalysts on solid supports (e.g., nanoparticles, carbon nanotubes, graphene, mesoporous silica, lamellar materials) with various applications (Hartmann, 2005; Ansari and Husain, 2012). Indeed, enzymes immobilization on a substrate provides several advantages: (i) the enzyme can be easily recovered after a reaction cycle and reused without a significant loss of material, (ii) the support can stabilize the enzyme and thus can increase its useful life and finally, (iii) the support ensures good dispersion of the enzyme and may increase the reaction yields (Katchalski-Katzir, 1993; Lei et al., 2002; Wang, 2006).

The low reactivity of clay minerals (they are relatively stable for wide pH ranges), abundance and their good compatibility with biomolecules explain their use as supports for enzymes immobilization (Fusi, et al., 1989; Mousty, 2004; Mbouguen et al., 2006; Mbouguen et al., 2007). Another key advantage is their ability to act as a local pH regulator through their amphoteric character. Thus, they are good candidates for the building of biological nanoreactors, suitable for reactions that are mostly very pH sensitive. As are the smectites, the fibrous minerals are well suited for the immobilization of enzymes (Sedaghat et al., 2009). The strategy for this immobilization is simple. It consists in dispersing natural or pretreated (acid or basic treated, or modified by a surfactant or organosilane) clay minerals in a solution of the enzyme. The resulting composite is then separated by centrifugation or filtration and dried. Stability tests are then performed under various experimental conditions to ensure good adhesion of the enzyme to the substrate.

Garcia-Segura et al. (1987) have immobilized urease on unmodified sepiolite and compared the catalytic activity of the immobilized enzyme to that of the free enzyme. It was clear from these studies that immobilization increases the thermal stability of the enzyme and its effectiveness in presence of ionic strength. After five cycles of use, the urease activity was retained. Most recently, it was demonstrated by immobilizing an enzyme (urease or cholesterol oxidase) on sepiolite previously modified by a phospholipid, its activity was increased (Wicklein et al., 2011). The role of the phospholipid was to improve compatibility with sepiolite, while ensuring an environment (mimicking living cells) suitable for enzymatic activity. According to the authors, the enzymes were reused several times without significant loss of activity.

Huang et al. (2008) studied the influence of the nature of palygorskite modification (acid treatment by adsorption of a cationic surfactant or functionalization with silane) on the amount of immobilized lipase and its catalytic activity for the hydrolysis of olive oil. If the modification by the surfactant yielded the larger enzyme immobilization, the best performance (activity and activity recovery) was obtained with the material resulting from the grafting of aminosilane. These results showed that high amounts of the biocatalyst on the support can inhibit active sites because of steric hindrance. These authors have shown subsequently that, when the enzyme was immobilized by chemical bonding (glutaraldehyde was used as a linker between the grafted aminosilane and the amine groups of lipase through $N=C-(CH_2)_3-C=N$ bonds), an improvement of the enzyme activity was observed, and the composite was reused several times with minimal loss of activity (Huang et al., 2009). Several other enzymes have been immobilized on fibrous clay minerals with significant biocatalytic activity. One can mention lipases of various origins (Cabezas etal., 1991; De Fuentes et al., 2001; Luna et al., 2012), α-chymotrypsin and α-amylase (Cabezas etal., 1991), catalase (Cengiz et al., 2012), phosphatase (Carrasco et al., 1995; Sedaghat et al., 2009), and linoleic acid isomerase (You et al., 2011).

This ability to form stable and efficient composite materials with enzymes has also been exploited in electrochemistry for the preparation of electrochemical biosensors. At the interface of such devices, enzymatic reactions enabled electroactive compounds (such as hydrogen peroxide) to be detected electrochemically (Wicklein et al., 2011). A palygorskite supported tyrosinase was described for the detection of phenol (Chen and Jin, 2010).

21.4 SEPIOLITE AND PALYGORSKITE TUNNELS AS NANOREACTORS FOR THE SYNTHESIS OF NANOMATERIALS

Different chemical compounds can be inserted in the sepiolite and palygorskite tunnels despite the steric restrictions imposed by the dimension of the section of the tunnels and by their chemical environment (Inagaki et al., 1990; Weir et al., 2000; Kuang and Detellier, 2004). These confined spaces can be used to perform chemical reactions on the inserted compounds.

21.4.1 SYNTHESIS OF POLYMERS CONFINED IN TUNNELS

Probably, the most obvious way to use the tunnels of sepiolite or palygorskite as reactors for organic synthesis is their use for the polymerization of monomers which were previously inserted. Although, intellectually attractive, this approach is not trivial to perform. Several practical problems were encountered in its implementation: in particular, the presences of zeolitic water molecules (occupying the available space), the difficulty of inserting the reagents, the control of the beginning of the reaction as well as the difficulty to confine the polymerization reaction exclusively inside the tunnels. The recovery of the products without destroying the reactor presents another drawback. However, Kitayama et al. (1997) reported the polymerization of pyrrole in sepiolite in the presence of iodine. The resulting composite (sepiolite with tunnels containing the polymer) showed electrical conductivity. Isoprene was also polymerized in the tunnels of sepiolite (Inagaki et al., 1995).

21.4.2   SYNTHESIS OF CARBON NANORODS

The possibility of synthesizing polymers inside the tunnels of sepiolite or palygorskite has induced the preparation of carbon with a structure based on these minerals, similarly to what has been done with swelling clays or other porous materials (Kyotani et al., 1988; Winans et al., 1995; Sandí et al., 1996; Kyotani, 2006; Nishihara and Kyotani, 2012). Sandí et al. (1999) used sepiolite as a template for the synthesis of carbon. The synthesis pathway is shown schematically in Figure 21.2. It consisted in filling the tunnels of sepiolite with gas phase ethylene or propylene prior to pyrolysis at 700°C under nitrogen. The inorganic template was then removed by concentrated HF solution. TEM images confirmed that the obtained carbon had the morphology and the size of the parent sepiolite fibres. Higher performances than those of graphite were obtained when this material was used as electrode for Li battery. A similar compound was obtained by Fernández-Saavedra et al. (2004) using acrylonitrile as a monomer for the in situ synthesis of carbon in sepiolite tunnels, followed by a calcination step under an inert atmosphere.

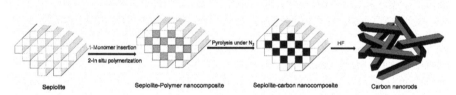

FIGURE 21.2   Synthesis of carbon nanorods using sepiolite as template.

21.4.3   GRAPHENE SYNTHESIS

The remarkable structure of sepiolite has also been exploited for the preparation of graphene. Many applications of graphene have been emerged since it was demonstrated that it was possible to isolate and therefore to produce this material. So far, it is still challenging to produce graphene in quantities that are sufficient enough for industrial applications. The usual chemical method used in laboratories is highly polluting since it requires a great amount of different chemicals and yields only low quality graphene in multiple steps. Recently, Ruiz-Hitzky et al. (2011) have prepared graphene-sepiolite nanocomposites. The strategy consisted first in depositing on sepiolite glucose ("the sweet way") or gelatine ("the jelly way"). Sucrose was transformed to caramel which adsorbed onto the surface of the mineral, and after a thermal treatment at 800°C under nitrogen atmosphere, a graphene-sepiolite nanocomposite were obtained. A similar nanocomposite was obtained from gelatine (Ruiz-Hitzky et al., 2011).

Palygorskite was modified by titanate and used as a catalyst support for the synthesis of multi-walled carbon nanotubes by chemical vapour deposition methods. Single crystalline iron carbide Fe_3C nano wires partially filled the carbon nanotubes (Cheng et al., 2003). Similar structures containing carbon nanotubes on sepiolite showed excellent properties for the adsorption of phenol (Nie et al., 2011).

21.5 CONCLUSION

The use of sepiolite and palygorskite is increasingly growing in the field of material chemistry, with important applications in organic synthesis. Their external surfaces have mainly been exploited because they are more accessible and can undergo chemical modifications, for example by grafting organosilanes. The modified minerals, particularly after being modified by metals or enzymes, can then undergo catalytic reactions with significant efficiency. Only few studies have so far reported the effective use of the tunnels of sepiolite and palygorskite as nanoreactors. The difficulties of introducing reagents and controlling the reaction conditions in their confined space cannot be underestimated. Future work should result in a better understanding and control of the tunnel environments in order to determine the ideal conditions for the insertion, stability and reactivity of strategic molecular compounds.

ACKNOWLEDGMENT

This work was financially supported by a Discovery Grant of the Natural Sciences and Engineering Research Council of Canada (NSERC). The Canada Foundation for Innovation and the Ontario Research Fund are gratefully acknowledged for infrastructure grants to the Centre for Catalysis Research and Innovation of the University of Ottawa.

KEYWORDS

- **Amphoteric**
- **Biocatalysts**
- **Fibrous minerals**
- **Nanoreactor**
- **Nanotubular minerals as templates**
- **Suzuki coupling reaction**
- **Zerovalent metals**

REFERENCES

1. Ansari, S. A.; and Husain, Q.; *Biotechnol. Adv.* **2012**, *30,* 512–523.
2. Aramendía, M. A.; Borau, V.; Corredor, J. I.; Jiménez, C.; Marinas, J. M.; Ruiz, J. R.; and Urbano, F. J. J.; *Colloid. Interface Sci.* **2000**, *227,* 469–475.
3. Aranda, P.; Kun, R.; Martin-Luengo, M. A.; Letaïef, S.; Dékány, I.; and Ruiz-Hitzky, E. *Chem. Mater.* **2008**, *20,* 84–91.
4. Bouna, L.; Rhouta, B.; Amjoud, M.; Maury, F.; Lafont, M. -C.; Jada, A.; Senocq, F.; and Daoudi, L.; *Appl. Clay Sci.* **2011**, *52,* 301–311.
5. Cabezas, M. J.; Salvador, D.; and Sinisterra, J. V. J.; *Chem. Technol. Biot.* **1991**, *52,* 265–274.
6. Carrasco, M. S.; Rad, J. C.; and Gonzalez-Carcedo, S.; *Bioresource Technol.* **1995**, *51,* 175–181.
7. Cengiz, S.; Çavaş, L.; and Yurdakoç, K.; *Appl. Clay Sci.* **2012**, *65–66,* 114–120.
8. Chen, J.; and Jin, Y.; *Microchim. Acta.* **2010**, *169,* 249–254.
9. Cheng, J.; Zhang, X.; Liu, F.; Tu, J.; Ye, Y.; Ji, Y.; and Chen, C.; *Carbon.* **2003**, *41,* 1965–1970.
10. Corma, A.; and Martín-Aranda, R. M. J.; *Catal.* **1991**, *130,* 130–137.
11. De Fuentes, I. E.; Viseras, C. A.; Ubiali, D.; Terreni, M.; and Alcántara, A. R. J.; *Mol. Catal. B-Enzym.* **2001**, *11,* 657–663.
12. Fang, X.; Liu, Z.; Hsieh, M. -F.; Chen, M.; Liu, P.; Chen, C.; and Zheng, N.; *ACS Nano.* **2012**, *6,* 4434–4444.
13. Fernández-Saavedra, R.; Aranda, P.; and Ruiz-Hitzky, E.; *Adv. Funct. Mater.* **2004**, *14,* 77–82.
14. Frost, R. L.; Xi, Y.; and He, H. J.; *Colloid. Interface Sci.* **2010**, *341,* 153–161.
15. Garcia-Segura, J. M.; Cid, C.; de llano, J. M.; and Gavilanes, J. G.; *Brit. Polym. J.* **1987**, *19,* 517–522.
16. Guggenheim, S.; and Krekeler, M. P. S.; The structures and microtextures of the palygorskite–sepiolite group minerals. In: Developments in Clay Science, (Developments in Palygorskite-Sepiolite Research). Galàn, E.; Singer, A.; eds. **2011**, *3,* 3–32.
17. Hartmann, M.; *Chem. Mater.* **2005**, *17,* 4577–4593.
18. He, X.; Tang, A.; Yang, H.; and Ouyang, J.; *Appl. Clay Sci.* **2011**, *53,* 80–84.
19. Huang, J.; Liu, Y.; and Wang, X. J.; *Mol. Catal. B-Enzym.* **2008**, *55,* 49–54.
20. Huang, J.; Liu, Y.; and Wang, X. J.; *Mol. Catal. B-Enzym.* **2009**, *57,* 10–15.
21. Inagaki, S.; Fukushima, Y.; Doi, H.; and Kamigaito, O.; Pore size distribution and adsorption selectivity of sepiolite. *Clay Miner.* **1990**, *25,* 99 105.
22. Inagaki, S.; Fukushima, Y.; and Miyata, M.; *Res. Chem. Intermediat.* **1995**, *21,* 167–180.
23. Jang, J.; and Ha, H.; *Langmuir.* **2002**, *18,* 5613–5618.
24. Katchalski-Katzir, E.; *Trends Biotechnol.* **1993**, *11,* 471–478.
25. Kitayama, Y.; Katoh, H.; Kodama, T.; and Abe, J.; *Appl. Surf. Sci.* **1997**, *121–122,* 331–334.
26. Kuang, W.; and Detellier, C.; *Can. J. Chem.* **2004**, *82,* 1527 1535.
27. Kyotani, T.; *Bull. Chem. Soc. Jpn.* **2006**, *79,* 1322–1337.
28. Kyotani, T.; Sonobe, N.; and Tomita, A.; *Nature.* **1988**, *331,* 331–333.
29. Lei, C.; Shin, Y.; Liu, J.; and Ackerman, E. J. J.; *Am. Chem. Soc.* **2002**, *124,* 11242–11243.
30. Lei, Z.; Zhang, Q.; Luo, J.; and He, X.; *Tetrahedron Lett.* **2005**, *46,* 3505–3508.
31. Letaief, S.; Grant, S.; and Detellier, C.; *Appl. Clay Sci.* **2011**, *53,* 236–243.
32. Liu, H.; Chen, T.; Chang, D.; Chen, D.; He, H.; and Frost, R. L. J.; *Mol. Catal. A-Chem.* **2012**, *363–364,* 304–310.
33. Luna, D.; Posadillo, A.; Caballero, V.; Verdugo, C.; Bautista, F. M.; Romero, A. A.; Sancho, E. D.; (...); and Calero, J.; *Int. J. Mol. Sci.* **2012**, *13,* 10091–10112.
34. Martín, N.; and Melo, F.; *React. Kinet. Catal. Lett.* **2006**, *88,* 27–34.
35. Melo, D. M. A.; Ruiz, J. A. C.; Melo, M. A. F.; Sobrinho, E. V.; and Martinelli, A. E. J.; *All. Compd.* **2002**, *344,* 352–355.

36. Melo, D. M. A.; Ruiz, J. A. C.; Melo, M. A. F.; Sobrinho, E. V.; and Schmall, M.; *Micropor. Mesopor. Mat.* **2000,** *38,* 345–349.
37. Nie, J. -Q.; Zhang, Q.; Zhao, M. -Q.; Huang, J. -Q.; Wen, Q.; Cui, Y.; Qian, W. -Z.; and Wei, F.; *Carbon.* **2011,** *49,* 1568–1580.
38. Nieto-Suárez, M.; et al. *J. Mater. Chem.* **2009,** *19,* 2070–2075.
39. Nishihara, H.; and Kyotani, T.; *Adv. Mater.* **2012,** *24,* 4473–4498.
40. Pecharromán, C.; Esteban-Cubillo, A.; Montero, I.; Moya, J. S.; Aguilar, E.; Santarén, J.; and Alvarez, A. J.; *Amer. Ceram. Soc.* **2006,** *89,* 3043–3049.
41. Rodríguez, A.; Ovejero, G.; Sotelo, J. L.; Mestanza, M.; and García, J.; *Ind. Eng. Chem. Res.* **2010,** *49,* 498–505.
42. Ruiz-Hitzky, E.; Darder, M.; Fernandes, F. M.; Zatile, E.; Palomares, F. J.; and Aranda, P.; *Adv. Mater.* **2011,** *23,* 5250–5255.
43. Salavati-Niasari, M.; *J. Mol. Catal. A-Chem.* **2009,** *310,* 51–58.
44. Salvador, R.; Casal, B.; Yates, M.; Martín-Luengo, M. A.; and Ruiz-Hitzky, E.; *Appl. Clay Sci.* **2002,** *22,* 103–113.
45. Sandí, G.; Carrado, K. A.; Winans, R. E.; Johnson, C. S.; and Csencsits, R. J.; *Electrochem. Soc.* **1999,** *146,* 3644–3648.
46. Sandí, G.; Winans, R. E.; and Carrado, K. A.; *J. Electrochem. Soc.* **1996,** *143,* L95–L98.
47. Sedaghat, M. E.; Ghiaci, M.; Aghaei, H.; and Soleimanian-Zad, S.; *Appl. Clay Sci.* **2009,** *46,* 131–135.
48. Shchukin, D. G.; and Sviridov, D. V.; *J. Photochem. Photobiol. C* **2006,** *7,* 23–39.
49. Shimizu, K. -I.; Maruyama, R.; Komai, S. -I.; Kodama, T.; and Kitayama, Y. J.; *Catal.* **2004,** *227,* 202–209.
50. Shuali, U.; Bram, L.; Steinberg, M.; and Yariv, S. J.; *Therm. Anal.* **1991,** *37,* 1569–1578.
51. Suárez, M.; and García-Romero, E.; Advances in the crystal chemistry of sepiolite and palygorskite. In: Developments in Clay Science, (Developments in Palygorskite-Sepiolite Research). Galàn, E.; Singer, A.; eds. **2011,** *3,* 33–65.
52. Vanscoyoc, G. E.; Serna, C. J.; and Ahlrichs, J. L.; *Amer. Miner.* **1979,** *64,* 215–223.
53. Wang, P.; *Curr. Opin. Biotechnol.* **2006,** *17,* 574–579.
54. Weir, M. R.; Facey, G. A.; and Detellier, C.; *Stud. Surf. Sci. Catal.* **2000,** *129,* 551 558.
55. Wicklein, B.; Darder, M.; Aranda, P.; and Ruiz-Hitzky, E.; *Appl. Mater. Interfaces.* **2011,** *3,* 4339–4348.
56. Winans R. E.; and Carrado, K. A.; *J. Power Sources.* **1995,** *54,* 11–15.
57. Yang, Y.; and Zhang, G.; *Appl. Clay Sci.* **2012,** *67–68,* 11–17.
58. Ye, Q.; Yan, L.; Wang, H.; Cheng, S.; Wang, D.; Kang, T.; and Dai, H.; *Appl. Catal. A-Gen.* **2012,** *431–432,* 42–48.
59. You, Q.; Yin, X.; Gu, X.; Xu, H.; and Sang, L.; *Bioprocess Biosyst. Eng.* **2011,** *34,* 757–765.
60. Zhao, D.; Zhou, J.; and Liu, N.; *Mater. Sci. Eng. A* **2006,** *431,* 256–262.
61. Zhao, D.; Zhou, J.; and Liu, N.; *Mater. Charact.* **2007,** *58,* 249–255.
62. Zhu, L.; Letaief, S.; Liu, Y.; Gervais, F.; and Detellier, C.; *Appl. Clay Sci.* **2009,** *43,* 439–446.

PART IX
MEDICAL AND HEALTH APPLICATIONS
OF NATURAL MINERAL NANOTUBES AND
THEIR HEALTH PROBLEMS

CHAPTER 22

MEDICAL AND HEALTH APPLICATIONS OF HALLOYSITE NANOTUBES

ELSHAD ABDULLAYEV

CONTENTS

22.1   INTRODUCTION

Nanoscale formulations allowed for significant progress in medicine. Design of functional smart nanocontainers for loading and controlled release of drugs and targeted delivery are in development (Brigger et al., 2002). Typically, biomimetic materials like polymers, lipids, polysaccharides, and proteins are used to create drug delivery systems. Inorganic tubular nanomaterials were also extensively studied for drug delivery, bone implants and, tissue scaffolds, and so on (Singhet al., 2008; Singh et al., 2010; and Luo et al., 2011). One such tubular nanomaterial is naturally available halloysite clay. Halloysite is a biocompatible material, which makes it prospective for medical applications and household products (Abdullayev and Lvov, 2013). Halloysite mixed with selectin protein was used for capturing leukemic and epithelial cancer cells (Hughes and King, 2010). The immobilization of enzymes within halloysite lumen may generate new functional materials exploiting nanoconfined reactions inside the pores. Bio-catalytic synthesis of vaterate within nanotube lumens has been demonstrated (Shchukin et al., 2005). All these features combined with the availability of halloysite from natural sources, biocompatibility and simplicity in processing makes it prospective material for medical applications.

22.2   HALLOYSITE BIOCOMPATIBILITY

Biocompatibility and biodegradability are the main requirements for the use of halloysite in medical and health applications. *In vitro* cell toxicity tests using human dermal fibroblast and breast cancer cells revealed that pristine halloysite mined from Dragon mine (Utah, USA) with over 95 percent purity is 50 times less toxic than the usual table salt (Veerabadran et al., 2009). Results on quantitative Caco-2/HT29-MTX cell viability, cytotoxicity, barrier permeability, and cytokine measurements indicate that halloysite exhibits a high degree of biocompatibility showing no toxicity below concentrations of 0.2 mg/mL (or 200×10^6 nanotubes/mL), though cells responded to exposure with differences in polarized proinflammatory cytokine release (Lai et al., 2013). Uptake of halloysite nanotubes by MCF-7 and HeLa cancer cells have been studied using confocal microscopy. Penetration and accumulation of halloysite within cell membrane do not prevent their proliferation (Figure 22.1). *In vitro* cytotoxicity test using these cells at 10^5 cells/mL population, 37°C and in 5 percent CO2, 95 percent relative humidity for 72 h revealed that majority of the cells survived at halloysite concentration of 0.1 mg/mL (Vergaro et al., 2010).

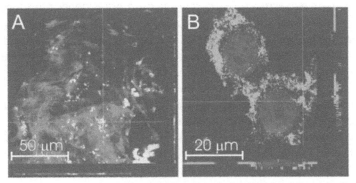

FIGURE 22.1   Confocal microscope images of the halloysite nanotubes uptake by MCF-7 cells and HeLa cells. (a) MCF-7 cell membrane (darker) with co-localized halloysite nanotubes (lighter) and, (b) localization of halloysite outside HeLa nuclei. Reproduced with permission from (Vergaro et al., 2010).Copyright 2010 American Chemical Society.

Human dermal fibroblasts were grown on multilayer films via layer-by-layer (LbL) assembly of the nanotubes with cationic polylysine and poly(ethyleneimine) on glass. Halloysite improved the adhesion of cells by maintaining their cellular phenotype (Kommireddy et al., 2005). Skibaet al. (2009) studied the effect of the addition of halloysite to a laying hens fodder, on the content of some egg yolk components. Hens were fed with a standard diet or with fodder enriched with halloysite. The results showed the ratio of $n6/n3$ polyunsaturated fatty acids was higher in eggs from hens that were fed a halloysite enriched diet. Moreover, halloysite caused a slight increase (by about 1–5%) of palmitic, palmitooleic acids, and vitamin A contents in the egg yolks.

*All these data provide strong evidence toward biocompatibility of halloysite. However, halloysite is not biodegradable and is not suitable for intravenous injections or oral administrations. Therefore, its application in medicine is limited to the biomedical devices, over skin treatments (such as scaffolds for wound healing, medications for healing skin diseases), bone implants (PMMA and calcium phosphate bone cements), dental fillers, and limited areas of the tissue engineering. Halloysite also has considerable potential for cosmetics applications as a controlled release agent (Suh et al., 2011).

22.3   HALLOYSITE FOR CONTROLLED DELIVERY OF PHARMACEUTICALS

Halloysite is an efficient container for active agents, especially, hydrophilic macromolecules including proteins, which are often too large to be immobilized in other

---

*HNTs are much shorter and thinner than asbestos microfibers, and fit the size window (0.5 1.5 μm), which is generally considered to be not toxic for silicate nanoparticles (Vergaro et al., 2010).

nanocontainers with smaller pores (e.g., Lvov and Price, 2008; Lvov and Abdullayev, 2013). For loading, halloysite is mixed with a saturated solution of the active agent. Then the suspension is stirred and sonicated for 10–20 min and transferred to a vacuum jar for 30 min. Air is removed from the tubule lumen under vacuum as indicated by slight fizzling of the solution. Once the vacuum is broken, the solution is sucked into pores. Solidification of the loaded substance within tube lumen is assumed to take place upon drying. Typical loading of active agents are ~10 wt %, which is close to the theoretical estimations (Abdullayev and Lvov, 2011). Loading and sustained release of cosmetic additives (Suh et al., 2011), antibacterials (Abdullayev et al., 2011), drugs (Veerabadran et al., 2007), proteins (Lvov and Price, 2008), and DNA (Shamsi and Geckeler, 2008) have been reported.

The first study on loading halloysite lumen with bioactive substances was performed using khellin (lypophilic vasodialator used for promoting hair growth), oxytetracycline HCl (water soluble antibiotic), and nicotinamide adenine dinucleotide (NAD). Khellin was loaded from its melt, oxytetracycline from saturated aqueous solution, and NAD from 5 percent aqueous polyvinylpyrrolidone solution. Release of oxytetracycline was completed in 30 h while khellin release extended over 500 h (Price et al., 2001). All these release rates were much longer than direct dissolution in water (few minutes) indicating nanopore-controlled diffusion. More extended release has been achieved for tetracycline from halloysite pores coated with chitosan. These capsules were mixed with polymers and have been proposed for dog teeth filling (Kelly et al., 2004). Such capsules can contain drugs for longer time with very low leakage, triggering the release, when the tube ends are exposed (e.g., in defects and/or microcracks).

The anti-angina drug diltiazem HCl was loaded within halloysite lumen from aqueous solutions using a vacuum suction pump to remove entrapped gases within halloysite lumen (Figure 22.2). Loading within halloysite lumen retarded the release rate to 1 h. An 80 percent initial burst was observed indicating that most of the drug was absorbed to the external surface rather than the inner lumen. In order to further retard the drug release, halloysite was coated with cationic chitosan by electrostatic surface adsorption. The burst in this case was significantly lower, about 30 percent. However, coating with chitosan also significantly reduced the drug loading efficiency (25.6% for the uncoated halloysite versus 7.8% at chitosan/halloysite ratio of 0.02). This formulation shows promise for oral administration of diltiazem hydrochloride, ensuring a prolonged exposure within the mouth (Levis and Deasy, 2003).

Halloysite is one of the few containers capable of encapsulating globular proteins with 2–6 nm proteins (Lvov and Price, 2008). This is particularly the case for tubules sharing a large diameter, for which the release rate of loaded proteins is rather slow; only 5 percent of catalase and glucose oxidase was released. The slow release rates were attributed to low diffusion coefficients of proteins coupled with stronger interactions with lumen walls (Figure 22.3).

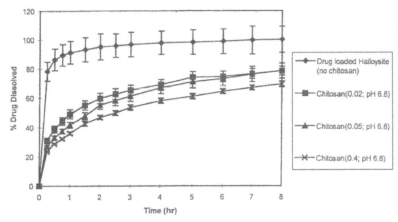

FIGURE 22.2   Release profiles of diltiazem HCl from halloysite coated with chitosan at varying chitosan/halloysite ratio, pH 6.8 and 37°C. Reproduced with permission from (Levis and Deasy, 2003). Copyright 2003 Elsevier.

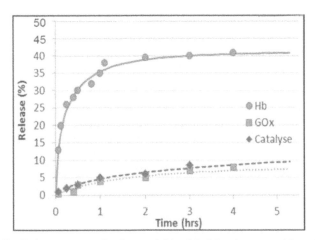

FIGURE 22.3   Release curves for hemoglobin (circle), glucose oxidase (squares) and catalase (diamonds) in water at pH 6.5. Reproduced with permission from (Abdullayev and Lvov, 2013).Copyright 2013 Royal Society of Chemistry.

Enzymes loaded into clay nanotubes are protected against loss of their activities. Loading into halloysite lumen did not inhibit the biocatalytic functionality of enzymes, allowing for selective biomimetic synthesis on halloysite external and internal walls. Biomimetic polymerization of aniline by positively charged hematin provided external coating of halloysite surface with polyaniline (Tierrablanca et al., 2010), while negatively charged urease allowed synthesis of vaterite within the nanotube lumen (Shchukin et al., 2005).

A drug delivery system based on halloysite-chitosan-poloxamer composite for treatment of animal periodontitis has been proposed by Levis and Deasy (2003). Halloysite was loaded with an antibiotic tetracycline, mixed with 0.2 percent solution of chitosan in acetate buffer (halloysite to chitosan ratio 1:0.114) and centrifuged. The sediment was dried and grinded. A 200 grams of such material was mixed with 1 mL of aqueous solution containing 20 percent poloxamer, 0.5 percent polyethylene glycol, and 1 percento cetylcyanoacrylate. This composite was capable of delivering antibiotic for up to 2.5 months and retained its syringeability over 9 months. *In vivo* study was performed by injecting this composite to the surgical pockets created in the mouths of dogs. Bacterial growth was inhibited by release of antibiotic as indicated by the reduction in colony forming units – CFU's (Figure 22.4).

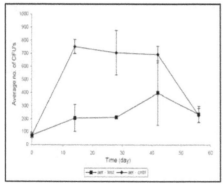 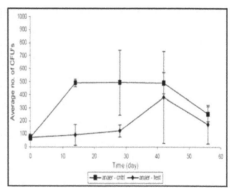

FIGURE 22.4   Inhibition of aerobic (left) and anaerobic (right) bacterial growth in dog mouth pockets after injection of halloysite-chitosan-poloxamer composite loaded with tetracycline. Reproduced with permission from (Kelly et al., 2004). Copyright 2004 Elsevier.

Halloysite loaded with tetracycline HCl was used as additives in polyvinyl alcohol and polymethylmethacrylate to provide functional composites with sustained drug release. Halloysite was mixed with the polymer by solution casting. Drug release from polymethylmethacrylate composite was reduced three times (about 7–8 h) by incorporating the drug in halloysite prior to producing the films (see Section 5 and Ward et al. (2010) for details).

22.4   HALLOYSITE FOR TISSUE ENGINEERING AND WOUND HEALING

In recent years, research on tissue engineering has progressed significantly due to its great potential to improve human health. Electrospinning of polymer solutions provides fibrous nets with diameters of fibers ranging from nanometers to a few micrometers (El-Refaie et al., 2002; Nitya et al., 2012; and Qi et al., 2010). This method uses high voltage to generate electrically charged jets of polymer solutions, which evaporate producing fibers of submicron diameters on the target material. Electrospinning technique was explored by El-Refaie et al. (2002) for the first time using polylactic acid and poly(ethylene-co-vinylacetate) fibers for the delivery of tetracycline HCl. Controllable drug release can be achieved by variations in the process parameters like applied voltage and polymer solution composition to obtain microfibers of desired diameter and porosity.

The potential application of halloysite-polycaprolactone (PCL) scaffolds as a wound healing material was explored by using nanotubes loaded with antiseptic agents. Release of Brilliant Green, well known antiseptic, from electrospun composites was extended to 5 days contrary to the 30 h for unmodified halloysite indicated earlier. About 20 percent of the initial drug burst is associated with the externally adsorbed molecules and is typical (Lvov and Abdullayev, 2013). Encapsulation of antibiotics in halloysite-PCL scaffolds may provide effective material for wound healing, and skin repair. In another study, halloysite-PCL scaffolds were explored for bone regeneration. Bioactivity of a scaffold material was improved significantly as evidenced by enhanced mineral deposition in simulated body fluid. All fibers were covered by an apatite-like layer after 21 days. Nanocomposite scaffolds exhibited excellent attachment, spreading, and proliferation of hMSC and osteoblast cells, indicating superior performance for bone tissue engineering (Nitya et al., 2012).

Porous scaffolds based on halloysite-poly(lactic-co-glycolic acid) (PLGA) composites have been made to explore the capability of delivering tetracycline hydrochloride and cell growth. Incorporation of halloysite significantly increased the yield strength of the fibrous mat produced by electrospinning of a composite containing 1 percent halloysite (4.2 MPa for pure PLGA vs. 6.6 MPa for halloysite-PLGA) and significantly retarded the release rate of the loaded drug. MTT cell viability essay was conducted on scaffolds using rat fibroblasts and results did not indicate any significant difference. This is probably associated with lower amounts of the clay used in this study. One percent of the halloysite was not sufficient to influence the surface of the PLGA fibers, which is critical for cell attachment and growth (Qi et al., 2010).

Incorporation of 10 percent halloysite within films made of polyvinyl alcohol significantly improved cell attachment and growth. Osteoblast-like cells grew on halloysite-PVA bionanocomposite films rather than on neat PVA films. Osteoblast filopodia extensions on PVA-10 percent halloysite composites were observed on 4th day of cell seeding. The increased cell adhesion properties are caused by both nanotopographic features and surface chemistry. Halloysite nanotubes impart nanoscale roughness on PVA film, causing enhanced osteoblast adhesion (Zhou et al., 2010). Similar results were also observed on surfaces modified by halloysite using by layer-by-layer (LbL) polyelectrolyte nanoassembly. Attachment and spreading of human dermal fibroblast cells were faster on substrates coated with halloysite clay compared with montmorillonite and silica nanoparticles. This was associated with nanoscale roughness caused by halloysite nanotubes as evidenced by an SEM microscopic study (Kommireddy et al., 2005). Another important feature of halloysite is the Si element in the surface of nanotube clay. It has been shown that incorporation of silicate ions into the hydroxyapatite structure can significantly improve its bioactivity. Silica has been shown to be important for bone formation by stimulating osteoblasts and osteoblast-like cells to secrete type I collagen and other biochemical markers (Zhou et al., 2010).

Recently, halloysite-chitosan biocomposite scaffolds have been demonstrated. Scaffolds were prepared by mixing halloysite and chitosan in aqueous acetic acid solution and freeze drying. Addition of halloysite resulted in increased compressive strength, modulus, thermal stability, and enhanced water uptake. Scaffolds have preserved their porous network with halloysite amounts as much as 80 wt% and showed enhanced bioactivity for NIH3T30 cells (cells grew on 80% halloysite scaffolds twice faster) (Liu et al., 2013).

22.5 HALLOYSITE FOR BONE CEMENTS AND IMPLANTS

Advances in surgical techniques and population longevity have drastically increased the need for and the number of dental and orthopedic procedures worldwide (Webb and Spencer, 2007). Dental and skeletal complaints are the major reason that most Americans see their physician, with over 7 million Americans seeking clinical intervention for dental and orthopedic disorders each year (McCaig, 2000; Cherry et al., 2008). The orthopedic implant industry, alone, was valued at 14.3 billion dollars in 2011 (www.reportlinker.com).

The surgical bone cement industry faces major challenges in meeting the changing consumer demands for more functional, bio-instructional, and longer-lasting products. Current treatment modalities of mixing antibiotics in commercial bone cement have significant limitations; (1) the addition of antibiotics to bone cement leads to a weakening of the cement, particularly, to a loss of mechanical strength,(2) sustained releases of the antibiotics from the current bone cements are available for only a short time period, providing no long-term protection against infection, and (3) mixing antibiotics intra-operatively into bone cement presents certain risk allergic reactions, cement mechanical failures, toxicity, and development of bacterial resistance.

The most common method chosen to secure prostheses, particularly in patients over 60 years of age, is the use of auto-polymerizing poly(methylmethacrylate) (PMMA) based bone cement. The cement is mixed in surgery until it becomes a doughy consistency when it can be inserted into the bone cavity and the completion of the polymerization takes place. This usually occurs in approximately 15 min. The polymerization process occurs as a result of the reaction between the benzoyl peroxide (BPO) in the polymer powder and N,N-dimethyl-p-toluidine in the monomer (DMPT). A radio pacifier such as barium sulphate or zirconium dioxide is also added to the powder component to enable the surgeon to view the cement in vivo. The polymerization of the MMA is highly exothermic and temperatures can exceed 80°C, which has detrimental effects for surrounding tissue (Harper, 1998). Addition of 10 percent halloysite by weight effectively reduced the temperature rise to 55 °C, which is safe for surrounding tissue if we consider that the duration of this state is less than 30 s. Nanotubes gavea reasonably good distribution of clay nanotubes within the polymer matrix (Figure 22.5(a)) (Wei et al., 2012).

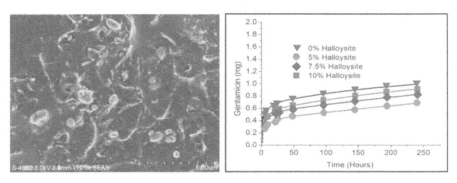

FIGURE 22.5 (a) Cross-sectional SEM image of halloysite/PMMA cement composites doped with 5 percent halloysite. Tubes and pre-crystallized PMMA particles are visible. (b) Gentamicin release from PMMA/halloysite composites. Total gentamicin loading is 2 mg.

The tensile strength of the composites increased from 20 to 29 MPa when 5 percent halloysite was added, while their flexural strength was slightly decreased. Optimal doping of 7 percent halloysite gave both a higher tensile strength and good flexural properties. Leakage of the commonly used bacteriocide, gentamicin, from PMMA-halloysite composites was slower than that from pure PMMA providing antibiotics supply over weeks (Figure 22.5(b)) (Wei et al., 2012).

An interesting phenomenon has been observed while studying the adhesive strength at the PMMA–bone interface. Three samples of PMMA cement: original (without additives), with 1 percent of free gentamicin (used as antibiotic) and with 7.5 percent of halloysite (loaded with 1% gentamicin) were used as the connecting glue for bovine femoral bone and pulled apart by a tensile tester after complete curing (Figure 22.6). For the halloysite/PMMA composite, failures occurred within the bone whereas it occurred at the cement–bone interface with the other two samples. The force required to break nanocomposite samples was 600 ± 20 N twice more than the original PMMA cement sample and the sample with free gentamicin (290 ± 20 N and 260 ± 30 N, respectively). An anchoring of needlelike halloysite tubes into pores existing on the bone surface (Figure 22.6(d)) is believed to be the cause of this phenomenon (Wei et al., 2012).

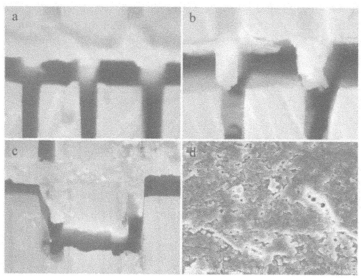

FIGURE 22.6 Images at PMMA composites–bone interface after fracturing with tensile tester: (a) original PMMA, (b) gentamicin-PMMA, (c) halloysite-PMMA, (d) SEM of cow femur bone surface. Reproduced with permission from (Wei et al., 2012). Copyright © 2012 WILEY-VCH Verlag GmbH & Co. KGaA, Weinheim.

Another type of commonly used material for bone cements is calcium phosphate salts because of their biocompatibility and osteoconductivity. Generally, these formulations contain calcium and phosphorus-based ingredients, which on mixing with water form workable and self-setting putty. The ingredients dissolve in the medium making it supersaturated with a desired calcium phosphate, which becomes reprecipitated inside the mass. The growth of the calcium phosphate phase as entangled crystallites helps the putty to retain its strength and shape (Camire et al., 2005). Addition of 10–20 wt % halloysite into β-tricalcium phosphate gave a reasonably good distribution of clay nanotubes within the cement matrix. Halloysite initiates non-isotropic crystallization of calcium phosphate yielding fibers of ca 50 μ lengths and 0.5 μ diameter (Figure 22.7). Formation of a microfibrous network significantly improved mechanical properties of the cement. Yield stress improved over 10 times from 0.33 MPa for the original cement to 3.5 MPa for the 10 percent halloysite-admixed cement. Elongation at the break improved 2.5 times and tensile modulus by 4 times.

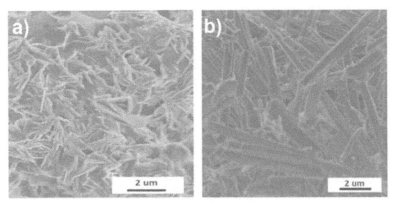

FIGURE 22.7   Scanning Electron Micrographs of the (a) original β tricalcium phosphate bone cement and (b) the cement with 10 percent halloysite.

22.6   HALLOYSITE FOR COSMETICS APPLICATIONS

Elongated hollow tubular structures have attracted many researchers from the cosmetics industry for their potential application in skincare products. The first study in this area was directed toward utilization of halloysite nanotube lumen for encapsulation and long lasting release of active agents such as: humectants, vitamins, fragrances, antimicrobials, antioxidants, and soothing agents, in cosmetic formulations. Such sustained release significantly improves the activity of these agents providing effective skin treatment. A publication of United States patent application 2007/0202061 A1 by Riedlinger and Corkery (2007) describes the loading and sustained release of glycerin and vitamins C and E from halloysite nanotubes. Consistent release has been obtained for up to 4 h in water suspensions.

A detailed study on the loading and release of glycerin was presented by Suh et al. (2011). Glycerin was loaded into halloysite samples from two natural deposits; Applied Minerals' Dragon Mine in USA and Imerys' clay deposit in New Zealand. Glycerin loading was accomplished by suspension of the halloysite in a 40 percent glycerin–water mixture. The suspension was sonicated for 1 h and placed into a vacuum jar for 20 min, which was then replaced by the atmosphere. The vacuum process was repeated 3–4 times to increase the loading efficiency. Then halloysite was separated from the solution by centrifugation and washed 2 times. Loading efficiencies of the glycerol were 19.0 and 2.3 percent for the USA and New Zealand halloysites, respectively. The release was much faster in the case of NZ halloysite, with an initial 60 percent burst and 80 percent within first 5 h. The USA sample did not display an initial burst and total release time exceeding 20 h, which is considered long enough for the humidifying effect. The difference in the release behavior has been attributed to a lower surface area and pore volume of New Zealand's halloysite (Figure 22.8).

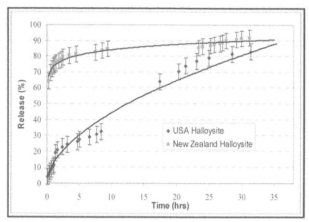

FIGURE 22.8   Glycerin release profile from halloysite nanotubes.

22.7   HALLOYSITE FOR BIOMEDICAL DEVICES

Nanoporous halloysite offers an attractive structure for the immobilization of enzymes. Urease-loaded halloysites have been used for biomimetic synthesis of $CaCO_3$ from water solutions containing $CaCl_2$ and urea. The urease-catalyzed reaction of $CaCO_3$ deposition occurs exclusively inside halloysite nanotubes and no $CaCO_3$ precipitate was found on the outer surface of the nanotubes or in solution after complete filling of the halloysite interior (Shchukin et al., 2005). On the other hand, the positively charged enzyme hematin, deposited on the tube exterior, reportedly provided biomimetic polymerization of aniline, coating halloysite surfaces (Tierrablanca et al., 2010).

Enzyme-loaded halloysites might potentially be useful for the fabrication of electrodes for biofuel cells and biosensors. Enzymatic fuel cells and sensors suffer from some limitations like lower power density and stability of enzymes, which has motivated scientists to search for a reliable substrate for enzyme immobilization. Careful electrode design is necessary for biosensors to improve their performance. The enzyme has to be immobilized as close to the electrode surface as possible to provide effective electron transport and retain its stability at least in the range of few months in order to meet consumer needs. Halloysite has the potential to provide multilayer coverage of an enzyme on an electrode surface by means of the layer-by-layer self-assembly method, which was shown to significantly improve current densities (typical current density for a traditional enzymatic electrodes lies in the interval of 50–200 $\mu A/cm^2$ and drastically increased to 800 $\mu A/cm^2$ upon multilayer coverage of enzymes) (Kamitaka et al., 2007). Furthermore, halloysite owns larger pores than traditional zeolites and carbon nanotubes, with diameters between 9 and 20 nm, which are comparable to the size of most enzymes. This property contributes to an effective immobilization without covalent bonding to a substrate (Zhai et al.,

2010). Glucose oxidase, immobilized on halloysite nanotubes, preserved its enzymatic activity even after 50 days of storage in phosphate buffered saline solution (Figure22.9). In another study, urease and α-amylase were also immobilized into halloysite nanotubes and retained 90 percent of their enzymatic activity after 15 days storage (Zhai et al., 2010). Enzymes being loaded into clay nanotubes are protected against proteolytic macromolecular agents while smaller molecular weight substrates (e.g., glucose) can pass through the tube openings, providing biocatalytic functionality for a longer time.

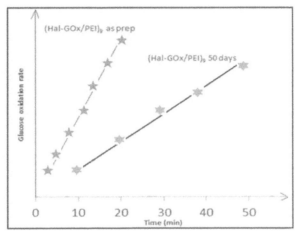

FIGURE 22.9   Enzymatic activity of halloysite–glucose oxidase thin film deposited on flat substrate.

A microfluidic device has been fabricated for capturing circulating tumor cells from the blood stream by coating the channel walls with a thin film made of P-selectin protein and halloysite nanotubes. The device showed enhanced capture of Colo205 colon carcinoma and KG1a leukemic cancer cells compared to the control device made without using halloysite. Nanotubes significantly increased the surface roughness (maximum surface roughness heights were 505 nm in films with halloysite vs. 30 nm in control with P-selectin only). High structures within microchannel walls presented significant amounts of P-selectin molecules farther into the flow for more efficiently capturing cells (El-Refaie et al, 2002). Similar results have also been demonstrated for liposome-halloysite-coated surfaces using MCF-7 and Colo205 cancer cells. The surface repelled healthy blood cells contrary to the cancerous cells due to the perfusion of the neutrophils. Liposomal doxorubicin was selectively delivered to cancer cells in solutions of cancer cells and neutrophils. The unique ability of microfluidic device made from halloysite-based coatings to attract cancer cells while repelling normal blood cells can reduce toxic non-specific effects and lower chemotherapeutic dosages required for cancer treatment (Mitchell et al.,

2012). This makes these devices promising in early detection and targeting the cancerous cells in diseased patients.

Silver-loaded halloysite has been applied as antimicrobial additives in coatings to prevent bacterial infection. Halloysite-based composite coating suppressed the growth of bacteria on paint surfaces. Ag^+ ions slowly leaked from the tubes and killed bacteria as indicated by the clear zone of inhibition in Figure 22.10(b). Contrary to the addition of bare silver oxide nanoparticles, halloysite core-shell structures provided better color stability for the coating (Figure 22.10(c), (d)) (Abdullayev et al., 2011).

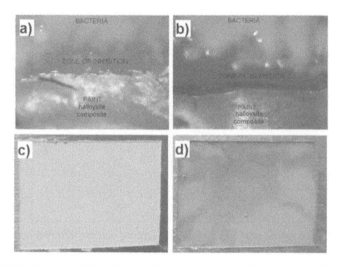

FIGURE 22.10 Images of the oil based alkyd paint doped with 5 percent halloysite loaded with silver after 1 week of exposure to *S. aureus* (a) and *E. coli* (b). Images of acrylic latex paint with 5 percent halloysite loaded with silver (c) and with 5 percent pure silver oxide (d). Reproduced with permission from (Abdullayev et al., 2011).Copyright 2011 American Chemical Society.

22.8 SUMMARY AND OUTLOOK

Halloysite minerals having tubular geometry are excellent materials for encapsulation of biologically active agents. Several pharmaceuticals like: diltiazem HCl, propranolol HCl, khellin, doxorubicin, resveratrol, and cosmetic agents like glycerol were successfully encapsulated by halloysite nanotubes, showing extended release rates in aqueous environments. This phenomenon drastically increases the biological activity of drug formulations due to the controllable release. Encapsulation of active cosmetics within halloysite lumen shows promise in the manufacturing of skincare crèmes with extended humidifying and nourishing effect. The biocompatible nature of halloysite has been reported in several studies on cancer cells.

Halloysite-polymer scaffolds find application in tissue engineering. Biocomposite scaffolds have been described using several polymers; PCL, PLGA, PVA, and chitosan. Halloysite was found to significantly improve mechanical properties and enhance cell adhesion. Sustained drug delivery from scaffolds shows promise in treatment of various diseases.

Halloysite was found to be an excellent substrate for immobilization of enzymes. This makes it a promising material for fabrication of biosensors and enzymatic fuel cells. A microdevice was fabricated by co-adsorption of P-selectin, poly-L-lysine, and halloysite for capturing cancer cells from the blood stream.

Finally, halloysite nanotubes can find applications in the improvement of artificial bone cements and dental fillers. Addition of halloysite into bone cement significantly improves its mechanical properties, which is crucial for bone implants. In addition, active agents that need to be embedded into bone cement (such as growth hormone or antibiotics) can effectively be encapsulated to provide long-term effects without the need for a high drug dose.

KEYWORDS

- Biomedical devices
- Biomimetic materials
- Halloysite-PCL scaffolds
- Nanopore-controlled diffusion
- Nanoporous halloysite
- Osteoconductivity
- Silver oxide nanoparticles
- Sustained drug release

REFERENCES

1. Abdullayev,E.; Sakakibara, K.; Okamoto,K.; Wei, W.; Ariga, K.; and Lvov, Y.;*ACS Appl. Mater.Interf.*2011,*3*, 4040–4046.
2. Abdullayev, E.; and Lvov, Y.;*J. Mater.Chem. B.*2013, 2894–2903.
3. Abdullayev, E.; and Lvov, Y.;*J. Nanosci.Nanotech.*2011,*11*, 10007–10026.
4. Brigger, I.; Dubernet, C.; and Couvreur, P.;*Adv. Drug Delivery Rev.*2002,*54*, 631–651.
5. Camire, C. L.;Saint-Jean, S. J.; Mochales, C.; Nevsten, P.; Wang, J. S.; Lidgren, L.; McCarthy, I.; and Ginebra, M. P.;*J. Biomed. Mater.Res. B Appl. Biomat.*2005,*76B*,424–431.
6. Cherry, D. K., Hing, E.; Woodwell, D. A.; and Rechsteiner E. A.; NationalAmbulatory MedicalCareSurvey: 2006 Summary. U.S.DepartmentofHealthandHumanServices;2008,*3*, 1–40.
7. El-Refaie, K.; Bowlin, G.; Mansfield, K.; Layman, J.; Simpson, D. G.; Sanders, E. H.; and Wnek. G. E.;*J. Contr. Release.*2002,*81*, 57–64.
8. Harper, E. J.;*Proc. Inst. Mec. Eng. H J. Eng. Med.*1998,*212*, 113–120.

9. Hughes, A.; and King, M.;*Langmuir.***2010,***26*, 12155–12159.
10. Kamitaka, Y.; Tsujimura, S.; Kataoka, K.; Sakurai, T.; Ikeda, T.; and Kano, K.;*J. Electroanal. Chem.* **2007,***601*, 119–124.
11. Kelly, H.; Deasy, P.; Ziaka, E.; and Claffey, N.;*Int. J. Pharm.***2004,***274*, 167–183.
12. Kommireddy, D.; Ichinose, I.; Lvov, Y.; and Mills,D.;*J Biomed.Nanotech.***2005,***1*, 286–290.
13. Lai, X.; Agarwal, M.; Lvov, Y.; Pachpande, C.; Varahramyan, K.; and Witzmann, F.;*J. Appl. Toxicity.***2013,***33*, DOI: 10.1002/jat.2858.
14. Levis, S.; and Deasy, P.;*Int. J. Pharm.***2003,***253*, 145–157.
15. Liu, M.; Wu, C.; Jiao, Y.; Xiong, S.; and Zhou, C.;*J. Mater.Chem. B***2013,***1*, 2078–2089.
16. Luo, X.; Matranga, C.; Tan, S.; Alba, N.; and Cui, X.;*Biomaterial.***2011,** *32*, 6316–6323.
17. Lvov, Y.; and Abdullayev, E.;*Prog.Polym. Sci.***2013,***38*, 1690–1719.
18. Lvov, Y.; and Price, R.;Halloysitenanotubulesa novel substrate for the controlled delivery of bioactive molecules. In: Bio-Inorganic Hybrid Nanomaterials. Ruiz-Hitzky, E.; Ariga, K.; Lvov, Y.;eds. Wiley: London, Berlin;**2008,**440–478.
19. McCaig L. F.;*Adv Data.***2000,***313*, 1–23.
20. Mitchell, M. J.; Castellanos, C. A.; and King, M. R.;*J. Nanomater.***2012,** DOI:10.1155/2012/831263
21. Nitya, G.; Nair, G. T.; Mony, U.; Chennazhi, K. P.; and Nair, S. V.;*J. Mater. Sci. Mater. Med.***2012,***23*, 1749–1761.
22. Price, R.; Gaber, B.; and Lvov, Y.;*J. Microencap.***2001,***18*, 713–723.
23. Qi, R.; Guo, R.; Shen, M.; Cao, X.; Zhang, L.; Xu, J.; Yu J.; and Shi, X.;*J. Mater. Chem.***2010,***20*, 10622–10629.
24. Riedlinger, M. D.; and Corkery, R. W.; Cosmetic Skincare Applications Employing Mineral-Derived Tubules for Controlled Release. US Patent Application 2007/0202061 A1, August 30,**2007,**Assigned to NaturalNano Inc.
25. Shamsi, M. H.; and Geckeler, K. E.;*e-Nanotech.***2008,***19*, 1–5.
26. Shchukin, D.; Price, R.; and Lvov, Y.;*Small.***2005,***1*, 510–513.
27. Singh, M.; et al *Nanoscale.***2010,***2*, 2855–2863.
28. Singh, M.; Shokuhfar, T.; Gracio, J.; Sousa, A.; Fereira, J.; Garmestani H.; and Ahzi, S.;*Adv. Func.Mater.***2008,***18*, 694–700.
29. Skiba, M.; Kulok, M.; Kolacz, R.; and Skiba, T.;The influence of halloysite supplementation in laying hens feeding on egg yolk lipid fraction. In:Proceedings of the 19th European Symposium on Quality of Poultry Meat, 13th European Symposium on the Quality of Eggs and Egg Products, **2009,** Turku, Finland.
30. Suh, Y.;Kil, D.; Chung, K.; Abdullayev, E.; Lvov, Y.;and Mongayt, D.;*J. Nanosci.Nanotech.***2011,***11*, 661–665.
31. Tierrablanca, E.; García, J. R.; Roman, P.; and Silva, R. C.;*Appl. Catal. A Gen.* **2010,***381*, 267–273.
32. Veerabadran, N.; Lvov, Y.; and Price, R.;*Macrom. Rap. Comm.***2009,***24*, 99–103.
33. Veerabadran, N.; Price, R.; and Lvov, Y.;*Nano.***2007,***2*, 215–222.
34. Vergaro,V.;Abdullayev,E.;Cingolani,R.;Lvov,Y.;andLeporatti,S.;*Biomacromolecule.***2010,***11*, 820–828.
35. Ward, C.; Song, S.; and Davis, E.;*J. Nanosci.Nanotech.***2010,***10*, 6641–6649.
36. Webb,J. C.; and Spencer, R. F.;*J. Bone Joint Surg. Brit.***2007,***89*, 851–857.
37. Wei, W.;Abdullayev, E.; Hollister, A.; Mills, D.; and Lvov.Y.;*Macromol.Mater. Eng.***2012,***297*, 645–653.
38. Www reportlinker.com, Orthopedic Implants-A Global Market Overview, **2011.**
39. Zhai, R.; Zhang, B.; Liu, L.; Xie, Y.; Zhang, H.; and Liu, J.;*Catal. Comm.***2010,***12*, 259–263.
40. Zhou, W.; Guo, B.; Liu, M.; Liao, R.; Bakr, A.; Rabie, M.; and Jia, D.;*J. Biomed.Mater.Res. A***2010,***93*, 1574–1587.41.

CHAPTER 23

MEDICAL AND HEALTH APPLICATIONS OF NATURAL MINERAL NANOTUBES

CÉSAR VISERAS, CAROLA AGUZZI, and PILAR CEREZO

CONTENTS

23.1   INTRODUCTION

This chapter addresses the issues related to health applications of a type of naturally occurring nanomaterial of great interest in academic and biomedical fields. Nanotechnology can be defined as the research and technology development at the atomic, molecular, or macromolecular levels, on the length scale in the 1–100 nm range. Nanomaterials are all nanoscale materials or materials that contain nanoscale structures internally or on their surfaces and include engineered and naturally occurring nano-objects, such as nanoparticles, nanoplates, and nanotubes (US National Nanotechnology Initiative. http://www nano.gov (accessed September 24, 2013)).

Natural mineral nanotubes include hollow tubular minerals with nanoscale diameters. There are several naturally occurring minerals with tubular morphologies (chrysotile, cylindrite, tochilinite, among others). Only some of these natural mineral nanotubes are interesting as raw materials for medical and health purposes on behalf of their demonstrated biocompatibility (safety) and efficacy: halloysite, imogolite, palygorskite, and sepiolite.

The first quality requirement of any substance intended for health care application is to demonstrate that is safe. Fibrous or tubular materials may show pathogenicity depending on their length, thickness, and biopersistence. The discovery of carbon nanotubes in the last decade of the twentieth century opened a new industrial revolution in modern technology. Nevertheless, carbon nanotubes and carbon nanofibers may pose some respiratory hazards (NIOSH: Occupational Exposure to Carbon Nanotubes and Nanofibers, US National Nanotechnology Initiative. http://www.nano.gov/node/1007 (accessed September 24, 2013)). Moreover, contamination due to toxic solvents and starting components in chemical synthesis as well as environmental concerns related to the accumulation of non-biodegradable residues have become relevant drawbacks of synthetic materials. Consequently, natural substances, and in particular, natural mineral nanotubes, are attracting growing attention in a variety of fields. Pharmaceutical technology is not an exception and natural source excipients are competing well with the synthetic materials (Ruiz-Hitzky et al., 2010; Ariga et al., 2012).

The present chapter is intended to provide an overview of the use of natural mineral nanotubes in medical and health care applications. The current state-of-art provides detailed knowledge on their structure, allowing deep characterisation and, if necessary, purification. Sometimes, properties of natural mineral nanotubes cannot achieve the desired objectives, requiring their modification or incorporation into polymeric matrices to obtain nanocomposites with improved properties compared to the individual components. Modified nanotubular minerals and nanocomposites are addressed in detail in other chapters of this book. All these improvements are notably prompting the availability of optimised and well characterized natural mineral nanotube derivatives, suitable for a wide range of biomedical applications.

Entrapping of drug molecules in natural mineral nanotubes is a helpful strategy to modify the rate and/or time and/or site of drug release but also to protect drugs against chemical and enzymatic degradation. Natural mineral nanotubes can be assembled with a large variety of polymers yielding nanostructured hybrid materials used to design nanoscale drug delivery systems (Viseras et al., 2008a, 2010; Abdullayev and Lvov, 2013). Tissue engineering offers a means to provide biocompatible replacement tissue with mechanical and functional integrity. A critical step in tissue engineering is the designing of scaffolds with structure, composition, physicochemical, mechanical, and biological features analogous to natural tissues. Many polymeric materials have been widely investigated but frequently their properties need to be improved by addition of inorganic fillers, as for example natural mineral nanotubes.

This chapter focus on drug and gene delivery, tissue engineering, use as drugs and pharmaceutical excipients, as well as future therapeutic and diagnostics trends of both natural mineral nanotubes and their derivatives (Table 23.1).

TABLE 23.1 Natural mineral nanotubes and their health care applications.

Mineral	Health Use	References
Halloysite	Anti-inflammatory	Cornejo-Garrido et al. (2012)
	Emulsifier	Wei et al. (2012a)
	Drug delivery	Price et al. (2001), Levis and Deasy (2003), Veerabadran et al. (2007), Lvov et al. (2008), Viseras et al. (2008b, 2009), Abdullayev and Lvov (2011), Aguzzi et al. (2013), Zhang et al. (2013)
	Cosmetics	Suh et al. (2011)
	Antimicrobial	Burridge et al. (2011)
	Diagnostics	Rawtani et al. (2013)
Halloysite functionalised	Drug delivery	Yah et al. (2012), Tan et al. (2013), Mitchell et al. (2012a, b), Mingyia et al. (2012), Vergaro et al. (2012)
	Gene delivery	Campos et al. (2011), Shi et al. (2011)
	Drug delivery	Kelly et al. (2004), Veerabadran et al. (2009), Forsgren et al. (2010), Abdullayev et al. (2011), Cavallaro et al. (2011a, b), Ghebaur et al. (2012), Qi et al. (2010, 2012, 2013), Wang et al. (2012), Abdullayev and Lvov (2013)
Halloysite nanocomposites	Tissue engineering	Qi et al. (2010, 2012, 2013), Chen et al. (2011), Nitya et al. (2012), Wang et al. (2012), Wei et al. (2012b), Liu et al. (2013c), Zhao et al. (2013)

	Bio-mimetic	Konnova et al. (2013)
Imogolite nano-composites	Gene delivery	Jiravanichanun et al. (2012)
	Tissue engineering	Ishikawa et al. (2010), Nakano et al. (2010), Teramoto et al. (2012)
Palygorskite-sepiolite	Drugs, excipients and cosmetics	Viseras et al. (2007), López-Galindo et al. (2011), de Gois da Silva et al. (2013)
	Drug delivery	Aguzzi et al. (2007), Viseras et al. (2010)
Sepiolite func-tionalised	Diagnostics	Erdem et al. (2012)
Sepiolite nano-composites	Drug delivery	Viçosa et al. (2009)
	Tissue engineering	Olmo et al. (1987), Wan and Chen (2011), Su et al. (2012)
	Immunotherapy	Ruiz-Hitzky et al. (2009)

23.2 HALLOYSITE

Halloysite was traditionally used as an additive in the manufacture of high quality dinnerware and also as molecular sieves in petroleum cracking, and other industrial uses related to its particular reactivity towards organic molecules and ions (Joussein et al., 2005). High surface area, pore dimensions of its lumen, relatively low price, and safety make this material very interesting also for biomedical applications (Du et al., 2010; Vergaro et al., 2011; Ma et al., 2012; Rawtani and Agrawal, 2012). Halloysite has been used as a particulate emulsifier in the preparation of polylactic-coglycolic acid microspheres via an emulsion solvent evaporation technique. In particular, halloysite nanotubes spontaneously localize at the oil/water interface and minimize the Helmholtz free energy, determining the stability and type of emulsion (Wei et al., 2012a). The use of halloysite as a drug with anti-inflammatory properties has also been described (Cornejo-Garrido et al., 2012).

Halloysite has been also used in drug delivery for drug encapsulation by chemisorption onto its polyanionic (silanol groups) faces via electrostatic interaction (Veerabadran et al., 2007; Lvov and Price, 2008; Lvov et al., 2008; Abdullayev and Lvov, 2011; and Zhang et al., 2013). Halloysite nanotubes allow the achievement of a modified release of the carried drugs when used alone (Price et al., 2001; Levis and Deasy, 2003) or combined with other excipients to optimise drug release patterns (Kelly et al., 2004; Forsgren et al., 2010; and Ghebaur et al., 2012). Among the drugs encapsulated onto halloysite nanotubes, 5-amino salicylic acid, an anti-inflammatory agent used in the therapy of ulcerative colitis and Crohn's disease, has been deeply investigated by our research group (Viseras et al., 2008b, 2009; Aguzzi et al., 2013). By means of equilibrium and kinetics studies of the adsorption process between the drug and the clay nanotubes, it was possible to demonstrate that the organic molecule was adsorbed as a result of two subsequent processes: rapid adsorption on the external clay mineral surface followed by slow adsorption inside the halloysite nanotubes (Viseras et al., 2008b). The structure of the loaded halloysite nanotubes obtained was detailed by solid state characterisation techniques, confirming the presence of the drug both at the surface and inside the halloysite tubes (Viseras et al., 2009). The mechanism and kinetics of release from such systems have been recently described as the sum of two first order processes, one of which was strongly controlled by diffusion of drug molecules from the interior of the nanotubes to the solid–liquid interface (Aguzzi et al., 2013).

Natural halloysite nanotubes do not always show adequate features for use as drug nanocarriers, necessitating a complete characterisation of their relevant properties (morphology, surface area, porosity, zeta potential, cation exchange capacity, etc.) (Pasbakhsh et al., 2013). Thermal and chemical treatments have been proposed to improve nanotube performance (Yuan et al., 2012, 2013). As for example, the inner-pores of halloysite have been opened and blocked as a result of different treatments, increasing the likelihood of drug retention (Joo et al., 2013). Bonding of silver nanoparticles to halloysite imparted antimicrobial properties to the resultant composites (Burridge et al., 2011). Similar hybrid systems have been also used to detect DNA damage aimed at many biological and medical applications (Rawtani et al., 2013). Functionalised halloysite nanotubes have been exploited for intracellular delivery of antisense oligonucleotides (Shi et al., 2011). Functionalization also led to improved drug immobilization and controlled release properties (Yah et al., 2012; Tan et al., 2013). Multifunctional halloysite nanotubes have been developed as nanovectors of doxorubicin (Mitchell et al., 2012a, b; Mingyia et al., 2012) and resveratrol (Vergaro et al., 2012) for anticancer drug delivery. The use of halloysite as an anticancer drug platform arises from the demonstrated capacity of its nanotubes to enhance cell adhesion of circulating cancer cells (Hughes et al., 2010, 2012). The clay has been also proposed as a transgene transmission reagent (Campos et al., 2011) because of its cellular uptake and cytocompatibility properties (Kommireddy et al., 2005; Vergaro et al., 2010; and Verma et al., 2012). Halloysite-DNA conjugates in which the clay particles were covered by the gene matter have also been developed and their dispersion properties evaluated (Shamsi and Geckeler, 2008). It is well established that the outer surface of halloysite is negatively charged while the inner lumen is positively charged in aqueous media and in a wide pH range (Lvov et al., 2008; Vergaro et al., 2010). Therefore a polyanion (such as DNA) should not wrap the nanotubes but instead fit in their cavity. The dispersion stability is therefore improved by electrostatic interactions as shown for small molecules adsorption (Cavallaro et al., 2012).

Besides drug and gene delivery, halloysite has also been proposed as a nanocarrier for controlled release of glycerol as a moisturizing agent intended for cosmetic applications (Suh et al., 2011).Organic/inorganic hybrid systems constituted by halloysite and different biodegradable polymers (alginate, chitin, chitosan, pectins, polylactic acid, poly-ε-caprolactone, polyvinyl alcohol, starch, among others) have been developed, obtaining new materials with improved properties and changed microstructures useful in health care (Liu et al., 2007, 2012a, b, 2013a, b; Zhou et al., 2010; Cavallaro et al., 2011a, 2013; Dong et al., 2011; Huang et al., 2012; Kasgoz et al., 2012; Schmitt et al., 2012; De Silva et al., 2013; Haroosh et al., 2013; Luo et al., 2013; and Prashantha et al., 2013). Layer by layer assembly (Ariga et al., 2011) and electrospinning (Dzenis et al., 2004) are among the techniques described to prepare these materials. Halloysite loaded with tetracycline has been incorporated into polyvinyl alcohol and polymethyl methacrylate films to reduce the initial burst effect observed in the release of the drug from the natural nanotubes alone (Ward et al., 2010). Diphenhydramine hydrochloride has been loaded into halloysite/polyvinyl alcohol nanocomposites to control drug delivery (Ghebaur et al., 2012). Biocompatible chitosan and gelatine multilayers were used for the encapsulation of halloysite tubes loaded with dexamethasone by means of layer by layer methodology (Veerabadran et al., 2009). Encapsulation of the drug in the mineral nanotubes, accomplished by the assembly of natural polymer shells, resulted in controlled release of the drug molecule.

Functional electrospun polymer nanotubes showed potential applications in the fields of tissue engineering, wound dressing, and drug-delivery (Wang et al., 2012). Composite electrospun nanofibers with tissue engineering and drug delivery applications were prepared with halloysite and polylactic-co-glycolic acid (Qi et al., 2010, 2012, 2013). The resultant systems showed a prolonged release profile of tetracycline by double encapsulation of the drug. Hemocompatibility of these halloysite nanocomposites was assessed, showing their suitability as artificial tissue/organ substitutes (Zhao et al., 2013). Nanocomposites of polymethylmetacrylate and halloysite have been used as bone cements doped with gentamicin providing slow release of the drug without compromising the composite mechanical strength while enhancing bone adhesiveness (Wei et al., 2012b). Halloysite was used as filler material to prepare polycaprolactone scaffolds with superior mechanical properties, substantial protein adsorption, enhanced mineralization, and osteogenic differentiation compared to the polymer scaffold without the clay (Nitya et al., 2012). Polyglycerol sebacate was filled with halloysite nanotubes to obtain elastomeric nanocomposites with optimal combination of compliance and a degradable profile compared with the pure counterpart, showing excellent resilience and satisfactory biocompatibility in vitro to be used as a soft tissue engineering material (Chen et al., 2011). Chitosan–halloysite nanotubes nanocomposite scaffolds were prepared by combining solution-mixing and freeze-drying techniques exhibiting a high porous structure and cytocompatibility with great potential for tissue engineering (Liu et al., 2013c).

As for future trends, mineral coating of biological cells can be used to obtain surface-functionalised 3-D structures (whole-cells) with biomimetic properties. Yeast cells have been used to obtain hollow halloysite based microcapsules ("cyborg cells") by layer-by-layer assembly of clay nanotubes and subsequent thermal decomposition of the cells (Konnova et al., 2013). These new assemblies may find applications in controllable cell growth.

23.3 IMOGOLITE

The copious advantages of organic/inorganic nanocomposites continuously stimulate researchers to find novel inorganic fillers able to improve properties of polymer matrices. In particular, imogolite is receiving considerable attention for its excellent adsorption properties and nanotubular shape. Reinforcement of polymeric materials with imogolite nanotubes has recently been described in the literature (Yamamoto et al., 2005a, b; Yah et al., 2010; and Shikinaka et al., 2011). These findings have been exploited to obtain collagen- (Nakano et al., 2010) and gelatine-based (Teramoto et al., 2012) nanocomposite hydrogels, showing improved properties and promising potential in pharmaceutical and tissue engineering development. Imogolite scaffolds exhibited good biocompatibility and enhancement effect of proliferation over mouse osteoblast-like cells (Ishikawa et al., 2010). Imogolite nanotubes have also found application in gene delivery. In particular, interaction between imogolite and DNA has been used to obtain hybrid hydrogels with a 3-D network structure (Figure 23.1) able to protect the gene matter against the degradation by nucleases but maintain their biological activity (Jiravanichanun et al., 2012). Besides these advances, modification and functionalization of imogolite nanotubes (Kang et al., 2010, 2011) will probably offer new perspectives in the biomedical field in the next year. As regards safety requirements, imogolite nanotubes must demonstrate the absence of potential toxicological effects toward humans. The possible cytotoxic and genotoxic effects of synthetic imogolite-like nanotubes have been evaluated, finding no cytotoxic activity in human fibroblast cells up to 0.1 mg/ml of nanoparticles whereas DNA damage depended on cell uptake and aspect ratio of the nanotubes (Liu et al., 2012c).

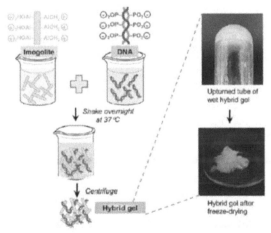

FIGURE 23.1 Schematic representation for the preparation of hybrid gel from imogolite and DNA. Adapted kind permission from (Jiravanichanun et al., 2012). Copyright (2013) American Chemical Society.

23.4   PALYGORSKITE AND SEPIOLITE

Palygorskite and sepiolite are natural fibrous clays, showing elongate structure with alternation of blocks and channels along the major axis direction of the clay particles. This morphology and small particle size give them unique properties (colloidal, high porosity, and surface area, great stability to high concentrations of electrolytes …) for several applications in the biomedical field. They are extensively used as drugs (antidhiarroeals, antacids, adsorbents of toxins, and exudates), excipients (diluents, glidants, disintegrants of solid dosage forms, as well as suspending and anti-caking agents in liquid and semisolid formulations) and cosmetics (adsorbents, cleaning agents, ultraviolet radiation filters, etc.) (Viseras et al., 2007; López-Galindo et al., 2011). Recent advances relate to the modification of palygorskite with lipophilic substances, resulting in wound healing activity with great potential to be exploited in dermopharmaceutical products (de Gois da Silva et al., 2013). Besides these applications, palygorskite and sepiolite are able to interact with a large number of drugs, being considered as promising nanoscale supports in drug delivery (Aguzzi et al., 2007; Viseras et al., 2010; and López-Galindo et al., 2011) and also in diagnostics to detect DNA and DNA–drug interactions (Erdem et al., 2012).

The high aspect ratio of palygorskite and sepiolite particles makes them excellent fillers for polymer matrix bionanocomposites both in drug delivery and tissue engineering. Nanocomposite hydrogels based on poly (vinyl alcohol) and sepiolite were prepared and assessed for their potential to improve the bioavailability of rifampicin in the therapy of tuberculosis (Viçosa et al., 2009). Sepiolite was incorporated into collagen matrices for artificial biological tissue applications (Olmo et al., 1987; Su et al., 2012). Natural fibrous sepiolite was also used as a template for growth of hydroxyapatite nanocrystals, giving a new class of nanocomposites to be exploited in tissue engineering and medical fields (Wan and Chen, 2011). An emerging aspect in nanocomposite applications concerns the development of vaccines in immunotherapy. In particular, bionanocomposites between sepiolite and xanthan gum were able to immobilise and preserve the antigenic activity of influenza virus particles through electrostatic interactions, resulting in potentially useful novel biosensor devices to enhance viral immune response (Ruiz-Hitzky et al., 2009).

23.5   FINAL REMARKS

Natural mineral nanotubes have gained great interest in the last years as materials for new advanced biomedical applications. They can be used as naturally occurring or functionalised materials both in drug and gene delivery and as inorganic fillers to obtain biopolymer-mineral nanohybrids. Such systems show improved features, being of particular relevance in the development of scaffolds for tissue engineering. Bionanocomposites can also act as multifunctional materials in diagnostics and immunotherapy, incorporating living entities or their fragments to achieve bio-mimetic properties. This is the new frontier of natural mineral nanotubes in health care applications.

ACKNOWLEDGMENTS

The authors thank the Spanish project CGL2010-16369 of the Ministerio de Ciencia e Innovación (MICINN) and the Andalusian group CTS-946.

KEYWORDS

- **Antimicrobial properties**
- **Dermopharmaceutical products**
- **Genotoxic effects**
- **Hemocompatibility**
- **Health care applications**
- **Imogolite**
- **Nanohybrids**
- **Polyanion**

REFERENCES

1. Abdullayev, E.; and Lvov, Y.; *J. Nanosci. Nanothecnol.* **2011**, *11*, 10007–10026.
2. Abdullayev, E.; Sakakibara, K.; Okamoto, K.; Wei, W.; Ariga, K.; and Lvov, Y.; *ACS Appl. Mater. Interfaces.* **2011**, *3(10)*, 4040–4046.
3. Abdullayev, E.; and Lvov, Y.; *J. Mater. Chem. B.* **2013**, *1*, 2894–2903.
4. Aguzzi, C.; Cerezo, P.; Viseras, C.; and Caramella, C.; *Appl. Clay Sci.* **2007**, *36(1–3)*, 22–36.
5. Aguzzi, C.; Viseras, C.; Cerezo, P.; Salcedo, I.; Sánchez-Espejo, R.; and Valenzuela, C.; *Colloid. Surf. B-Biointerfaces.* **2013**, *105*, 75–80.
6. Ariga, K.; Lvov, Y. M.; Kawakami, K.; Ji, Q.; and Hill, J. P.; *Adv. Drug Del. Rev.* **2011**, *63*, 762–771.
7. Ariga, K.; Ji, Q.; McShane, M. J.; Lvov, Y. M.; Vinu, A.; and Hill, J. P.; *Chem. Mater.* **2012**, *24*, 728 737.
8. Burridge, K.; Johnston, J.; and Borrmann, T.; *J. Mater. Chem.* **2011**, *21*, 734–742.
9. Campos, V. F.; Marques Moura de Leon, P.; Rossi Komninou, E.; Dellagostin, O. A.; Deschamps, J. C.; Kömmling Seixas, F.; and Collares, T.; *Theriogenology.* **2011**, *76*, 1552–1560.
10. Cavallaro, G.; Lazzara, G.; and Milioto, S.; *Langmuir.* **2011a**, *27(3)*, 1158–1167.
11. Cavallaro, G.; Donato, I.; Lazzara, G.; and Milioto, S.; *J. Phys. Chem. C* **2011b**, *115(42)*, 20491–20498.
12. Cavallaro, G.; Lazzara, G.; and Milioto, S.; *J. Phys. Chem. C* **2012**, *116(41)*, 21932–21938.
13. Cavallaro, G.; Gianguzza, A.; Lazzara, G.; Milioto, S.; and Piazzese, D.; *Appl. Clay Sci.* **2013**, *72*, 132–137.

14. Chen, Q.-Z.; Liang, S.-L.; Wang, J.; and Simon, G. P.; *J. Mech. Behav. Biomed. Mater.* **2011,** *4,* 1805–1818.
15. Cornejo-Garrido, H.; et al. *J. Appl. Clay Sci.* **2012,** *57,* 10–16.
16. De Gois da Silva, M. L.; *J. Therm. Anal. Calorim.* **2013,** DOI 10.1007/s10973-012-2891-4.
17. De Silva, R. T.; Pasbakhsh, P.; Goh, K. L.; Chai, S. P.; and Ismail, H.; *Polym. Test.* **2013,** *32,* 265–271.
18. Dong, Y.; Chaudhary, D.; Haroosh, H.; and Bickford, T.; *J. Mater. Sci.* **2011,** *46,* 6148–6153.
19. Du, M.; Guo, B.; and Jia, D.; *Polym. Int.* **2010,** *59,* 574–582.
20. Dzenis, Y. *Science.* **2004,** *304,* 1917–1919.
21. Erdem, A.; Kuralay, F.; Çubukçu, H. E.; Congur, G.; Karadeniza, H.; and Canavar, E.; *Analyst.* **2012,** *137,* 4001–4004.
22. Forsgren, J.; Jämstorp, E.; Bredenberg, S.; Engqvist, H.; and Strømme, M.; *J. Pharm. Sci.* **2010,** *99(1),* 219–226.
23. Ghebaur, A.; Garea, S. A.; and Iovu, H. *Int. J. Pharm.* **2012,** *436,* 568–573.
24. Haroosh, H. J.; Dong, Y.; Chaudhary, D. S.; Ingram, G. D.; and Yusa, S. *Appl. Phys. A.* **2013,** *110(2),* 433–442.
25. Huang, D.; Wang, W.; Kang, Y.; and Wang, A. *Polym. Compos.* **2012,** DOI 10.1002/pc.22302, 1693–1699.
26. Hughes, A. D.; and King, M. R. *Langmuir.* **2010,** *26(14),* 12155–12164
27. Hughes, A. D.; Mattison, J.; Western, L. T.; Powderly, J. D.; Greene, B. T.; and King, M. R. *Clin. Chem.* **2012,** *58(5),* 846–853.
28. Ishikawa, K.; Akasaka, T.; Yawaka, Y.; and Watari, F.; *J. Biomed. Nanotechnol.* **2010,** *6(1),* 59–65.
29. Jiravanichanun, N.; Yamamoto, K.; Kato, K.; Kim, J.; Horiuchi, S.; Yah, W.-O.; Otsuka, H.; and Takahara, A.; *Biomacromolecules.* **2012,** *13(1),* 276–281.
30. Joo, Y.; Sim, J. H.; Jeon, Y.; Leed, S. U.; and Sohn, D.; *Chem. Commun.* **2013,** *49,* 4519–4521.
31. Joussein, E.; Petit, S.; Churchman, J.; Theng, B.; Righi, D.; and Delvaux, B.; *Clay Minerals.* **2005,** *40(4),* 383–426.
32. Kang, D. Y.; Zang, J.; Wright, E. R.; McCanna, A. L.; Jones, C. W.; and Nair, S.; *Acs Nano.* **2010,** *4,* 4897–4907.
33. Kang, D. Y.; Zang, J.; Jones, C. W.; and Nair, S.; *J. Phys. Chem. C* **2011,** *115,* 7676–7685.
34. Kasgoz, H.; Durmus, A.; Kasgoz, A.; and Aydin, I.; *J. Macromol. Sci. Part A-Pure Appl. Chem.* **2012,** *49(1),* 92–99.
35. Kelly, H.; Deasy, P.; Ziaka, E.; and Claffey, N.; *Int. J. Pharm.* **2004,** *274,* 167–183.
36. Konnova, S. A.; et al. *Chem. Commun.* **2013,** *49,* 4208–4210.
37. Kommireddy, D. S.; Ichinose, I.; Lvov, Y. M.; and Mills, D. K.; *J. Biomed. Nanotechnol.* **2005,** *1,* 286–290.
38. Levis, S.; and Deasy, P.; *Int. J. Pharm.* **2003,** *253,* 145–157.
39. Liu, M; Guo, B.; Du, M.; and Jia, D.; *Appl. Phys. A* **2007,** *88,* 391–395.
40. Liu, M; Li, W.; Rong, J.; and Zhou, C.; *Colloid Polym. Sci.* **2012a,** *290,* 895–905.
41. Liu, M.; Zhang, Y.; Wu, C.; Xiong, S.; and Zhou, C.; *Int. J. Biol. Macromol.* **2012b,** *51,* 566–575.
42. Liu, W.; et al. *Chem. Res. Toxicol.* **2012c,** *25,* 2513–2522.
43. Liu, M.; Zhang, Y.; and Zhou, C.; *Appl. Clay Sci.* **2013a,** 75–76, 52–59.
44. Liu, M.; Zhang, Y.; Li, J.; and Zhou, C.; *Int. J. Biol. Macromol.* **2013b,** *58,* 23–30.
45. Liu, M.; Wu, C.; Jiao, Y.; Xiong, S.; and Zhou, C.; *J. Mater. Chem. B.* **2013c,** *1,* 2078–2089.
46. López-Galindo, A.; Viseras, C.; Aguzzi, C.; and Cerezo, P.; Pharmaceutical and cosmetic uses of fibrous clays. In: Developments in Palygorskite-Sepiolite Research. A New Outlook on These Nanomaterials. Galan, E.; Singer, A.; eds. Amsterdam: Elsevier; **2011,** 299–324.

47. Luo, B.-H.; Hsu, C.-E.; Li, J.-H.; Zhao, L.-F.; Liu, M.-X.; Wang, X.-Y.; and Zhou, C.-R.; *J. Biomed. Nanotechnol.* **2013,** *9(4),* 649–658.

48. Lvov, Y.; Shchukin, H.; Möhwald, H.; and Price, R.; *ACS Nano.* **2008,** *2(5),* 814–820.

49. Lvov, Y.; and Price, R.; Halloysite nanotubules a noble substrate for the controlled delivery of bioactive molecules. In: Bio-Inorganic Hybrid Nanomaterials. Ruiz-Hitzky, E.; Ariga, 50. K.; and Lvov, Y.; eds. Berlin, London: Wiley; **2008,** 440–478.

51. Ma, Z.; Wang, J.; Gao, X.; Ding, T.; and Qin, Y.; *Prog. Chem.* **2012,** *24(2),* 275–283.

52. Mingyia, G.; Aifei, W.; Faheema, M.; Wenxiub, Q.; Haoa, R.; Yingjie, G.; and Guangshan, Z.; *Chin. J. Chem.* **2012,** *30,* 2115–2120.

53. Mitchell, M. J.; Chen, C. S.; Ponmudi, V.; Hughes, A. D.; and King, M. R.; *J. Control. Rel.* **2012a,** *160,* 609–617.

54. Mitchell, M. J.; Castellanos, C. A.; and King, M. R.; *J. Nanomat.* **2012b,** article number 831263.

55. Nakano, A.; Teramoto, N.; Chen, G.; Miura, Y.; and Shibata, M.; *J. Appl. Polym. Sci.* **2010,** *118,* 2284–2290.

56. Nitya, G.; Nair, G. T.; Mony, U.; Chennazhi, K. P.; and Nair, S. V.; *J. Mater. Sci. Mater. Med.* **2012,** *23,* 1749–1761.

57. Olmo, N.; Lizarbe, M. A.; and Gavilanes, J. G.; *Biomaterials.* **1987,** *8,* 67–69.

58. Prashantha, K.; Lecouvet, B.; Sclavons, M.; Lacrampe, M. F.; and Krawczak, P.; *J. Appl. Polym. Sci.* 2013, DOI: 10.1002/APP.38358, 1895–1903.

59. Price, R.; Gaber, B.; and Lvov, Y.; *J. Microencapsul.* **2001,** *18,* 713–723 .

60. Pasbakhsh, P.; Churchman, G. J.; and Keeling, J. L.; *Appl. Clay Sci.* **2013,** *74,* 47–57.

61. Qi, R.; Guo, R.; Shen, M.; Cao, X.; Zhang, L.; Xu, J.; Yu, J.; and Shi, X. *J. Mater. Chem.* **2010,** *20,* 10622–10629.

62. Qi, R.; Cao, X.; Shen, M.; Guo, R.; Yu, J.; and Shi, X.; *J. Biomat. Sci.-Polym. E.* **2012,** *23(1–4),* 299–313.

63. Qi, R.; Guo, R.; Zheng, F.; Liu, H.; Yu, J.; and Shi, X.; *Colloid. Surf. B-Biointerf.* **2013,** *110,* 148–155.

64. Rawtani, D.; and Agrawal, Y. K.; *Rev. Adv. Mater. Sci.* **2012,** *30,* 282–295.

65. Rawtani, D.; Agrawal, Y. K.; and Prajapati, P.; *Bionano Sci.* **2013,** *3(1),* 73–78.

66. Ruiz-Hitzky, E.; Darder, M.; Aranda, P.; Martín del Burgo, M. A.; and del Real, G.; *Adv. Mater.* **2009,** *21,* 4167–4171.

67. Ruiz-Hitzky, E.; Aranda, P.; Darder, M.; and Rytwo, G.; *J. Mater. Chem.* **2010,** *20,* 9306–9321.

68. Schmitt, H.; Prashantha, K.; Soulestin, J.; Lacrampe, M. F.; and Krawczak, P.; *Carbohyd. Polym.* **2012,** *89,* 920–927.

69. Shamsi, M.; and Geckeler, K.; *Nanotechnology.* **2008,** *19,* 1–5.

70. Shi, Y.-F.; Tian, Z.; Zhang, Y.; Shen, H.-B.; and Jia, N.-Q.; *Nanoscale Res. Lett.* **2011,** *6,* article number 608, 1–7.

71. Shikinaka, K.; Koizumi, Y.; Osada, Y.; and Shigehara, K.; *Polym. Adv. Technol.* **2011,** *22,* 1212–1215.

72. Su, D.; Wang, C.; Cai, S.; Mu, C.; Li, D.; and Lin, W.; *Appl. Clay Sci.* **2012,** *62,* 41–46.

73. Suh, Y. J.; Kil, D. S.; Chung, K. S.; Abdullayev, E.; Lvov, Y. M.; and Mongayt, D.; *J. Nanosci. Nanothecnol.* **2011,** *11(1),* 661–665.

74. Tan, D.; Yuan, P.; Annabi-Bergaya, F.; Yu, H.; Liu, D.; Liu, H.; and He, H.; *Micropor. Mesopor. Mat.* **2013,** *179,* 89–98.

75. Teramoto, N.; Hayashi, A.; Yamanaka, K.; Sakiyama, A.; Nakano, A.; and Shibata, M.; *Materials* **2012,** *5,* 2573–2585.

76. US National Nanotechnology Initiative. http://www.nano.gov accessed September 24, 2013.

77. US National Nanotechnology Initiative. http://www.nano.gov/node/1007 accessed September 24, 2013.

78. Veerabadran, N.; Price, R. R.; and Lvov, Y.; *Nano J.* **2007,** *2,* 215–222.
79. Veerabadran, N.; Mongayt, D.; Torchilin, V.; Price, R. R.; and Lvov, Y. M.; *Macromol. Rapid Commun.* **2009,** *30,* 99–103.
80. Vergaro, V.; Abdullayev, E.; Lvov, Y.M. Zeitoun, A.; Cingolani, R.; Rinaldi, R.; and Leporatti, S.; *Biomacromolecules.* **2010,** *11,* 820–826.
81. Vergaro, V.; et al. *Adv. Drug Del. Rev.* **2011,** *63,* 847–864.
82. Vergaro, V.; Lvov, M. V.; and Leporatti, S.; *Macromol. Biosci.* **2012,** *12,* 1265–1271.
83. Verma, N. K.; Moore, E.; Blau, W.; Volkov, Y.; and Babu, P. R.; *J. Nanopart. Res.* **2012,** *14,* 1137–1147.
84. Viçosa, A. L.; Gomes, A. C. O.; Soares, B. G.; and Paranhos, C. M.; *Express Polym. Lett.* **2009,** *3(8),* 518–524.
85. Viseras, C.; Aguzzi, C.; Cerezo, P.; and Lopez-Galindo, A.; *Appl. Clay Sci.* **2007,** *36(1–3),* 37–50.
86. Viseras, C.; Aguzzi, C.; Cerezo, P.; and Bedmar, M. C.; *Mater. Sci. Tech.-Lond.* **2008a,** *24(9),* 1020–1026.
87. Viseras, M.T.; Aguzzi, C.; Cerezo, P.; Viseras, C.; and Valenzuela, C.; *Micropor. Mesopor. Mat.* **2008b,** *108(1–3),* 112–116.
88. Viseras, M.-T.; Aguzzi, C.; Cerezo, P.; Cultrone, G.; and Viseras, C.; *J. Microencapsul.* **2009,** *26(3),* 279–286.
89. Viseras, C.; Cerezo, P.; Sanchez, R.; Salcedo, I.; and Aguzzi, C.; *Appl. Clay Sci.* **2010,** *48(3),* 291–295.
90. Wan, C.; and Chen, B.; *Nanoscale.* **2011,** *3,* 693–700.
91. Wang, S.; Zhao, Y.; Shen, M.; and Shi, X.; *Therapeutic Delivery.* **2012,** *3(10),* 1155–1169.
92. Ward, C. J.; Song, S.; Davis, and E. W.; *J. Nanosci. Nanotechno.* **2010,** *10(10),* 6641–6649.
93. Wei, Z.; Wang, C.; Liu, H.; Zou, S.; and Tong, Z.; *J. Appl. Polym. Sci.* **2012a,** *125,* E358–E368.
94. Wei, W.; Abdullayev, E.; Hollister, A.; Mills, D.; and Lvov, Y. M.; *Macromol. Mater. Eng.* **2012b,** *297,* 645–653.
95. Yah, W. O.; Yamamoto, K.; Jiravanichanun, N.; Otsuka, H.; and Takahara, A.; *Material.* **2010,** *3,* 1709–1745.
96. Yah, W. O.; Takahara A.; and Lvov, Y. M.; *J. Am. Chem. Soc.* **2012,** *134,* 1853–1859.
97. Yamamoto, K.; Otsuka, H.; Wada, S.-I.; Sohn, D.; and Takahara, A.; *Polymer.* **2005a,** *46,* 12386–12392.
98. Yamamoto, K.; Otsuka, H.; Wada, S.-I.; Sohn, D.; and Takahara, A.; *Soft Matter.* **2005b,** *1,* 372–377.
99. Yuan, P.; Southon, P. D.; Liu, Z.; and Kepert, C. J.; *Nanotechnology.* **2012,** *23(37),* article number 375705.
100. Yuan, P.; Tan, D.; Annabi-Bergaya, F.; Yan, W.; Fan, M.; Liu, D.; and He, H.; *Respiratory Care.* **2013,** *58(3),* 561–573.
101. Zhang, Y.; Chen, Y.; Zhang, H.; Zhang, B.; and Liu, J.; *J. Inorg. Biochem.* **2013,** *118,* 59–64.
102. Zhao, Y.; Wang, S.; Guo, Q.; Shen, M.; and Shi, X.; *J. Appl. Polym. Sci.* **2013,** *127(6),* 4825–4832.
103. Zhou, W. Y.; Guo, B.; Liu, M.; Liao, R.; Rabie, A. B. M.; and Jia, D.; *J. Biomed. Mater. Res. A* **2010,** *93(4),* 1574–1587.

CHAPTER 24

THE ANTI-INFLAMMATORY PROPERTIES OF DIFFERENT NATURALLY-OCCURRING HALLOYSITES

JAVIERA CERVINI-SILVA, ANTONIO NIETO-CAMACHO, and MARÍA TERESA RAMÍREZ-APÁN

CONTENTS

24.1 INTRODUCTION

Halloysite is a clay mineral of the kaolin group sharing chemical and physical properties attractive to the areas of biotechnology, pharmaceutical, and medical research (Price et al., 2001; Veerabadran et al., 2007; Vergano et al., 2010; Cornejo-Garrido et al., 2012; and Oh Yah et al., 2012), and for developing technologies related to cancer cell separation, bone implants, cosmetics, control drug delivery (Levis and Deasy, 2002; Hughes and King, 2010; and Oh Yah et al., 2012), dentistry and bone tissue scaffolds (Kelly et al., 2004), vehicle drug delivery systems (Lvov et al., 2008), enzymatic nanoscale reactors (Shchukin et al., 2005), stem cell attachment and proliferation devices (Kommireddy et al., 2006), and adhesion of human dermal fibroblasts, with fibroblasts maintaining their cellular phenotype and spreading (Kommireddy et al., 2006).

A related study showed that halloysite **CLA-1** presented anti-inflammatory properties comparable to indomethacin as evidenced by determination of the mieloperoxidase (MPO) enzymatic activity, a specific marker for migration and cellular infiltration, with edema reaching maximum levels only after 4 h (Cornejo-Garrido et al., 2012). Halloysite **CLA-1** was concluded to offer anti-inflammatory properties, so has potential health benefits, unlike kaolinite, which has been shown to be lethal to rats on prolonged exposure in high concentrations (Westiaux and Daniel, 1990). More recent work (Cervini-Silva et al., 2013) compared halloysite from three additional locations and reported an apparent inverse relation between their anti-inflammatory activity, on the one hand, and surface area, and lumen space, on the other. The present chapter compared the anti-inflammatory and cytotoxic activity of halloysites from eight different locations, including four studied elsewhere (Cornejo-Garrido et al., 2012; Cervini-Silva et al., 2013).

24.2 MATERIALS AND METHODS

24.2.1 HALLOYSITE SPECIMENS

Halloysite specimens were provided as a gift by John Keeling (Resource Evaluation and Planning Section Geological Survey, Minerals and Energy Resources, Primary Industries and Resources, South Adelaide, Australia) and Jock Churchman (Soil and Land Systems, School of Earth and Environmental Sciences, The University of Adelaide, Australia). Listed in Table 24.1 are the halloysite specimens used in this study, varying in surface morphology depending on the availability and incorporation of structural Fe (Churchman and Theng, 1984; Novo et al., 1986; and Keeling et al., 2011). The crystallization of long tubes (e.g., **MB**) occurs when the availability of Fe is low; decreases in length (e.g., **KA**) or width (e.g., **TA**) take place with increasing availability of Fe; and the formation of stubby, tubular particles (e.g., **TP**) is favored upon the increased incorporation of Fe. The crystallization of some non-tubular forms (e.g., **TP**) prevails when the incorporation of Fe is relatively high. Spheroidal forms (e.g., **OP**) can contains either low or high contents of Fe, but are most likely to adopt this morphology as a result of their conditions of formation. They particularly occur as a result of the rapid dissolution of volcanic glass (Joussein et al., 2005).

TABLE 24.1   Selected halloysite specimens.

Sample Code	Sample Locality	Parent Material (Map Reference)	Mode of Formation
JA	~45 km SE of Perth, Western Australia	Granite and dolerite	Lateritic weathering
KA	Mt. Parakiore, Northland, New Zealand	Dacite (N20/797044)	Hydrothermal action
TA	~8 km W of Ngaruawahia, Waikato, New Zealand	Andesite (N56/570461)	Weathering
PA	Siberia, ~85 km NW of Kalgoorlie, Western Australia	Arkean greenstone	Weathering
OP	~2 km W of Opotiki, Bay of Plenty, New Zealand	Rhyolite (N78/681198)	Weathering
MB	Matauri Bay, Northland, New Zealand	Rhyolite and Dacite (N11/398770)	Low temperature hydrothermal alteration, then weathering. Differential weathering along cracks in phonolite boulders.
TP	~10 km W of Te Puke, Bay of Plenty, New Zealand	Rhyolite and andesite	Weathering and/or hydrothermal action
CLA-1	Maralinga, Barton locality, South Australia	Rhyolite and andesite	Weathering and/or hydrothermal action.

Data for halloysite **JA** was taken from Churchman and Gilkes (1989); halloysites **KA**, **TA**, **OP**, **MB**, and **TP** from Churchman and Theng (1984); and halloysite **PA** from Norrish (1995).

Two specimens are of particular interest: **PA** and **CLA-1**. The former halloysite specimen was found in a pit 8 m under a thick lateralitic cap and had a very pale white blue translucent appearance and high moisture content in the field (Norrish, 1995). It shows an atypical morphology with long tubes up to 30 μm in length that have very thin walls, with a uniform internal diameter of *ca.* 187 Å (Norrish, 1995; Cervini-Silva et al., 2013). The other notable specimen was collected from the **CLA-1** site of Maralinga, the Barton locality in South Australia and contains *ca.* 0.5 million tons of halloysite. Halloysite specimens from the **CLA-1** site were found to be up to 4-μm in length.

24.2.2   MOUSE-EAR EDEMA TESTS

Adult male Wistar rats (200–250 g) were provided by the Instituto de Fisiología Celular, UNAM. Adult male Wistar rats were approved by the Animal Care and Use Committee (NOM-o62-ZOO-1999). They were maintained at 25°C on a 12/12 h light–dark cycle with free access to food and water.

24.2.2.1   TPA METHOD

The incubation of ear edema in mice was achieved by using a method reported by Merlos et al. (1991). Briefly, a group of six male CD1 mice was anesthesized using Sedaphorte (32 mg kg ^{1} of sodium barbital, intraperitoneal). Under complete anesthesia, 10 μL of a 0.25 mg mL ^{1} ethanolic solution of 12-*O*-tetradecanoylphorbol-13-acetate (TPA) was typically applied on both faces of the mouse right ear; thus, 5-μL aliquots were applied to each face. To confirm full TPA absorption, the mouse ears were left in contact with the TPA solution for 15 min before proceeding to the next experimental step. Parallel reference experiments were conducted in the left ear were only ethanol was applied.

Halloysite dispersions prepared by adding 1 mg halloysite to 20 μL of 1:1 water: acetone aliquots of 10 μL of the freshly prepared halloysite dispersion were applied to both faces of the right ear.

A parallel set of experiments to study the effect of indomethacin was conducted using a second group of animals. A solution that contains 1 mg of indomethacin dissolved in 1:1 ethanol: acetone was prepared. Twenty microliter aliquots of indomethacin solution were spread on the right ear faces such that *ca.* 10 μL was applied to each face. In this case, a group of animals that served as control were exposed to the vehicle solutions that namely, either 1:1 water: acetone or 1:1 ethanol: acetone.

After the application of the dispersions containing halloysite, indomethacin, or vehicle, the animals were sacrificed in a CO_2 chamber. Then 7-mm diameter plugs were removed from each ear.

The edematous response was determined from measured plug mass difference. The edema inhibition (EI) was calculated according to:

$$EI(\%) = \frac{A-B}{A}(x100) \tag{1}$$

Where A and B correspond to mass determined for samples exposed to TPA only and TPA plus indomethacin or halloysite.

24.2.2.2 MPO METHOD

Quantification of mieloperoxidase (MPO) enzymatic activity is used in inflammation models as an enzymatic marker specific for migration and cellular infiltration (De Young et al., 1989). This is particularly the case for neutrophiles bearing high contents of MPO (Bradley et al., 1982). In all cases, the enzymatic activity of MPO was determined from right-ear biopsies either 4 or 24 h after exposure to TPA-containing samples as described above.

The enzymatic activity of MPO was quantified colorimetrically using a BioTek micro-plate reader (ELx808) at $ca.$ 450 nm. The activity of MPO was expressed as optical density per biopsy (OD/Biopsy). All experiments were conducted in quadruplicates.

24.2.2.3 CELL VIABILITY AND CYTOTOXICITY

The cell viability was determined by the reduction of tetrazolium as induction step and the MTT bromide assay (Mosmann, 1983). Cytotoxicity [where cytotoxicity (%) ≡ 1-cell viability (%)] in halloysite suspensions (100 mg mL[1]) was also determined.

24.3 RESULTS AND DISCUSSION

24.3.1 MOUSE-EAR EDEMA TESTS

24.3.1.1 TPA METHOD

In an earlier study, applying 2.5 µg ear[1] TPA brought about a 16 mg edema only 4 h after induction which remained invariant for 24 h (Cornejo-Garrido et al., 2012). By applying 1 mg indomethacin inhibited the edema production in a significant manner ($p \leq 0.01$). At $t = 4$ h, the mass of edema decreased from 16 to 2.9 mg, which corresponded to 82 percent inhibition. After 24 h, the mass decreased from 16 to 3.3 mg; that is, 79 percent inhibition. Applying 1 mg ear[1] halloysite **CLA-1** for 4 h decreased the edema mass, from 16 to 12 mg ($p < 0.01$), that is, 22 percent inhibition, and from 16 to 3.6 mg (*ca.* 77% inhibition after 24 h). Clearly, the anti-inflammatory activity of halloysite **CLA-1** depended on exposure time (Cornejo-Garrido et al., 2012). At longer exposure times, the activity of halloysite **CLA-1** was comparable to that of indomethacin.

In the current study, the anti-inflammatory activity of halloysite **PA, OP, MB,** and **TP** at t = 4 h or **PA** and **OP** at t = 24 h surpassed that for indomethacin (Table 24.2, top). Of all samples, halloysite **OP** shows that more effective and prolonged anti-inflammatory activity, with increases by 8.5 percent at $4 \leq t \leq 24$ h. For the rest of the samples, the anti-inflammatory activity was highest at t = 4 h (Table 24.2).

TABLE 24.2   Time-dependent anti-inflammatory activity of halloysite specimens as determined by the TPA and MPO models.

	TPA Model				
	t = 4 h		t = 24 h		
	Edema (mg)	EI (%)	Edema (mg)	EI (%)	DEI (%)
Control	11 ± 3.15	-	9.8 ± 1.6	-	-
Indomethacin	2.6 ± 0.6**	76.55**	2.6 ± 0.6**	73.8**	+2.78
	Halloysites				
JA	5 ± 1.45*	54.8*	5 ± 1.45**	49.4**	+5.4
KA	5.9 ± 1.8*	46*	5.9 ± 1.8**	39.7**	+6.3
TA	4.8 ± 0.9*	55.8*	4.8 ± 0.9**	50.6**	+5.2
PA	2.4 ± 0.5**	78.4**	2.4 ± 0.5**	75.8**	+ 2.6
OP	2.9 ± 0.6**	79.4**	1.3 ± 0.5**	86.8**	7.4
MB	3.2 ± 0.4**	77.3**	3.1 ± 0.6**	68.3**	+9
TP	2.5 ± 0.2**	82.4**	2.9 ± 0.5**	70.4**	+12
CLA-1	12.5 ± 0.7**	22.4**	-	-	-
	MPO Model					
	t = 4 h			t = 24 h		
	DO$_{450nm}$ biopsy	Edema (mg)	EI (%)	DO$_{450nm}$ biopsy	Edema (mg)	EI (%)	DEI (%)
Basal	0.04 ± 0.009	-	-	0.04 ± 0.004	-	-	-
Control	0.23 ± 0.05	14.2 ± 1	-	0.5 ± 0.03	9.8 ± 1.6	-	-

TABLE 24.2　*(Continued)*

	Edema (mg)		EI (%)	Edema (mg)		EI (%)	DEI (%)
	TPA Model						
	t = 4 h			**t = 24 h**			
Indomethacin	0.05 ± 0.005**	1.03 ± 0.24**	92.7**	-	2.6 ± 0.6**	73.8**	+19
	Halloysites						
JA	0.27 ± 0.07	9.4 ± 0.6**	34.2**		5 ± 1.045	49.4**	15.2
KA	0.06 ± 0.007**	6.6 ± 0.9**	53.4**	0.4 ± 0.07	5.9 ± 1.8*	39.7**	+13.7
TA	0.009	7.8 ± 0.9**	45**	0.5 ± 0.07	4.8 ± 0.9**	50.6**	5.65
PA	0.14 ± 0.04	7.7 ± 1.1**	45.8**	0.4 ± 0.07	2.4 ± 0.5**	75.8**	30.05
OP	0.06 ± 0.002**	2.9 ± 0.6**	79.4**	0.3 ± 0.03	1.3 ± 0.5**	86.8**	7.44
MB	0.045 ± 0.01**	3.2 ± 0.4**	77.3**	0.4 ± 0.06	3.1 ± 0.6**	68.3**	+9.02
TP	0.09 ± 0.02**	2.5 ± 0.2**	82.4**	0.4 ± 0.08	2.9 ± 0.55**	70.4**	+12.02
CLA-1	0.12 ± 0.02**	-	50.2**	-	-	-	-

Values with $p \leq 0.05$ (*) and $p \leq 0.01$ (**) were considered to differ statistically from control experiments. All results were analyzed using the t student test. In all experiments, the dose corresponded to 1 mg ear-1. Specimens were sieved, and the ≤ 0.05 mm size fraction was separated and used as received. ΔEI (%) $\equiv$ EI(%)(4h)　EI(%)(24 h). Experiments were conducted in heptaplicates average values with standard error. Experiments were conducted in quintuplicates, average values with standard error. Data was taken from Cornejo-Garrido et al. (2012), Table 24.1, bottom.

24.3.1.2　MPO METHOD

The quantification of MPO was highest for the TPA-bearing samples. The presence of halloysite **CLA-1** after 4 h led to a 50 percent reduction of MPO (Cornejo-Garrido et al., 2012). Longer exposure times significantly increased the accumulation of MPO. By contrast, the effect of indomethacin on the accumulation of MPO was not dependent on the exposure time, and remained constant at *ca.* 90 percent. Upon TPA application, infiltration and pro-inflammatory effects in neutrophiles are temporarily shuffled processes (De Young et al., 1989). Thus, the activity of halloysite **CLA-1** with time was tested next (Table 24.2, bottom). Edema reached maximum levels after 4 h while cellular migration was noted to be low. By contrast, after 24 h, an opposite trend in reactivity was noted, thereby confirming the anti-inflammatory properties of **CLA-1**.

The extent to which halloysite inhibited the number of neutrophiles that infiltrated the inflammation site varied from one sample to the other. Notably, halloysites **TP** and **OP** provoked an inhibition by 82.4 percent (4 h) and 86.83 percent (24 h) more relative to samples containing TPA only. Diffractograms for halloysites **TP** and **OP** (not shown) evidenced a high degree of purity relative to the rest of the samples, suggesting that, like clay–water hydration, migration, and cellular infiltration was constrained by varying access to active surface sites. Furthermore, secondary surfaces may occlude active sites when present. Since several of the halloysites studied herein showed higher EI values at longer reaction times (Tables 24.2), we do not discard the physical separation between secondary surfaces and halloysite with time as plausible. A consequence is the enabling of active surface sites, leading to a higher effectiveness of migration and cellular infiltration.

The question whether halloysite is able to stabilize radical intermediates was evaluated by using the 2,2 -diphenil-1-picrrylhydrazyl radical (DPPH) as radical scavenger (Table 24.3(a) in Cornejo-Garrido et al., 2012). The optical density determinations in 100-mM DPPH reference solutions and halloysite **CLA-1** containing dispersions revealed no statistically significant difference, indicating that the halloysite **CLA-1**-DPPH molecular interaction is weak. EPR data for AAPH-DMPO and AAPH-DMPO halloysite **CLA-1** dispersions showed quartet signals with intensities and hyperfine-coupling constants of 1:2:2:1 and $a_N = 1.48$ mT and $a_H = 1.48$ mT, respectively, indicating the formation of the alcoxy radical adduct DMPO-OR (Krainev and Bigelow, 1996; Cornejo-Garrido et al., 2012). Peak intensities for dispersions with AAPH-DMPO (no halloysite) and AAPH-DMPO-halloysite **CLA-1** (either 3,000 or 30,000 ppm halloysite) showed no significant difference. Therefore, the presence of halloysite **CLA-1** did not contribute to the stabilization of DPPH-daughter radical(s). No oxidant activity associated with the formation of alcoxy free radicals after AAPH decomposition (Krainev and Bigelow, 1996) became apparent.

24.3.1.3 CELL VIABILITY AND CYTOTOXICITY

Cell viability values (%) varied from one halloysite to the other: **OP** (93.6 ± 6.4), **PA** (92.3 ± 6.0), **TA** (91.2 ± 6.6), **JA** (84.9 ± 7.8), **MB** (84.8 ± 6.2), **TP** (80.8 ± 4.7), and **KA** (77.2 ± 6.8).

24.3.1.4 RELATIONSHIP TO PROPERTIES OF HALLOYSITE

Since Cervini-Silva et al. (2013) found that surface area and lumen space tended to govern anti-inflammatory activity, Table 24.3 presents this data where available for the halloysites studied herein. Since the data was not available for all specimens, Table 24.3 also presents average lengths and widths of specimens from the same localities that are gleaned from other studies. The relative values of each of these indicators of available surface for the different halloysites are compared in Table 24.3 with values for anti-inflammatory activity and cytotoxicity that were measured in this study.

TABLE 24.3 Relative values of some physicochemical properties of halloysites and relation to anti-inflammatory activities and cytotoxicities.

Specific Surface Area (Cumulative, from Adsorption Isotherm (m^2g^{-1})

PA (88[*]) > **CLA-1** (77[*]) > **OP** (42[**], 169[***†]) ~ **JA** (47[*], 56[*]) > **TP** (36[*], 35[**]) > **MB** (23[*]); **KA, TA** not measured

Lumen Space (%)[*]

PA (47) > **CLA-1** (34) > **JA** (29) > **TP** (19) > **MB** (11); **OP, KA TA** not measured

Tube Length[§]/Particle Diameter[§‡] (µm)

TP (0.27) < **KA** (0.33) < **OP** (0.42) < **TA** (0.51) < **MB** (0.67); **JA, PA** not measured

Tube Width[§]/Particle Diameter[§‡] (µm)

TP (0.1) = **TA** (0.1) < **KA** (0.2) = **MB** (0.2) < **OP** (0.42); **JA, PA** not measured

Anti-Inflammatory Activity (TPA and MPO Models, 24 h) (mg Edema)

OP (1.3[¶]) > **PA** (2.4[¶]) > **TP** (2.9[¶]) > **MB** (3.1[¶]) > **CLA-1** (3.3[¶¶]) **TA** (4.8[¶]) > **JA** (5[¶]) > **KA** (5.9[¶])

Anti-Inflammatory Activity (MPO Model, 24 h) (mg Edema)

OP (1.3[¶]) > **PA** (2.4[¶]) > **TP** (2.9[¶]) > **MB** (3.1[¶]) > **CLA-1** (3.3[¶¶]) > **TA** (4.8[¶]) > **JA** (5[¶]) > **KA** (5.9[¶¶])

Cell Viability (%)[¶]

OP (93.6) > **PA** (92.3) > **TA** (91.2) > **JA** (84.9) ~ **MB** (84.8) > **TP** (80.8) > **KA** (77.2); **CLA-1** not measured

[*]from Pasbakhsh et al. (2013)
[**]from Churchman et al. (1995)
[§]from Churchman and Theng (1984)
[†]specimen may include allophone
[‡]particle diameter for spheroidal halloysite (**OP**)
[¶]from this study
[¶¶]from Cornejo-Garrido (2012)

It was found that **OP** and **PA** halloysites had similarly high anti-inflammatory activities and cytotoxicities to each other. Whereas, **KA** halloysite had low values for both anti-inflammatory activity and cytotoxicity. The **OP** and **PA** halloysites both showed among the highest surface areas, along with **CLA-1** (examined by Cornejo-Garrido et al., 2012). **PA** also had the largest lumen space. By contrast, **MB**, **TP**, and **JA** all had among the smallest values for surface areas and lumen space. **MB** had intermediate values for both anti-inflammatory activity and cytotoxicity. **JA** tended to have one of the lowest values for both anti-inflammatory activity and cytotoxicity. **KA**, with the lowest values measured for both anti-inflammatory activity and cytotoxicity, had tubes that were especially short and only of intermediate width. It should be noted that **OP** and **TP** were not primarily tubular in morphology.

24.4   CONCLUSIONS

This study with 7 halloysite specimens confirms Cervini-Silva's (2013) conclusion that there is a direct relationship between the surface area of the halloysites and their anti-inflammatory activity and cytotoxicity.

ACKNOWLEDGMENTS

The authors thank Lic. María del Rocío Galindo Ortega and Natalia Alfonsina Mendoza Mena (UAM-Cuajimalpa); M. in Sc. Claudia Rivera Cerecedo and Héctor Malagón Rivero (Bioterio, Instituto de Fisiología Celular, UNAM); for technical assistance. The comments of two anonymous reviewers contributed significantly to improve the original version of this manuscript. This project was supported in part by Universidad Autónoma Metropolitana Unidad Cuajimalpa.

KEYWORDS

- Anti-inflammatory properties
- Cell viability values
- Cellular infiltration
- Cytotoxicity
- Diffractograms
- Heptaplicates average values

REFERENCES

1. Bates, T. F.; *Am. Mineral.* **1959**, *44*, 78–114.

2. Bradley, P. P.; Priebat, D. A.; Christensen, R. D.; Rothstein, G.; *J. Invest. Dermatol.* **1982,** *78,* 206–209.
3. Cervini-Silva, J.; Nieto-Camacho, A.; Palacios, E.; Montoya, A.; Gómez-Vidales, V.; Ramírez-Apán, M. T.; *Colloids Surf. B: Biointerf.* **2013,** *111,* 651–655.
4. Cornejo-Garrido, H.; et al. *J. Appl. Clay Sci.* **2012,** *57,* 10–16.
5. Churchman, G. J.; Davy, T. J.; Aylmore, L. A. G.; Gilkes, R. J.; and Self, P. G.; *Clay Minerals.* **1995,** *30,* 89–98.
6. Churchman, G. J.; and Gilkes, R. J.; *Clay Minerals.* **1989,** *24,* 589–590.
7. Churchman, G. J.; and Theng, B. K. G.; *Clay Minerals.* **1984,** *19,* 161–175.
8. Davies, D.; and Cotton, R.; *Br. J. Indus. Med.* **1983,** *40,* 22–27.
9. De Young, L. M.; Kheifets, J. B.; Bailaron, S. J.; and Young, J. M.; *Agents Actions.* **1989,** *26,* 335–341.
10. Hughes, A. D.; and King, M. R.; *Langmuir.* **2010,** *26,* 12155–12164.
11. Joussein, E.; Petit, S.; Churchman, J.; Theng, B.; Righi, D.; and Delvaux, B.; *Clay Minerals.* **2005,** *40,* 383–426.
12. Kelly, H.; Deasy, P.; Ziaka, E.; and Claffey, N.; *Int. J. Pharm.* **2004,** *274,* 167–183.
13. Keeling, J. L.; Pasbakhsh, P.; and Churchman, G. J.; *Proc. Int. Conf. Mineral.* **2011,** 1–9.
14. Kommireddy, D.; Sriram, S.; Lvov, Y.; and Mill, D.; *Biomaterial.* **2006,** *27,* 4296–4303.
15. Levis, S. R.; and Deasy, P. B.; *Int. J. Pharm.* **2002,** *243,* 125–134.
16. Lvov, Y.; Shchunkin, D. G.; Möhwald, H.; and Prince, R. R.; *ACS Nano.* **2008,** *2,* 814–820.
17. Merlos, M.; Gomez, L. A.; Giral, M.; Vericat, M. L.; Garcia-Rafarell, J.; and Forn, J.; *Br. J. Pharmacol.* **1991,** *104,* 990–994.
18. Mosmann, T.; *J. Immunol. Methods.* **1983,** *65,* 55–63.
19. Norrish, K.; In Clays Controlling the Environment: Proceedings of the 10th International Clay Conference. Churchman, G. J.; Fitzpatrick, R. W.; Eggleton, R. A.; eds. Melbourne: CSIRO Publishing; **1995,** 275–284.
20. Novo, S.; Pinto, A.; Abrignani, M. G.; Galati, D.; and Strano, A.; *Am. J. Nephrol.* **1986,** *6(Suppl. 1,* 87–90.
21. Oh Yah, W.; Takanara, A.; Lvov, Y. M.; *J. Am. Chem. Soc.* **2012,** *134,* 1853–1859.
22. Pasbakhsh, P.; Churchman, G. J.; and Keeling, J. L.; *Appl. Clay Sci.* **2013,** *74,* 47–57.
23. Price, R. R.; Gaber, B. P.; and Lvov, Y. M.; *J. Microencapsul.* **2001,** *18,* 713–722.
24. Shchukin, D.; Price, R.; Sukhorukov, G.; and Lvov, Y.; *Small.* **2005,** *1,* 510–513.
25. Stumm, W.; and Morgan, J. W.; Aquatic Chemistry. John Wiley & Sons. 3rd edition, **1996.**
26. Veerabadran, N.; Price, R.; and Lvov, Y. M.; Nano: Brief Reports and Reviews. **2007,** *2,* 115–120.
27. Vergano, V.; Abdullayev, E.; Lvov, Y. M.; Zeitoun, A.; Cingolani, R.; Rinaldi, R.; and Leporatti, S.; *Biomacromolecule.* **2010,** *11,* 820–826.
28. Westiaux, A.; and Daniel, H.; In: Health Related Effects of Phyllosilicates. Bignon, J.; ed. NATO ASI Series, G21, Springer, Berlin.

CHAPTER 25

HEALTH EFFECTS OF CARBON NANOTUBES AND SOME COMPARISONS WITH NATURAL MINERAL NANOTUBES

MARIE-CLAUDE JAURAND

CONTENTS

25.1   INTRODUCTION

Carbon nanotubes (CNTs) are an important new class of technological materials that have numerous novel and interesting properties. These properties allow a large spectrum of technological applications, including in biomedicine. They are nanoparticles (defined as a particle with one dimension lower than 100 nm), and are the subject of concern as to whether they have the ability to generate adverse effects on human health. Today, the development of nanotechnologies crosses paths with the road of the legacy of a terrible past use of nanofibers, namely asbestos. Industrial and economical development of nanotechnology requires urgency on the part of the political and scientific communities to consider issues relating to safety of these materials (Jaurand et al., 2009; Jaurand and Pairon, 2011).

CNTs are formed as a single or several sheets of graphene rolled to form a hollow cylinder. There are several classes of CNTs: Single-walled (SWCNT), double-walled (DWCNT), and multi-walled carbon nanotubes (MWCNT). Their dimensions are in the nanometric range for the width and length, but the length may reach several micrometers or dozen of micrometers. They belong to the so called High Aspect Ratio Nanoparticles (HARN). The similarities in shape between HARN and asbestos, addressing the question as whether exposure to HARN may cause similar adverse health effects, has focused on the health risks of CNTs exposure, and development of research to study their biological effects (Tran et al., 2008). CNTs are not the sole type of elongated and thin particles to present some similarities with asbestos, as can be seen in this volume for other nanotubular structures. However, CNTs have received great interest in the toxicological field, especially since the publication of similar biological effects (inflammation and mesothelioma induction) between long CNTs and asbestos when they were injected in the peritoneal cavity of mice (Poland et al., 2008; Takagi et al., 2008), following many years of their use.

In the following lines, some key mechanisms in asbestos toxicity will be first remembered. Thereafter, the potential health effects of CNTs will be summarized, on the basis of the results of cell and animal studies carried out with CNTs. Finally, it will be briefly discussed whether other nanotubular particles could have the same effects as CNTs, and finally, whether the data available in the literature on the biological effects of nanotubular particles have already been described. The health effects are discussed here, mostly as a consequence of exposure by the inhalation route, because most studies have been devoted to this route of exposure. However, it is worth mentioning that it is not an unique route of possible exposure, in view of the various applications of nanotubular particles. Oral and dermal exposures may occur in technological applications, as well as systemic exposure in the field of biomedical applications, considering the research on the health effects of CNTs; it must be also mentioned that the ecotoxicological impact of CNTs is the subject of present concern and of a number of investigations (Jackson et al., 2013).

25.2 SUMMARY OF KEY MECHANISMS IN FIBER TOXICITY

25.2.1 ASBESTOS CHARACTERISTICS

Asbestos has received abundant applications dating to antiquity, due to its remarkable properties of resistance to thermal and chemical breakdown, tensile strength, and fibrous shape. This fibers crystallize with the asbestiform habit. In this morphology, asbestiform crystals have grown to yield long and thin fibers together, which can be easily separated into very thin fibers (fibrils). This morphology gives the mineral special features including a high aspect ratio (length/diameter ratio), and favorable mechanical properties (including strength, flexibility, and durability).

Two families of asbestos have been used in industry (IARC, 1987; IARC, 2012). Chrysotile, from the serpentine group, is a magnesium phyllosilicate (Mg_3 Si_2 $O_5(OH)_4$); Al^{3+} or Fe^{2+} may substitute for Si^{4+} or Mg^{2+}. It is composed of sheets of $Mg(OH)$ and SiO_2. The sheets are curved and chrysotile has the structure of a hollow tube because of differences in the sizes of atoms. Amphibole asbestos fibers (amosite, crocidolite, and anthophyllite) are double-chain silicates containing a variety of cations including Fe^{2+}, Fe^{3+}, Mg^{2+}, Al^{3+}, Ca^{2+}, and Na^+. Asbestos is well known for its adverse health effects, characterized by the induction of lung, pleural, and peritoneal cancer and fibrosis (IARC, 2012). Later on, other types of fibers were found to induce cancer in humans. Erionite is a fibrous form of zeolite found to induce pleural and peritoneal cancer (mesothelioma) in areas in Turkey (Baris and Grandjean, 2006; Jasani and Gibbs, 2012). Fluoro-edenite is a non-asbestifom form of amphiboles associated to an enhanced risk of mesothelioma in Italy (Comba et al., 2003).

25.2.2 FIBER PARAMETERS THAT MODULATE TISSUE AND CELL RESPONSES

Among the fiber characteristics that play a role in the toxicity of asbestos, fiber dimensions are a critical parameter. It has been demonstrated that long and thin fibers are toxicologically more active than short fibers (Sanchez et al., 2009; Huang et al., 2011). These different findings entailed the definition of a specific status for elongated nanoparticles "HARN," defining a criterion to consider for investigating their potential toxicity. Although not a unique criterion that plays a role in the mechanism of action of particles, HARN stands out as a very important toxicological parameter (Tran et al., 2008).

Determinants of pathogenicity have been identified from the *in vitro* and *in vivo* studies on the mechanism of action of asbestos fibers. The fiber toxicology paradigm defines dimensions: length, width, or diameter, and also aspect ratio, as important. These parameters influence both fiber deposition and fiber retention. Briefly, it can be remembered that following inhalation, fibers are deposited in the lung, and may be eliminated by clearance pathways. They include the muco-ciliary escalator, uptake by macrophages and lymphatic clearance (Andujar et al., 2011; Mossman et al., 2011). Particle deposition is dependent on the aerodynamic diameter (D_{ae}) of the particle. For fibers, D_{ae} is controlled by the diameter, but the length impacts little on D_{ae} (Tran et al., 2008). Therefore, although maximum deposition in the alveolar region of the lung is about 5 μm (D_{ae}) for globular particles, long fibers can be also deposited because of their thinness. One can find details in a report discussing whether high aspect ratio nanoparticles should raise the same concerns as with asbestos fibers (Tran et al., 2008).

Because of the occurrence of malignant mesothelioma in asbestos-exposed subjects, fiber translocation from the lung to the pleura has been investigated. Both human and animal studies have demonstrated the presence of fibers in the pleura (Boutin et al., 1996; Everitt et al., 1997; Mitchev et al., 2002; Muller et al., 2002; and Broaddus et al., 2011).

Biopersistence defines the time of fiber retention in the lung. This notion has been introduced as a determinant to assess the hazard of fiber exposure, persistent fibers being more active than non-biopersistent fibers. It is clear that biopersistence is dependent of the clearance system and is partly determined the fiber length, as long fibers are normally cleared more slowly than short fibers that are usually cleared by macrophages. It is also determined by the potential of the fibers to dissolve in the biological milieu (durability), resulting in breakage of long fibers into shorter fibers (Sanchez et al., 2009).

The fibrous shape is a cause of the biological effects of particles. This was demonstrated with the use of globular particles of the same composition as the elongated counterparts. Experiments carried out with asbestos fibers have demonstrated that length is a key parameter in the pathogenicity of fibers. This evidence arose from comparative studies in which cells or animals were exposed to long and short fibers. The cut-off was generally 5 μm of length. It is based on the World Health Organization (WHO) definition, as the current regulations focus on the long asbestos fibers (Length: $L \geq 5$ μm, Diameter: $D < 3$ μm, and L/D ratio > 3) (NIOHS, 1994 (Baron, 2001 No. 4949; WHO/EURO, 1985)). The value of 5 μm as size limit of length is not based on scientific evidence, but is used in the metrological characterization of the samples in toxicological studies. Although there is a general agreement to consider that fibers less than 5 μm of length are likely to be non-carcinogenic, one cannot exclude a low potential of these fibers as discussed elsewhere (AFSSET, 2009). Data obtained in studies comparing the biological effects of short and long fibers showed that long fibers are more active than short fibers. Several types of *in vitro* cell systems showed inflammatory responses, oxidative stress, activity of regulatory pathways, and genotoxicity (Jaurand, 1997; Huang et al., 2011). Concerning *in vivo* studies, animals were exposed via inhalation, intra-tracheal, or intra-cavitary routes (pleural, peritoneal). Both higher rates of lung fibrosis and of lung cancer and pleural mesotheliomas have been observed in animals exposed to long fibers than in those exposed to short fibers (IARC, 1987; IARC, 2012). It is noteworthy that similar results were found with other types of fibers, such as man-made mineral fibers, confirming a role of the particle shape and dimensions in mediating the cell responses (IARC, 1988; IARC, 2002).

While length is a very important parameter, it does not appear to be the unique fiber parameter that modulates the biological effects. The surface chemistry, especially the presence of Fe^{2+} and Fe^{3+} or other ions of transition metals can catalyze the formation of reactive oxygen species (ROS), which can be associated with biological effects including lipid peroxidation, oxidative DNA damage, and activation of intracellular signaling pathways fibrotic and carcinogenic effects (Shukla et al., 2003; Manke et al., 2013).

25.2.3 MECHANISM OF FIBER TOXICITY

Fiber uptake is an important mechanism in asbestos toxicity. In the presence of long fibers, macrophages may be unable to engulf all the fiber, resulting in so called frustrated phagocytosis. This process is associated with inflammation and oxidative stress and damage in the surrounding tissue (Donaldson et al., 2010). It seems linked to the inflammasome activation (Dostert et al., 2008).

Concerning carcinogenesis, asbestos fibers cause a wide variety of DNA and chromosomal damages, as demonstrated by *in vitro* studies (Jaurand, 1997; Huang et al., 2011). *In vivo*, few studies have been carried out. DNA breakage and DNA base oxidation may be generated by ROS and more stable clastogenic factors. Although DNA repair cellular systems exist, DNA lesions may be misrepaired or remain unrepaired, leading to mutations. Mutations can also affect chromosomes, either their structure (breakage, deletions, amplifications, translocations) or number (aneuploidy and polyploidy). Asbestos fibers provoke numerous chromosomal abnormalities (numerical and structural abnormalities, chromosomal missegregation, alteration of cytokinesis, loss of heterozygozytie, non-homologous recombination, centrosome duplication) (Huang et al., 2011). These alterations may be linked to ROS activity and impairment of mitosis.

25.3 POTENTIAL HEALTH EFFECTS OF CNTS

25.3.1 HEALTH RISK OF CNTS

So far, there is no epidemiological data on the effects of CNTs on human health, and the health risk remains undefined. Our present knowledge on the potential health risk of CNTs comes from experimental studies. Based on reviews of animal dose-response data relevant to assessing the potential of non-malignant adverse respiratory effects of CNTs, NIOSH (National Institute for Occupational Safety and Health) proposed limit levels of exposure for CNTs. In 2010, a first draft proposed a recommended exposure limit (REL) of 7 mg per cubic meter. The last 2013 REL proposes an exposure limit for workers of 1 mg per cubic meter of elemental carbon (respirable mass for 8-h-time-weighted average concentration), and that efforts should be made to reduce exposure as much as possible (NIOHS, 2010).

25.3.2 EXPERIMENTAL STUDIES INVESTIGATING THE BIOLOGICAL EFFECTS OF CNTS

Experimental studies have considered the biological effects of CNTs using both *in vivo* animal experiments and *in vitro* cell systems. Most studies were developed to evaluate the effects via the inhalation route. However, inhalation studies need sophisticated and expansive protocols. CNTs administration was often performed by intra-tracheal instillation or pharyngeal aspiration. These modes of administration are not physiological, and extrapolation to inhalation procedure is made with caution. However, results provide useful information on the lung and lung cells responses to CNTs and can be compared to the findings of inhalation studies. In studies comparing two routes of exposure, pharyngeal aspiration, and inhalation (Shvedova et al., 2008; Shvedova et al., 2013), SWCNTs showed significantly greater potency (inflammation, fibrotic and genotoxic effects) after inhalation than after pharyngeal aspiration. Authors conclude that the difference most likely reflects the degree of agglomeration of SWCNTs. SWCNTs delivered upon suspension for pharyngeal aspiration tend to form agglomerates while the aerosol was more dispersed (Shvedova et al., 2013). Several recent reviews describe the potential health effects of CNTs, focusing on different aspects quoted in the present paper accordingly.

All sorts of CNTs (SWCNTs, DWCNTs, and MWCNTs) have been tested to investigate their biological effects. It seems presently difficult to draw clear-cut conclusions on the relative effects of the different sorts of CNTs as contradictory data are sometimes reported with the same type of CNT. These inconstancies can be related to different intrinsic characteristics of the CNT tested, concerning dimensions, functionalization, and/or treatment prior to exposure to cells or animals. It must be noted that when used in suspension for cultured cells or intratracheal instillation exposures, dispersants (dipalmitoyl phosphatidyl choline, dimethyl sulfoxide, natural surfactant, proteins…) are generally used to avoid or minimize the cluster or aggregate formation of CNTs. These treatments can modulate the cell responses (Herzog et al., 2009). Unless it does not alter the CNT structure, ultrasound sonication is an alternative.

25.3.2.1 IN VIVO EFFECTS IN THE LUNG

Experimental studies carried out in mice, rats, and guinea pigs have shown that administration of SWCNTs or MWCNTs cause inflammation, granulomas, and fibrosis in the lungs, by intra-tracheal instillation. Granulomatous lesions were associated with the presence of agglomerates of CNTs. Inflammation is expressed as an increase in neutrophils, eosinophils, lymphocytes, and macrophages recovered in bronchoalveolar lavage. Inflammation is associated with the increase in the amount of pro-inflammatory cytokines including TNF-α, IL1-β of and IL-8 (Aschberger et al., 2010; Manke et al., 2013; and Rodriguez-Yanez et al., 2013). Fibrosis located on granulomatous lesions is likely to be due to particles that are less aggregated or are isolated. The results of these studies suggest that CNTs can produce similar inflammatory effects as observed after exposure to asbestos. Despite some data indicating a possible attack of CNTs by enzymatic activity (Shvedova et al., 2012), CNTs are generally considered as biopersistent, suggesting that persistent inflammation may sustain inflammation and the fibrotic process, and cell and tissue damages (Donaldson et al., 2013). Allergic effects have been reported after intra-tracheal instillation of MWCNTs) in Aschberger et al. (2010).

One study investigated the profile of gene expression in the mouse lung following pharyngeal aspiration MWCNT. Several genes were differentially expressed in CNTs-exposed mice in comparison with control mice. They include genes previously identified as prognostic biomarkers in lung cancer (Pacurari et al., 2011). Pathway analysis revealed that several genes belong to pathways involved in carcinogenesis (Pacurari et al., 2011). Further studies will be likely developed to determine gene expression in CNTs-treated cells. They will be helpful to compare the effects of different types of CNTs.

25.3.2.2   TRANSLOCATION TO THE PLEURA

Because of the known role of translocation in pleural pathogenesis of asbestos, the translocation of CNTs to the pleura is the subject of researches. There is a general agreement to consider that asbestos reach the pleura, and that a direct interaction between cells and asbestos occurs. This does not exclude the possibility of the occurrence of paracrine mechanisms via the production of inflammatory and growth factors by neighbor cells or cells circulating in the pleural cavity. In rodents, long CNTs have been shown to translocate to the subpleural regions of the lungs (Ryman-Rasmussen et al., 2009; Mercer et al., 2010; Porter et al., 2010; Mercer et al., 2011; Mercer et al., 2013; and Porter et al., 2013) and to induce inflammation, frustrated phagocytosis, and granulomas similar to asbestos fibers following intraperitoneal injection (Poland et al., 2008). Xu et al. (2012) also investigated whether MWCNTs administered into the lung by intrapulmonary spraying induce mesothelial lesions in F344 rats. MWCNTs were found in the cell pellets of the pleural cavity lavage, mostly in macrophages, confirming that CNTs may reach the pleural space. Moreover, hyperplastic proliferative lesions of the visceral mesothelium were observed (Xu et al., 2012). Cooperation between macrophages and mesothelial cells for production of inflammatory factors has been reported in *in vitro* cell systems (Murphy et al., 2012). In a review, Stella (2011) discusses the experimental evidence and open issues on pleural inflammation from CNTs.

25.3.2.3   IN VIVO GENOTOXICITY, CARCINOGENICITY, AND REPROTOXICITY

Several studies have focused on the genotoxic effects of CNTs (for reviews see: (Jaurand et al., 2009; Singh et al., 2009; Schulte et al., 2012; and Van Berlo et al., 2012). They reported DNA breakage in lung cells after exposure to SWCNTs, and chromosome damage in bone marrow cells in mice, after intraperitoneal injection. DNA base oxidation was reported following treatment with SWCNTs and MW-CNTs. Mutations in the proto-oncogene *K-ras* were found in mice exposed to SW-CNTs by inhalation (Shvedova et al., 2008; Shvedova et al., 2013). Mutations were also detected in lung cells from transgenic *gpt delta* mice exposed to MWCNTS by intratracheal instillation. Genotoxicity studies have demonstrated that CNTs induced chromosomal aberrations, micronucleus formation aneuploidy, and DNA damage in rats after intratracheal instillation, and in mice bone marrow cells after intraperitoneal injection, indicating chromosome breakage, and missegregation of part whole chromosomes, and interference with the development mitosis (Muller et al., 2008a; Patlolla et al., 2010; and Kato et al., 2013).

Direct evidence of carcinogenicity has been reported after intracavitary injection of MWCNTs in rodents. Takagi et al. (2008) first observed malignant mesotheliomas after intraperitoneal injection into p53$^{+/}$ mice. Later, Sakamoto et al. (2009) also observed mesotheliomas after intrascrotal injection of the same type of MWCNT in wild-type Fischer-344 rats. Two studies focused on reprotoxicity and teratogenicity. A study of male mice exposed to MWCNTs by repeated intravenous administration showed that CNTs could cause reversible damage to the testes without affecting the fertility of animals (Bai et al., 2010). Intraperitoneal or intra-tracheal instillation of MWCNTs in pregnant ICR mice resulted in various types of external and skeletal abnormalities in fetuses (Fujitani et al., 2012).

In contrast, an absence of carcinogenic potential in rats after intraperitoneal injection of MWCNTs was reported in other studies. Muller et al. (2009) did not found peritoneal mesothelioma following injection of short (0.7 μm) and very thin (11.3 nm) MWCNTs in a 2-year study (Muller et al., 2009). Nagai et al. (2011) used five different types of MWCNTs and showed that thick (145 nm) or tangled (15 nm) forms of MWCNTs presented a lower risk of mesothelial carcinogenesis in rats (Nagai et al., 2011). The different findings are consistent with the hypothesis that a lower risk of carcinogenesis is related to the size of the MWCNTs, but does not exclude other parameters playing a role in view of the other differences between the samples. The aggregation state of CNTs is considered as a parameter that modulates the biological effects.

25.3.2.4   IN VITRO EFFECTS IN CELL CULTURES

Most *in vitro* studies have shown that eukaryotic cells internalize CNTs. Phagocytosis is an important step in the mechanism of action of fibers. With asbestos, DNA damage is associated with fiber internalization. As mentioned above, incomplete phagocyosis of fibers may entail enhanced inflammation and oxidative stress (Donaldson et al., 2010), and internalized fibers may lead to abnormal mitosis, chromosome missegregation, and aneuploid chromosome number (Jaurand, 1997; Huang et al., 2011). CNTs uptake has been reported in a variety of cell types. Several mechanisms of particle uptake exist; phagocytosis, endocytosis, and penetration through the cytoplasmic membrane. Raffa et al. (2010) found that CNTs uptake was dependent on the degree of dispersion, the formation of supramolecular complexes, and the nanotube length. Phagocytosis occurred for aggregates and bundles of CNTs ≥ 1 μm of length, endocytosis for supramolecular structures, and passive diffusion for dispersed submicronic CNTs (Raffa et al., 2010). Nagai et al. (2011) investigated internalization of MWCNTs by human mesothelial cells. They observed that penetration of MWCNTs was dependent on nanotube diameter. Thin MWCNTs (diameter ~50 nm) were internalized and showed cytotoxicity *in vitro* and subsequent inflammation and carcinogenicity *in vivo*, while thick or tangled MWCNTs (diameter: ~150 nm and ~2–20 nm, respectively) were less toxic.

Several types of cells have been used in cytotoxicity assays, as previously done with asbestos. The cytotoxic effect of CNTs seems to depend on time and the exposure concentration and the cell type. Slowdown of cell proliferation has also been demonstrated in different cell types. Cells exposed to CNTs respond by an oxidative stress and production of inflammatory cytokines and chemokines, and activation of pro-inflammatory transcription factors. These effects are mediators of fibrosis (Manke et al., 2013; Rodriguez-Yanez et al., 2013).

Genotoxicity has been investigated using mutagenicity assays in prokaryotes and eukaryotes. No mutagenicity was observed in studies carried out in prokaryotes. It is likely due to the limitations of this assay to test particulate matter because of the inability of prokaryotes to internalize CNTs, as it has been discussed with asbestos. In these systems, a mutagenic effect should be likely generated by soluble substances or molecules diffusing in the culture medium. Several studies have been carried out with eukaryotic cells. Results showed a clastogenic potential of SWCNTs (DNA breakage) highlighted by the comet assay, the induction of micronuclei, alterations in the mitotic process, centrosome fragmentation, and formation of aneuploid cells (Sargent et al., 2012; Van Berlo et al., 2012). MWCNTs also induced DNA breaks, micronuclei, centrosome disruption, chromosomal aberrations, and/or abnormal chromosome segregation in different cell types, demonstrating a genotoxic potential in eukaryotes (Sargent et al., 2010; Sargent et al., 2012; and Van Berlo et al., 2012). The mechanism of genotoxicity remains to be clarified; it has been related to the formation of reactive oxygen by cellular function and to alteration of mitosis (Singh et al., 2009; Van Berlo et al., 2012) although oxidative stress alone does not seem to account for genotoxicity (Manshian et al., 2013). Sargent et al. (2010) suggested a possible interaction with microtubules as the basis of the missegregation of chromosomes.

A long-term study was carried out to determine the ability of single- and multi-walled CNTs to induce phenotypes related to neoplastic-like transformation, and protein and gene profile expression in small airways epithelial cells (Wang et al., 3014). Results were consistent with the expression of neoplastic-like criteria (increased cell proliferation, anchorage-independent growth, invasion, and angiogenesis). Gene and protein expression analyses showed that single- and multi-walled CNTs shared similar signaling signatures, which were distinct from asbestos (Wang et al., 2014). In a short-term assay, a mouse fibroblasts cell line (Balb/3T3) was treated with un-functionalized and functionalized (-NH($_2$), -OH and -COOH groups) MWCNTs (Ponti et al., 2013). Genotoxicity and transformation assays revealed that MWCNTs induced morphologically transformed colonies but no cytotoxicity or genotoxicity (micronucleus formation) (Ponti et al., 2013).

25.3.2.5   OTHER BIOLOGICAL EFFECTS

A few studies have reported the effects of CNTs on the cardiovascular system of mice after pharyngeal aspiration or intra-tracheal instillation. They suggest that CNTs cause effects that may predispose to, or accelerate, atherogenesis (Li et al., 2007; Castranova, 2011; and Manke et al., 2013).

Little information concerning the oral route is available. In an acute oral toxicology assay, no pathological abnormalities were found in mice shortly after exposure (see reviews: (Aschberger et al., 2010; Johnston et al., 2010)). Two studies of genotoxicity performed by gavage in mice or rats reported data determined 24 h post-exposure. One showed an oxidation of DNA in the lung and liver cells (Folkmann et al., 2009) and the other did not find urinary excretion of mutagenic substances (Szendi and Varga, 2008).

Similarly, little information is available on the effects of CNTs via the dermal route. One study reported the production of free radicals, oxidative stress, and inflammation causing dermal toxicity by SWCNTs in mic. Other studies even indicate that CNTs are not irritating to the skin (see reviews (Aschberger et al., 2010; Johnston et al., 2010)).

25.3.3   DETERMINANTS OF CNT TOXICITY

From the different studies carried out in animals and in cultured cell systems, the biological effects and mechanisms of action of CNTs are likely to depend on their physico-chemical characteristics, including dimensions (length, diameter), shape or morphology (aggregation state), and degree of purity (presence of impurities or catalyst residues). In addition to the intrinsic properties of CNTs their characteristics may be changed by interactions with the biological medium, in particular the adsorption of biomolecules. Other parameters are likely to modify the response to CNTs, such as surface properties and functionalization. In a study comparing the potential of functionalized (carboxyl groups on the nanotube surface) and non-functionalized MWCNTs, in bone marrow cells of mice, functionalized CNTs exerted a higher level of structural chromosomal aberrations and micronucleated cells, and DNA damage in leukocytes compared to non-functionalized form of MWCNT (Patlolla et al., 2010). Lung acute MWCNTs toxicity (*in vivo* inflammation in rat lung, *in vitro* genotoxicity using the micronucleus assay) has been related to the presence of defects at the MWCNTs surface, as toxicity was reduced upon heating but restored upon grinding (Muller et al., 2008b).

The impact of surface state on interaction with biological molecules and subcellular structures is a matter of research in the field of biomedical research. This aspect has not been taken into consideration in the present paper. Although human exposure for biomedical applications is different from potential exposures of workers and consumers, these studies are of great interest to explain the interactions between CNTs and biological molecules, subcellular and cellular structures. They will increase our knowledge of the mechanisms of action of CNTs and will be complementary for toxicological issues.

The mechanism of action CNTs deserves thorough and fundamental studies to improve our understanding of their mechanism of action. They include exposure assessment, toxicokinetics, systemic effects, and a better definition of the cell and tissue specific responses (Simko and Mattsson, 2010). Further fundamental studies are needed to characterize the physico-chemical properties of CNTs that affect their biological effects.

As mentioned above, a wide diversity of CNTs have been used in toxicological assays. Although their characteristics were reported in several studies, it is not yet possible to determine the role of the different physico-chemical parameters. So far, differences between samples concern purity (especially the presence of metallic impurities from catalysis) size, state of agglomeration, and surface state (oxidation or functionalization) (Johnston et al., 2010).

25.4   BIOLOGICAL EFFECTS OF POTENTIAL INTEREST IN NANOTUBULAR PARTICLES TOXICOLOGY

Our present state of knowledge on the biological effects to consider in order to defines the health risk potency of CNTs. It seems that the HARN paradigm implies that these particles have a pathological potential for human health, and that the effects are modulated by physico-chemical properties, especially dimensions, and the state of aggregation. The expected effects concern inflammation and the response to stress in the form of oxidative stress and cellular stress, including mitosis impairment. These reactions consist in defence mechanisms that become deleterious when they cannot return to a homeostasis state. Shvedova et al. (2012) focused on oxidative stress. ROS may be produced by the cells but also generated by oxidant-generating properties of the particles, as reported in Fe-containing asbestos fibers. However, CNTs may quench rather than generate oxygenated free radicals (Fenoglio et al., 2008; Crouzier et al., 2010). Other sources of ROS in the cells are the NADPH oxidase and the mitochondria.

One important mechanism of damage induced by CNTs consists in the physical interference of CNTs with cellular and subcellular constituents. Similarity between SWCNT and microtubules, interaction of CNTs with centrosomes and cytosqueletal actin may have consequences in cell mobility and cell division (mitosis) and alter to cell functions.

25.5 COMPARISON BETWEEN CNTS AND ASBESTOS

Both CNTs and asbestos produce oxidative stress, inflammation, activation of signaling pathways, and genotoxicity. Lung fibrosis is observed in rodents exposed by intra-tracheal instillation or inhalation to CNTs or asbestos, and mesotheliomas are found by intraperitoneal injection of both fiber types. With both types of particles a size-dependent effect has been reported, with longest particles being the most active. These findings make CNTs close to asbestos in assuming their potential effects on human health. Despite these similarities, differences exist, and several issues remain to be investigated. The ability of CNTs to form clusters and aggregates is a specific feature of CNTs, and its role in the mechanism of action deserves further studies. Some issues are presently discussed, such as the CNTs capability of produce ROS from surface reactions, their biopersistence, and a better definition of the size limit for "long" and "short" CNTs (see (Sanchez et al., 2009; Donaldson et al., 2013; and Pacurari et al., 2010) for reviews).

Whether CNTs can provoke adverse health effects remain to be established. Regarding the route of exposure by inhalation, and fibrosis and cancer as the suspected diseases, the mechanism of action of asbestos can be a paradigm to assess the health effects. The mechanism of action of asbestos is linked to fiber characteristics, and cell and tissue responses; it can provide some insights to assess the CNTs effects. For asbestos, high aspect ratio, thinness, length, and bio-persistence (clearance and durability) are identified as the most important fiber parameters. Deposit of asbestos in the lung can cause oxidative stress, inflammation and genotoxicity, and fiber translocation to the pleura, with similar biological effects at the pleural level. CNTs share a number of similarities with asbestos fibers. Tables 25.1 and 25.2 summarize the CNTs effects. It is worth noting that these summaries draw general and not definitive conclusions, since some contradictory data are found in the literature on the effects and properties of CNTs. However, they emphasize the similarities between CNTs and asbestos.

TABLE 25.1   Summary of biological effects of CNTs.

Human	No Specific Data
Animal experiments (rat, mice) by inhalation, intra-tracheal instillation, pharyngeal aspiration	Recruitment of inflammatory cells. Early effects (within weeks). Production of molecules involved in inflammation and fibrogenesis. Granulomas, thickening of alveolar wall, fibrosis. ROS production (contradictory results). Internalization by alveolar macrophages, epithelial cells. Translocation (interstitium, lymph nodes, pleura). Genotoxicity. Persistence of CNTs (contradictory results).
Other routes of exposure: intra-cavitary (Intraperitoneal, IPer; intrapleural, IPl.)	Inflammation. Fibrosis, granulomas—length-dependent responses (IPer. IPl.). Mesotheliomas (IPer.).
Studies with cells in culture (macrophages, epithelial cells, mesothelial cells...)	Phagocytosis or no uptake; passive transfer (contradictory results). Involvement of oxidative stress response (metal-dependent or independent). ROS production. Genotoxicity.

TABLE 25.2   Genotoxicity of CNTs.

In vivo (inhalation, intratracheal instillation, pharyngeal aspiration)	Enhancement in the percentage of cells with micronucleus (pneumocytes II) ; DNA breakage ; DNA oxidation. Mutations in K-*ras* in lung cells, *gpt* in *gpt*-delata mice.
In vivo, after intra-peritoneal injection	Chromosomal aberrations in bone marrow cells, micronuclei, DNA breakage.
Studies with cells in culture (Prokaryotes: bacterias. Eukaryotes: lymphocytes, epithelial cells, fibroblasts, macrophage cell lines; from human, rat, mouse or hamster)	Bacteria: No mutagenicity. Eukaryotes: DNA damage (mutations, breakage, base oxidation); chromosome damage (breakage, bridges, lagging chromosomes, aneuploidy); mitosis impairment (interaction with mitotic spindle centrosome fragmentation, impairment of cytokinesis).

25.6   BRIEF SUMMARY OF NANOTUBULAR PARTICLES TOXICOLOGY

Other naturally occurring minerals of nanotubular structure are used in many industrial applications. Halloysite is clay forming multi-layered hollow cylinders. This mineral is considered as biocompatible, has a high mechanical strength and easy natural availability; it is (or can be) used for drug entrapment, in the pharmaceutical field, and also in medicine (Abdullayev and Lvov, 2011; Rabiskova, 2012).

Zeolite has received some medical applications, as an adjunct to traditional methods of surgical local hemorrhage control, and the efficacy and safety of these agents are currently under investigation (Recinos et al., 2008). Erionite is the fibrous counterpart of natural zeolites. Erionite exposure is an example of nonasbestos-mediated cause of mesothelioma. This observation has been made in Turkey where exposure to this type of fiber is highly prevalent (Jasani and Gibbs, 2012).

25.7   CONCLUSIONS

CNTs have different morphologies (number of walls, diameter, structural defects, residual catalysts, etc.) and literature data are difficult to interpret because of this large diversity. First, the characteristics of the CNTs used in cellular and animal experiment are not always well characterized; second, when used in suspension, their properties may change. Nevertheless, from our present knowledge, it can be proposed that morphology and surface chemistry (surface functionalization, etc.) impurities (presence of trace metals), dimensions (length and diameter), stability in biological milieu, and degree of agglomeration influence the effects.

Table 25.3 summarizes the comparisons of intrinsic properties of CNTs and asbestos, and show that some CNTs parameters are consistent with a toxic potential, similar to that of asbestos fibers. It is therefore, important to consider the similarities in order to prevent potential diseases, prior to have consistent epidemiological studies. Nevertheless, from literature results, some questions arise concerning (i) the size limits accounting for the effects, since the spectrum of dimensions of CNTs is different from that of asbestos; (ii) the ability of CNTs to form clusters and aggregates that modulate their effects; (iii) the possibility of surface reactivity modifications by functionalization; (iv) the capability to generate ROS and to capture inflammatory-mediated ROS; and (v) the possibility of biological degradation.

TABLE 25.3 Comparative intrinsic properties between CNTs and asbestos fibers.

Shape	Similarities: High aspect ratio.
Chemistry	Different, but may contain associated metals potentially active.
Dimensions	Similarities: Large spectrum of dimensions: Nano size, up to micrometers of length.
Particulate state	Both similar and different (aggregation) aspects.
Surface reactivity	Similarities: Interactions with biological molecules.
	Differences: Hydrophobicity, but functionalization may change surface state.
	ROS production: Capture of ROS
Durability (stability in biological milieu)	Similarities: Generally considered as durable but data on low durability, enzymatic biodegradation.

Exposure to carbon nanotubes can occur at any time during the life cycle of products containing them (during their manufacture, transport, use, or disposal) and producers and consumers can be exposed. Most studies on human health risk potency have concerned exposure by inhalation. Meanwhile investigations on the ecotoxicological impact of CNTs are well under development.

Our knowledge of the human health risk needs to be improved by investigating other routes of exposure, by a better characterization of CNTs, and the development of large-scale molecular studies of the cell responses to CNTs. It is also important to perform studies to better define of the circumstances, types, and levels of human exposure.

KEYWORDS

- **Asbestiform crystals**
- **Biopersistence**
- **Extrapolation**
- **Fiber toxicity**
- **Granulomatous lesions**
- **Macrophages**
- **Pathogenicity of fibers**
- **Phagocytosis**

REFERENCES

1. Abdullayev, E.; and Lvov, Y.; *J. Nanosci. Nanotechnol.* **2011**, *11*, 10007–10026.
2. AFSSET ‡Les fibres courtes et les fibres fines d'amiante. Prise en compte du critère dimensionnel pour la caractérisation des risques sanitaires liés à l'inhalation d'amiante, Agence Française de Sécurité Sanitaire de l'Environnement et du Travail, **2009**.
3. Andujar, P.; Lanone, S.; Brochard, P.; and Boczkowski, J.; *Rev. Mal. Respir.* **2011**, *28*, e66–75.
4. Aschberger, K.; et al. *Crit. Rev. Toxicol.* **2010**, *40*, 759–790.
5. Bai, Y.; et al. *Nat. Nanotechnol.* **2010**, *5*, 683–689.
6. Baris, Y. I.; and Grandjean, P.; *J. Natl. Cancer Inst.* **2006**, *98*, 414–417.
7. Boutin, C.; Dumortier, P.; Rey, F.; Viallat, J. R.; and Devuyst, P.; *Am. J. Respir. Crit. Care Med.* **1996**, *153*, 444–449.
8. Broaddus, V. C.; Everitt, J. I.; Black, B.; and Kane, A. B.; *J. Toxicol. Environ. Health B Crit. Rev.* **2011**, *14*, 153–178.
9. Castranova, V.; *J. Occup. Environ. Med.* **2011**, *53*, S14–17.
10. Comba, P.; Gianfagna, A.; and Paoletti, L.; *Arch. Environ. Health.* **2003**, *58*, 229–232.
11. Crouzier, D.; et al. *Toxicology.* **2010**, *272*, 39–45.
12. Donaldson, K.; Murphy, F. A.; Duffin, R.; and Poland, C. A.; *Part Fibre Toxicol.* **2010**, *7*, 5.
13. Donaldson, K.; Poland, C. A.; Murphy, F. A.; Macfarlane, M.; Chernova, T.; and Schinwald, A.; *Adv. Drug Deliv. Rev.* **2013**, *65*, 2078–2086.
14. Dostert, C.; Petrilli, V.; Van Bruggen, R.; Steele, C.; Mossman, B. T.; and Tschopp, J.; *Science.* **2008**, *320*, 674–677.
15. Everitt, J. I.; et al. *Environ. Health Perspect.* **1997**, *105(Suppl. 5)*, 1209–1213.
16. Fenoglio, I.; et al. *Chem. Res. Toxicol.* **2008**, *21*, 1690–1697.
17. Folkmann, J. K.; Risom, L.; Jacobsen, N. R.; Wallin, H.; Loft, S.; and Moller, P.; *Environ. Health Perspect.* **2009**, *117*, 703–708.
18. Fujitani, T.; Ohyama, K.; Hirose, A.; Nishimura, T.; Nakae, D.; and Ogata, A. J.; *Toxicol. Sci.* **2012**, *37*, 81–89.
19. Herzog, E.; Byrne, H. J.; Davoren, M.; Casey, A.; Duschl, A.; and Oostingh, G. J.; *Toxicol. Appl. Pharmacol.* **2009**, *236*, 276–281.
20. Huang, S. X.; Jaurand, M. C.; Kamp, D. W.; Whysner, J.; and Hei, T. K.; *J. Toxicol. Environ. Health B Crit. Rev.* **2011**, *14*, 179–245.
21. IARC *IARC Monogr. Eval. Carcinog. Risks Hum. Suppl.* **1987**, *7*, 1–440.
22. IARC *IARC Monogr. Eval. Carcinog. Risks Hum.* **1988**, *43*, 1–300.
23. IARC *IARC Monogr. Eval. Carcinog. Risks Hum.* **2002**, *81*, 1–381.
24. IARC *IARC Monogr. Eval. Carcinog. Risks Hum.* **2012**, 100 C.
25. Jackson, P.; et al. *Chem. Cent. J.* **2013**, *7*, 154.
26. Jasani, B.; and Gibbs, A.; *Arch. Pathol. Lab Med.* **2012**, *136*, 262–267.
27. Jaurand, M. C.; *Environ. Health Perspect.* **1997**, *105*, 1073–1084.
28. Jaurand, M. C.; Renier, A.; and Daubriac, J.; *Part Fibre Toxicol.* **2009**, *6*, 16–29.
29. Jaurand, M. C.; and Pairon, J. C.; In: Nanoethics and Nanotoxicology. Houdy, P.; Lahmani, M.; Marano, F.; eds. Springer: Heidelberg Dordrecht London New York; **2011**, 3–35 p.
30. Johnston, H. J.; et al. *Nanotoxicology.* **2010**, *4*, 207–246.
31. Kato, T.; et al. *Nanotoxicology.* **2013**, *7*, 452–461.
32. Li, Z.; et al. *Environ. Health Perspect.* **2007**, *115*, 377–382.
33. Manke, A.; Wang, L.; and Rojanasakul, Y.; *Toxicol. Mech. Methods.* **2013**, *23*, 196–206.
34. Manshian, B. B.; et al. *Nanotoxicology.* **2013**, *7*, 144–156.
35. Mercer, R. R.; et al. *Part Fibre Toxicol.* **2010**, *7*, 28.
36. Mercer, R. R.; et al. *Part Fibre Toxicol.* **2011**, *8*, 21.
37. Mercer, R. R.; et al. *Part Fibre Toxicol.* **2013**, *10*, 38.

38. Mitchev, K.; Dumortier, P.; and De Vuyst, P.; *Am. J. Surg. Pathol.* **2002,** *26,* 1198–1206.
39. Mossman, B. T.; Lippmann, M.; Hesterberg, T. W.; Kelsey, K. T.; Barchowsky, A.; and Bonner, J. C.; *J. Toxicol. Environ. Health B Crit. Rev.* **2011,** *14,* 76–121.
40. Muller, J.; et al. *Carcinogenesis.* **2008a,** *29,* 427–433.
41. Muller, J.; et al. *Chem. Res. Toxicol.* **2008b,** *21,* 1698–1705.
42. Muller, J.; Delos, M.; Panin, N.; Rabolli, V.; Huaux, F.; and Lison, D.; *Toxicol. Sci.* **2009,** *110,* 442–448.
43. Muller, K. M.; Schmitz, I.; and Konstantinidis, K.; *Respiration.* **2002,** *69,* 261–267.
44. Murphy, F. A.; Schinwald, A.; Poland, C. A.; and Donaldson, K.; *Part Fibre Toxicol.* **2012,** *9,* 8.
45. Nagai, H.; et al. *Proc. Natl. Acad. Sci. USA.* **2011,** *108,* E1330–1338.
46. NIOHS http://www.cdc.gov/niosh/pdfs/97-162-f.pdf, **1994.**
47. NIOHS Occupational exposure to carbon nanotubes and nanofibres. *Depart. Health Hum. Serv. Cent. Dis. Control Prev.* **2010,** *65.*
48. Pacurari, M.; Castranova, V.; and Vallyathan, V.; *J. Toxicol. Environ. Health A.* **2010,** *73,* 378–395.
49. Pacurari, M.; et al. *Toxicol. Appl. Pharmacol.* **2011,** *255,* 18–31.
50. Patlolla, A. K.; Hussain, S. M.; Schlager, J. J.; Patlolla, S.; and Tchounwou, P. B.; *Environ. Toxicol.* **2010,** 608–621.
51. Poland, C. A.; et al. *Nat. Nanotechnol.* **2008,** *3,* 423–428.
52. Ponti, J.; et al. *Nanotoxicology.* **2013,** *7,* 221–233.
53. Porter, D. W.; et al. *Toxicology.* **2010,** *269,* 136–147.
54. Porter, D. W.; et al. *Nanotoxicology.* **2013,** *7,* 1179–1194.
55. Rabiskova, M.; *Ceska. Slov. Farm.* **2012,** *61,* 255–260.
56. Raffa, V.; Ciofani, G.; Vittorio, O.; Riggio, C.; and Cuschieri, A.; *Nanomedicine (Lond).* **2010,** *5,* 89–97.
57. Recinos, G.; Inaba, K.; Dubose, J.; Demetriades, D.; and Rhee, P.; *Ulus. Travma. Acil. Cerrahi. Derg.* **2008,** *14,* 175–181.
58. Rodriguez-Yanez, Y.; Munoz, B.; and Albores, A.; *Toxicol. Mech. Methods.* **2013,** *23,* 178–195.
59. Ryman-Rasmussen, J. P.; et al. *Nat. Nanotechnol.* **2009,** *4,* 747–751.
60. Sanchez, V. C.; Pietruska, J. R.; Miselis, N. R.; Hurt, R. H.; and Kane, A. B.; *Wiley Interdiscip Rev. Nanomed Nanobiotechnol.* **2009,** *1,* 511–529.
61. Sargent, L. M.; Reynolds, S. H.; and Castranova, V.; *Nanotoxicology.* **2010,** *4,* 396–408.
62. Sargent, L. M.; et al. *Mutat. Res.* **2012,** *745,* 28–37.
63. Schulte, P. A.; et al. *Am. J. Ind. Med.* **2012,** *55,* 395–411.
64. Shukla, A.; Ramos-Nino, M.; and Mossman, B.; *Int. J. Biochem. Cell Biol.* **2003,** *35,* 1198–1209.
65. Shvedova, A. A.; et al. *Am. J. Physiol. Lung. Cell Mol. Physiol.* **2008,** *295,* L552–565.
66. Shvedova, A. A.; Pietroiusti, A.; Fadeel, B.; and Kagan, V. E.; *Toxicol. Appl. Pharmacol.* **2012,** *261,* 121–133.
67. Shvedova, A. A.; et al. *Am. J. Physiol. Lung Cell Mol. Physiol.* **2013,** In Press.
68. Simko, M.; and Mattsson, M. O.; *Part Fibre Toxicol.* **2010,** *7,* 42.
69. Singh, N.; et al. *Biomaterials.* **2009,** *30,* 3891–3914.
70. Szendi, K.; and Varga, C.; *Anticancer. Res.* **2008,** *28,* 349–352.
71. Takagi, A.; et al. *J. Toxicol. Sci.* **2008,** *33,* 105–116.
72. Tran, C. L.; et al. Report on Project CB0406 2008, IOM, Research Consulting Services, http://www.iom-world.org.
73. Van Berlo, D.; Clift, M. J.; Albrecht, C.; and Schins, R. P.; *Swiss Med. Wkly.* **2012,** *142,* w13698.
74. Wang, L.; Stueckle, T. A.; Mishra, A.; Derk, R.; Meighan, T.; Castranova, V.; and Rojanasakul, Y.; *Nanotoxicology.* **3014,** *8,* 485–507.
75. WHO/EURO World Health Organisation Geneva, Regional Office, Coipenhagen, **1985.**
76. Xu, J.; et al. *Cancer Sci.* **2012,** *103,* 2045–2050.

PART X
FINAL REMARKS

CHAPTER 26

CURRENT TRENDS IN RESEARCH AND APPLICATION OF NATURAL MINERAL NANOTUBES

G. JOCK CHURCHMAN and POORIA PASBAKHSH

CONTENTS

26.1   MINERALS STUDIED

Halloysite, palygorskite, sepiolite, and chrysotile are the four major natural mineral nanotubes which are discussed in this book. Of all the possibilities, these four types have been studied the most and have offered the most possibilities for practical applications. There are some other mineral species that are nanotubular but which have received only limited attention in the literature for possible applications. These include the hydrous aluminosilicate, imogolite, and the zeolite, erionite. They each receive occasional mention in this volume but extra information on them may be sourced elsewhere, for example, Guimarães et al. (2007) and Brigatti et al. (2013), for imogolite; Dogan and Dogan (2008), for erionite.

26.2   RECORDED TRENDS IN PUBLICATIONS ON NANOTUBULAR MATERIALS

Undoubtedly, the major focus of this present volume has been on applications. Its collation of the information from especially the most recent literature has revealed that, as nanotubular materials, halloysite, palygorskite, and sepiolite particularly have found or have potential for application in a wide variety of fields and most commonly in polymer, manufacturing, chemical and construction industries, medicine, and the environment. The trends in number of publications for these three natural mineral nanotubes and also chrysotile and their applications are summarized in Table 26.1 (data were collected from Scopus in December 2013). Table 26.1 also locates the chapters in this book which deal with each of the topics identified in this survey of the literature. In addition, the numbers of publications on carbon nanotubes are given in Table 26.1 for comparison.

26.3   COMPARISONS BETWEEN DIFFERENT NATURAL MINERAL NANOTUBES

Table 26.1 and other information discussed in the book enable comparisons to be made between the usefulness and applicability of the various natural mineral nanotubes: (1) with non-tubular nanominerals, (2) with one another, and (3) with carbon nanotubes. In the first place, it was pioneering work on polymer nanocomposites using the non-tubular but relatively ubiquitous smectitic minerals, mostly as bentonites, that provided the starting point for demonstrations of some superior properties for these nanocomposites when natural mineral nanotubes such as HNTs are used in their manufacture. This topic is well-covered herein.

Few, if any, comparisons have yet been made between the different mineral species comprising nanotubes for their relative efficiencies for the different applications, whether in nanocomposites, as vehicles for transport or slow release of incorporated species, or as templates for chemical reactions. Summarizing the broad thrust of

TABLE 26.1 Summary of the publication records, their major research areas and applications based on the data obtained from scopus

Type of Mineral	Number of Papers Before 2005	Number of Papers from 2005 to 2013	Major Fields of Research in the Literature	Related Research and/or Applications	Chapters of This Book
Halloysite nanotubes	0	414	Earth and Planetary Science; Material Science; Environmental Science; Chemistry; Agricultural and Biological sciences; Engineering; ChemicalEngineering; Physics	Mineralogy, chemistry, geology, identification and nomenclature	1,2,5, 15, 18
Halloysite	714	757		Reinforcing filler for polymer nanocomposites	8, 10, 12, 13, 14
				Packaging	
				Protective coating	
				Concrete and cement	8, 10
				Wound dressing, inflammatory, drug delivery system, and tissue engineering scaffolds	15, 19,20
					19, 20
				Controlled release	8, 11, 22, 24
				Adsorption of contaminants	
				Nanotemplating and nanoscale reaction vessels, support for catalyst immobilization, heterogeneous catalysts	22, 23, 24
					15,19, 20
					8, 15, 19

TABLE 26.1   *(Continued)*

Type of Mineral	Number of Papers Before 2005	Number of Papers from 2005 to 2013	Major Fields of Research in the Literature	Related Research and/or Applications	Chapters of This Book
Palygorskite	615	670	Earth and Planetary Science Environmental Science Chemistry Material Science Engineering, Agricultural and Biological sciences Chemical Engineering	Mineralogy, chemistry, geology, identification and nomenclature	1,3,6,16,17
				Reinforcing filler for polymer nanocomposites	9
				Packaging	9
				Wound dressing, inflammatory, drug delivery system, and tissue engineering scaffolds	9, 23
				Controlled release	23
				Adsorption of contaminants	
				Nanotemplating and nanoscale reaction vessels, support for catalyst immobilization, heterogeneous catalysts	17
					21
Sepiolite	957	927	Earth and Planetary Science Chemistry Material Science Chemical Engineering Environmental Science Engineering	Mineralogy, chemistry, geology, identification and nomenclature	1,3,6,16
				Reinforcing filler for polymer nanocomposites	9
				Packaging	9
				Wound dressing, inflammatory, drug delivery system, and tissue engineering scaffolds	9, 23
				Controlled release	23
				Adsorption of contaminants	
				Nanotemplating and nanoscale reaction vessels, support for catalyst immobilization, heterogeneous catalysts	23
					21

TABLE 26.1 (Continued)

Type of Mineral	Number of Papers Before 2005	Number of Papers from 2005 to 2013	Major Fields of Research in the Literature	Related Research and/or Applications	Chapters of This Book
Chrysotile	2765	905	Medicine Environmental Science Earth and Planetary Science Pharmacology, Toxicology and pharmaceutics Biochemistry, Genetics and Molecular	Mineralogy, Chemistry, Geology, identification and nomenclature Reinforcing filler for polymer nanocomposites Nanotemplating and nanoscale reaction vessels, support for catalyst immobilization, heterogeneous catalysts	1,4,7 NA NA
Carbon nanotubes	12594	71887	Material Science Physics Engineering Chemistry ChemicalEngineering Biochemistry, Genetics and Molecular	NA	NA

the various chapters on application in the book, Table 26.1 reveals that halloysite, palygorskite, and sepiolite, at least, often share many similar applications. However, the fundamental properties of the different species, which are covered in chapters on structure and mineralogy, can differ substantially, so that it is likely that some mineral types have better properties, or sets of properties, for some applications than for others. It would be unwise for us to conclude the specific applications best suited to each mineral type given both the detailed information that can be found throughout the book and also the range of characteristics that each mineral type can encompass. Nonetheless, it may be that, if size of opening (lumen) is of most relevance to an application, halloysite could generally provide a wider range of possibilities than either palygorskite or sepiolite. On the other hand, these two latter minerals tend to have higher surface areas and more reactive surfaces than halloysite. Their charge characteristics also differ, with sepiolite especially and palygorskite often having higher cation exchange capacities than halloysite (Wilson, 2013). On the other hand, the asymmetry of the halloysite layer structure and composition means that halloysite can present both positive and negative charges, on their different surfaces, at the same time. There is scope for experimental comparisons of the different nanotubular mineral species for properties that are appropriate to each type of application.

According to the Scopus analysis (Table 26.1), the amount of research on "halloysite nanotubes" has increased significantly since 2005 compared to other nanotubular minerals. Most of this interest to halloysite nanotubes has referred to their potential applications in polymer reinforcement and medicine but various other applications for halloysite such as polymer composites, electronic components, cosmetics, and other personal care products have been identified recently. It may also be that much research that is applicable to nanotubular applications has not been tied specifically to "halloysite nanotubes," but has involved only "halloysite" as such. Both nomenclatures are covered in Table 26.1. Nomenclature is particularly important for halloysite because not all examples of this mineral type are tubular. In reality, nonetheless, very few publications on halloysite prior to 2005 focus specifically on nanotubular applications.

Studies on palygorskite and sepiolite have also shown an ascending trend over the past 10 years. When it comes to chrysotile; however, most of the research (34%) on this mineral has been related to medicine due to the concerns about long-term ill-health and especially cancer that has possibly been caused by the widespread use of chrysotile asbestos materials in the past. As a result of the concentration on health issues coupled with concerns about its future use, rather than on the potential for useful applications, the number of the studies of this mineral has tended to stagnate since 2005.

Another area which shows promise for further comparative studies, is that between different specimens of each mineral type. The chapters in this book describing geology, as well as those on mineralogy and nomenclature show clearly that as natural minerals with a geological origin, the characteristics of the nanotubes such as, for example, their tube lengths and widths, vary widely between occurrences. Some specimens will present better characteristics for some applications but not such good properties for other applications as a result. The recognition of the range of properties encompassed by each mineral species name means that studies using specimens from more than one locality or type of origin thereby gain value for generalizations about that mineral. Many studies of natural mineral nanotubes have used just one specimen and many possible specimens of each mineral type have not been studied very much, if at all, for their nanotubular characteristics, interactions, and applications.

26.4 POSSIBLE MANUFACTURE OF NANOTUBES FROM NON-NANOTUBULAR MINERALS

Whereas halloysite exhibits a range of morphologies, sepiolite and palygorskite are necessarily fibrous, albeit that their dimensions vary between occurrences.On the other hand, while sepiolite and palygorskite most commonly differ from each other in chemical composition, halloysite may also be compared with kaolinite, which has essentially the same chemical composition – interlayer water aside – but is invariably platy in morphology (e.g., Wilson, 2013). Very early work by Gastuche et al. (1954) and also Hope and Kittrick (1964) showed that some tubular material was produced when kaolinites were boiled in nitrobenzene to break hydrogen bonds between layers, bring about delamination, and allow the natural curling tendency of kaolin layers to take effect. More recently, intercalation of large cationic or polar species into the interlayer region of kaolinites, sometimes but not always followed by one or more steps of intercalation, has been shown to also bring about delamination followed by curling of the thin layers into tubes (Singh and Mackinnon 1996; Matsuik et al. 2009; and Kuroda et al. 2011). Kaolinite is more widespread and more abundant in deposits worldwide than halloysite (Galán and Ferrell, 2013). Kaolinite also exhibits a wide range of particle sizes in its different occurrences (Wilson, 2013). Hence there is a potentially worthwhile field of study in the manufacture of HNTs, tailor-made for tube length and lumen size – among other characteristics – from kaolinite precursors, as has already been recognized by Matsuik et al. (2009) and Kuroda et al. (2011).

26.5 COMPARISONS WITH CARBON NANOTUBES

In the past 20 years, Carbon nanotubes (CNTs) have attracted a great deal of attention in the literature and have also been used extensively. There is scope for studies comparing natural mineral nanotubes with CNTs in fields they both may have applications. In one respect, that of cost, natural mineral nanotubes, which are usually mined quite cheaply, will often have an advantage over CNTs, which require complex manufacturing processes. It may be time to assess the performance of CNTs against their promise, especially for engineering, and to compare them in this light and for relative value for cost with natural mineral nanotubes. In another important respect, considerable evidence is discussed herein for the possibility that some CNTs may be carcinogenic. On the other hand, the apparent non-toxicity of the particular nanotubular minerals, halloysite, sepiolite, and palygorskite allow their safe ingestion by animals generally and human beings in particular and hence their pharmaceutical and wider medical use.

26.6   FUTURE PROSPECTS: SUMMARY

Undoubtedly, natural mineral nanotubes and especially those of halloysite, palygorskite, and sepiolite offer much promise for applications in the fields of medicine, bionanocomposites, tissue engineering, and contaminant removal due to their unique sizes and dimensions, pore spaces, and mechanical, chemical, and rheological properties. Studies and applications of natural mineral nanotubes should continue to advance in both numbers and scope.

KEYWORDS

- Biochemistry
- Carcinogenic
- Cation exchange capacities
- Environmental science
- Medicine
- Nanominerals
- Polymer composites

REFERENCES

1. Brigatti, M.F.; Galán, E.; and Theng, B.K.G.; Structure and mineralogy of clay minerals. In: Handbook of Clay Science.2nd edition, Part A, Fundamentals.Bergaya, F.; Lagaly, G.;eds. Elsevier, Amsterdam;**2013**,21–81 p.
2. Dogan, A.U.; and Dogan, M.;*Environ.Geochem.Health.***2008**,*30,* 355–366.
3. Galán, E.; and Ferrell, R.E.; Genesis of clay minerals. In: Handbook of Clay Science.2nd edition, Part A, Fundamentals.Bergaya, F.; Lagaly, G.;eds.Elsevier, Amsterdam;**2013**,83–126 p.
4. Gastuche, M.C.; Delvigne, J.; and Fripiat, J.J.; Transactions of the Fifth International Congress of Soil Science. Leopoldville;**1954**,*2,* 439–456.
5. Guimarães, L.; Enyashin, A.N.; Frenzel, J.; Heine, T.;Duarte, H.A., and Seifert, G.;*ACS Nano.***2007**,*1,* 362–368.
6. Hope, E.W.; and Kittrick, J.A.;*Am.Mineral.***1964**,*49,* 859–866.
7. Kuroda, Y.; Ito, K.; Itabashi, K.; and Kuroda, K.;*Langmuir.***2011**,*27,* 2028–2035.
8. Matsuik, J.; Gawel, A.; Bielánska, Osuch, W.; and Bahranowski, K.;*Clays Clay Minerals.***2009**,*57,* 452–464.
9. Singh, B.;and Mackinnon, I.D.R.;*Clays Clay Minerals.***1996**,*44,* 825–834.
10. Wilson, M.J.; Rock-forming minerals.Sheet Silicates, Clay Minerals.London:The Geological Society;**2013**, 3C.

INDEX